Short-Distance Phenomena in Nuclear Physics

NATO Advanced Science Institutes Series

A series of edited volumes comprising multifaceted studies of contemporary scientific issues by some of the best scientific minds in the world, assembled in cooperation with NATO Scientific Affairs Division.

This series is published by an international board of publishers in conjunction with NATO Scientific Affairs Division

A Life Sciences **B Physics**	Plenum Publishing Corporation New York and London
C Mathematical and Physical Sciences	D. Reidel Publishing Company Dordrecht, Boston, and London
D Behavioral and Social Sciences **E Applied Sciences**	Martinus Nijhoff Publishers The Hague, Boston, and London
F Computer and Systems Sciences **G Ecological Sciences**	Springer Verlag Heidelberg, Berlin, and New York

Recent Volumes in Series B: Physics

Volume 97 —Mass Transport in Solids
edited by F. Bénière and C. R. A. Catlow

Volume 98 —Quantum Metrology and Fundamental Physical Constants
edited by Paul H. Cutler and Amand A. Lucas

Volume 99 —Techniques and Concepts in High-Energy Physics II
edited by Thomas Ferbel

Volume 100—Advances in Superconductivity
edited by B. Deaver and John Ruvalds

Volume 101—Atomic and Molecular Physics of Controlled Thermonuclear Fusion
edited by Charles J. Joachain and Douglass E. Post

Volume 102—Magnetic Monopoles
edited by Richard A. Carrigan, Jr., and W. Peter Trower

Volume 103—Fundamental Processes in Energetic Atomic Collisions
edited by H. O. Lutz, J. S. Briggs, and H. Kleinpoppen

Volume 104—Short-Distance Phenomena in Nuclear Physics
edited by David H. Boal and Richard M. Woloshyn

Short-Distance Phenomena in Nuclear Physics

Edited by

David H. Boal

Simon Fraser University
Burnaby, British Columbia, Canada

and

Richard M. Woloshyn

TRIUMF
Vancouver, British Columbia, Canada

Plenum Press
New York and London
Published in cooperation with NATO Scientific Affairs Division

Proceedings of the Pacific Summer Institute,
held August 23-September 3, 1982,
at Pearson College on Vancouver Island, British Columbia, Canada

Library of Congress Cataloging in Publication Data

Main entry under title:

Short-distance phenomena in nuclear physics.

(NATO advanced science institutes series. Series B, Physics; v. 104)
"Proceedings of the Pacific Summer Institute, held August 23-September 3, 1982, at Pearson College on Vancouver Island, British Columbia, Canada."
Organized by the Theoretical Physics Division of the Canadian Association of Physicists.
"Published in cooperation with NATO Scientific Affairs Division."
Includes bibliographical references and index.
1. Quarks—Congresses. 2. Gluons—Congresses. 3. Nuclear structure—Congresses. I. Boal, David H. II. Woloshyn, Richard M. III. Canadian Association of Physicists. Theoretical Physics Division. IV. North Atlantic Treaty Organization. Scientific Affairs Division. V. Series.
QC793.5.Q2522S48 1983 539.7′21 83-17773
ISBN 0-306-41494-5

A Division of Plenum Publishing Corporation
233 Spring Street, New York, N.Y. 10013

Printed in the United States of America

PREFACE

Each summer, the Theoretical Physics Division of the Canadian Association of Physicists organizes a summer institute of two weeks duration on a current topic in theoretical physics. This volume contains the lectures from the Pacific Summer Institute held at Pearson College on Vancouver Island, B.C. (Canada) from August 23 to September 3, 1982. The Institute was titled "Progress in Nuclear Dynamics: Short-Distance Behavior in the Nucleus".

The primary source of funds for the Institute came from NATO through its Advanced Study Institute programme. Significant financial support is also gratefully acknowledged from TRIUMF, Simon Fraser University, Natural Sciences and Engineering Research Council of Canada, and Atomic Energy of Canada Ltd.

The topic of the school was the role of the substructure of hadrons—quarks and gluons—in nuclear physics. This includes not only the effects which may be observed in specific nuclear states, such as form factors at large momentum transfer, or the presence of hidden color components in the ground states of few nucleon systems, but also effects which may be observed in the nuclear matter continuum: the phase transition from normal nuclear matter to a plasma of quarks and gluons. The current status of the long distance phenomenology of the nucleus—the interacting boson approximation and the role of π's and Δ's in nuclear structure, is also reviewed. Because it was the intention of the organizers of the Institute to make both nuclear and particle physicists who are interested in this topic fluent in each other's language, the lectures begin with a review of quantum chromodynamics and its implementation in particle physics.

The Institute was organized by a committee whose membership included, in addition to DHB and RMW, Prof. B. Castel (Queen's University, Kingston), and Prof. K.S. Viswanathan (Simon Fraser University). Secretarial help was provided by TRIUMF in the person of Ms. I. Duelli, who also undertook the arduous task of typing the proceedings. Topics and speakers for the Institute were suggested by an International Advisory Committee, whose members were Profs. S.D. Drell (SLAC), J. Hüfner (Heidelberg), F.C. Khanna (Chalk River),

C. Mahaux (Liège), B. Mottelson (NORDITA), G.R. Satchler (Oak Ridge), D. Sprung (McMaster), I. Talmi (Weizmann), and D. Wilkinson (Sussex).

We thank as well the staff at Pearson College, for helping to make the operation of the Institute as smooth as it was. Lastly, someone should be thanked for the weather: two weeks without rain to enjoy the scenery and marine life of Canada's West Coast.

Vancouver, B.C.
April, 1983

David H. Boal, Director
Richard M. Woloshyn, Co-director

CONTENTS

QCD AS A BASIS FOR QUARK AND NUCLEAR FORCES

F.E. Close

Rutherford Appleton Laboratory
Chilton, Oxon OX11 OQX
England

INTRODUCTION AND GENERAL REMARKS ON QUARKS

Quarks are constituents of nucleons, which are in turn the building blocks of nuclei. Now that high energy physicists believe that they have a viable theory of quark forces, at least within nucleons, an obvious question is whether we can exploit the theory to better understand the nuclear forces.

Progress in this direction will be described elsewhere in these proceedings. My job is to give you a first introduction to quarks in nucleons and the ideas behind the quantum chromodynamic (QCD) theory of their interactions. As the ultimate aim is to confront nuclear dynamics I shall begin with a comparison of nuclei and nucleons and a discussion of length or energy scales in physics. This will reveal ways in which investigations of nucleons are just higher energy action replays of nuclear physics experiments, but will also highlight certain crucial differences. Before getting into this, I shall briefly review the present knowledge about the basic building blocks: quarks and leptons.

The original evidence for quark constituents in nucleons and nuclear matter arose from the discovery of a multitude of nucleon resonance states and related strongly interacting particles (hadrons) through the 1950's and 60's. The observation of systematic regularities among their properties enabled them to be grouped into families exhibiting regular patterns (known as the 'Eightfold Way'). This was analogous to Mendeleev's historic formulation of the Periodic Table of the atomic Elements. The Periodic Table emerges naturally once the atoms are recognized to be composites of a few basic varieties of matter, (electrons and nuclei, supplemented by the Pauli principle: similarly, the Eightfold way emerges if hadrons are built from quarks).

Originally (1964) three varieties, or "flavors", of quark were all that were needed to build all hadronic matter. These "up, down and strange" quarks have electrical charges that are fractions 2/3 or -1/3 of a proton's charge. This distinctive feature should make isolated quarks eminently noticeable—yet none such has ever been seen. To understand why quarks appear to be "permanently" confined is a central unsolved problem in QCD theory of quark forces—and one which I shall not discuss!

Since 1974 two more flavors of quark have been found—these are the charmed quark and the bottom quark, again with charges of +2/3 and -1/3 respectively. Thus (u = up, d = down etc.)

$$\begin{array}{ll} \text{charge} = +2/3: \\ \text{charge} = -1/3: \end{array} \begin{pmatrix} u \\ d \end{pmatrix} \begin{pmatrix} c \\ s \end{pmatrix} \begin{pmatrix} ? \\ b \end{pmatrix}$$

the "?" representing the suspicion that a sixth quark, the 'top' (t) with charge = 2/3, exists to complete a third pair. This is reinforced by the fact that there appear to be six leptons (particles that do not experience strong nuclear forces and hence are not bound in nuclei), three charged and three neutral ("neutrinos"):

$$\begin{array}{ll} \text{charge} = -1: \\ \text{charge} = \ \ 0: \end{array} \begin{pmatrix} e^- \\ \nu_e \end{pmatrix} \begin{pmatrix} \mu^- \\ \nu_\mu \end{pmatrix} \begin{pmatrix} \tau^- \\ \nu_\tau \end{pmatrix} .$$

A tantalizing property of these doublets of quarks and leptons will be noted later on.

That quarks are genuine physical entities inside nucleons was verified around 1970 by scattering high energy electrons from proton and neutron targets. Just as in 1911 the violent collisions between α-particle beams and atoms had led Rutherford to conclude that atoms have a compact nuclear core, so did violent deflections of high energy electrons show that the proton's charge is not diffusely spread throughout its 1 fermi size but is instead located on three quarks within. The proton and neutron seem to be essentially

$$p^{+1} = u^{2/3} \ u^{2/3} \ d^{-1/3}$$
$$n^{0} = d^{-1/3} \ d^{-1/3} \ u^{2/3} \ .$$

In the subsequent decade no phenomenon has been seen inconsistent with the following dogma:

(1) Isolated quarks (Q) or antiquarks ($\bar{Q}$) do not exist.
(2) Quarks cluster in threes (QQQ) or with antiquarks ($Q\bar{Q}$).

The QQQ systems are known as 'baryons' (e.g. nucleons); $Q\bar{Q}$ are 'mesons' (e.g. pions).

NUCLEI AND NUCLEONS: LENGTH SCALES

Although the presence of nuclei was first inferred from historic α-particle scattering, more recently electron scattering has been important in resolving nuclear and nucleon structure. Nucleons in nuclei are resolved by inelastic scattering of 100 MeV electrons, typically momentum transfers by the exchanged photon being

$$Q^2 \equiv +\vec{q}^{\,2} - \nu^2 \lesssim 1 \text{ GeV}^2 .$$

($\vec{q}, \nu$ being the virtual photon's three-momentum and lab frame energy.) Scattering electrons from nucleons at 100 *GeV* resolves a *quark* substructure.

Different realms of physical phenomena have been revealed as a consequence of the change in the energy of the beam or, equivalently, the smaller scales of microscopic distance that the more energetic smaller beam can probe. Nature seems to be layered—characteristic phenomena are important at distinct length scales. To resolve atomic, nuclear, nucleon, quark structure requires probes with the intrinsic resolving powers of less than 10^{-8} cm, 10^{-12}, 10^{-13}, 10^{-16} cm, which is equivalent to probes whose energies are of the order hundreds of keV, MeV, GeV and TeV respectively.

When the probe's momentum is "tuned" to the characteristic length scale of some layer of matter then interesting phenomena are seen when the momentum or energy of the probe is slightly altered. Resonances are excited, the system breaks up and ultimately a new layer of shorter distance phenomena is seen.

As one varies the energy of the beam through a range that resolves distances less than one scale but still much larger than the next shorter distance scale, then dimensionless physical quantities (such as certain structure functions F_2—see Ref. 1) will not alter. The physics seems to be invariant under the change of scale. We say that the particles from which the beam scatters are "pointlike". More precisely we should say that their substructure has not yet been resolved. When the energy is increased sufficiently to resolve this next layer, manifest violations of scale invarinace will be seen.

This is indeed the case for nuclei. Scattering electrons off of α-particles between $Q^2 = 0.08$ GeV^2 and 0.1 GeV^2 reveals a dramatic change in the cross section for $e\alpha \to e\alpha$ but the quasi-elastic peak, arising from $ep \to ep$ inside the α-particle, stays unaltered. What is happening is that over this range of Q^2 the protons are seen as effectively pointlike particles—(their inner structure is not resolved) whereas the internal makeup of the α-particle is directly seen. (See Ref. 1 for a more detailed discussion.)

To see this scaling invariance on nuclei has required

$Q^2 > R^{-2}$ (nucleus); $<R^{-2}$ (proton constituents)
ν > Binding energy (protons quasi-free).

If we perform an analogous experiment at much higher energies then the proton substructure is resolved: quark constituents are seen. A similar scale invariance is observed when

$Q^2 > R^{-2}$ (proton) (< "R^{-2} quark")

(the reason for the quotation marks will emerge later). However, the astonishing feature is that $\nu \not>$ binding energy of quarks because quarks appear to be "permanently" confined. That they are, however, (almost) free is a property of the forces between them which the QCD theory predicts should tend to vanish at asymptotically high energies (so-called "asymptotic freedom").

Precise measurements over the last decade have shown that there are small violations of scale invariance in the data. This may be due to a new discrete substructure in the quarks—(a nucleus-nucleon-quark replay)—or it may be due to the nature of the forces among the quarks. QCD makes specific predictions for the nature of the scaling violation in this latter scenario.[2]

COLOR

QCD is a theory of quark forces similar to QED. Whereas QED arises from the presence of electrical charges, so QCD arises from "color" charges. Particles that possess color will feel the QCD forces, particles that do not contain it will not experience them.

We now believe that the strongly interacting hadrons contain color—carried by their quarks—and that this is the source of the strong nuclear force at 1 fermi dimensions. At very short dimensions this color force is rather feeble—hence the apparent freedom of quarks in the very high energy experiments of the last section.

The evidence that quarks carry color, is discussed by Llewellyn Smith.[2] I refer you to his lectures and I will now take color for granted and show how it affects quark combinations and yields a theory of their forces.

Building Colorless Hadrons

The simplest QED atom is positronium where the opposite (electrical) charges of e^- and e^+ mutually attract. Mesons are somewhat analogous in that the "charges" (yet to be defined) of quark and antiquark are opposite and unlike charges attract.

The overall scale of relative size between positronium and quarkonium is $R^{-1} \sim 10^6$. Therefore electromagnetic effects in quarkonium should be of order 10^6 bigger than in atoms, hence O(MeV). The mass difference of $\pi^+ - \pi^0$ is indeed of this order. The excitation spectrum of $Q\overline{Q}$ is O(100 MeV) between orbital levels and $\vec{L}\cdot\vec{S}$ or $\vec{S}\cdot\vec{S}$ splittings are O(10 MeV).[3] This hints at an interquark force that is rather stronger than in QED.

The existence of baryons is the first clue that the interquark forces are profoundly different to QED. That a pair of quarks will attract a third quark forming QQQ clusters hints that "QQ behave like $\overline{Q}$". This property emerges naturally if quarks can carry any of three colors and if color acts as the source of a force by analogy with electrical charge generating electromagnetic forces. Suppose that "Like colors repel, unlike (can) attract". If the three colors form the basis of an SU(3) group then they cluster to form "white" systems—the singlets of SU(3). This attraction of three *different* colors implies that the Ω^-, in particular, is built from three *distinct* strange quarks, s_R, s_Y, s_B for colors R(ed), Y(ellow), B(lue). In the absence of color the strange quarks in Ω^-(sss) would have been identical, in violation of the Pauli principle. (This Pauli paradox with uncolored quarks was historically one of the first hints that something like color must exist as a property of quarks.)[4] We will investigate the powerful effects of the Pauli principle for colored quarks later on.

First some more comments on "white" states.

If quark (Q) and antiquark ($\overline{Q}$) are 3 and $\overline{3}$ of color SU(3), then combining up to three together gives SU(3) multiplets of the following dimensions[1]

$$QQ = 3 \times 3 = 6 + \overline{3}$$

$$Q\overline{Q} = 3 \times \overline{3} = 8 + 1 \quad .$$

(Note the $\overline{3}$ in the QQ system—this is the explanation of the qualitative remark that "two quarks act like an antiquark"—the "acting like" is in color space.) The $Q\overline{Q}$ contains a singlet—the physical mesons.

$$QQ\overline{Q} = 15 + 6 + 3 + 3$$

$$QQQ = 10 + 8 + 8 + 1 \quad .$$

Note the singlet in QQQ—the physical baryons.

For clusters of three or less only $Q\overline{Q}$ and QQQ contain color singlets and, moreover, these are the only states realized physically. Thus are we led to the hypothesis that only color singlets can exist free in the laboratory; in particular the quarks will not

Table 1. Color Content in Some Representations of SU(3)

$\bar{3}_{anti}$	$\bar{6}_{sym}$
RB - BR	RB + BR
RY - YR	RY + YR
BY - YB	BY + YB
	RR
	BB
	YY

exist in isolation. (It is an interesting question whether or not other color singlets, such as Q^6, $QQ\bar{Q}\bar{Q}$ exist. See Refs. 3,5.)

Effects of Color Inside Hadrons

Spin flavor correlations. To have three quarks in color singlet:

$$1 \equiv \frac{1}{\sqrt{6}}\left[(RB - BR)Y + (YR - RY)B + (BY - YB)R\right]$$

any pair is in the $\bar{3}$ and is antisymmetric. Note that $3 \times \bar{3} = 6 + 3$. These are shown in Table 1. Note well: *Any Pair is Color Antisymmetric*. The Pauli principle requires total antisymmetry and therefore any pair must be: *Symmetric in all else* ("else" means "apart from color").

This is an important difference from nuclear clusters where the nucleons have no color (hence are trivially *symmetric* in color!). Hence for nucleons Pauli says: *Nucleons are Antisymmetric in Pairs* and for quarks: *Quarks are Symmetric in Pairs*. If we forget about color (color has taken care of the antisymmetry and won't affect us again), then

i. Two quarks can couple their spins as follows

$$\begin{Bmatrix} S=1: & \text{symmetric} \\ S=0: & \text{antisymmetric} \end{Bmatrix}$$

ii. Two u,d, quarks similarly form isospin states

$$\begin{Bmatrix} I=1: & \text{symmetric} \\ I=0: & \text{antisymmetric} \end{Bmatrix}$$

iii. In the ground state, L=0 for all quarks; hence the orbital state is trivially symmetric. Thus for pairs in L=0 we have

$$\left\{\begin{matrix} S=1 & \text{and} & I=1 & \text{correlate} \\ S=0 & \text{and} & I=0 & \text{correlate} \end{matrix}\right\} .$$

Thus the Σ^0 and Λ^0 which are distinguished by their ud quarks being I=1 or 0 respectively also have the ud pair in spin = 1 or 0 respectively:

$$\left\{\begin{matrix} \Sigma^0[(ud)_{I=1}s] \leftrightarrow [(ud)_{S=1}s] \\ \Lambda^0[(ud)_{I=0}s] \leftrightarrow [(ud)_{S=0}s] \end{matrix}\right\} .$$

Thus the spin of the Λ^0 is carried entirely by the strange quark.

These correlations between flavor and spin cause interesting flavor-dependent effects to arise as a product of spin-dependent forces (e.g. the $\vec{S}\cdot\vec{S}$ force from the magnetic contact interaction between quarks). Thus can the u and d flavors have different momentum distributions in the proton as revealed in deep-inelastic scattering.

This is also the source of the Σ-Λ mass difference. The $\vec{S}\cdot\vec{S}$ interaction acts between all possible pairs; thus

$$\Sigma^0[(ud)_1 s]: \quad \langle\vec{S}\cdot\vec{S}\rangle_1 + \langle\vec{S}\cdot\vec{S}\rangle_{s,1} \quad \text{and} \quad \Lambda^0[(ud)_0 s]: \quad \langle\vec{S}\cdot\vec{S}\rangle_0 .$$

Note $\langle\vec{S}\cdot\vec{S}\rangle$ between a spinless diquark and anything vanishes; hence the absence of $\langle\vec{S}\cdot\vec{S}\rangle_{s,0}$. Now $\langle\vec{S}\cdot\vec{S}\rangle_0 = -3\ \langle S\cdot S\rangle_1$, (see p. 91 of Ref. 1). Further, if $m_s = m_{u,d}$ the Σ and Λ become mass degenerate and so in this limit $\langle\vec{S}\cdot\vec{S}\rangle_{s,1} = -4\ \langle\vec{S}\cdot\vec{S}\rangle_1$. For unequal masses of u and s the magnetic interaction scales as the inverse mass.[1,6] Hence finally

$$\Sigma^0 \sim \langle\vec{S}\cdot\vec{S}\rangle_1 \left\{1-4\,\frac{m_u}{m_s}\right\} .$$

$$\Lambda^0 \sim \langle\vec{S}\cdot\vec{S}\rangle_0 \left\{-3\right\} .$$

Then with $m_s > m_u$ we find $m_\Sigma > m_\Lambda$ as observed. Increasing m_s/m_u enhances the effect (e.g. for the charmed analogues $\Sigma_c[(ud)c]$ and $\Lambda_c[(ud)c]$ the splitting will be larger—again observed).

Magnetic moments. A very beautiful demonstration of symmetry at work is the magnetic moment of two similar sets of systems of three, *viz.*

$$\left\{\begin{matrix} N & ; & P \\ ddu; & & uud \end{matrix}\right\} \qquad \mu_P/\mu_N = -3/2$$

and the nuclei

$$\left\{\begin{matrix} H^3 \; ; & He^3 \\ NNP; & PPN \end{matrix}\right\} \qquad \mu_{He}/\mu_H = -2/3 \quad .$$

The Pauli principle for nucleons requires He^4 to have *no* magnetic moment:

$$\mu[He^4;\; P^{\uparrow}P^{\downarrow}N^{\uparrow}N^{\downarrow}] = 0 \quad .$$

Then

$$He^3 \equiv He^4 - N$$

$$He^3 \equiv He^4 - P$$

and so

$$\frac{\mu_{He^3}}{\mu_{H^3}} = \frac{\mu_N}{\mu_P} \quad .$$

To get at this result in a way that will bring best comparison with the nucleon-three quark example, let's study the He^3 directly. $He^3 = PPN$: PP are flavor symmetric, hence spin antisymmetric i.e. S=0. Thus

$$[He^3]^{\uparrow} \equiv (PP)_0 \; N^{\uparrow} \tag{1}$$

and so the PP do not contribute to its magnetic moment. The magnetic moment (up to mass scale factors) is

$$\mu_{He^3} = 0 + \mu_N \quad .$$

Similarly,

$$\mu_{H^3} = 0 + \mu_P \quad .$$

Now let's study the nucleons in analogous manner.

The proton contains uu flavor symmetric and *color antisymmetric*; thus the spin of the "like" pair is symmetric (S=1) in contrast to the nuclear example where this pair had S=0. Thus coupling spin 1 and spin 1/2 together, the Clebsches yield

$$p^{\uparrow} = \frac{1}{\sqrt{3}}\,(uu)_0 d^{\uparrow} + \sqrt{\frac{2}{3}}\,(uu)_1 d^{\downarrow} \tag{2}$$

[contrast Eq. (1)], and (up to mass factors)

$$\mu_P = \frac{1}{3}\,(0 + d) + \frac{2}{3}\,(2u - d) \quad .$$

Suppose that $\mu_{u,d} \propto e_{u,d}$ then

$$\mu_u = -2\,\mu_d$$

so

$$\frac{\mu_P}{\mu_N} = \frac{4u-d}{4d-u} = -\,\frac{3}{2}$$

(the neutron follows from proton by replacing $u \leftrightarrow d$).

I cannot overstress the crucial, hidden role that color played here in getting the flavor-spin correlation right. We can extend this discussion to the full baryon octet. This is described in Ref. 7.

Color as the Source of a Field Theory: Molecular and Nuclear Forces

QED is a renormalizable gauge field theory of Abelian electric charge, (mathematically a U(1) for one charge). The spin 1/2 charged electron necessarily requires a massless vector boson—the photon—to maintain local gauge invariance. The coupling of electron and photon is uniquely fixed: $e\bar{\psi}\gamma_\mu\psi A^\mu$.

QCD is a renormalizable gauge theory of non-Abelian color. The three-fold color generates an SU(3) theory. The spin 1/2 colored quark necessarily requires massless vector bosons—gluons—to maintain local gauge invariance. The coupling of a quark and gluon is uniquely fixed: $g\bar{\psi}\gamma_\mu\psi G^\mu$. The three colors of quark leads to $8(=3^2-1)$ varieties of colored gluons.

Gluons themselves have color (contrast photons which do *not* carry the charge though they couple to charge). Gluons couple not just to quarks but also to one another. Thus the manner in which the color force spreads across space differs from the behavior of electromagnetic forces. This is surely at the root of color confinement though the explicit demonstration of this is still awaited.

Opposite electric charges mutually attract to form net uncharged systems. Analogously colored objects attract to form net uncolored systems. Parallels between atomic and nuclear systems *vis a vis* QED and QCD are

		QED electric charge	QCD color
Feel the force	Have manifest charge	e^- Z^+ Na^+ Cl^-	quarks GLUONS
	contain it hidden	atoms molecules	nucleons nuclei
Don't feel force	totally neutral	ν^0 PHOTON	$\begin{pmatrix}\nu^0\\ e^-\end{pmatrix}$ leptons

(weak interactions can generate a non-zero charge radius for ν^0 via $\nu^0 \to e^- W^+ \to \nu^0$; but we are presently considering a QED-QCD world alone).

Now let's look inside atoms and hadrons.

Inside atoms there is a Coulomb potential, α/r, and relativistic corrections due to one (transverse) photon exchange which generates the famous Fermi-Breit Hamiltonian. This latter splits the energy levels of states with parallel or antiparallel spins: e.g. the $^3S_1$ and $^1S_0$ levels in hydrogen—the $^3S_1$ is pushed up in energy while the $^1S_0$ is pulled down.

Inside hadrons the single gluon exchange generates an analogous α_s/r and Fermi-Breit interaction. This will shift the $^3S_1$ $Q\overline{Q}$ energy up relative to $^1S_0$—as empirically seen—vector mesons are more massive than their pseudoscalar counterparts.

Although the long-range interquark potential is unlike the electromagnetic, the Fermi-Breit is manifested as a perturbation. We shall not go into this in detail but it is interesting to see how color is necessary to get the *sign* of the hyperfine shifts correctly.

The magnetic interaction H_I between charges e_1e_2 is proportional to $-e_1e_2 \ \vec{\sigma}_1 \cdot \vec{\sigma}_2$. Now $<\sigma_1 \cdot \sigma_2> = +1(S_{total} = 1)$; $-3(S_{total} = 0)$ and so for opposite charges $(-e_1e_2 > 0)$ H_I is an upward shift for S=1, downward for S=0. This is indeed the case in hydrogen and quarkonium where Q and $\overline{Q}$ carry opposite colors ("$-e_1e_2 > 0$").

Between two quarks in a baryon, in the absence of color, $-g_1g_2 < 0$ (g being the "strong charge" giving rise to the interquark forces). Thus in baryons one might have expected that the high spin clusters shift *down*, not up. In nature, however, they shift up—as in mesons. It is the color that causes this to happen. Heuristically the QQ acting "like $\overline{Q}$" generates an effective sign flip—in color space baryons are effectively $3 \times \overline{3}$ like mesons—thus the direction of their energy shifts is the same as for mesons: high spin clusters go up, low spin clusters come down, (more details can be found in Refs. 1 and 2).

Before studying the details of non-Abelian theories there is one final comparison to make—namely the forces between "neutral" clusters. These may be classified in three types: constituent exchange (covalent), imbalanced charge distribution leading to residual "van der Waals" forces, and ionic forces. These are summarized in Table 2, which is again self-explanatory.

The color SU(3) symmetry is the only *exact* internal symmetry that occurs at quark level. It was once thought that flavor was a broken symmetry. [Do *not* confuse SU(3) color with the old, misidentified, SU(3) of flavor. We now know that there are at least five flavors, u,d,s,c,b—there may be even more. There are *only three* colors—that is why *three* quarks cluster to form baryons.]

Table 2. Classification of Forces between "Neutral" Clusters

	Covalent	van der Waals	Ions
Atoms Molecules	electron exchange	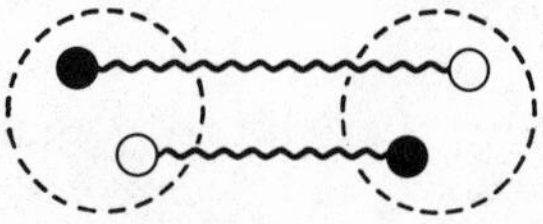	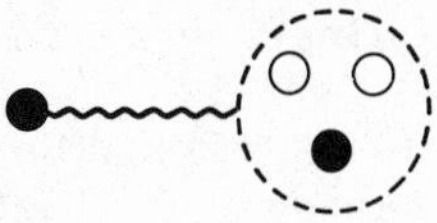
Hadrons Nuclei	quark exchange	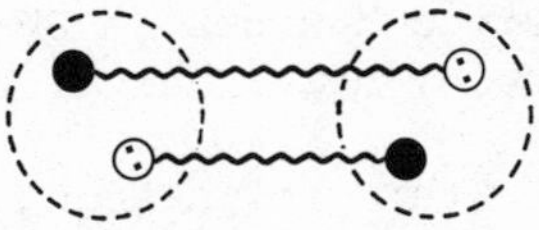	Not > 1 fm. Maybe in dense quark matter
	at low energies (macrosocopic > 1 fm) color-QCD hidden		

NON-ABELIAN THEORIES: VERTICES

Qualitative Features

U(1): In QED the photon couples to electric charge: thus there are vertices for its coupling to electrons and quarks but not to neutrinos. In particular it will not couple directly to other photons as these have zero charge.

SU(2): Suppose that charged photons existed, $\gamma^+\gamma^0\gamma^-$ which could be assigned to the regular $(3 = 2^2-1)$ representation of an SU(2) group whose fundamental representation contains two fields, U,D, with charges ±1/2 (if we call this SU(2) "charged isospin" then U and D are $I_3 = \pm 1/2$ states). The following vertices exist

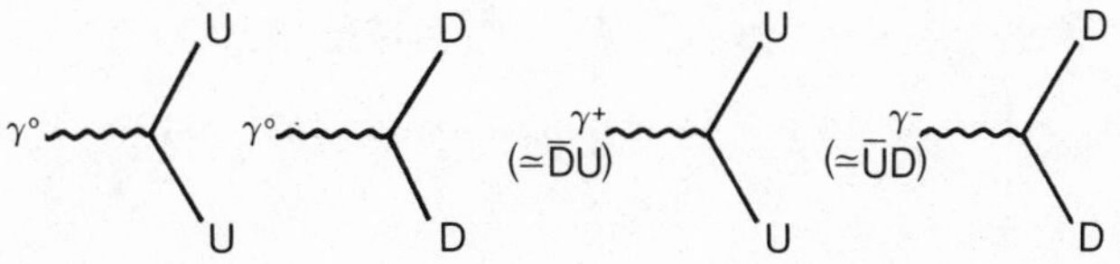

and because γ^0 couples to charge there will also be a "three boson vertex"

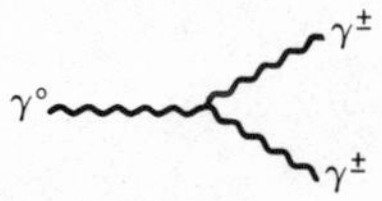

Nature exploits this in an SU(2) × U(1) sense where the doublets have electrical charges shifted from ±1/2 by an amount called the "weak hypercharge" [essentially the U(1) contribution][8] e.g.

$$\begin{pmatrix} \nu^0 \\ e^- \end{pmatrix} = \begin{pmatrix} +1/2 & -1/2 \\ -1/2 & -1/2 \end{pmatrix} \longrightarrow \text{weak hypercharge of } -1/2 \quad .$$

The SU(2) × U(1) manifests itself with *four* bosons ($W^+Z^0W^-\gamma^0$) as against *three* in SU(2).

SU(3): The above immediately generalizes to three fundamental charges and eight associated "photons". The use of charge here is loose and not totally correct. In the sense of conserved quantities there are *two* conserved charges in $SU(3)_c$: the (third component of) color-isospin and color hypercharge.

Nature exploits this in that quarks appear to carry any of three colors. By analogy with the heuristic SU(2) example, gluons will couple to colored quarks but in addition can couple to one another—a three gluon vertex will exist.

Boson-Fermion Vertices in SU(N)

U(1): The space time and charge coupling structure of the electron photon vertex is $<|ie\gamma_\mu|>$. Let's generalize to SU(2).

SU(2): For γ^0 coupling to U,D with charge ±1/2

U(1) 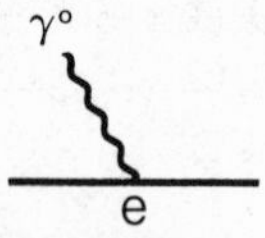

$\longrightarrow \langle | ie\gamma_\mu | \rangle$

SU(2) 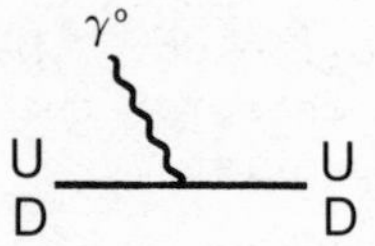

$\longrightarrow$ if $\gamma_\mu \langle | +1/2 | \rangle$ or if $\gamma_\mu \langle | -1/2 | \rangle$.

If we summarize the U and D in the fundamental SU(2) doublet

$$2 \equiv \begin{pmatrix} U \\ D \end{pmatrix}$$

then

$$< \gamma^0 > = \text{if } \gamma_\mu \frac{\tau_3}{2} \; ; \quad \tau_3 \equiv \begin{pmatrix} 1 & 0 \\ 0 & -1 \end{pmatrix}$$

and for the charged photon

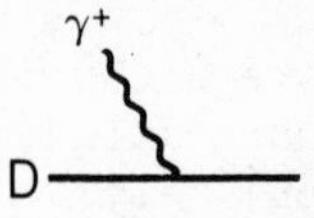

$$= \text{if } \gamma_\mu \left\langle U \left| \frac{\tau^+}{2} \right| D \right\rangle ; \quad \frac{\tau^+}{2} \equiv \begin{pmatrix} 0 & 1 \\ 0 & 0 \end{pmatrix} \quad .$$

In general

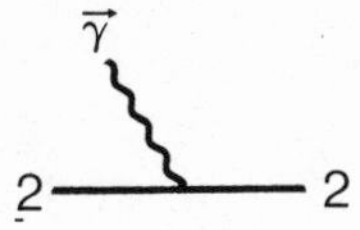

$= \; i f \gamma_\mu \vec{\tau}/2$ where $\vec{\tau}$ are the 2×2 traceless Pauli matrices.

SU(3): The natural extension is

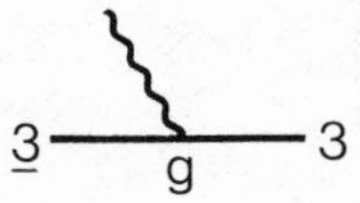

$\rightarrow \; i g \gamma_\mu \vec{\lambda}/2$ where $\vec{\lambda}$ are any of eight traceless 3×3 matrices.

If we look down a gluon line and see it convert into Q_3 and $\overline{Q_{\bar{3}}}$ then there are *a priori* nine combinations. Of these six carry color and three have no net color. Of these one is "color blind", $R\bar{R} + B\bar{B} + G\bar{G}$, and in matrix notation acting on the basis $\begin{pmatrix} R \\ B \\ G \end{pmatrix}$ it is $\begin{pmatrix} 1 & & \\ & 1 & \\ & & 1 \end{pmatrix}$.

This is not traceless and is a singlet of SU(3) $(\bar{3} \times 3 = 8 + 1)$. The coupling strength of this entity is independent of the 8 of gluons in which we are interested. The two remaining octet members are the combinations $\frac{1}{\sqrt{2}}(R\bar{R} - B\bar{B})$ and $\frac{1}{\sqrt{6}}(R\bar{R} + B\bar{B} - 2G\bar{G})$.

Gauge Invariance: QED, QCD and the Three Gluon Vertex

The three gluon vertex in QCD is quantitatively constrained by gauge invariance. If the boson polarization is ε in one gauge and the boson absorption matrix element is $\varepsilon \cdot M$ then in another gauge where $\varepsilon' = \varepsilon + cq$, for momentum q and arbitrary c, $\varepsilon \cdot M \equiv \varepsilon' \cdot M$ if $q \cdot M = 0$. Gauge invariance requires that observables are unaltered in going from one gauge to another, thus we can test gauge invariance by demanding that $q \cdot M = 0$.

In QED the Compton amplitude

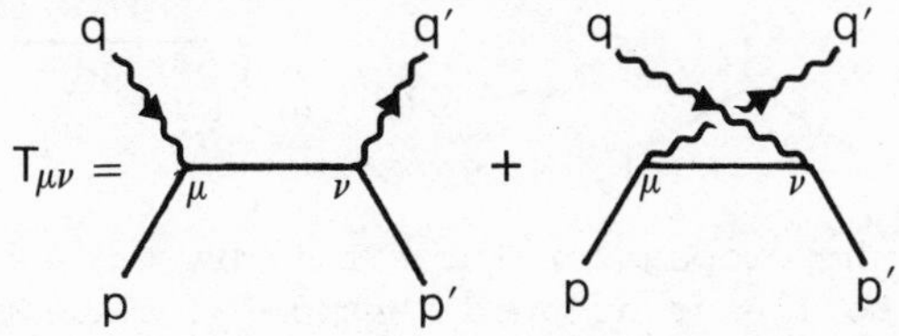

is gauge invariant as may be verified by explicitly checking that

$$q_\mu T_{\mu\nu} q'_\nu = 0 \; .$$

In a non-Abelian theory the colors of boson and fermion must also be included. Writing

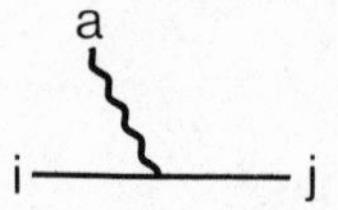

$$= \bar{\psi}(-ig\gamma_\mu (t^a)_{ij})\psi$$

the Compton amplitude becomes

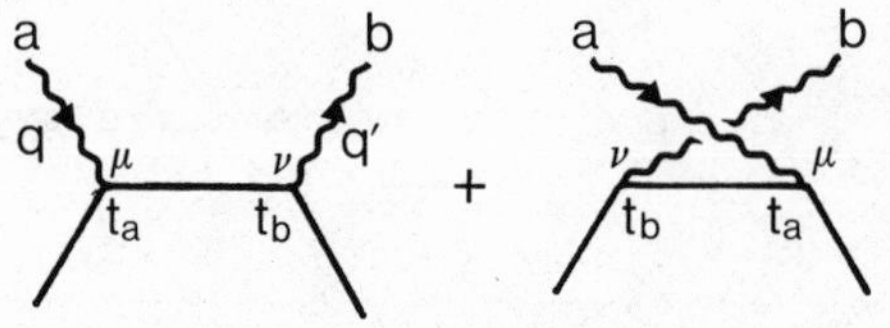

which gives

$$q_\mu T_{\mu\nu} q'_\nu = \bar{u}(p')\left\{\left[-ig\not{q}'t^b \frac{i}{\not{p}+\not{q}} (-ig\not{q}t^a)\right] + \left[-ig\not{q}t^a \frac{i}{\not{p}-\not{q}} (-ig\not{q}'t^b)\right]\right\} .$$

Rewrite $\not{q} \equiv (\not{p}+\not{q}) - \not{p}$ in the first parenthesis (s-channel). Then $\not{p}u(p) = 0$ by Dirac's equation for m=0. Cancel the $\not{p}+\not{q}$ in numerator and denominator. Play a similar game for the u-channel piece and we find

$$ig^2\bar{u}(p')\left\{\not{q}'[t_a,t_b]_{ij}\right\}u(p) \neq 0 .$$

In the Abelian case the commutator vanished yielding a zero (gauge invariance). In the non-Abelian case the s and u diagrams are not gauge invariant.

We have seen heuristically why a three gluon vertex should exist and so we expect that a t-channel diagram will occur:

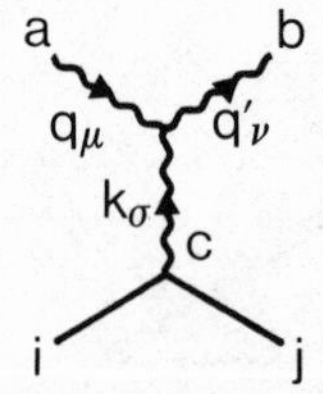

$$\leftarrow g_{\sigma\sigma'}/k^2$$
$$\leftarrow -ig\gamma_{\sigma'}(t_c)_{ij}$$
$$\Bigg\} \frac{-ig\gamma_\sigma(t_c)_{ij}}{k^2}$$

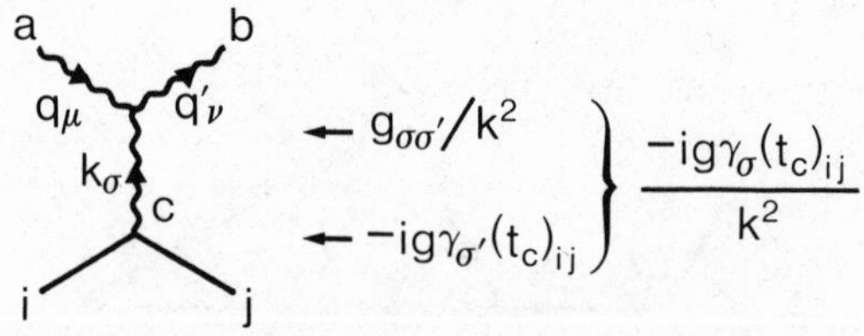

($k_\sigma k_{\sigma'}$ terms in the propagator for arbitrary gauges do not affect the calculation so I have dropped them.) We can immediately see some features that the three gluon vertex must have

i. $O(g)$ so that overall $O(g^2)$ obtains.
ii. Involves momenta to kill k^{-2} propagator.
iii. Contains $g_{\sigma\nu}$ *etc.* to give γ_ν and hence $\not{q}'$
iv. Depends on abc, (call this C_{abc}).

The answer is

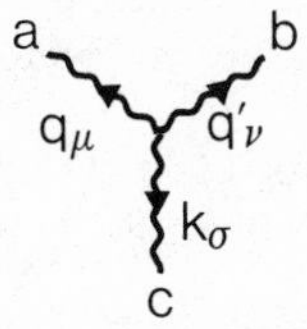

$$-gC_{abc}\left[(q-k)_\nu g_{\mu\sigma} + (k-q')_\mu g_{\nu\sigma} + (q'-q)_\sigma g_{\mu\nu}\right]$$

for all momenta outgoing. The s,t,u sum will yield zero for $q_\mu T_{\mu\nu} q'_\nu$ if

$$[t_a, t_b] = C_{abc}\, t_c$$

which is the essential structure of a Lie algebra [in SU(3) $C_{abc} \equiv f_{abc}$].

I have devoted much attention to the three gluon vertex because proof that it exists and contributes to observables would be an important contribution to confirming QCD. Similar arguments applied to gluon-gluon Compton scattering lead to a four gluon vertex. I shall not detail it here (see Ref. 9).

RENORMALIZATION AND SCALING VIOLATION

We have seen the basic rationale behind the quark-gluon vertices, albeit heuristically. These ideas can be used to study q-g interactions as perturbative corrections to the naive quark-parton model of deep inelastic phenomena where the quarks are treated as quasi-free. But if QCD is the correct theory of Nature, then this quasi-free behavior should emerge naturally in the theory, not be put in by hand. We shall see that that is indeed the case. Anticipating the successful outcome let's begin by describing some general features of theories like QED and QCD and gradually make contact with physically interesting applications.

We saw in the introduction how scaling and its violation are related to the absence or presence of inherently important mass scales in the physics of interest.

If we concentrate on u,d quark flavors then

$$m_{u,d} << m_\pi,\ m_P;\quad m_{gluon} = 0\ . \tag{3}$$

So we can consider a limit where all the fields are massless: there is no explicit mass scale in the theory. If an electron is scattered from such a system of free quarks and gluons then the (dimensionless) structure function of deep inelastic scattering $F_2(X,Q^2)$ would be invariant under a change of momentum scale (be Q^2 independent). This is the essence of the quark parton model.

Let them interact—as the coupling is dimensionless there is still no mass scale in the theory and so one might still expect scaling to obtain. However, there is a *hidden* running mass scale in QCD (and QED) connected with renormalization. This has the consequence that at any Q^2 mass scales are always important and a (logarithmic in Q^2) violation of scaling occurs.

Lagrangians and Renormalization

The equations of motion (i.e. physics) arise from minima of the action (good old classical mechanics taken over to field theory). The action is related to a Lagrangian $\mathcal{L}$.

$$\text{Equations of motion} \longleftrightarrow \partial A = 0 \qquad (4)$$

$$\updownarrow \qquad\qquad \updownarrow$$

$$\partial_\mu \frac{\partial \mathcal{L}}{\partial(\partial_\mu \phi)} = \frac{\partial \mathcal{L}}{\partial \phi} \qquad A \equiv \int \mathcal{L}\, d^4x \quad . \qquad (5)$$

Thus it is convenient to write down a Lagrangian, connected to phyics by the above steps.

Lagrangians should contain the field(s) of interest and be Lorentz invariant. There is nothing magic about them—they are constrained to give sensible physics as output: you put in what you want to get out. If you find renormalization philosophically disturbing, remember this point later.

I shall not labor the well known features of QED: *viz.* that local gauge invariance under

$$\psi(x) \rightarrow e^{i\theta(x)}\psi(x) \qquad (6)$$

is not satisfied by $\overline{\psi}(x)\gamma_\mu\partial^\mu\psi(x)$ alone and that one necessarily needs to introduce a massless gauge field (photon)

$$A_\mu(x) \rightarrow A_\mu(x) + \frac{1}{g}\partial_\mu\theta(x) \qquad (7)$$

and that

$$\mathcal{L} = \overline{\psi}(i\not{\partial} - ig\not{A})\psi - \frac{1}{4}F_{\mu\nu}F^{\mu\nu} \quad . \qquad (8)$$

The generalization (Yang and Mills; Shaw)[10] to SU(N) non-Abelian theory is

$$\psi(x) \rightarrow \exp\left[i\theta^a(x)\cdot\lambda^a\right]\psi(x) \qquad a = 1\ldots(N^2 - 1) \quad . \qquad (9)$$

The resulting modification in the gauge field (check by analogous steps to familiar QED) is

$$A^a_\mu(x) \rightarrow \underbrace{\left[A^a_\mu(x) + \frac{1}{g}\,\partial_\mu \theta^a(x)\right]}_{\text{trivial generalization}} \quad + \quad \underbrace{\left[f^{abc}\,A^b_\mu\,\theta^c\right]}_{\text{new piece}} \tag{10}$$

where $f_{abc}\,\lambda^c \equiv [\lambda^a, \lambda^b]$.

As in QED we must include the free boson field in the total Lagrangian. Introduce

$$F^a_{\mu\nu} \equiv \underbrace{\left[\partial_\mu A^a_\nu - \partial_\nu A^a_\mu\right]}_{\text{trivial generalization}} \quad - \quad \underbrace{\left[g f^{abc}\,A^b_\mu\,A^c_\nu\right]}_{\text{new from commutator}} \tag{11}$$

Since $a = 1 \ldots (N^2-1)$ we see N^2-1 massless gauge fields replace the single photon of the U(1) theory, QED. The Lagrangian is formally the same as in QED but with the new difinition of $F_{\mu\nu}$ above.

Note that $F_{\mu\nu}F_{\mu\nu}$ contains interaction terms

i. $g f^{abc}(\partial_\mu A^a_\nu) A^b_\mu A^c_\nu \equiv$  (12)

ii. $g^2 f^{abc}\,A^b_\mu A^c_\nu\,f^{ade}\,A^d_\mu A^e_\nu$ 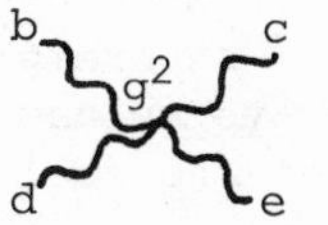 (13)

which is the four-gluon vertex.

This is the full content of the theory at the basic Lagrangian level. The previous approach presupposed the gauge invariance and showed that multi-gluon vertices were needed in a particular case. The Lagrangian approach is more fundamental and assures us that this is the whole story. (In this introduction I will not delve into possible extra CP violating contributions proportional to the so-called θ-parameter.)

The original Lagrangian generates quantum loops which have infinite amplitude e.g. the electron and photon propagators have divergent contributions  and 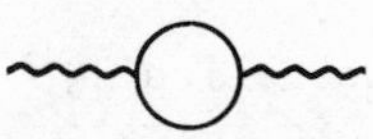

But physics is finite. This means that we weren't smart enough in our original guess for $\mathcal{L}$.

If for every + ∞ we could find a − ∞ we might end up with a finite, physically sensible theory. Let's try

$$\mathcal{L} = \bar{\psi}(x)(i\nabla\!\!\!/ - eA\!\!\!/)\psi(x) - \frac{1}{4}F_{\mu\nu}F^{\mu\nu} - B\bar{\psi}(x)(i\nabla\!\!\!/ - eA\!\!\!/)\psi(x) + CF_{\mu\nu}F^{\mu\nu} . \tag{14}$$

Then at second order the diagrams generated would be:

i.  diverges, but this divergence is cancelled by

−B  ;

ii.  diverges; cancelled by −B 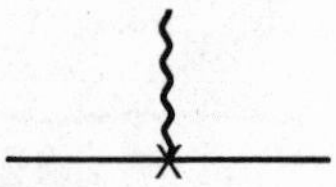 (that the same B appears is due to gauge invariance).[11]

iii.  diverges; cancelled by C  .

The B and C have contributions to each order of α

$$B = B_1\alpha + B_2\alpha^2 + \ldots$$
$$C = C_1\alpha + C_2\alpha^2 + \ldots \tag{15}$$

The important point is that the counter terms have the same form as terms already in $\mathcal{L}$. It is not trivial that finiteness obtains at each and every order: that it does can be proved—the theory is said to be "renormalizable". An example of non-renormalizability is if at each order new counter-terms had to be invoked. For example in a Lagrangian with $\lambda\phi^6$ then the following divergent diagram arises

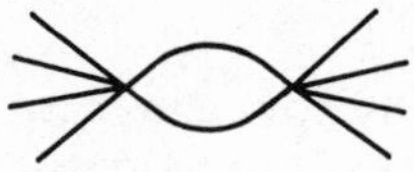

The counter term 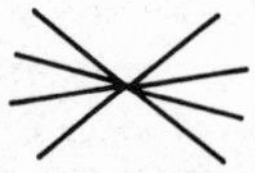 is of form $\lambda\phi^8$ a structure that was not in the original. One can add this in. This in turn generates

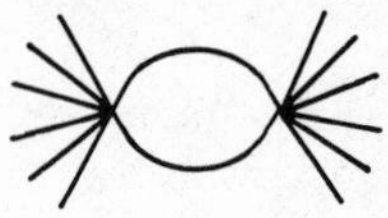 necessitating a $\lambda\phi^{12}$ counter term and so on.

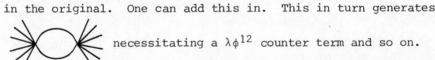

Specifically, if we start with "unrenormalized" quantities ψ_u, F_u, e_o in terms of which

$$\mathcal{L} = \bar{\psi}_u(i\nabla\!\!\!/ - e_o A\!\!\!/_u)\psi_u - \frac{1}{4}F_uF_u \tag{16}$$

(dropping Lorentz indices for convenience), then infinite amplitudes would obtain. If instead we use rescaled ("renormalized") quantities

$$\begin{aligned} \psi &\equiv (1-B)^{-1/2}\ \psi_u \equiv Z_1^{-1/2}\ \psi_u \\ A &\equiv (1-C)^{-1/2}\ A_u \equiv Z_3^{-1/2}\ A_u \\ e &\equiv (1-C)^{+1/2}\ e_o \equiv Z_3^{1/2}\ e_o \end{aligned} \tag{17}$$

(the $Z_{1,3}$ are infinite!) then

$$\mathcal{L} = (1-B)\ \bar{\psi}(i\partial\!\!\!/ - eA\!\!\!/)\psi - (1-C)\ \frac{1}{4}\ FF$$

and perturbation theory with these "physical" fields and couplings will yield finite answers (multiplied by infinities which are uniformly absorbed in the Z factors and which never appear in physical observables).

These remarks may be better understood by an explicit example. Consider an $O(\alpha)$ calculation of the photon propagator $D(q^2)$

$$D_u(q^2) \underset{\substack{(\text{suppressing}\\ \mu,\nu\ \text{indices})}}{=\!=\!=\!=\!=} \int \langle T[A_u(x)\ A_u(0)]\rangle\ e^{iq\cdot x}dx$$

$$= \frac{1}{q^2}\left[1 + \frac{\alpha}{3\pi}\log\ (-q^2/M^2) + \ldots\right]$$

(18)

(this is derived in Bjorken and Drell).[11] The quantity M is a cut-off introduced to regulate the infinite divergence. If we choose

$$Z_3 \equiv (1-C) = 1 - \frac{\alpha}{3\pi}\log M^2/\mu^2 \tag{19}$$

where μ^2 is arbitrary but finite then $D_u = Z_3 D_R(q^2,\mu^2,\alpha)$,

$$D_R(q^2,\mu^2,\alpha) = \frac{1}{q^2}\left[1 + \frac{\alpha}{3\pi}\log\ (-q^2/\mu^2) + \ldots\right] \tag{20}$$

which is indeed finite as promised!

Various comments should be made at this point and we shall study them in greater detail subsequently. First of all *we see that a mass scale*, μ, *has appeared in the theory even though none was manifest in the original Lagrangian*. This was the motivation behind the discussion that started this section. This is the price that we pay for finiteness: namely the introduction of an *arbitrary* mass scale μ (known as the "renormalization point"). The above is not a unique scheme for rendering Green functions finite. This particular approach is known as the "MOM" or "*momentum*" scheme and is such that if we put $q^2 = -\mu^2$ then

$$D(q^2, q^2, \alpha) = \frac{1}{q^2} \tag{21}$$

which is the canonical free photon propagator. One can generalize to arbitrary Green functions (details in Ref. 9) and define the MOM scheme by demanding that renormalized propagators and vertices recover their canonical form when $q^2 = -\mu^2$.

Logs in Perturbation Theory

The appearance of logs poses a problem for the naive application of perturbation theory. If we calculate perturbatively we would typically find the following, e.g.

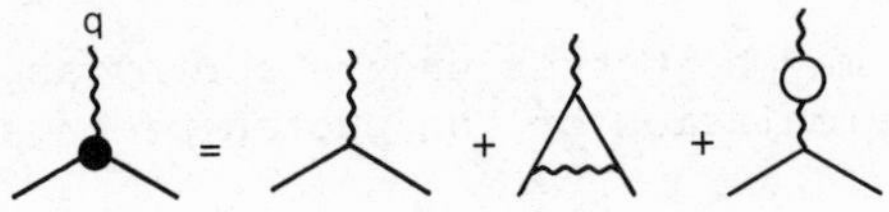

$$A(q^2) = \sum_n \alpha^n f_n(q^2)$$

where

$$f_o = \alpha_o$$
$$f_1 = \alpha_1 \log(-q^2/\mu^2) + b_1$$
$$f_n = \alpha_n[\log(-q^2/\mu^2)]^n + b_n(\log)^{n-1} + \ldots \tag{22}$$

Although α may be small, $\log q^2/m^2$ and hence $\alpha \log q^2/m^2$ could be large, thus a perturbative expansion in the $f_i(q^2)$ would break down.

In QED this is not a problem in practice. If $\mu^2 = m_e^2$ then with $\alpha \simeq 1/137$ we see that $\alpha \ln(-q^2/m^2) << 1$ so long as $-q^2 < m_e^2 \exp(137)$ (which is always true!). In QCD, however, it is a real problem. First of all $\alpha/\pi \simeq O(1/10)$ at moderate Q^2 and it grows as Q^2 decreases. And so the leading logarithms (at least) in each order are quite comparable. Thus it is in practice desirable to reorganize the series into the form

$$\begin{aligned} A(q^2) &= \sum \alpha^n F_n[\alpha \log(-q^2/\mu^2)] \\ &= \sum_n \alpha_n[\alpha \log(-q^2/\mu^2)]^n + \alpha \\ &\quad + \alpha \sum b_n[\alpha \log(-q^2/\mu^2)]^{n-1} + \ldots \end{aligned} \tag{23}$$

The perturbation expansion is now column by column in Eq. (22) rather than row by row. The problem is now to sum the leading logarithms.

We show how this is done in Ref. 9. We here merely quote the answer in order to motivate the immediate development in the lectures. If you chose $\mu^2 \equiv -q^2$ then the log would vanish and the naive perturbation expansion in α ("row by row") would be recovered. Since μ is arbitrary this is quite valid—so where did the log problem go to? "There is no such thing as a free lunch" as any American knows, so there must be a price to pay; some hidden charge. The clue is to ask—"but what if I chose a different μ", or equivalently, "what if I now changed q^2"? The answer is that the naive perturbation in α can be performed, but α is now an *effective* Q^2 dependent quantity

$$\frac{\alpha(Q_1^2)}{\alpha(Q_1^2)} \sim \log(Q_2^2)/\log(Q_1^2) \quad . \tag{24}$$

Thus the dimensional, or momentum, dependence appears in $\alpha(Q^2)$.

The apparatus that is used to quantify this miracle is the renormalization group equation. Although it is necessary to study it in order to formalize the procedure, it is possible to show that α is Q^2-dependent without going into the RGE. The approach will be heuristic in that some points will be glided over; proof of their validity will only be seen when the full RGE treatment is employed.[2]

$\alpha(Q^2)$: A COUPLING CONSTANT WHICH ISN'T A CONSTANT

Doubtless you are all familiar with the notion that in QCD the coupling "constant" is not constant but instead has a Q^2 dependence, tending to zero as $Q^2 \to \infty$. This is the phenomenon of asymptotic freedom[12] that has generated so much excitement in that it:

i. explains why quarks appear to be rather weakly bound when viewed at large Q^2;

ii. has given a new perspective on our ideas of the intrinsic "strengths" of the fundamental interactions. As we shall see, α in QED is not (always) 1/137, the "strong" interactions are not always strong. These discoveries have paved the way for an exciting line of investigation;

iii. if the strengths of the fundamental interactions are energy dependent, does there exist some extreme energy where all interactions have the same strength? If so, can one create a unified theory of all the forces which is exact at this "grand unified mass scale?"

I shall first demonstrate that even in QED α is not a constant, and then proceed to QCD to show the similarities and crucial differences.

$\alpha(Q^2)$ in QED

The Coulomb amplitude between two infinite mass sources, to and including the one loop correction, follows immediately from Eq. (20). For massless particles in the loop we find

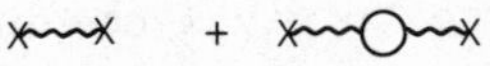

$$= \frac{e^2}{q^2}\left[1 + \frac{\alpha}{3\pi}\log(-q^2/\mu^2)\right] Z_3(\infty) \quad . \tag{25}$$

Now, the electron-positron have small but *non-zero* masses. While Eq. (25) is fine for $m^2 = 0$ or for $q^2 \gg m^2$ it is not good enough for $q^2 \ll m^2$. The effect of including the lepton masses is to modify the logarithm thus[11]

$$\log(-q^2/\mu^2) \to c\int dx\; f(x)\; \log\left[\frac{-q^2 f(x) + m^2}{\mu^2 f(x) + m^2}\right] \quad . \tag{26}$$

[Note that if there were only massless charged leptons available for the loop, then the $\log(-q^2/\mu^2)$ would be correct and an infrared divergence would arise as $q^2 \to 0$. In QCD this is a real problem because there is the possibility of massless colored gluons in the loop, with consequent infrared divergence.]

Now we are in a position to study three particular regions of q^2.

i. $-q^2 \ll m^2 \to 0$. This yields the Coulomb amplitude at large distances: Eq. (25) as $q^2 \to 0$ then yields (the log $\to$ 0)

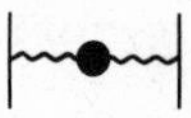

$$= \frac{e^2}{q^2} Z_3 \equiv \frac{e_R^2}{q^2} \tag{27}$$

where e_R^2 is the renormalized (charge).[2] It is *this* quantity ($\div 4\pi$) that is 1/137 as $q^2 \to 0$, *not* the e^2 in the Dirac equation.

The quantity Z_3 will be present in any photon exchange amplitude. The divergences which naively arise in such amplitudes will disappear if they are written in terms of the renormalized quantities.

ii. $-q^2 \to m^2 \neq 0$. Expanding out the logarithm reveals that the amplitude becomes

$$\frac{e^2}{Q^2}\left[1 + \frac{\alpha}{3\pi}\left(\frac{Q^2}{5m^2}\right)\right] \tag{28}$$

(where $Q^2 \equiv -q^2 > 0$). Rewriting this in terms of the renormalized charge, to this order, yields ($Z_3 e^2 \equiv e_R^2$)

$$\text{Amplitude} = \frac{e_R^2}{Q^2}\left[1 + \frac{\alpha_R}{15\pi}\frac{Q^2}{m^2} + \ldots\right] \equiv \frac{e_{\text{eff}}^2(Q^2)}{Q^2} \quad . \tag{29}$$

Note that the positive sign causes the interaction strength to increase (which fits in with the heuristic picture described earlier for the dielectric and vacuum polarization). Such an effect has been detected experimentally in the -27 Mc s^{-1} "Uehling" contribution relative to the Lamb shift in hydrogen.

iii. $-q^2 >> m^2$. In this case we recover the starting equation

$$\text{Amp} = \frac{Z_3 e^2}{q^2}\left[1 + \frac{\alpha}{3\pi}\log(-q^2/\mu^2)\right] . \tag{30}$$

Writing $Z_3 e^2 \equiv e_R^2$, then to this order we see that

$$\alpha_{\text{eff}}(Q^2) \sim \alpha\left[1 + \frac{\alpha}{3\pi}\log(-q^2/\mu^2)\right] = \tag{31}$$

and so α_{eff} rises as Q^2 rises.

We can immediately generalize this to the case of an infinite sequence of single loops, thus

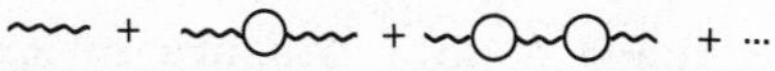

$$\alpha_{\text{eff}}(Q^2) = \alpha\left[1 + \frac{\alpha}{3\pi}\log + \left(\frac{\alpha}{3\pi}\log\right)^2 + \left(\frac{\alpha}{3\pi}\log\right)^3 + \ldots\right] \tag{32}$$

Therefore

$$\alpha_{\text{eff}}(Q^2) = \alpha/\left(1 - \frac{\alpha}{3\pi}\log(-q^2/\mu^2)\right) . \tag{33}$$

In the sense of Eqs. (22) and (23) these are the leading logarithms, namely the terms of the form $\alpha^n \log^n$. The

$$\alpha_{\text{eff}}(Q^2) = \alpha \sum_n \left(\frac{\alpha \log}{3\pi}\right)^n$$

and, in the sense of Eq. (23), has summed the leading logarithms. Thus

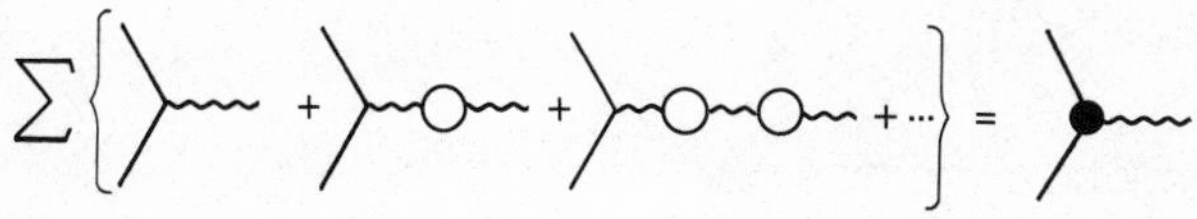

Leading logs; all order in perturbation theory

$\alpha_{\text{eff}}(Q^2)$ 1st order in "leading log improved" perturbation theory. (34)

$\alpha(Q^2)$ in QCD

The rise in $\alpha_{\text{eff}}(Q^2)$ in QED at large Q^2 was due to the charge screening by e^-e^+ loops. In QCD the virtual quark loop has the same effect and tends to increase α. The presence of a three gluon vertex generates a new topology in QCD:  which tries to decrease α. There is a competition and the gluons win if there are not too many quark flavors.

Explicit calculation reveals that in lowest order

$$\alpha_{eff}(Q^2) \sim \alpha\left[1 - \frac{\alpha}{4\pi} B \log(-q^2/\mu^2)\right] \tag{35}$$

(where $B \sim 11 - 2/3$ number of quark flavors). On summing all bubbles, as in QED, this exponentiates to

$$\alpha_{eff}(Q^2) \sim \alpha/1 + \frac{\alpha}{4\pi} B \log(-q^2/\mu^2) \quad . \tag{36}$$

From all of this discussion we see that α is not fundamental but is a *function*. It is conventional to summarize this functional dependence to leading order by writing

$$\alpha_{eff}(Q^2) = \frac{\pi}{B \log(Q^2/\Lambda^2)} + \ldots = \frac{12\pi}{(33-2F)\log Q^2/\Lambda^2} + \ldots \tag{37}$$

Naively $\alpha(Q^2 = \Lambda^2) \to \infty$ and hence Λ is a measure of the scale where the interaction becomes strong.

Quantitatively it is not so precise because Eq. (37) is only a leading order expression and breaks down as α becomes large [i.e. as $(Q^2 \to \Lambda^2)$].

This all suggests that although the interquark forces are large at small Q (large distances) they are indeed feeble at short distances. At high energies it may be possible to successfully make perturbative calculations in $\alpha_s(Q^2)$ and confront data with QCD directly. This is discussed by Llewellyn Smith.[2]

Although no one has yet managed to generate a unified quantitative description of both short- and long-range properties of quark forces, we can make some qualitative remarks on quarks and nuclear forces.

QUARKS AND NUCLEAR FORCES

Earlier we drew analogies between QED and QCD; covalent atomic multi-gluon exchanges. Let's look at these in slightly more detail in order to bring out an important point about color analogues of van der Waals forces.

i. Covalent: Colored quarks are confined in colorless clusters or a distance of about 1 fermi. Denote this by a dotted region.

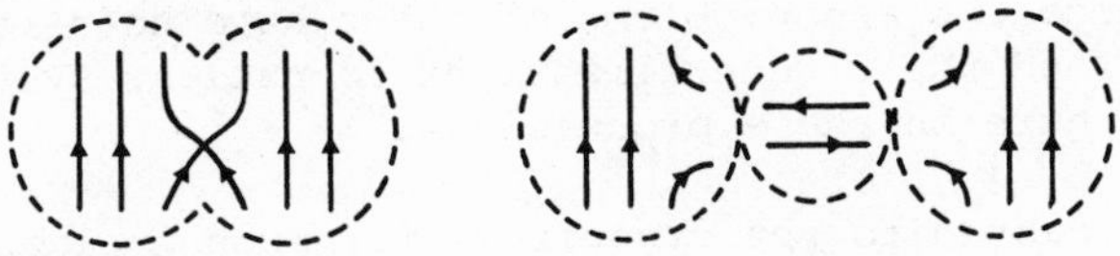

Short distance Long distance

At short distance the two confining bags overlap and quarks can be freely exchanged by the hadrons. This is the motivation behind the successful calculations of energy and angular distributions of large angle hadron scattering (see Brodsky in these proceedings).[13]

At large distances the exchanged quarks would have to escape from the confining domain. They must therefore cluster into their own colorless system. The result is the $q\bar{q}$ meson exchange familiar to nuclear physicists.

A unified description of these two extremes is not yet to hand. However, the qualitative features are clear and have an important moral for the van der Waals force analogue involving gluon exchange.

ii. van der Waals: Analogous to the quark exchange contrast two regions of distance scale for gluon exchanges.

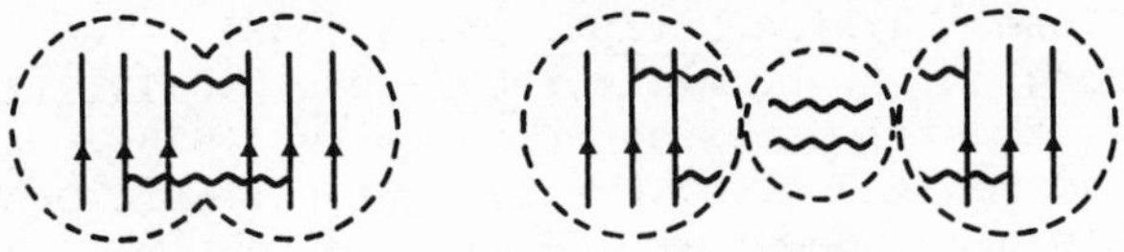

Short distance Long distance

At short distances perturbative QCD applies. The two gluon exchange is perturbatively calculable as in QED. It is wrong to extrapolate this to large distance and propose that van der Waals color forces will be manifested in low energy NN or nuclear forces. Just as the quarks cluster into color singlet systems at large distance, so will two gluons form color singlet "glueballs".

Thus there may be a component in nuclear forces arising from exchange of the lightest glueball. This may be hard to find as glueballs, if they exist, are probably over 1 GeV in mass.[14] This may generate a new short-range Isospin zero exchange force in nuclei. However, the naive van der Waals behavior is not to be expected.

iii. Multiquark Clusters: Within $q\bar{q}$ or qqq clusters there has been good progress in understanding the energy shifts arising from one gluon exchange. This generates a color analogue of the Fermi-Breit Hamiltonian of QED which may be treated as a perturbation about the zeroth order energies arising from the dominant confining potential.

The successes and problems of these have been reviewed elsewhere[6,7,15] and their generalizations to nuclear systems are discussed elsewhere in these proceedings.[3,5]

I am indebted to the organizers for such a pleasant shool and to Ingrid Duelli for typing this manuscript.

REFERENCES

1. F. E. Close, "Introduction to Quarks and Partons," Academic Press, London (1979). This also contains a detailed bibliography and list of references.
2. C. H. Llewellyn Smith, these proceedings.
3. M. Harvey, these proceedings.
4. O. W. Greenberg, Phys. Rev. Lett. 13:598 (1964).
5. H. J. Lipkin, these proceedings.
6. A. De Rújula, H. Georgi and S. L. Glashow, Phys. Rev. D12:147 (1975); N. Isgur and G. Karl, Phys. Lett. 72B:109 (1977); Phys. Rev. D18:4187 (1978).
7. F. E. Close in: "Quarks and Nuclear Forces," Springer Verlag, Heidelberg (1982).
8. H. Fritzsch, Physica Scripta 25:119 (1982); J. D. Bjorken, Physica Scripta 25:69 (1982).
9. F. E. Close, Physica Scripta 25:86 (1982).
10. C. N. Yang and R. L. Mills, Phys. Rev. 96:191 (1954); R. Shaw, Ph.D. thesis, Cambridge University (1955).
11. J. D. Bjorken and S. D. Drell, "Relativistic Quantum Mechanics", McGraw Hill, New York (1964).
12. G. 't Hooft, (unpublished, 1972); H. D. Politzer, Phys. Rev. Lett. 30:1346 (1973); D. J. Gross and F. Wieczek, Phys. Rev. Lett. 30:1343 (1973).
13. S. J. Brodsky, these proceedings.
14. F. E. Close "QCD and the search for glueballs," Rutherford Lab Report RL-82-041 (1982).
15. F. E. Close and R. H. Dalitz, in: "Low and Intermediate Energy Kaon-Nucleon Physics," D. Reidel, Dordrecht (1981).

WHY BELIEVE IN QCD?

C.H. Llewellyn Smith

Department of Theoretical Physics
1 Keble Rd.
Oxford OX1 3NP, England

OVERVIEW OF EVIDENCE

Introduction

I will divide the evidence for QCD into *a priori* evidence and *a posteriori* evidence, obtained by comparing specific QCD predictions with data.[1] The former consists of the very strong evidence, reviewed below, that hadrons are made of colored quarks which interact through the exchange of vector (spin one) gluons. Evidently the force must be color dependent, since we only see a single π meson rather than the nine different color combinations which can be made from a tricolored quark and antiquark. The only sensible (renormalizable, unitary etc.) theory of vector gluons coupled to color is QCD [or a color gauge theory based on SO(3) but this turns out to fail the *a posteriori* tests and will not be discussed further].

The *a posteriori* evidence can be divided into short distance and long distance tests of QCD. The former will be considered in detail in later sections. Some remarks about long distance tests are made at the end of this section where references to the literature are given.

A Priori Evidence

The most convincing evidence that quarks have three "color" degrees of freedom is that:

i. Color is needed to solve the "spin statistics" problem of the quark model for baryons, allowing (e.g.) the $\Delta^{++}(1232)$ state to be made of $u^{\uparrow}u^{\uparrow}u^{\uparrow}$ in S-waves with a symmetric orbital wave function without violating the Pauli principle by letting the 3 quarks have different colors.

ii. Color is needed to account for $\sigma(\overline{e}e \to \text{hadrons})$. Consider the dimensionless ratio

$$R = \frac{\sum_F \sigma(\overline{e}e \to F)}{\sigma(\overline{e}e \to \overline{\mu}\mu)} \quad .$$

Since the electromagnetic current couples only to quarks, the numerator is given by

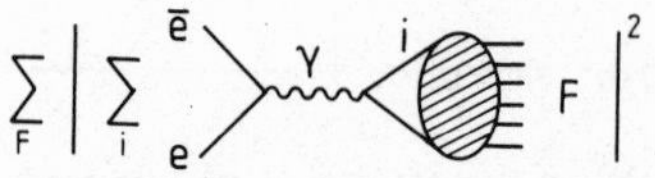

where i runs over all types of quarks. It has long been thought that at high energies the quark-hadron transition does not change the total cross section and can be ignored (see following section) and that interquark interactions can also be ignored in the spirit of the impulse approximation (this is only really justified in QCD). In this case, the numerator and denominator in R are the same apart from the fact that the charge ($Q_i|e|$) of a quark differs from the charge (e) of the muon and we have

$$R = \sum_i Q_i{}^2 \quad .$$

Well above the threshold for b-quark production this gives

$$R = N_c\left(Q_u^2 + Q_d^2 + Q_s^2 + Q_c^2 + Q_b^2\right)$$

$$= N_c \; 11/9$$

where N_c is the number of colors. The data (Fig. 1) support the idea that R is energy dependent well away from thresholds for new flavor production and give $N_c = 3$.

iii. It is needed to account for $\Gamma(\pi^0 \to \gamma\gamma)$. By considering a Ward identity (an exact relation which follows from the symmetries of the theory), it is possible to show[3] that in the chiral limit (defined below) the amplitude

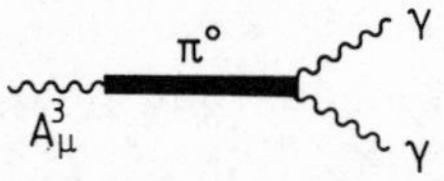

is given by a bare quark triangle diagram which is proportional to N_c. Here $A_\mu{}^3$ is the third component of the axial isospin current:

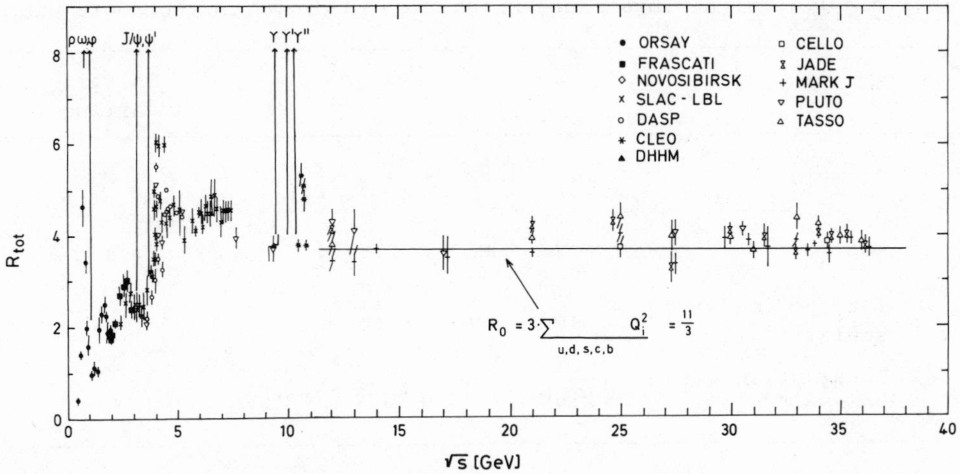

Fig. 1. $R = \sigma(\bar{e}e \rightarrow \text{hadrons})/\sigma(\bar{e}e \rightarrow \bar{\mu}\mu)$ as a function of the center of mass energy $\sqrt{s}$. Data from PETRA groups are plotted together with some lower energy data, compiled in Ref. 2. Only statistical errors are shown. The solid line is the prediction for u, d, s, c and b quarks with three colors.

$$\vec{A}_\mu = \bar{\psi}\gamma_\mu\gamma_5\vec{T}\psi$$

with

$$\psi = \begin{pmatrix} u \\ d \end{pmatrix} ,$$

where u and d are up and down quark fields. The coupling of $A_\mu{}^3$ to π^0 is equal, by an I-spin rotation, to the $\pi^\pm$ decay constant f_π, which is measured by $\Gamma(\pi^+ \rightarrow \mu\nu)$. Hence

$$\Gamma(\pi^0 \rightarrow \gamma\gamma) f_\pi^2 = \text{const } N_C \ .$$

Putting in the measured value of f_π gives

$$\Gamma(\pi^0 \rightarrow \gamma\gamma) = 7.87 \left(\frac{N_C}{3}\right)^2 \text{ eV} \ .$$

Comparison with the experimental value of 7.95 ± 0.55 eV requires $N_C=3$.

Strong evidence that gluons have spin one is provided by the success of predictions obtained by assuming an exact chiral symmetry.[4] As we shall see, these predictions work to 10% and it seems clear that in the limit of exact chiral symmetry the world would not change much. In such a world the force between quarks can only be vector in nature (assuming that the gluons do not carry weak and electro-

Table 1. Chiral Invariance Properties of Terms in the Lagrangian

Term		Chiral Invariance?
Kinetic	$\bar{\psi}\gamma_\mu\psi = \bar{\psi}_L\gamma_\mu\psi_L + \bar{\psi}_R\gamma_\mu\psi_R$	✓
Mass	$m\bar{\psi}\psi = m\bar{\psi}_L\psi_R + m\bar{\psi}_R\psi_L$	x
Interaction with scalar field	$\phi\bar{\psi}\psi = \phi\bar{\psi}_L\psi_R + \phi\bar{\psi}_R\psi_L$	x
Interaction with vector field	$A^\mu\bar{\psi}\gamma_\mu\psi = A^\mu\bar{\psi}_L\gamma_\mu\psi_L + A^\mu\bar{\psi}_R\gamma_\mu\psi_R$	✓

magnetic charges, as required by many successes of the quark model). Define

$$\psi = \begin{pmatrix} u \\ d \end{pmatrix} = \psi_L + \psi_R$$

with

$$\psi_{L,R} = \left(\frac{1 \mp \gamma_5}{2}\right) \psi \quad .$$

For massless quarks, $\psi_L(\psi_R)$ represents particles with helicity $-1/2(+1/2)$. Ordinary I-spin conservation corresponds to invariance under a simultaneous and equal isospin rotation on ψ_L and ψ_R. It leads to conservation of the current

$$\vec{V}_\mu = \bar{\psi}\gamma_\mu\vec{T}\psi \quad ,$$

whose charged components generate Fermi transitions. This leads to CVC. Chiral symmetry is invariance under *independent* I-spin rotation on ψ_L and ψ_R. It leads to conservation of the axial isospin current

$$\vec{A}_\mu = \bar{\psi}\gamma_\mu\gamma_5\vec{T}\psi \quad ,$$

whose charged components generate Gamow-Teller transitions. Before we consider the implications of chiral symmetry, we record in Table 1 the behavior of various possible terms in the Lagrangian under chiral rotations, which follow from elementary properties of γ_5. Only a vector gluon interacting with massless quarks is allowed in a renormalizable theory with exact chiral symmetry.

If the axial current is conserved, $(\partial^\mu\vec{A}_\mu = 0)$ then the matrix elements of the axial contribution to neutron β-decay

$$\langle P(p')|A_\mu^+(0)|N(p)\rangle = \overline{V}(p')(\gamma_\mu\gamma_5 g_A + q_\mu\gamma_5 f_p)\, V(p)$$

where $q = p-p'$, must satisfy

$$q^\mu\langle P|A_\mu|N\rangle = 0 = \overline{V}(p')\left(-2Mg_A(q^2) + q^2 f_p(q^2)\right) V(p) \quad .$$

At $q^2 = 0$, this equation requires that *either* $M_{n,p} = 0$, which we reject as it is remote from the real world, *or* $f_p \xrightarrow{q^2\to 0} \frac{2Mg_A}{q^2}$. One of the many contributions to the amplitude is

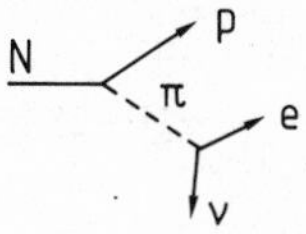

which gives

$$f_p = \frac{g_{pn\pi} f_\pi}{q^2 - M_\pi{}^2} + \ldots$$

We therefore see that $\partial^\mu \vec{A}_\mu = 0$ would be possible if the pion were massless and

$$-2Mg_A + g_{pn\pi}\, f_\pi = 0 \quad .$$

This relation, which is known as the Goldberger-Treiman relation, works to 8%. Furthermore, M_π is small compared to M_ρ, ρ being the

Table 2. Tests of Predictions in Chiral Limit

	Prediction in chiral limit	Data
$\frac{M_\pi{}^2}{M_\rho{}^2}$	0	0.03
$1-\frac{2Mg_A}{f_\pi g_{pn\pi}}$	0	0.08 ± 0.01
$M_\pi^2\ a_{\pi N}^{1/2}$	0.16	0.17 ± 0.05
$M_\pi^2\ a_{\pi N}^{3/2}$	-0.078	-0.088 ± 0.004
λ^0	0.021 ± 0.003	0.019 ± 0.004
$\Gamma(\pi^0 \to \gamma\gamma)$	7.87 eV	7.95 ± 0.55 eV

next lightest state made of u and d quarks. Assuming that chiral symmetry is broken by quark masses, it turns out that $M_\pi^2 \propto m_{u,d}$ so the correct measure of chiral symmetry breaking is

$$\frac{M_\pi^2}{M_\rho^2} = 0.03 \ .$$

Many exact relations can be obtained in the chiral limit. Those which have been tested by accurate data are shown in Table 2, where $a^I_{\pi N}$ are pion-nucleon scattering lengths for isospin I and λ^0, the slope parameter in the Dalitz plot for $K \to \pi e \nu$. The success of so many predictions cannot be an accident. We conclude that the interquark force is mediated by vector gluons which couple to color. The only viable renormalizable field theory of this type is QCD.

A Posteriori Evidence

Consider a test charge placed in a dielectric as shown in Fig. 2. To define the charge, we construct a hypothetical sphere of radius R and calculate

$$Q(R) = \int \vec{E} \cdot d\vec{s} \ .$$

Q(R) will vary with R, being the free space value Q for R small compared to the intermolecular spacing, but being reduced to Q/ε for large R because polarization produces a net flow of negative (screening) charge into the sphere. The idea that charges vary with distance is also true in quantum field theory. In QED, for example, the force between an electron and proton which is given by the diagrams

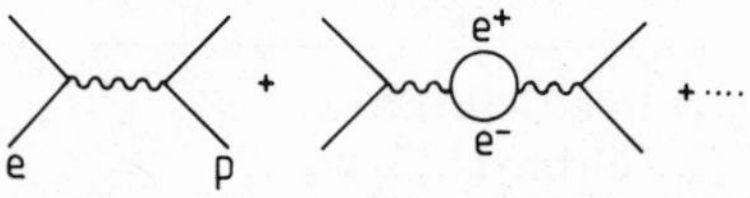

can be written $\frac{\alpha(r)}{r}$, or in momentum space $\frac{\alpha(q^2)}{q^2}$ where q^2 is the four-momentum transfer squared. The "running" fine structure constant $\alpha(r)$ or $\alpha(q^2)$ is equal to α for $r = \infty$ or $q^2 = 0$. For small q, it is given by

$$\alpha(q^2) = \alpha - \frac{\alpha}{15\pi} \frac{q^2}{M^2} + \ldots$$

The second term is very well tested as it contributes -27 MHz to the Lamb shift.

In QED, $\alpha(r)$ increases as $r \to 0$. In QCD, however, $\alpha_s(r) \equiv \frac{g^2(r)}{4\pi}$ (where g_s is the QCD coupling constant in units with $\hbar = c = 1$) *decreases*—a phenomenon known as asymptotic freedom,[5] to which we

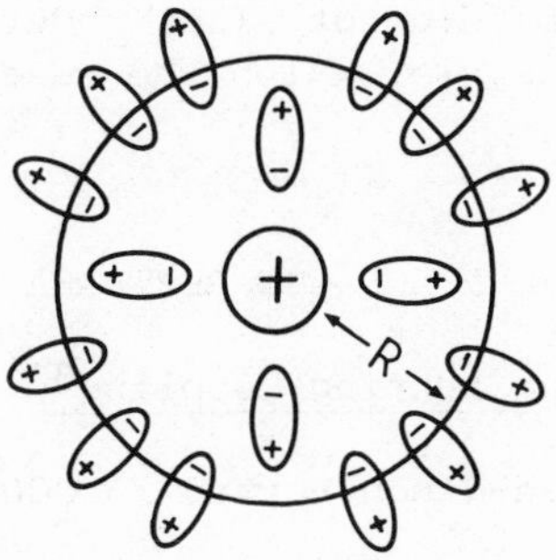

Fig. 2. Test charge placed in a dielectric.

shall return in the next section. Non-Abelian gauge theories are the only asymptotically free theories. The success of the parton model, or impulse approximation picture, in deep inelastic scattering ($eN \to eX$ etc.) shows that the strong force is indeed weak at short distance and it is obvious that it is large at long distance. In fact asymptotic freedom not only underwrites the parton model as a zeroth order approximation but allows us to predict deviations from the original naive impulse approximation which are confirmed semi-quantitatively by the data, as we shall see in the third section.

At large distances, QCD also seems capable of explaining the data. First, if we assume that the long range force between quarks has the simplest possible non-trivial color dependence (which is the same as that in one gluon exchange, although the long range potential is certainly *not* due to one gluon exchange), it turns out that it is most attractive in the color singlet channel.[6] There is therefore hope of explaining the fact that only color singlets have been seen. Color singlets can only be made from three quarks or a quark plus antiquark (or multiples thereof). Thus confined color would in turn explain the failure to observe states such as 4q or qq, which was originally one of the most puzzling features of the quark model. Furthermore, QCD inspired ideas turn out to work very well in the constituent quark model. In particular the assumption that the spin dependent potential is short range and, invoking asymptotic freedom, can therefore be abstracted from one gluon exchange, explains[7] why $M_\rho > M_\pi$, $M_\Delta > M_N$, $M_\Sigma > M_\Lambda$ and $\langle\gamma^2\rangle_N < 0$.

In the last few years there have been very interesting attempts to solve QCD numerically in an approximation in which fields are only defined on discrete points in a fine grained space-time lattice. Lattice QCD provides strong evidence for quark confinement and reproduces qualitatively many of the observed features of the low-energy spectrum, albeit in an approximation in which virtual quark loops are neglected. In the coming years it seems likely that lattice QCD will be used to investigate most aspects of low energy hadron structure. Time did not allow a detailed discussion of this topic in these lectures, for which the reader is referred to the literature.[8]

THE RENORMALIZATION GROUP AND QCD PERTURBATION THEORY

Renormalization and the Running Coupling[9]

Consider the three-gluon vertex in QCD, ignoring quarks *pro tem.* for simplicity:

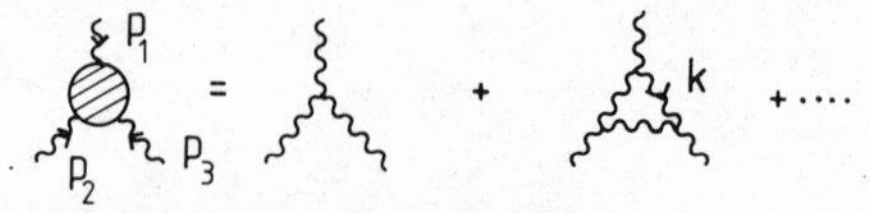

At the symmetric point $p_i^2 = -2p_i \cdot p_j (i \neq j) = p^2$, it is characterized by an invariant function $g(p^2)$ which can be calculated from the bare coupling g_o thus

$$g(p^2) = g_o + g_o^3 \int I(p,k) d^4k + \ldots$$

where the two terms correspond to the two diagrams above. In $\hbar=c=1$ units, g_o is dimensionless and I has dimensions M^{-4}. It is therefore not surprising that I behaves as k^{-4} for $k \to \infty$ and the integral diverges. Going to Euclidean space ($d^4k = d\Omega k^3 dk$) the integral is proportional to something like

$$\int_p^\infty \frac{dk}{k} \quad .$$

Even if we cut off the integral at the mass of the Universe, which seems reasonable, this seems absurd. It implies that $g(p^2)$ is equally sensitive to physics in the range of one to two solar masses, where we clearly do not know the correct form of the integrand, and the range p to 2p. This may be true, in the same limited sense that the properties of water depend on QCD! However, we can study hydrodynamics without particle physics by introducing a few phenomenological parameters (density, viscosity etc.), which are determined in principle by atomic physics. Atomic parameters in turn depend on parameters controlled by nuclear physics etc. In renormalizable field theory the same idea works. If we express $g(p^2)$ in terms of the value at a neighbouring scale μ, the result is finite

$$g(p^2) = g(\mu^2) + g^3(\mu^2) \int \left[I(p,k)\big|_{p^2} - I(p,k)\big|_{\mu^2} \right] d^4k + \ldots$$

and the dependence on high energy physics drops out, the integral being controlled by $p \lesssim k \lesssim \mu$. This is the basic idea of renormalization and works to all orders in perturbation theory. Measurable quantities are finite when expressed in terms of $g(\mu^2)$, which is used to parametrize the theory.

In non-renormalizable theories, the dependence on unknown physics at remote scales cannot be removed in this way. Physics at one scale depends essentially on all other scales. We could not understand anything without understanding everything! In renormalizable theories, however, we can understand one layer of the onion at a time and physicists can make a living.

Given that $g(p^2)$ is a finite function of $g(\mu^2)$, we can differentiate the relation with respect to $\ln p^2$ and set p^2 equal to μ^2 to obtain

$$\frac{\partial g(\mu^2)}{\partial \ln\mu^2} = \beta[g(\mu^2)] \quad .$$

This is a basic renormalization group equation which states in words that the change of g with μ depends only on g at the same scale. Provided g is small, β can be calculated perturbatively

$$\beta = -cg^3 + O(g^5) \quad .$$

In non-Abelian gauge theories c is positive i.e., g decreases as μ increases. This is asymptotic freedom. In QCD explicit calculation gives

$$\frac{\partial \alpha_s(\mu^2)}{\partial \ln\mu^2} = -\left(\frac{33-2F}{12\pi}\right)\alpha_s^2(\mu^2) + O(\alpha_s^3)$$

in terms of $\alpha_s \equiv g^2/4\pi$, where F is the number of quark flavors (assumed here to be massless for simplicity). Integration gives

$$\alpha_s(\mu^2) = \alpha_s(\mu_0^2)\left(1 + \frac{(33-2F)}{12\pi}\ln \mu^2/\mu_0^2\right)^{-1} + \ldots$$

Provided $F \leqslant 16$, we have asymptotic freedom. In QED the coefficient 33, which is due to gluon-gluon interactions, is absent and α increases for $\mu^2 \to \infty$.

Even if g is large for small μ^2, so that β is incalculable, it

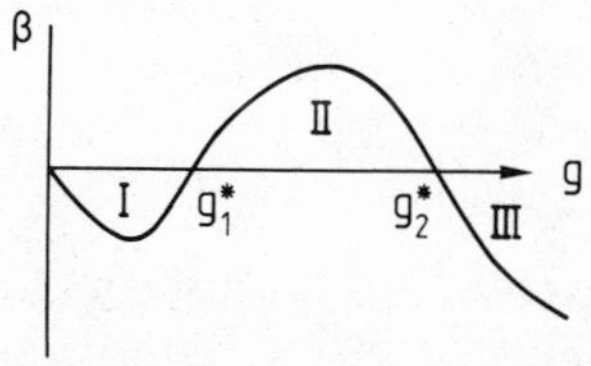

Fig. 3. An example of the possible behavior of the function β.

will become small asymptotically provided β is negative through its range. Conversely it will grow indefinitely for small μ^2. Lattice calculations suggest that this happens in QCD. If, however, β changes sign—as shown in Fig. 3, there will be three phases,

$$\begin{array}{llll} \text{I} & g \xrightarrow{\mu^2\to\infty} 0, & g \xrightarrow{\mu^2\to 0} g_1^* \\ \text{II} & g \xrightarrow{\mu^2\to\infty} g_2^* & g \xrightarrow{\mu^2\to 0} g_1^* \\ \text{III} & g \xrightarrow{\mu^2\to\infty} g_2^* & g \xrightarrow{\mu^2\to 0} \infty & . \end{array}$$

To find out which phase is realized, it would be necessary to calculate the vacuum energy.

At large μ^2, α_s is parametrized in terms of $\alpha_s(\mu_0^2)$ for some μ_0^2. Alternatively we can introduce a scale λ writing

$$\alpha_s(\mu^2) = \frac{\alpha_s(\mu_0^2)}{1 + c\alpha_s(\mu_0^2)\ln \mu^2/\mu_0^2} + \ldots = \frac{1}{c\ln \mu^2/\lambda^2} + \ldots$$

(Beyond leading order λ is defined slightly differently.) Roughly speaking λ characterizes the energy at which strong interactions become strong. The fact that λ has dimensions means that *for massless quarks QCD actually has no parameters!* Given λ we can calculate M_p/λ in principle, making $\alpha_s(M_p^2/\lambda^2)$ calculable! Alternatively, if we like we can choose M_p as our unit of mass making λ calculable.

The Renormalization Group

As discussed above, we must choose a scale μ to define $\alpha_s(\mu)$ which is the expansion parameter in perturbative calculations. The choice of μ does not matter i.e.,

$$\mu \frac{d}{d\mu} \text{Observable} = 0 \quad ,$$

or

$$\left(\frac{\partial}{\partial \ln \mu^2} + \beta \frac{\partial}{\partial g}\right) \text{Observable} = 0 \quad .$$

This is the renormalization group equation or RGE. At first it might be thought that if μ is chosen sufficiently large so that $\alpha_s(\mu^2) << 1$, perturbation theory would always be justified. In fact, this is not the case as terms of the form

$$(\alpha_s(\mu)\ln \mu/E)^n$$

are found, were E is a typical energy in the problem (the term of the form $\int_p^\mu \frac{dk}{k} = \ln\mu/p$ encountered in the last section is the simplest example of the way such terms are generated). However, we can remove such logs by setting $\mu=E$ (equivalently the logs can be summed using the RGE). If E is large, so that $\alpha_s(E) \ll 1$, perturbation theory might be useful.

Large E is therefore a necessary condition for the utility of perturbation theory but it is not sufficient. In most amplitudes, calculation generates terms such as

$$\left(\alpha_s(E)(\ln E/m)^a\right)^n$$

with $a=1$ or 2, where m is a quark mass, which obviously spoil perturbation theory. There are two mathematical circumstances in which perturbation theory may be useful, whose physical significance will be discussed below:

i. There is no dependence on $\ln m$ or equivalently the result is "mass finite" for $m \to 0$. A dimensionless observable $\hat{O}$ satisfies

$$\hat{O} \xrightarrow{E\to\infty} f\left(\alpha_s(E)\right) + O\left(\frac{m^2}{E^2}\right)$$

in this case, where f can be calculated perturbatively for large E.

ii. The E and m dependence factorize i.e.,

$$\hat{O} \xrightarrow{E\to\infty} A\left(\alpha_s(\mu), E/\mu\right) B\left(\alpha_s(\mu), m/\mu\right) + O\left(\frac{m}{E}\right) .$$

The renormalization group equation then reads

$$\frac{1}{A}\,\mu\,\frac{dA}{d\mu} = -\frac{1}{B}\,\mu\,\frac{dA}{d\mu} = 2\gamma\left(\alpha_s(\mu)\right)$$

where the separation constant γ is known as an anomalous dimension. The equation for A can then be written

$$\left(-\frac{\partial}{\partial \ln E^2} + \beta\frac{\partial}{\partial g} - \gamma\right) A = 0 \quad ,$$

so that the renormalization group gives the asymptotic energy dependence of $\hat{O}$, although its absolute value, which depends on B, is not determined.

Conditions for the Use of Perturbation Theory

In the last section we found that the utility of perturbation theory is controlled by the dependence on the quark mass m. Amplitudes are only sensitive to m when virtual particles become almost real [so that propagators $\sim(p^2-m^2)^{-1}$ blow up] and can propagate over large distances. Quantities which are sensitive to m therefore involve long distance physics and we could not expect to be able to calculate them perturbatively since

(a) $\alpha_s(r)$ is large for large r and
(b) large distance effects obviously depend on how hadrons are made of quarks and gluons.

We can now reconsider the two cases in which perturbation theory may apply:[10]

i. Mass finite quantities are completely insensitive to m for large E and may therefore be calculable in terms of quarks and gluons, ignoring the way they turn into hadrons. In processes with no incoming hadrons, quantities in which the contributions of all states which become degenerate as $m \to 0$ are included turn out to be mass finite (this is a consequence of the KLN theorem).[11] The most important examples are $\sigma(\bar{e}e \to \text{hadrons})$ and $\sigma(\bar{e}e \to \text{jets})$. In the former case, the picture is

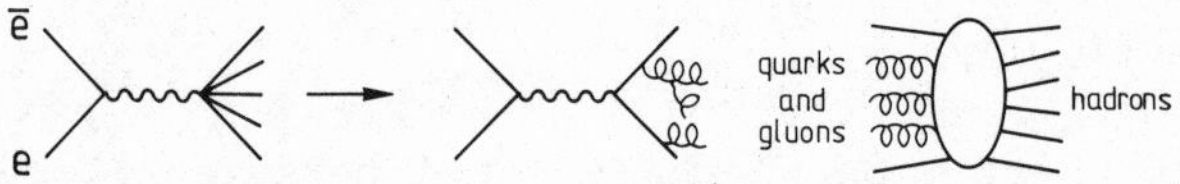

It is argued that m independence means that the total cross section is determined by the first step, which is calculable in QCD because of asymptotic freedom, and is independent of the second. It is not possible to really prove this without solving QCD but it is true in two-dimensional QCD and it works experimentally!

ii. If the m and E dependence factorize, the energy dependence is controlled by short distances and it is argued that it may be independent of how hadrons are made of quarks and gluons. Factorization occurs in all processes which admit a parton model interpretation in which soft partons do not play an essential role.[12] The classic example is inclusive deep inelastic scattering $eN \to eX$. The picture is

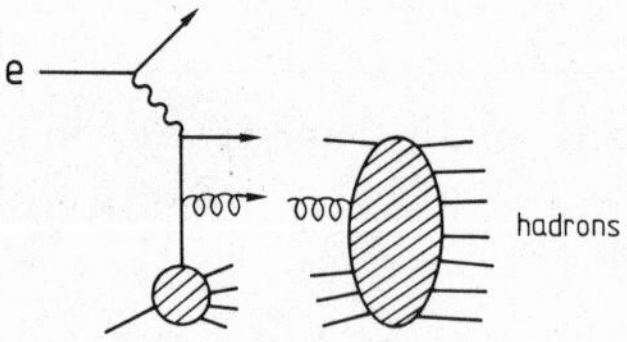

In perturbation theory, there are no mass singularities associated with the final state which suggests that, as in $\bar{e}e \to X$, the final quark + gluon → hadron transition has no influence on the cross section. Singularities associated with the incoming particles factorize, suggesting that although we cannot calculate the distribution of quarks in a hadron perturbatively, we can calculate how that distribution changes with E. As E increases, regions of high energy phase space open up which are controlled by short distance physics. Again this cannot really be proven but it is true in two-dimensional QCD and in potential models.

HIGH ENERGY TESTS OF QCD

Electron-Positron Annihilations

As discussed above, it is thought that to $O(m^2/E^2)$ the total cross section can be calculated perturbatively from the diagrams

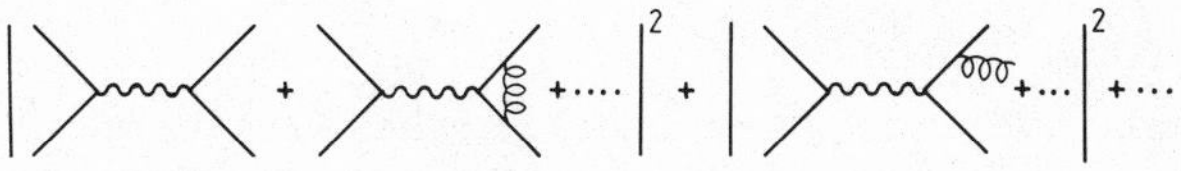

which give

$$\sigma = \sigma_{Born}\left(1 + \frac{\alpha_s(E)}{\pi} + \dots\right) .$$

We have already seen (Fig. 1) that this idea works well. Although the total cross section is mass finite, this is not true of individual diagrams. In fact the lowest order contribution to $\bar{e}e \to q\bar{q}g$ contains $\alpha_s(E)\ln E/m$ and is of the same order as σ_{Born}. However, this contribution is cancelled by virtual corrections to $\bar{e}e \to q\bar{q}$. Thus σ_{Born} $(\bar{e}e \to q\bar{q})$ gives the asymptotic answer although $\bar{e}e \to q\bar{q}$ is *not* the dominant process. In fact $\sigma(\bar{e}e \to q\bar{q})$ is zero! [as is $\sigma(\bar{e}e \to \bar{\mu}\mu)$]. Colored (electrically charged) particles like to radiate when accelerated and the probability of scattering without radiating any gluons (photons) is zero when summed to all orders (order by order it contains infrared divergences). Perturbatively, the processes which build up σ_{tot} are ones in which the initial qq pairs are accompanied by almost parallel moving hard gluons and quarks or soft gluons. In the diagram for $\bar{e}e \to q\bar{q}g$ above it is easy to see that the log e/m contribution comes from a region where the angle θ between the quark and gluon is less than some fixed small value for $E \to \infty$, since for $\theta \to 0$ the virtual quark becomes real and the propagator blows up for $m \to 0$.

Thus perturbatively the leading [in $\alpha_s(E)$] contribution to σ is due to events in which all the energy is in two narrow jets. This feature should survive the soft long range confining forces which cannot induce large momentum transfers. Indeed, it is clearly seen in the high energy data from PETRA and PEP, where most events have an obvious two jets structure.[13]

On heavy quark resonances strong deviations from the two jet structure are expected and seen. At the $b\bar{b}$ bound state $\Upsilon(^3S)$ the $b\bar{b} \to$ hadrons transition is mass finite when the masses of the lighter quarks vanish and can therefore probably be calculated perturbatively in terms of $\alpha_s(m_b)$, the leading contribution being given by

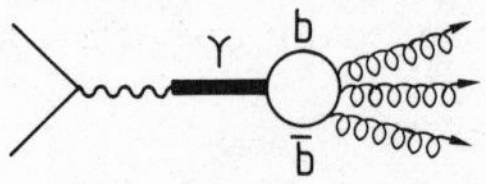

where the three gluons should be interpreted as jets (one gluon is forbidden by color, two by Yang's theorem). For every event, a quantity called thrust is constructed defined by

$$T = \frac{\text{Max} \sum_i |p^i_{\parallel}|}{\sum_i |p^i|}$$

where the sum runs over all particles in the event and the axis relative to which $p_{\parallel}$ is measured is chosen to maximize this expression. An event with two pencil jets could give T=1 whereas an isotropic event would give T=1/2. The data[14] (Fig. 4) show that

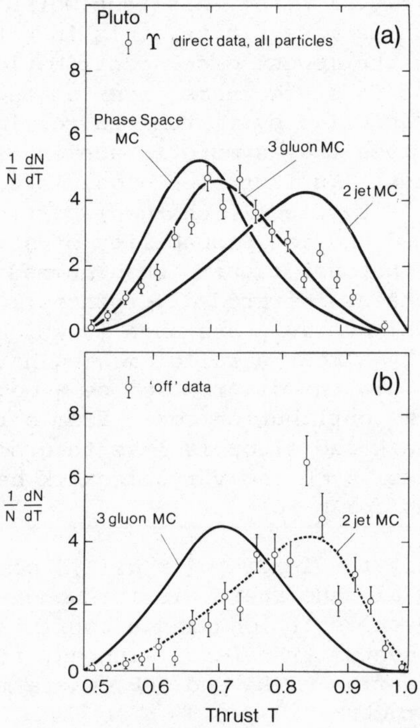

Fig. 4. Thrust distributions (a) on the Υ(1s) and (b) in the continuum close to the Υ measured by the PLUTO group. Predictions based on phase space, $q+\bar{q}$ jets and $q+\bar{q}+g$ jets are shown.

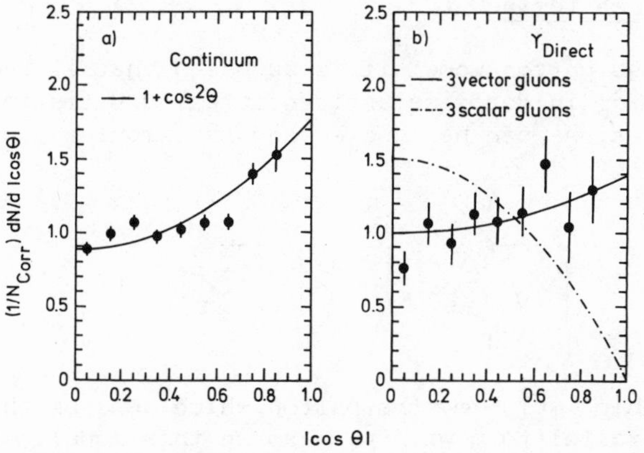

Fig. 5. Distribution of the thrust axis with respect to the beam axis (a) in the continuum close to the ϒ and (b) on the ϒ(1s), measured by the LENA group,[15] compared to the expected distributions for vector and scalar gluons.

events on the ϒ are much more isotropic than in the continuum. In fact when a Monte Carlo program is constructed to describe individual jets, it is found (Fig. 4) to give a good fit to the continuum assuming two jets and the resonance data assuming three. Furthermore, whereas the distribution of the thrust axis relative to the beam direction is well described by $1 + \cos^2\theta$ for the continuum, as expected for spin 1/2 quarks, the resonance data[15] (Fig. 5) has the distribution expected for vector gluons but is inconsistent with the distribution which would be expected for scalar gluons.

In the continuum, the α_S/π correction to the Born term value of R is due to hard gluon bremsstrahlung ($\bar{e}e \to q\bar{q}g$) and virtual corrections to $\bar{e}e \to q\bar{q}$. Again the quarks and gluons should be interpreted as jets so QCD predicts that a fraction of order α_S/π of all events should have a three jet structure. Such events are seen.[13] A careful analysis shows clearly that they really are genuine three jet events and not statistical fluctuations in a distribution of two jet events. Their characteristics are exactly as expected on the basis of QCD and, as in the case of ϒ decay, exclude scalar gluons. The frequency of three jet events gives $\alpha_S[(35\text{ GeV})^2] = O(0.15)$. It is hard to obtain a precise number as the exact result depends on the details of the Monte Carlo program used to generate hadronic jets from quarks and gluons. However, the fact that $\alpha_S \sim 0.15$ is obtained is good news. Given that, roughly speaking, λ_{QCD} represents the energy at which strong interactions become strong we would expect $10\text{ MeV} < \lambda_{QCD} < 1\text{ GeV}$ corresponding to $0.10 < \alpha_S(35\text{ GeV}) < 0.23$ and a value outside this range would have been very hard to understand (unfortunately this insensitivity to λ_{QCD} means that λ_{QCD} is very hard to measure).

Deep Inelastic Scattering

In the naive parton model it is supposed that if the momentum transfer $Q^2 = \vec{q}^2 - q_0^2$ in deep inelastic scattering satisfies $Q^2 > Q_0^2$ then all interactions can be ignored and the impulse approximation can be used:

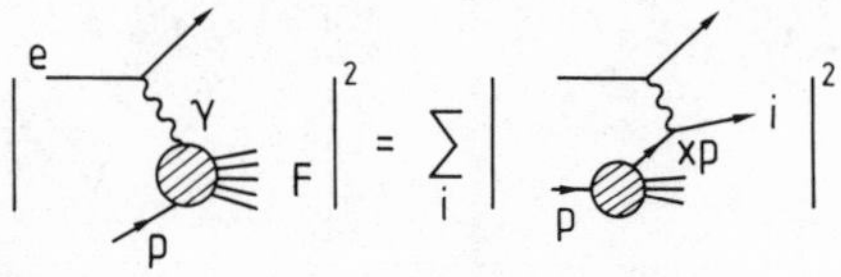

In a frame in which $|\vec{p}| \to \infty$, the parton which absorbs the virtual photon moves parallel to $\vec{p}$ with $\vec{p}_i = x\vec{p}$ in this model. For Q^2 large enough that the parton mass can be neglected, we have

$$p_i^2 \simeq (xp+q)^2 \simeq 0$$

which gives $x = Q^2/2q\cdot p$. Because there is no intrinsic scale, the total cross section for absorbing transverse or longitudinal photons becomes

$$\sigma_{T,L} \to \frac{1}{Q^2} f_{T,L}(x)$$

for $Q > Q_0$. By comparing σ_L and σ_T we can measure the spin of the partons, which turns out to be 1/2. Furthermore, by using neutrino data it is possible to show that there are three more fermions than antifermions in the nucleon. These results were important in establishing the reality of quarks. Finally, $f_T(x)$ measures the distribution of quarks in the nucleon and shows that approximately 50% of the proton's momentum is carried by neutral particles. This was the first evidence for gluons.[16]

In renormalizable field theories there is no intrinsic scale Q_0 beyond which interactions are negligible and $f_{T,L}$ must depend on Q^2. In QCD it turns out that the logarithmic decrease of α_s only leads to a weak Q^2 dependence, which agrees well with what is observed, and the parton results quoted above survive except that the quark and gluon momentum distributions depend (weakly) on Q^2.

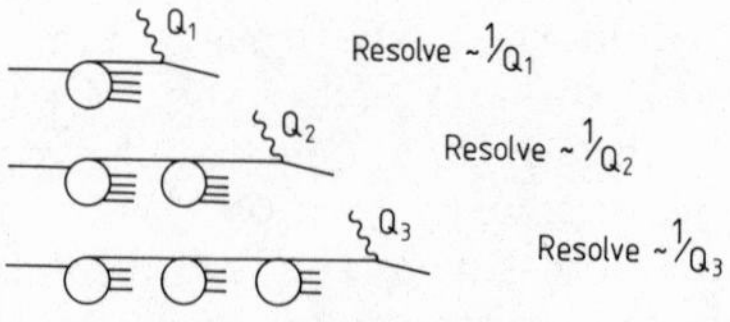

Fig. 6. Increase in resolution with increasing Q.

We will now develop an intuitive picture of deep inelastic scattering in renormalizable field theories, which can be justified by more formal considerations.[17] When virtual photons with $Q_1 = \sqrt{Q_1^2}$ are scattered from a target they can resolve constituents of size $O(1/Q_1)$ (for fixed $x \neq 0$) but are not sensitive to structure on a smaller scale. When Q is increased, however, substructure may be revealed. Further increases may reveal further substructure. This is shown schematically in Fig. 6. This process is easiest to visualize if there is a relatively sharp transition between length scales, as occurs in the successive resolution of molecules, atoms, nuclei, nucleons and quarks. In order to discuss scattering with momentum Q_N which resolves the Nth layer of constituents, we need to know the effective Hamiltonian H_N which describes the $(N-1)^{th}$ layer in terms of the Nth layer but we do not need to know anything about substructure at the $(N+1)^{th}$ and deeper levels e.g. at the stage at which quasi-elastic scattering off single nucleons dominates, nucleons are the relevant degrees of freedom—we need the Hamiltonian which describes nuclear forces and to understand nuclear structure, but we can ignore the fact that nucleons are made of quarks.

When the level of quarks and gluons is reached the degrees of freedom do not change as Q is increased, as far as we know, but the process of successive resolution continues as increasingly virtual $q\bar{q}$ pairs and gluons are resolved. At any scale Q we must use the appropriate effective coupling constant $g^2(Q^2)$ in the Hamiltonian. We consider three possibilities as $Q^2 \to \infty$:

i. $g^2 \to (M^2/Q^2)^p$ with $p > 0$. For $Q^2 \gg M^2$ interactions could be neglected, the structure resolved would not change with Q^2 and there would be exact scaling in this case. However, this can only happen if the coupling constant is proportional to a positive power of the dimensional parameter M, which is impossible for interactions involving fermions (for non-renormalizable interactions the power would be negative giving $p < 0$). In theories involving fermions, exact scaling can only be achieved by artificially introducing some damping factor or cutoff, which necessarily violates Lorentz invariance, locality or some other cherished property.

ii. $g^2 \to 1/\ln Q^2$—asymptotic freedom

iii. $g^2 \to g_*^2 \neq 0$. This fixed point behavior occurs in all cases except non-Abelian gauge theories (unless $g^2 \to \infty$ or wanders around in closed limit cycle loops in coupling constant space, which is a possibility if there are several coupling constants). In this case the relation of one layer to the next becomes universal as Q^2 is increased in a sense which will be discussed below.

Keeping a discrete index N to label Q_N for simplicity, we can define probability distributions for partons of type i at the Nth level $\left(F_N^i(x)\right)$ and the probability $f_N^{ij}(y)$ of resolving parton type j with momentum $y\vec{k}$ at the Nth level inside parton type i with momentum $\vec{k}$ at the $(N-1)^{th}$ level:

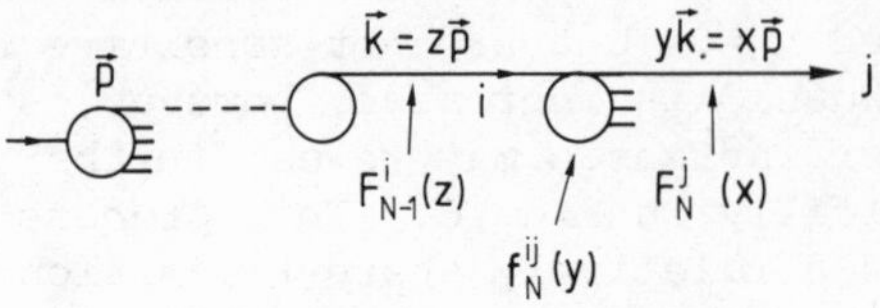

According to this picture:

i. Momentum conservation reads

$$\sum_i \int x\, F^i_N(x)\,dx = 1$$

for all N. However, since the number of partons increases with N we expect

$$\langle x\rangle_N \equiv \sum_i \int x^2\, F^i_N(x)\, dx \xrightarrow[N\to\infty]{} 0\ .$$

Thus $\sum_i x\, F^i_N$ will behave as shown in Fig. 7. Since the electromagnetic function F_2 is given by $F_2(Q_N) = \sum_i Q_i^2\ x\, F^i_N(x)$, we expect it to increase at small x and decrease at large x as Q^2 increases in all field theories—exactly the trend seen in the data.

ii. The distributions F^i_N must be consistent with the conservation of B, I_3, Q etc. for all N and all the parton sum rules and inequalities will therefore be satisfied.

iii. F^i_N is related to F^j_{N-1}

$$F^i_N(x) = \sum_j \iint dydz\ F^j_{N-1}(z)\ f^{ji}_N(y)\ \delta(x-yz)$$

$$= \sum_j \int_x^1 dy\ F^j_{N-1}(x/y)\ \frac{f^{ji}_N(y)}{y}\ .$$

Dropping the labels i and j for simplicity and taking moments of this equation we obtain

$$\frac{\int_0^1 x^p\, F_N(x)\,dx}{\int_0^1 z^p\, F_{N-1}(z)\,dz} = \int_0^1 y^p\ f_N(y)\,dy\ .$$

Specific models will tell us something about f and we therefore see why it is the behavior of the moments of structure functions which

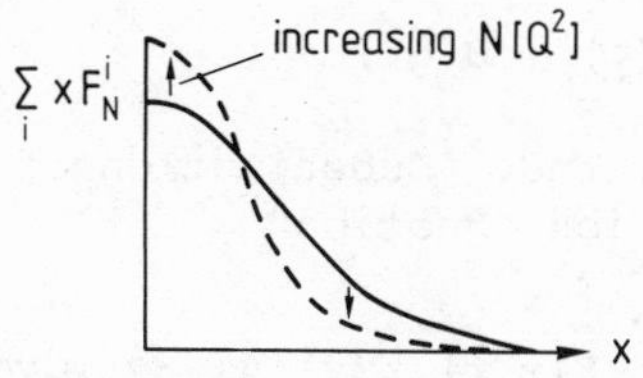

Fig. 7. Behavior of the structure function with increasing Q^2.

is predicted in field theories. First consider fixed point theories in which $g_N^2 \to g_*^2$. If the ratio $Q_{N+1}/Q_N = \Lambda$ is chosen to be independent of N, f then its moments

$$\int y^p \ f(y) dy \equiv q_p$$

will be independent of N for large Q^2. Since

$$\frac{Q_N^2}{Q_0^2} = \Lambda^{2N}$$

we have

$$N = \frac{1}{2} \frac{\ln Q_N^2/Q_0^2}{\ln\Lambda}$$

and

$$\frac{\int x^p \ F_N(x) dx}{\int x^p \ F_o(x) dx} = (q_p)^N = \left(\frac{Q^2}{Q_0^2}\right)^{\frac{\ln q_p}{2 \ln\Lambda}} .$$

With just one species of parton momentum conservation gives $q_1 = 1$ so the first moment scales but $q_{p>1} < 0$ and higher moments decrease like powers of Q^2 in theories with fixed points.

Turning to asymptotically free theories, we set $Q_{N-1}^2 = Q^2$ and consider the case that

$$Q_N^2 = Q^2(1+\varepsilon)$$

with $\varepsilon << 1$. The kernel f must have the form

$$f = \delta(1-y) + \varepsilon K(y,Q^2) + O(\varepsilon^2) \quad .$$

For large Q^2, K can be calculated perturbatively in powers of $g^2(Q^2) = (C)/(\ln Q^2/\lambda_{QCD}^2)$. Since K is dimensionless and the only dimensional parameter is λ_{QCD} which enters only through g, necessarily

$$K(y,Q^2) = g^2(Q^2)\, k(y) + O(g^4)$$

where $k(y)$ can be calculated. Substitution of f into the equations above yields the "evolution equation"

$$\frac{dF(x,t)}{dt} = g^2(t) \iint F(z,t)\, k(y)\, \delta(x-yz)\,dy\,dz$$

$$= g^2(t) \int_x^1 dy\, F(x/y,t)\, \frac{k(y)}{y}$$

where $t = \ln Q^2/\lambda^2$. Taking moments of this equation:

$$\frac{dM_n(t)}{dt} = g^2(t)\, k_n M_n(t)$$

where

$$M_n(t) \equiv \int_0^1 x^n\, F(x,t)\,dx$$

and

$$k_n = \int_0^1 y^n\, k(y)\,dy$$

is known explicitly. The solution of this equation is:

$$\frac{M_n(t)}{M_n(t_o)} = \left(\frac{t}{t_o}\right)^{ck_n} .$$

With just one species of parton, momentum conservation requires the first moment of the kernel f to be one, giving $k_1 = 0$, but higher moments of f must be less than one so $k_{n>1} < 0$. Hence M_1 scales but higher moments decrease as *calculable* powers of $\ln Q^2$ in asymptotically free theories.

We must now face the fact that the kernel f is actually a matrix f_{ij} for the probability of finding a parton of type j in a parton of type i giving rise to coupled evolution equations for the different species of parton. The most annoying consequence of this coupling is that the neutral gluons, which do not contribute to lepto-production and are not meaured at Q^2, nevertheless generate effects at $Q^2 + \delta Q^2$ since they can dissociate into $q\bar{q}$ pairs:

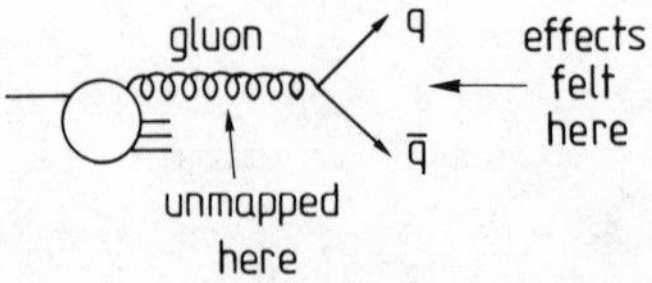

Table 3. Tests of QCD from Neutrino Experiments

n,m	Measured slope	QCD (2nd order)	Scalar gluons
3,4	1.38 ± 0.06	1.32	1.08
4,5	1.20 ± 0.03	1.20	1.04
3,5	1.68 ± 0.11	1.57	1.12

Hence, in general, lepto-production data at two different values of Q^2 are needed as boundary values to determine the quark and gluon distributions for all Q^2. However, if we consider the difference of the distributions for two different flavors [e.g. $u(x,Q^2) - d(x,Q^2)$] or the difference of a quark and an antiquark distribution, the effects of gluon dissociation cancel out. Examples of structure functions to which the gluon does not contribute, which are known as non-singlets, are $\sigma^{ep}-\sigma^{en}$ and $\sigma^{\nu}-\sigma^{\bar{\nu}}$, which measures the difference of q and $\bar{q}$. Their evolution is determined by the kernels $k_{qq} = k_{\bar{q}\bar{q}}$ which are flavor independent because quark masses are negligible at large Q^2. The moments M_n^{NS} of non-singlet structure functions therefore do satisfy the simple equation above with $k_n \to k_n^{NS}$. From this equation it follows that

$$k_m \ln M_n(t) = k_n \ln M_m(t) + \text{const.}$$

so a plot of $\ln M_n(t)$ against $\ln M_m(t)$ should yield a straight line with slope k_n/k_m predicted by QCD. The data do indeed seem to lie on straight lines.[18]

Using data with $Q^2 > 5$ GeV^2, the CDHS neutrino experiment[19] has measured the slopes for various n and m and compared them to the predictions of QCD and a "straw man" theory with scalar gluons (Table 3). QCD is clearly favored (other groups have also checked these predictions but with lower Q^2 cuts, making the significance less clear).

Non-singlet tests are cleanest theoretically. However, non-singlets involve differences and therefore suffer from bigger errors than singlets. Singlet data have generally been fitted using QCD by assuming some parametrization of the gluon distribution at some initial Q^2. The quality of the fits is good. For a recent review see Ref. 20.

SUMMARY

In $\bar{e}e$ annihilation and inclusive deep inelastic scattering QCD provides a semi-quantitative description of the data. There are many other processes which provide tests of QCD in principle, e.g. $eN \to e + jets$, $pp \to \mu\bar{\mu}X$, $\gamma\gamma \to$ large p_T, $e\gamma \to eX$, $pp \to$ large p_T, $e^+e^- \to \pi X$ $eN \to e\pi X$. In all cases the data agree qualitatively with the predictions of QCD although theoretical ambiguities and experimental errors have made real quantitative tests impossible up to now.

Because α_s is not very small at available energies, there is nothing which tests QCD in the way that (g-2) tests QED. However, taken altogether the success is impressive. Recalling the strong *a priori* evidence and the encouraging results of lattice calculations, we see that there are excellent reasons for believing in QCD.

REFERENCES

1. Since these lectures are of an introductory nature many of the references are to reviews, where further references may be found, rather than to the original literature. For reviews of QCD see e.g. W. Marciano and H. Pagels, Phys. Rep. 36:137 (1978); G. Altarelli, Phys. Rep. 81:1 (1982); C.H. Llewellyn Smith, Phil. Trans. Roy. Soc. 304:5 (1982); H. D. Politzer, to be published in Proc. XXI International Conference on High Energy Physics, Paris 1982.
2. R. Felst, in: "Proc. 1981 Symposium on Lepton and Photon Interactions at High Energy," W. Pfeil, ed., University of Bonn, Bonn (1981).
3. S. L. Adler, in: "Lectures on Elementary Particles and Quantum Field Theory," S. Deser, M. Grisaru and H. Pendelton, eds., MIT Press, Cambridge (1970); R. Jackiw, in: "Lectures on Current Algebra and its Applications," by S. Teiman, R. Jackiw and D. Gross, Princeton University Press, Princeton (1972).
4. V. De Alfaro et al., "Currents in Hadron Physics", North-Holland, Amsterdam (1973); H. Pagels, Phys. Rep. 5:219 (1975).
5. D. J. Gross and F. Wilczek, Phys. Rev. Lett. 30:1343 (1973); H. D. Politzer, Phys. Rev. Lett. 30:1346 (1973).
6. R. P. Feynman, in: "Weak and Electromagnetic Interactions at High Energy," R. Balian and C. H. Llewellyn Smith, eds., North Holland, Amsterdam (1977).
7. A. De Rújula, H. Georgi and S. L. Glashow, Phys. Rev. D12:147 (1975); N. Isgur, in: "Proc. Madison Conference on High Energy Physics," L. Durand and L. G. Pondrom, eds., Madison (1980).
8. C. Rebbi, to be published in "Proc. XXI International Conference on High Energy Physics," Paris 1982, and references therein.
9. The discussion in this section is simplified e.g. I have ignored external line corrections and subtleties such as gauge dependence. For a proper discussion see any modern book on quantum field theory, for example, C. Itzykson and J.-B. Zuber, "Quantum Field Theory," McGraw Hill, New York (1980).

10. See Refs. 1 and A. H. Mueller, Phys. Rep. 73:237 (1981); Yu. L. Dokshitser et al., Phys. Rep. 58:269 (1980) and E. Reya, Phys. Rep. 69:195 (1981). For a more detailed discussion from a similar point of view to that followed in these lectures see C. H. Llewellyn Smith, to be published in Proc. 1981 NATO Banff Summer Institute.
11. T. Kinoshita, J. Math. Phys. 3:650 (1962); T. D. Lee and M. Nauenberg, Phys. Rev. 133:B1349 (1964).
12. G. T. Bodwin, S. Brodsky and G. P. Lepage, Phys. Rev. Lett. 47:1799 (1981) and S. Brodsky, this volume, have discovered effects which may spoil the stronger assumption, which has often been made, that the function which absorbs the m dependence is "universal" i.e. totally process independent.
13. For a general review of jet production in e^+e^- annihilation, both on and off resonances, see K. H. Mess and B. H. Wiik, DESY preprint 82-011 (1982).
14. Ch. Berger et al., Phys. Lett. 78B:176 (1978) and 82B:449 (1979).
15. J. Bienlein, in: "Proc. 1981 Symposium on Lepton and Photon Interactions at High Energy", W. Pfeil, ed., University of Bonn, Bonn (1981); B. Niczyporuk et. al., Zeit. Phys. C9:1 (1981).
16. R. P. Feynman, "Photon-Hadron Interactions", Benjamin, Reading (1972); F. E. Close, "An Introduction to Quarks and Partons", Academic Press, London (1978); C. H. Llewellyn Smith, in: "Hadron Structure and Lepton-Hadron Interactions", M. Levy et al., eds., Plenum, New York (1979).
17. J. Kogut and L. Suskind, Phys. Rev. D9:697 and 3391 (1974).
18. P. Bosetti et al., Nucl. Phys. B140:1 (1980); H. Anderson et al., Phys. Rev. Lett. 40:1061 (1978); J. G. H. de Groot et al., Zeit. Phys. C1:143 (1979).
19. H. Wahl, *loc. cit.*
20. F. Eisele, to be published in "Proc. XXI International Conference on High Energy Physics," Paris 1982.

THE SUCCESSES AND FAILURES OF THE CONSTITUENT QUARK MODEL

H.J. Lipkin*

Weizmann Institute of Science
Rehovot, Israel

INTRODUCTION—HISTORY OF THE CONSTITUENT QUARK MODEL

As I mentioned in my 1978 Banff lectures[1] a nonrelativistic constituent quark model has been remarkably successful in describing many phenomena in hadron physics.[1-7] However, there are other areas where the model has been spectacularly unsuccessful.[8-10] George Zweig used to say that the quark model[11,12] gives an excellent description of half the world.

The model has no sound theoretical basis. In the early days there was no clue to the underlying theory. Today we believe that the underlying theory is QCD and that hadrons are composed of quarks and gluons. However, the equations of QCD are so complicated that no one has been able to solve them to derive hadron spectroscopy and dynamics. The constituent quark model can now be considered as an intermediate phenomenological model which fits the experimental data and will hopefully be derived from QCD.

Our approach considers the model as a possible bridge between QCD and the experimental data and examines its predictions to see where these succeed and where they fail. Pinpointing the successes and failures may give us clues to the connection with the underlying theory. We also attempt to improve the model by looking for additional simple assumptions which give better fits to the experimental data. But we avoid complicated models with too many *ad hoc*

---

*Also at Argonne National Laboratory, Argonne, Ill. 60439 and Fermi National Laboratory, Batavia, Ill. 60510.

assumptions and too many free parameters; these can fit everything but teach us nothing. We also attempt to look beyond the simple *ad hoc* assumptions to see whether QCD gives any indication for a possible justification.

Everyone has his own version of the quark model. We define our constituent quark model by analogy with the constituent electron model of the atom and the constituent nucleon model of the nucleus. In the same way that an atom is assumed to consist only of constituent electrons and a central Coulomb field and a nucleus is assumed to consist only of constituent nucleons, hadrons are assumed to consist only of their constituent valence quarks with no bag, no glue, no ocean, nor other constituents. Although these constituent models are oversimplified and neglect other constituents we push them as far as we can. Atomic physics has photons and vacuum polarization as well as constituent electrons, but the constituent model is adequate for calculating most features of the spectrum when finer details like the Lamb shift are neglected. Similarly, constituent nucleon models are used extensively in nuclear spectroscopy even though we know that at some stage the contributions of mesons, isobars, exchange currents, etc. must be also taken into account.

The simple nonrelativistic constituent quark model has had remarkable success in describing the low-mass spectroscopy of the quark-antiquark and three quark systems.[1-7] Most recently it has given excellent descriptions of quarkonium systems like charmonium and the Υ states.[5,6] However, for detailed properties of multiquark systems the model has failed almost completely and given no predictions which have been verified by experiment.[8-10] In this talk, we shall try to understand how the model can be so successful in the quark-antiquark and three quark systems and fail for almost everything else.

We first review the history of the constituent quark model. This can be conveniently divided into three stages.

The Simple Nonrelativistic Quark Model

The simple nonrelativistic quark model[11,12] (no color) was first introduced to explain the quantum numbers occurring in the low-lying meson and baryon spectrum. This model was then extended to treat all possible properties of hadrons by making the simplest dynamical assumptions possible.[2,3] Additive single quark operators were used wherever possible to describe observable properties of hadrons, including weak,[13] electromagnetic[14,15] and strong[16,17] interaction matrix elements. Nucleon-antinucleon annihilation was described as quark rearrangement.[18] Hadron mass splittings were described by a nuclear shell model approach with two-body interactions.[19,20] The results obtained were in surprising agreement with experiment for all of hadron spectroscopy. However, there were several outstanding open problems.

The statistics problem. Although quarks were expected to obey Fermi statistics, the wave functions required to fit the baryon spectrum were symmetric in all the known degrees of freedom rather than antisymmetric.

The saturation problem. Quarks had not been observed and were assumed to have a high mass. Mesons and baryons were thus strongly bound systems indicating the presence of strong attractive forces both in the quark-antiquark and quark-quark systems. There was no explanation for why only quark-antiquark and three quark systems should be strongly bound. This paradox is illustrated most dramatically by comparing the well known deuteron stripping reaction with its analog in hadron physics

$$d + {}^{3}H \rightarrow n + \alpha \quad , \tag{1a}$$

$$M + B \not\rightarrow q + \bar{q}qqq \quad . \tag{1b}$$

In a collision between a deuteron and a triton, the proton can be stripped from the deuteron and combined with the triton to make an alpha particle with considerable energy release. Since the nucleon-nucleon force is strongly attractive, the proton in the deuteron is much more strongly attracted by the three nucleons in the triton than by the single neutron in the deuteron and the transfer releases energy. In the analogous case of a meson-baryon collision (1b) one would expect the antiquark in the meson to be much more strongly attracted to the three quarks in the baryon than to the single quark in the meson. However, the antiquark stripping reaction (1b) does not occur.

The stripping reaction (1b) might somehow be ruled out by a general principle forbidding production of states having fractional charge. But it would not forbid production of multiquark states with integral charges which are also not seen. There is no bound

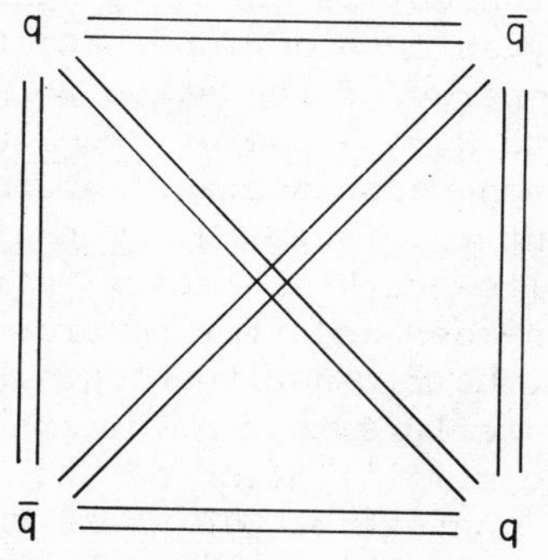

Fig. 1. Binding in a four quark state.

state of two pions with charge 2 and a mass below two pion masses. Such multiquark states with integral charge and baryon number are now called exotics. They might still be found as high mass resonances, but there are clearly no stable low-mass bound states. Their absence could not be explained in any simple way by the naive constituent quark model without the color degree of freedom. If the quark-quark and quark-antiquark forces are attractive in all channels as is implied by the existence of meson and baryon bound states with all possible quantum numbers, then the four-quark state shown in Fig. 1 should be bound much more strongly than two separated quark-antiquark pairs.

The meson-baryon problem. There was no simple description of the forces required to make both mesons and baryons. A vector interaction like electrodynamics would give attractive quark-antiquark forces and repulsive quark-quark forces and would not bind baryons. A scalar or tensor interaction would give attractive and equal quark-quark and quark-antiquark forces and lead to a diquark spectrum identical to the meson spectrum. Combinations of vector and scalar or tensor interactions to make the quark-quark force attractive but weaker than the quark-antiquark forces could explain the difference between mesons and baryons. But this seemed contrived and not very convincing.

The free quark problem. There was no reasonable explanation for the failure to find free quarks.

The solution of the statistics problem was found[21] in 1964. The additional internal degree of freedom now called color enabled quarks to satisfy Fermi statistics with wave functions symmetric in all the previously known degrees of freedom and antisymmetric in color.

In 1968 some simple features of the $N \to \infty$ limit[22] where N is the number of colors were noted as a possible explanation for how hadrons could be bound states of quarks but free quarks would not be created in hadron-hadron collisions. If quark-antiquark pairs were bound into mesons by an interaction characterized by a coupling constant g, the binding energy of the state we now call the color singlet is proportional to Ng^2. However, the quark-quark scattering cross section and the meson-meson scattering cross section would be proportional to g^2 without the factor N. Thus in the limit where $N \to \infty$ but Ng^2 remains finite quark-antiquark pairs would be bound into mesons but the meson-meson scattering cross section which might break the mesons up into their constituent quarks would go to zero. This explanation for the failure to find quarks is not considered seriously today. But the simplifiation of the large N limit has been rediscovered in the context of QCD where the decoupling of hadron-hadron interactions in this limit makes it a useful starting point for expansions in strong interaction dynamics.

The Global Color Nonrelativistic Quark Model

The color degree of freedom was explicitly introduced into phenomenological dynamical quark models in 1973 to solve the saturation and meson-baryon problems.[23] The use of a non-Abelian gauge theory with confinement had not yet been proposed. At the 1972 Batavia conference, Murray Gell-Mann in his summary[24] presented a strong case for the color degree of freedom based primarily on the electromagnetic and weak interactions and suggested that some kind of vector gluons were responsible for the strong interactions. However, there was no suggestion that the gluons were colored or that the color degree of freedom was in any way essential for the strong interactions. There was no hint of asymptotic freedom, infra-red slavery or non-Abelian gauge theories with confinement.

The global color model[23] considered a two-body interaction which would be produced by the exchange of an octet of colored massive vector bosons.[25] The bosons, like the quarks, would have to be massive to explain the failure to observe them experimentally. Confinement was not understood at that time and the only mechanism to explain the failure to observe such particles was by giving them a high mass. The color-exchange Yukawa interaction solved the saturation and meson-baryon problems. The quark-antiquark and three quark systems behaved like neutral atoms, there were no strongly bound exotics and the quark-quark interaction was exactly half of the quark-antiquark interaction

$$V(qq) = \frac{1}{2} V(q\bar{q}) \quad . \tag{2}$$

This is exactly the relation required to bind both mesons and baryons.

In a model where quarks were very massive and hadrons had a very low mass on the quark mass scale, the expression (2) could explain the existence of mesons and baryons by setting the strength of the matrix element of the interaction (2) to be approximately equal to the quark mass. Then

$$\langle V(qq)\rangle \approx \frac{1}{2} \langle V(qq)\rangle \approx -M_q \tag{3a}$$

$$\langle V(q\bar{q})\rangle \approx -2M_q \tag{3b}$$

$$3\,\langle V(qq)\rangle \approx -3M_q \tag{3c}$$

where M_q is the quark mass. The interaction between the two quarks in a diquark thus cancels only one quark mass of the two body system and leaves the diquark with essentially the same high mass as the quark. The quark-antiquark interaction which cancels two quark masses leaves mesons down at low mass and the three quark-quark interactions in a baryon cancel the three quark masses to give low-lying baryons. However, there was still no justification for a nonrelativistic picture nor for the use of the interaction (2) which comes from one gluon exchange in a strong interaction model where multiple gluon exchanges are not easily neglected.

The model gave no strongly bound exotic state. With potentials having a reasonable spatial variation, there was no possibility of getting a lower energy than that of two spatially separated mesons. The two-body interactions are attractive and described by the expressions (2) and (3) only for the color singlet quark-antiquark configuration found in mesons and for the color antitriplet quark-quark configuration found in baryons. The interactions in the quark-antiquark color octet configuration and in the quark-quark color sextet configuration are both repulsive. These repulsive channels play no role in meson and baryon states but are responsible for the saturation in the multiquark states. The attractions and repulsions cancel in calculating the force between a color singlet hadron and an external quark in the same way that the Coulomb attractions and repulsions cancel in the force from a neutral atom on an external charged particle.

The QCD Motivated Nonrelativistic Quark Model

After the introduction of asymptotic freedom and confinement in non-Abelian gauge theories led to the development of QCD,[4,26] De Rújula, Georgi and Glashow[27] (DGG) introduced ideas from QCD into the nonrelativistic quark model. They attributed the spin dependence of the two-body interaction to the spin dependent part of a one gluon exchange interaction. This explained for the first time the sign of the hyperfine splittings; e.g. why the Δ is heavier than the nucleon, and related the magnitudes of the hyperfine splittings to the quark masses. With this model it was possible to obtain two independent relations between the strange and nonstrange quark masses in terms of experimental hadron masses[27,28] and to use them to predict the Λ magnetic moment.[1,27,29] Both of these predictions agreed exactly with one another and with the experimental value of the Λ moment. The original prediction by DGG in 1975 is particularly impressive because it was made *before* the magnetic moment had been measured precisely.

In the multiquark sector, this model gave complete nonsense. The saturation feature of the global color model remained. Forces between color singlet hadrons were much weaker than those that bind the quarks into hadrons. But the introduction of confining potentials led to unphysical long range forces which were in disagreement with experiment[8-10] and rapidly changing color[30] correlations between spatially separated systems which violated causality. Multiquark baryonium states were predicted and not found experimentally but created considerable confusion as one candidate after another was shown to be only a statistical fluctuation.[31]

The situation can be summed up by saying that the nonrelativistic constituent quark model with input from QCD gives a very good phenomenological description of the quark-antiquark and three quark systems but breaks down in the multiquark sector. Arguments explaining this breakdown are presented in the last section.

In the next section we examine the possible basis from QCD of one application of the simple additive quark model with single-quark operators, the calculation of hadron total cross sections. Following this section we look beyond the approximation of single quark operators for phenomenological contributions from two-quark operators in total cross sections and magnetic moments. In the last section we analyze the difference between the quark-antiquark and three quark systems where the model succeeds and the multiquark systems where the model fails and show how gluon dynamics can make the difference. Somehow it is possible to replace the gluon field by an effective interaction in the quark-antiquark and three quark systems. But the gluon dynamics plays an essential role in the multiquark system and the constituent quark picture is no longer adequate.

WHY ARE TOTAL CROSS SECTIONS ADDITIVE?

The additive quark model (AQM) for high energy scattering[17] has been remarkably successful in fitting and predicting experimental data, despite the absence of a satisfactory QCD derivation. It was first introduced on the basis of an impulse approximation as shown in Fig. 2. This *ad hoc* assumption was never justified but accepted as reasonable. However, it appeared to be completely wrong after the advent of QCD. Single gluon exchange shown in Figs. 3(a) and 3(b) is a two-quark operator which violates the impulse approximation of the AQM when the two gluons are emitted by two different quarks in the same hadron as in Fig. 3(b). A two-gluon exchange model for the Pomeron contribution to scattering proposed independently by Nussinov[32] and Low[33] explained some dynamic features, but did not relate meson and baryon couplings in a simple way. However, it turns out that the combinatorial factors in two gluon and three-gluon exchange vertices do indeed satisfy the additivity assumption of the AQM in lowest order, even though there is no impulse

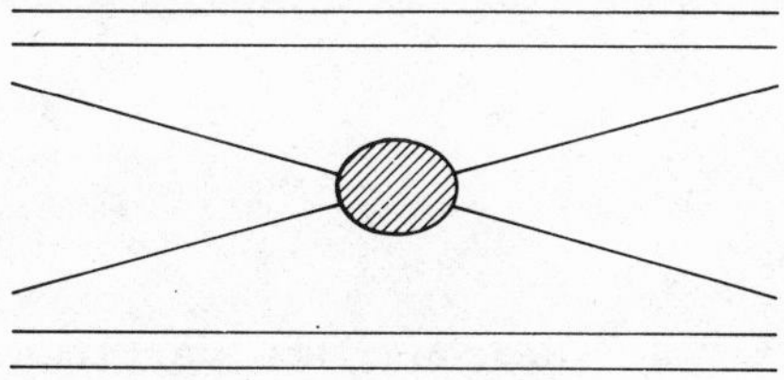

Fig. 2. The additive quark model for hadron scattering.

(a) (b)

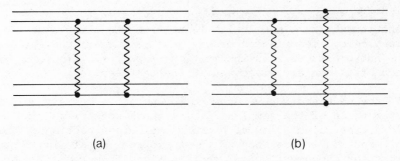

Fig. 3. Two gluon exchange diagrams.

approximation and two-quark and three-quark operators play an important role.[34,35] The color algebra allows the multiquark operators to be reduced to single quark operators and gives the simple quark-counting rules relating meson and baryon couplings. The problems never solved in the original impulse approximation approach remain; e.g. ignoring the differences between mesons and baryons in form factors, masses and momentum fractions carried by an active quark. However, the additional difficulty of obviously wrong combinatorial factors arising from two-quark and three-quark operators is resolved.

The quark counting factor was first noted by Gunion and Soper.[34] We follow the general algebraic derivation of Ref. 35 which obtains the result explicitly from the color algebra of the vertices describing gluon emission from color singlet hadrons shown in Figs. 4(a) and 4(b). These vertices have a very simple color structure because of the factorization of the color degree of freedom which allows all color factors to be expressed in terms of generators of the SU(3) color group.

The emission of a gluon by a single quark is described in the quark space by the matrix element of a color octet operator between two color triplet states. By the Wigner-Eckart theorem, this matrix element factorizes into the product of a reduced matrix element of a color-independent operator and an SU(3) Clebsch-Gordan coefficient

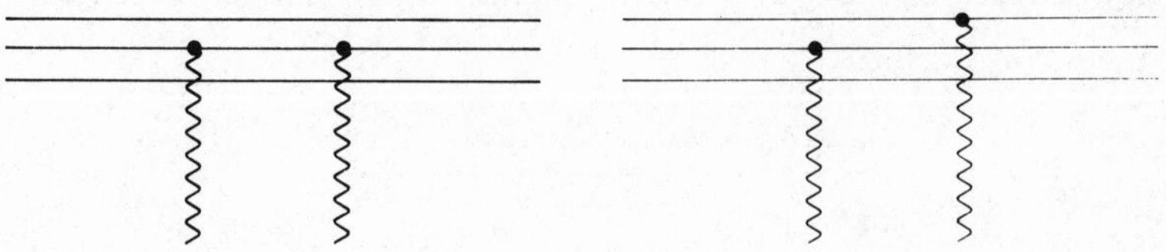

(a) (b)

Fig. 4. Hadron-gluon vertices.

which is the same for any color octet operator and is therefore proportional to the matrix element of the corresponding generator of SU(3) color. Thus we can write

$$\langle (G^a) q_i' | V | q_i \rangle = \langle q_i' \| A_i \| q_i \rangle \langle q_i' | F_i^a | q_i \rangle \tag{4}$$

where F_i^a denotes one of the eight SU(3) color generators for the ith quark or antiquark and a is a color index, G^a denotes a gluon with the color quantum number a, q_i' and q_i denote any two states of the ith quark or antiquark and the double-barred reduced matrix element of the color-indpendent operator A_i describes all the other degrees of freedom except color. The generator F_i^a is identical to the λ^a matrix[4,8,26-36] for quark states. For antiquarks $F^a = (-\lambda^a)^*$.

Color factorization also occurs in the meson and baryon wave functions with a color-independent factor in the baryon wave function totally symmetric under permutations of the quarks. The reduced matrix elements of products of two color-independent single quark operators A_i and B_j are thus independent of the quark indices i and j.

$$\langle H \| A_i B_j \| H \rangle = M_1 \delta_{ij} + M_2 - M_1 \tag{5}$$

where M_1 and M_2 are independent of the indices i and j. The matrix element (5) has the value either M_1 or M_2 depending only upon whether i and j are the same or different, but not on their explicit values.

Consider the vertex $\langle (GG) H | V | H \rangle$ describing the emission from a color singlet hadron H of two gluons in an overall color singlet state. Two types of contributions arise. The gluons can be emitted either by the same quark in the hadron H as in Fig. 4(a) or by two different quarks as in Fig. 4(b).

$$\langle (GG) H | V | H \rangle = \sum_a \left\{ f_1 \sum_i \langle H | F_i^a F_i^a | H \rangle + f_2 \sum_{j \neq i} \sum_i \langle H | F_i^a F_j^a | H \rangle \right\} . \tag{6}$$

The coefficients f_1 and f_2 denote all the additional dynamical and kinematic factors which do not depend upon the color degree of freedom obtained by evaluating matrix elements of color-independent operators.

The second term on the right hand side of Eq. (6) is manifestly non-additive and depends quadratically on the quark number. However, a hidden n-dependence in the color coupling factors exactly cancels the non-additive combinatorial factor to give an overall result proportional to the quark number. This can be shown by using the following identities for color couplings with any color singlet quark-antiquark or three-quark hadron $|H\rangle$.

$$\sum_i F_i^a |H\rangle = 0 \tag{7a}$$

$$\sum_a F_i^a F_i^a |H\rangle = n_H c |H\rangle \tag{7b}$$

where c is the eigenvalue of the quadratic Casimir operator for SU(3) color for a single quark state and n_H is the number of constituents in hadron $\underline{H}$; i.e. $n_H = 2$ for mesons and 3 for baryons. We first rewrite Eq. (6) as

$$\langle (GG)H|V|H\rangle = \sum_a (f_1 - f_2) \sum_i \langle H|F_i^a F_i^a|H\rangle + f_2 \sum_j \sum_i \langle H|F_i^a F_j^a|H\rangle . \tag{8}$$

The two-body terms in Eq. (8) can be seen to vanish by the identity (7a) that the sum of any SU(3) color generator F_i^a over all quarks is a generator of the total SU(3) color group which annihilates any color singlet state. The remaining term is evaluated with the identity (7b). Thus

$$\langle (GG)H|V|H\rangle = \sum_a (f_1 - f_2) \sum_i \langle H|F_i^a F_i^a|H\rangle = (f_1 - f_2) n_H c \ . \tag{9}$$

This result (9) shows that the two-gluon emission vertex for any hadron H is proportional to the quark number n_H. A similar result was obtained for the three-gluon vertex. Any model for Pomeron exchange such as that of Ref. 2 in which the coupling of the Pomeron to the hadron H is via two or three gluons will give the AQM ratio of 3/2 for baryons to mesons for the imaginary part of the forward scattering amplitude and the total cross section. This includes not only the contributions from the simple gluon-exchange diagrams but also from more complicated ladder exchanges which couple to the external hadrons via a two-gluon or three-gluon vertex.

The inclusion of four-gluon exchanges breaks the additive quark model. That additivity cannot be saved by manipulation of color factors is easily seen in the example of one possible four-gluon diagram with two gluons coupled to a color singlet and coupled to a single quark and the remaining two gluons coupled to any two quarks and coupled to a color singlet as in the two gluon vertex (6). This contribution then factorizes into two two-gluon vertices, one behaving like the first term on the right hand side of Eq. (6) and the second like the entire right hand side. The result is a quadratic function of quark numbers.

BEYOND THE ADDITIVE QUARK MODEL

The assumption that all possible processes are described by single-quark operators must break down at some level. The question arises whether it makes sense at all to attempt to describe higher order effects by two-quark operators, or whether the whole model should be discarded at the 15% level where additivity breaks down. This question was investigated[37] in 1974 with the aim of looking for some signal in the discrepancies at the 10% level between AQM predictions and experimental data on total cross sections which were sufficiently precise to look at 1% physics. Since Regge exchange was the popular mechanism for high energy scattering at that time, the dominant mechanism described by a flavor-dependent two-quark operator was a double exchange involving a Pomeron exchange and an exchange of the f-meson Regge trajectory. The simplest test of this model gave new relations between hadron total cross sections which were in remarkable agreement with experiment. The situation was characterized by the following remark by one of my colleagues: "I do not believe a word of this crazy model. But the numbers are impressive. You must find a better explanation." For eight years a better explanation has been sought but not found. Instead all that has been found are more and more impressive numbers, showing agreement with new data on hadron nucleon total cross sections at higher energies and new channels and on real parts of forward amplitudes with the simple predictions of this "two-component Pomeron model" which adds a second component described by a two-body operator to the simple quark-counting first component.[37]

The basic feature of the data described by this model is the simultaneous breaking of quark model additivity and SU(3) symmetry. These two effects are then seen to be empirically related and both described by a single mechanism. The additive component of the total cross section due to Pomeron exchange was assumed to be universal and a pure SU(3) singlet with no symmetry breaking. All the strangeness dependence of the Pomeron component as well as the meson-baryon difference came from a single non-additive second component which enhanced contributions from nonstrange quarks by an amount depending upon the total number of quarks in the hadron. The SU(3) breaking appeared as an *enhancement of the contribution from the nonstrange quarks, rather than as a suppression of the contribution of the strange quarks.*

The most recent success of this model is in the hyperon-nucleon cross sections[38] which had not been measured when the model was first proposed. In the AQM the strangeness dependence is universal in mesons and baryons and attributed to the difference at the quark level between the scattering amplitudes of the strange and nonstrange quarks. The difference between $\sigma(\Xi N)$, $\sigma(\Sigma N)$ and $\sigma(NN)$ must then be equal to the difference between $\sigma(KN)$ and $\sigma(\pi N)$.[39]

$$\sigma(pp) - \sigma(\Sigma p) = \sigma(\Sigma p) - \sigma(\Xi p) = \sigma(\pi^- p) - \sigma(K^- p) \quad . \tag{10}$$

The two-component Pomeron model, on the other hand, attributes the strangeness dependence to a second order effect which is a quadratic function of the quark numbers, rather than a linear function. The change in total cross section when a nonstrange quark is replaced by a strange quark is larger in baryons than in mesons by a factor of 3/2.[40,41]

$$\sigma(pp) - \sigma(\Sigma p) = \sigma(\Sigma p) - \sigma(\Xi p) = (3/2)\left\{\sigma(\pi^- p) - \sigma(K^- p)\right\} \quad . \tag{11}$$

This difference by a factor of 3/2 between the predictions (10) and (11) of the AQM and of the two-component Pomeron model has now been tested experimentally. The prediction (11) from the two-component Pomeron model agrees with experiment.

The general approach of this two-component Pomeron model pinpoints certain features of the experimental data which have a simple physical interpretation. This is most clearly demonstrated by inverting the relations between hadron-nucleon and quark-nucleon amplitudes and obtaining the contributions of strange and nonstrange quarks to the experimental baryon-nucleon and meson-nucleon cross sections.

$$\sigma(nN)_B = (1/6)\left\{\sigma(pp) + \sigma(pn)\right\} \tag{12a}$$

$$\sigma(nN)_M = (1/4)\left\{\sigma(\pi^- p) - \sigma(K^- p) + \sigma(\pi^+ p) - \sigma(K^- n) + \sigma(K^+ p) + \sigma(K^+ n)\right\} \tag{12b}$$

$$\sigma(sN)_B = (1/6)\left\{\sigma(\Sigma^- p) + \sigma(\Sigma^- n) + \sigma(\Xi^- p) + \sigma(\Xi^- n) - \sigma(pp) - \sigma(pn)\right\} \tag{12c}$$

$$\sigma(sN)_M = (1/4)\left\{\sigma(K^- p) - \sigma(\pi^- p) + \sigma(K^- n) - \sigma(\pi^+ p) + \sigma(K^+ p) + \sigma(K^+ n)\right\} \tag{12d}$$

where $\sigma(nN)_B$, $\sigma(nN)_M$, $\sigma(sN)_B$ and $\sigma(sN)_M$ denote the contributions from nonstrange and strange quarks to the isospin averaged baryon-nucleon and meson-nucleon scattering cross sections respectively as calculated from the AQM and the conventional duality assumption of equality of the contributions from strange quarks and antiquarks is used to eliminate antiquark contributions from Eqs. (12b) and (12d).

$$\sigma(sN)_M = \sigma(\bar{s}N)_M \tag{13}$$

The AQM predicts the equality of the corresponding quark-nucleon contributions to baryon and meson cross sections. Substituting the relations (12) gives two sum rules which can be tested against experimental data:

The nonstrange sum rule,

$$\sigma(nN)_B = \sigma(nN)_M \tag{14a}$$

$$(1/6)\left\{\sigma(pp) + \sigma(pn)\right\} = $$
$$= (1/4)\left\{\sigma(\pi^- p) - \sigma(K^- p) + \sigma(\pi^+ p) - \sigma(K^- n) + \sigma(K^+ p) + \sigma(K^+ n)\right\} \tag{14b}$$

$$12.9 \pm 0.01 \text{ mb.} = 11.2 \pm 0.05 \text{ mb.} \tag{14c}$$

and the strange sum rule,

$$\sigma(sN)_B = \sigma(sN)_M \tag{15a}$$

$$(1/6)\{\sigma(\Sigma^-p)+\sigma(\Sigma^-n)+\sigma(\Xi^-p)+\sigma(\Xi^-n)-\sigma(pp)-\sigma(pn)\} = $$
$$= (1/4)\{\sigma(K^-p)-\sigma(\pi^-p)+\sigma(K^-n)-\sigma(\pi^+p)+\sigma(K^+p)+\sigma(K^+n)\} \tag{15b}$$

$$7.7 \pm 0.1 \text{ mb.} = 7.75 \pm 0.05 \text{ mb.} \tag{15c}$$

The experimental data quoted are taken at 100 GeV/c momentum, where there are both new data on hyperon-nucleon cross sections and previous data on the other hadronic cross sections available.

The strange sum rule (15) is seen to be in excellent agreement with experiment, while there is strong disagreement with the non-strange sum rule (14). The 15% discrepancy is significant and shows that the contribution from *strange* quarks to the hadron-nucleon cross sections is the same in mesons and baryons but that the contribution from *nonstrange* quarks is greater in baryons than in mesons. *This indication that strange quark contributions are somehow simpler than nonstrange contributions is a significant and recurrent feature of the data which has no explanation from first principles.* The model also relates the SU(3) symmetry breaking which produces the difference between the strange and nonstrange contributions (14) and (15) to the breaking of additivity in the sum rule (14). The assumption that a single two-quark mechanism explains both effects gives a new relation between these quantities

$$\sigma(nN)_B - \sigma(sN)_B = (3/2)\{\sigma(nN)_M - \sigma(sN)_M\} \tag{16a}$$

$$(1/3)\{\sigma(pp)+\sigma(pn)\}-(1/6)\{\sigma(\Sigma^-p)+\sigma(\Sigma^-n)+\sigma(\Xi^-p)+\sigma(\Xi^-n)\} =$$
$$= (3/4)\{\sigma(\pi^-p)-\sigma(K^-p)+\sigma(\pi^+p)-\sigma(K^-n)\} \tag{16b}$$

$$5.15 \pm 0.07 \text{ mb.} = 5.2 \pm 0.1 \text{ mb.} \tag{16c}$$

Here the SU(3) symmetry breakings in the baryon and meson sectors are compared and shown to be related in the exact number predicted by the two-component Pomeron model. This prediction has now been strikingly confirmed by the new hyperon-nucleon data. It is just the factor 3/2 appearing in Eqs. (16a) and (11) that makes the difference between the two predictions (10) and (11). The analogous sum rule from the AQM differs from (16a) by not having the factor of 3/2 on the right hand side. The value of 3.46 ± 0.08 mb. obtained without the factor 3/2 is in strong disagreement with the value 5.15 ± 0.07 mb. on the left hand side.

The two-component Pomeron model was extremely successful in fitting data available at the time and has successfully predicted a large quantity of data from subsequent experiments. The striking

Table 1. Theoretical and Experimental Values of Baryon Magnetic Moments

Baryon	SU(3) symmetric model	Experimental value[44,45]	Standard broken SU(3)	Two comp. broken SU(3)
proton	1.83	2.793	2.79	2.75
neutron	-1.22	-1.913	-1.86	-1.83
Λ	-0.61	-0.614 ± 0.005	-0.61	-0.61
Σ^+	1.83	2.33 ± 0.13	2.67	2.24
Σ^-	-0.61	-0.89 ± 0.14	-1.09	-0.81
Ξ^-	-0.61	-0.75 ± 0.06	-0.50	-0.61
Ξ^0	-1.22	-1.25 ± 0.014	-1.44	-1.22
$R(p,\Sigma^+,\Xi^-)$	0	2.76 ± 0.85	0.38	2.50
$R(\Xi,\Lambda)$	1.0	1.09 ± 0.03	1.06	1.0
$R'(\Xi,\Lambda)$	1.0	1.12 ± 0.02	1.0	1.0
$R''(\Xi,\Lambda)$	1.0	1.22 ± 0.1	0.82	1.0

agreement between prediction and experiment shown in Eqs. (15)-(16) is particularly impressive since no data were available at this energy and no hyperon-nucleon cross sections had been measured at all when the model was first proposed. The success of these sum rules seems to indicate that the breaking of SU(3) and additivity are mysteriously related and that the corrections to the simple SU(3)-symmetric AQM *affect only the contributions of the nonstrange quarks*.

The model has no convincing derivation from first principles. The original double exchange picture fails to explain the observed energy dependence, which differs from predictions from Pomeron-f double exchange. The success of the sum rules has motivated a search for an alternative mechanism to give a contribution with the same dependence on quantum numbers as Pomeron-f exchange but a different energy dependence. So far this search has been unsuccessful.

At this point one might look for further clues in other properties of hadrons where the AQM breaks down. Baryon magnetic moments have been treated with an additive quark model where SU(3) breaking is introduced by suppressing the additive contribution of the strange quarks.[14,42] However, it now appears that additivity is also broken.[43] It may be that here also the SU(3) breaking mechanism is better described as a non-additive enhancement of the non-strange quark contribution, rather than a suppression of the additive strange quark contribution. We therefore examine the present status of baryon magnetic moments from this point of view. The essential details of the problem are shown in Table 1.

Included in Table 1 are four functions of the magnetic moments which project out certain physically interesting features of the data.

$$R(p,\Sigma^+,\Xi^-) = -3\{\mu(p) - \mu(\Sigma^+)\}/\{\mu(\Xi^o) - \mu(\Xi^-)\} \quad (17a)$$

$$R(\Xi,\Lambda) = \{\mu(\Xi^o) + \mu(\Xi^-)\}/3\mu(\Lambda) \quad (17b)$$

$$R'(\Xi,\Lambda) = \{\mu(\Xi^o) + 2\mu(\Xi^-)\}/4\mu(\Lambda) \quad (17c)$$

$$R''(\Xi,\Lambda) = \mu(\Xi^-)/\mu(\Lambda) \quad (17d)$$

The expressions (17a) and (17b) were defined in Ref. 43. The expressions (17c) and (17d) are modified versions of (14b) which are somewhat more sensitive to disagreements with the model predictions.[46]

Motivated by the suggestion that the AQM works better for strange quarks we begin our analysis from the unorthodox SU(3) symmetry limit in which all quarks have the mass of the strange quark and use the Λ magnetic moment as input. The magnetic moment of a baryon with a configuration denoted by (1,2;3), where quarks 1 and 2 have the same flavor, is then given in the static model with L=0 SU(6) wave functions by

$$\tilde{\mu}(1,2;3) = (2q_1 + 2q_2 - q_3)\ \mu(\Lambda) = (4q_1 - q_3)\ \mu(\Lambda) \quad , \quad (18)$$

where $\tilde{\mu}$ denotes the magnetic moment in the SU(3) limit where all quarks have the mass of the strange quark, q_i denotes the electric charge of quark i, and $q_1 = q_2$. This SU(3)-symmetric form is scaled to give the correct Λ moment.

The predictions of the standard broken-SU(3) model are obtained from this limit by enhancing the contributions of the nonstrange quarks by a factor which fits the proton moment, while leaving the strange quark contributions unchanged.

$$\begin{aligned}\mu(1,2;3) &= (2q_1+2q_2-q_3)\mu(\Lambda) + (2q_1x_1+2q_2x_2-q_3x_3)\{(1/3)\mu(p)-\mu(\Lambda)\} \\ &= (4q_1-q_3)\mu(\Lambda) + (4q_1x_1-q_3x_3)\{(1/3)\mu(p)-\mu(\Lambda)\} \quad (19)\end{aligned}$$

where x_i is defined to be 1 if quark i is a nonstrange quark and zero if it is a strange quark and $x_1 = x_2$. The symmetry breaking terms are all proportional to x_i which vanishes for strange quarks. Thus the scaling of the strange quark contribution to the moment remains unchanged. Equation (19) can be rewritten

$$\begin{aligned}\mu(1,2;3) &= (2q_1+2q_2-q_3)(1/3)\mu(p) - \\ &- \{2q_1(1-x_1) + 2q_2(1-x_2) - q_3(1-x_3)\}\{(1/3)\mu(p)-\mu(\Lambda)\} \\ &= (4q_1-q_3)(1/3)\mu(p) - \{4q_1(1-x_1) - q_3(1-x_3)\}\{(1/3)\mu(p)-\mu(\Lambda)\} \quad (20)\end{aligned}$$

This is the conventional form in which the SU(3) symmetry breaking appears as a suppression of the strange quark contribution rather than an enhancement of the nonstrange contribution. The SU(3)-symmetric term is scaled to give the correct proton moment and the symmetry breaking terms are all proportional to $(1-x_i)$ which vanishes for nonstrange quarks.

If the comparatively small discrepancy in the Ξ^- moment is neglected, the predictions in the SU(3) symmetry limit shown in Table 1 are seen to fit the Ξ moments reasonably well and the nucleon moments very badly. The introduction of symmetry breaking fits the nucleons well, but at the price of a much stronger disagreement in the Ξ sector. The Σ^+ and Σ^- moments are midway between the two predictions, but the large error on the Σ^- moment allows a fit to either within two standard deviations. Furthermore, the discrepancy in the difference between the proton and Σ^+ moment remains and is nearly unaffected by the symmetry breaking.

This paradox suggests that SU(3) symmetry breaking is not a simple phenomenon and goes beyond enhancing the magnetic moment of the nonstrange quarks relative to that of the strange quarks by the same factor in all hadrons. Nonstrange quark moments seem to be quenched in strange hadrons[43] (or equivalently enhanced in nonstrange hadrons), perhaps by pion exchange.[47,48] However, the results tabulated in Table 1 show a very large enhancement of about 50% is needed to fit the nucleon data, with no enhancement required for the Ξ's. This suggests that all SU(3) breaking comes from a new dynamical mechanism like pion exchange, described by a two-body operator enhancing the nonstrange contributions mainly in the nucleon, with no additional breaking from the quark mass difference. This is difficult to believe; yet it corresponds exactly to the model described above for high energy scattering. The common feature in the total cross section and magnetic moment data that the *nonstrange* quarks break the SU(3) symmetry rather than the strange quarks may be significant.

This point motivated a "two-component" model[46] with one fully SU(3)-symmetric component given by Eq. (18) and the second component breaking SU(3) with a non-additive enhancement of the contributions of the nonstrange quarks by an *ad hoc* factor chosen to fit the nucleon moments $1+(1/4)N$, where N is the number of additional nonstrange quarks in the baryon; i.e. $N=0,1$ and 2 in the Ξ, Σ and nucleon respectively.

$$\begin{aligned}\mu(1,2;3) &= (2q_1+2q_2-q_3)\mu(\Lambda) + \\ &+ \left\{2q_1x_1(x_2+x_3) + 2q_2x_2(x_3+x_1) - q_3x_3(x_1+x_2)\right\}\left\{\mu(\Lambda)/4\right\} \\ &= (4q_1-q_3)\mu(\Lambda) + \left\{2q_1x_1(x_1+x_3) - q_3x_3x_1\right\}\left\{\mu(\Lambda)/2\right\} \qquad (21)\end{aligned}$$

The simultaneous breaking of SU(3) symmetry and additivity is evident in this relation, since the symmetry-breaking terms all contain two-quark products $x_i x_j$. This contrasts with Eq. (19) where all symmetry breaking terms are linear in the x_i's. The predictions of this model are listed in Table 1 as "Two comp. broken SU(3)".

The empirical formula (21) has no theoretical basis. Its superiority over the standard model, which also has only one SU(3)-breaking parameter, shows that the data are parametrized better by non-additive rather than additive enhancement of the nonstrange quark contribution. The factor (1/4) is chosen to fit the well-known approximate enhancement factor of 3/2 for the nucleon. This corresponds in the standard model to a quark mass ratio of 3/2. With parameters fixed in both models by fitting the nucleon and Λ, the significant test comes in the Σ moments. The additive standard model (19) predicts the same nonstrange enhancement in all baryons and gives the same enhancement in the Σ as in the nucleon in disagreement with experiment. The data show that the nonstrange enhancement needed for the Σ is roughly half that needed for the nucleon, which agrees with the prediction of the non-additive two-component model, Eq. (21).

Models for hadron structure and dynamics must eventually describe all properties including masses, magnetic moments and scattering cross sections. The standard broken-SU(3) model predicts the Λ magnetic moment from the proton moment and hadron mass differences. This clue to hadron structure should not be easily discarded in correcting the model to fit other hyperon moments. It is interesting and puzzling that both the magnetic moments and the total cross sections are fit reasonably well by a two-component description in which both additivity and SU(3) symmetry are broken only by a second component which enhances nonstrange contributions. Unfortunately there is no simple relation between the two "second components" in Eqs. (16) and (21). Further experimental work in hyperon physics may help to clarify these paradoxes. A model which explains why strange quarks seem to have a simpler behavior than nonstrange quarks would be very interesting.

PROBLEMS OF THE CONSTITUENT QUARK MODEL FOR MULTIQUARK STATES

Models with only quark degrees of freedom and effective interactions are analogous to conventional atomic models with only electrons, nuclei and a Coulomb interaction. But the gluon of non-Abelian QCD with its color charge and nonlinear self coupling is very different from the neutral photon. The color current carried by the gluon field plays an essential role in ensuring local conservation of color charge and local gauge invariance.

The charge of an electron cannot be changed by photon emission nor shielded by the surrounding cloud of virtual photons. Charge

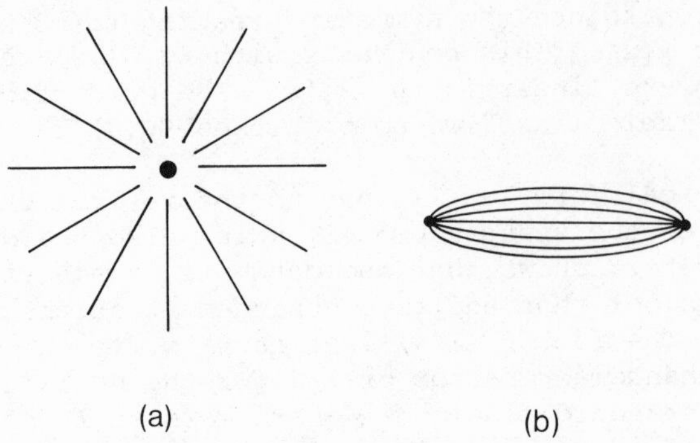

Fig. 5. (a) Lines of force in QED; (b) lines of force in QCD.

exchange does not occur between electrons and nuclei in atoms. Maxwell's equations are linear and the lines of force from a point charge radiate isotropically in all directions as shown in Fig. 5(a). An additive static Coulomb interaction proportional to gauge-invariant c-number charges and uniquely determined by the positions and charges of all particles is adequate for atomic physics. The contribution from the Coulomb field of a nucleus to the force on an atomic electron is independent of the positions of the other electrons. All effects of screening of the nuclear Coulomb field are included by simply adding the Coulomb field of all charged particles.

The color charge of a quark is changed by gluon emission and shielded by virtual gluons. Color exchange occurs in interacting multiquark systems and depends upon gluon dynamics. The color charge of a quark is a gauge-dependent dynamical variable subject to quantum fluctuations. A quark-antiquark pair at two different space points cannot be in a color singlet state in all gauges. A local gauge transformation at the position of the quark changes its color without changing the color of the antiquark. The QCD field equations are nonlinear and the lines of color force are confined to strings or flux tubes as shown in Fig. 5(b). The forces on a quark due to its interactions with two other quarks at different space points are not additive, because of the nonlinear couplings in the gluon field between them. The color energy of a given configuration is not completely determined by the locations and color charges of the constituent quarks. It also depends upon the configuration of the strings or flux tubes connecting these constituents.

Consider a multiquark system containing two quarks and an antiquark as shown in Figs. 6(a) and 6(b). A flux tube may connect the antiquark with either quark while another flux tube connects the

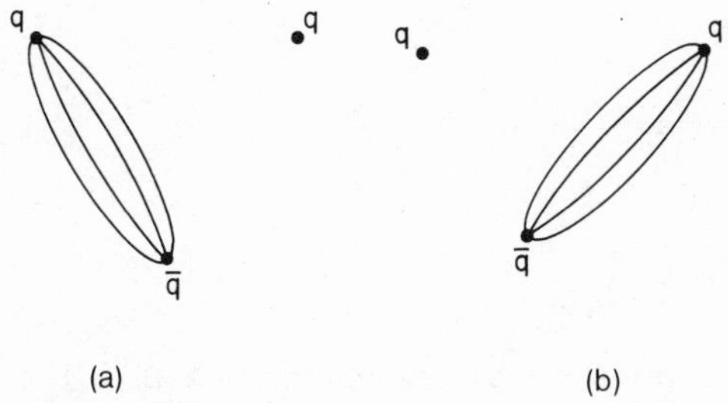

Fig. 6. Flux tubes in the $qq\bar{q}$ system.

remaining quark to the rest of the system. The force on the antiquark depends not only on the positions of all other constituents, but also on whether its flux tube connects it to one quark or the other as shown in Fig. 6. The dynamics of the gluon field may cause the flux tube to jump back and forth between the two configurations. An adequate description of these dynamics may not be possible in a model with only quarks and effective static interactions and no flux tubes. Greenberg and Hietarinta[49] have argued that the string degrees of freedom must be included in any phenomenological model describing the long range forces in multiquark systems.

A potential model with colored quarks interacting via a phenomenological color-exchange force motivated by one-gluon-exchange was first introduced before QCD to explain the saturation of interquark forces in hadrons and the relation between meson and baryon spectra.[23] Only global color symmetry was assumed, with massive gluons giving a short-range Yukawa interaction and no gauge theory nor confinement. This model was later connected with QCD by hand-waving arguments[10] without noting the inconsistency introduced by the additional requirement of local gauge invariance and color charge conservation in a constituent colored quark model with no explicit description of the gluon degrees of freedom.

This inconsistency can be seen[9] by applying a local gauge transformation to a color-singlet quark-antiquark wave function,

$$|1\rangle = \sum_a \bar{q}_a(x) q_a(y) \tag{22}$$

where x and y are two different space points and <u>a</u> is a color index. The state $|1\rangle$ is manifestly a color singlet under a global SU(3) color transformation.

We now apply a *local* SU(3) color gauge transformation which acts only at the point y and not at the point x and is described by a unitary matrix U_{ab} in color space chosen to have all its diagonal elements vanish

$$|1\rangle \rightarrow \sum_{ab} U_{ab}\bar{q}_a(x)q_b(y) = |8\rangle \tag{23a}$$

where

$$U_{aa} = 0 \ . \tag{23b}$$

The state $|8\rangle$ is a pure octet under global SU(3) color and has no singlet component. Since all physical hadron states in QCD are assumed to be color singlets and color octet states do not exist, an allowed local gauge transformation (23) which transforms the color singlet state $|1\rangle$ into the color octet state $|8\rangle$ indicates an inconsistency with local gauge invariance.

One might try to avoid this inconsistency by choosing a particular gauge before using the wave function (22). However, the following physical argument suggests that the inconsistency results from the neglect of the role of the gluon current in ensuring local conservation of color and probably cannot be eliminated by choosing an appropriate gauge. The global color singlet wave function $|1\rangle$ has a quark and an antiquark at two different space points with instantaneously correlated color fluctuations. The quark and antiquark each have equal probabilities of being red, blue or green, but the antiquark must be green when the quark is green. In a complete QCD description color is locally conserved and is exchanged between the quark and antiquark only via gluon exchange. The wave function $|1\rangle$ must have a "string" of gluon current connecting the points x and y to conserve color and preserve local gauge invariance. This difficulty does not arise in Abelian QED with neutral photons where electric charge is trivially conserved in photon exchange.

We can now see the implications of this inconsistency for the colored constituent quark model which disregards the contribution of the gluon field to the color current. The model fails consistently in cases where localized color densities occur in the quark sector and gluon currents necessary for current conservation must introduce additional dynamic effects like screening. All phenomenological successes of the model occur when color and space are separated; the failures occur when they are not. The model succeeds in the quark-antiquark and three-quark systems where the meson and baryon wave functions factorize into a color factor and a factor depending upon all other degrees of freedom. Color completely separates from space and all dynamics are described by a color-independent effective interaction. The color degree of freedom enters only in two ways:

i. The allowed states for three-quark baryons are required to be *symmetric* in the other degrees of freedom in accordance with Fermi statistics for colored quarks in a color singlet state.[21]

ii. A color factor related to the eigenvalue of a color SU(3) Casimir operator appears in relating quark-quark interactions in baryons to quark-antiquark interactions in mesons.[23]

These successes suggest that whenever color and space factorize the inconsistency (23) is unimportant and the dynamics of the gluonic degrees of freedom can either be neglected or somehow absorbed into an effective constituent quark wave function without gluons. This occurs rigorously when $x=y$ and there is no string or flux tube because all quarks are at the same point. The transformation (23) is then not allowed and the state $|1\rangle$ remains a color singlet under all gauge transformations. The effective constituent quark picture might still hold for a wave function which includes terms with $x=y$ and where color and space factorize, so that all the color tranformation properties are determined by the portion of the wave function with $x=y$ and there are no correlations between color and space.

The model has failed when wave functions without factorization of color and space are used to describe multiquark systems and correlations between color and space play an essential role; e.g. in "color chemistry" predictions of unobserved "baryonium" states with color-space correlations[31] and the calculation of long range Van der Waals forces between separated hadrons[1,8,10] arising from the admixture into the two-hadron wave function of states describing spatially separated color-octet pairs coupled to an overall color singlet. Such correlations are not gauge invariant in a model which considers only quark degrees of freedom. In QCD with confinement any local color-octet density in the quark sector is screened by gluons and the oversimplified constituent quark picture which does not include gluon dynamics cannot be valid.

The predictions of exotic multiquark hadrons bound by color hyperfine interactions[50] with color-spin correlations but no color-space correlations have not yet conclusively succeeded or failed. No such multiquark states have yet been convincingly identified, and further experimental tests are of great interest.[51] The four-quark model for scalar mesons[50-52] bound by hyperfine interactions explains properties of the low-lying scalar mesons δ and S^* which are otherwise very mysterious. A zero-range hyperfine interaction which acts only between a pair of quarks at the same point produces bound states analogous to the deuteron with two quark-antiquark clusters separated by a distance of several cluster radii.[52] These states can be described as two color-singlet clusters, bound by an effective short-range potential obtained from a hyperfine interaction, with no unphysical color-space correlations.

This picture might be tested by experiments producing the δ and S^* on nuclear targets. The reaction

$$K^- + p \rightarrow \Lambda + M^0 \tag{24}$$

where M^o is a neutral meson should have an A dependence when the proton is bound in a nucleus which depends on the *size* of the meson M^o. If the δ and S* are large four-quark states they should be absorbed much more strongly in a nuclear target than $q\bar{q}$ states like ρ^o, ω, ϕ, f, A_2 and f′ which should be copiously produced in the same reactions. Studying the $\eta\pi$ spectrum in such reactions would enable direct comparison of the A-dependence of δ and A_2 production, while the $\pi\pi$ spectrum would compare S* production with ρ^o and f^o production.

One might even expect that stripping reactions analogous to (1) would occur between a deuteron-like S* and δ and a nucleus

$$(\delta^o,S^*) + (Z,A) \rightarrow {}_\Lambda(Z,A) + K^o \tag{25a}$$

$$(\delta^o,S^*) + (Z,A) \rightarrow {}_\Lambda(Z-1,A) + K^+ \tag{25b}$$

where ${}_\Lambda(Z,A)$ denotes the hypernucleus with charge Z and baryon number A. Comparing the reactions (24) and (25) might lead to production of hypernuclei with double strangeness.

$$K^- + (Z,A) \rightarrow (\delta^o,S^*) + {}_\Lambda(Z-1,A) \rightarrow {}_{\Lambda\Lambda}(Z-1,A) + K^o \tag{26a}$$

$$K^- + (Z,A) \rightarrow (\delta^o,S^*) + {}_\Lambda(Z-1,A) \rightarrow {}_{\Lambda\Lambda}(Z-2,A) + K^+ \ . \tag{26b}$$

We now use the string picture to see the role of gluon dynamics in more detail. The original global SU(3) potential model[23] takes a two-body potential, introduces a color factor appropriate for a one-gluon-exchange Yukawa interaction and applies it everywhere to multiquark systems. This is clearly unjustified in a gauge theory. However, the potential model has a number of interesting properties, of which some can be expected to hold in a correct description and others to break down completely.

The potential used has the form[8]

$$H_I = -\sum_{i>j} F_i \cdot F_j \ V(|x_i - x_j|) \tag{27}$$

where F_i denotes one of the eight SU(3) color generators for the ith quark or antiquark.

In the color-singlet quark-antiquark state used in meson spectroscopy the color coupling of a quark and an antiquark to a color singlet is unique and factors out. The interaction (27) can be made identical (by hand) to the credible effective two-body potential used in charmonium and obtained from QCD considerations such as lattice gauge calculations. This picture has a string between the quark and antiquark as shown in Fig. 7(a) and the effective potential is obtained by summation over all string configurations.

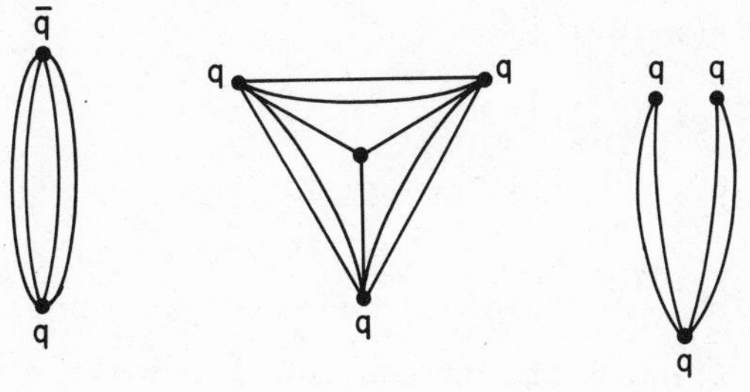

Fig. 7. Flux configurations in mesons and baryons.

In the three-quark color singlet state used in baryon spectroscopy, described in Fig. 7(b), the color coupling is also unique and factors out. The interaction given by Eq. (27) for this case has the correct asymptotic form for the case where two quarks are very close together as shown in Fig. 7(c). The force between a color triplet quark and a point-like color antitriplet diquark is the same as the quark-antiquark force as expected from QCD. Thus this interaction may be taken as a crude approximation to the correct QCD description of baryon spectroscopy.

For multiquark systems containing more than three constituents the asymptotic behavior of the interaction (27) for states describable as two separated point-like clusters agrees with the expectations from QCD. There is no strong force between two point-like color singlet clusters and the triplet-antitriplet force is identical to the quark-antiquark force. For more complicated spatial configurations the color couplings for an overall singlet state are no longer unique and factorization of color no longer occurs. The interaction (27) is a nontrivial matrix in color space[1,23] with matrix elements depending upon the spatial degrees of freedom. It is no longer positive definite and can produce pathologies with confining potentials which always have "anticonfining" components.[8] Furthermore, there is not even a one-to-one correspondence between the configurations allowed by QCD and those allowed by the model based on the interaction (27).

The difficulty can be seen explicitly by examining the system of two quarks and two antiquarks located at the four corners of a tetrahedron. Some possible color couplings for the four particles are:

$$|A\rangle = |\{(13)_1;\ (24)_1\}_1\rangle \tag{28a}$$

$$|B\rangle = |\{(13)_8;\ (24)_8\}_1\rangle \tag{28b}$$

$$|C\rangle = |\{(14)_1;\ (23)_1\}_1\rangle \tag{28c}$$

$$|D\rangle = |\{(14)_8;\ (23)_8\}_1\rangle \tag{28d}$$

$$|E\rangle = |\{(12)_{3^*};\ (34)_3\}_1\rangle \tag{28e}$$

$$|F\rangle = |\{(12)_6;\ (36)_{6^*}\}_1\rangle \tag{28f}$$

where particles 1 and 2 are quarks and 3 and 4 are antiquarks and $|\{(ij)_n;\ (km)_{n^*}\}_1\rangle$ denotes a color singlet state with particles i and j coupled to the representation n of SU(3) color and particles k and m coupled to the representation n*.

In a description with strings or flux tubes each of the six configurations (28) is described by drawing strings connecting the four particles in different ways. Figures 8(a), 8(b) and 8(c) show the configurations for the states $|A\rangle$, $|C\rangle$ and $|E\rangle$ respectively. The wave function $|\{(ij)_n;\ (km)_{n^*}\}_1\rangle$ denotes configurations with strings joining the pairs of particles (ij) and (km) and additional strings joining these two strings when the pairs (ij) and (km) are not color singlets. Each of these configurations is linearly independent of the other five. This can be seen, for example, by noting that the state $|A\rangle$ described by Fig. 8(a) has flux tubes along the edges of the tetrahedron joining particles 1 and 3 and joining particles 2 and 4 and no flux elsewhere, while the four states (28c)-(28f) have no flux along the edges (13) and (24) and all their flux tubes elsewhere. Thus the state $|A\rangle$ cannot be expressed as a linear combination of these four states. The quark description with the interaction (27) has only two independent color couplings.[1,23] The states (28) are linearly dependent in this description and span a Hilbert space of only two dimensions in the color degree of freedom. The state $|A\rangle$ for example is a linear combination of the states $|C\rangle$ and $|D\rangle$ or of the states $|E\rangle$ and $|F\rangle$. Thus there is no simple way to eliminate the QCD strings to obtain an effective interaction of the form (3). Essential information is lost in using a two-dimensional Hilbert space for string dynamics whose description require a space of at least six dimensions.

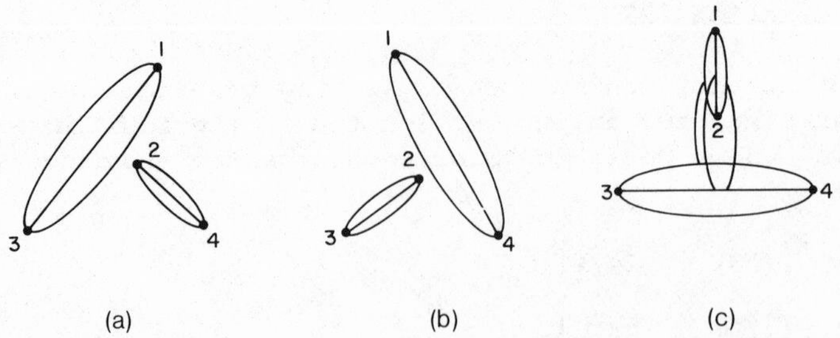

Fig. 8. Flux configurations in the $qqq\bar{q}$ system.

The quark-antiquark and three-quark systems have unique color couplings and simple flux tube configurations shown in Figs. 7(a) and 7(b), which can probably be replaced by an effective potential. Other multiquark systems no longer have a unique color coupling and several flux tube configurations are equally probable as shown in Fig. 8. For these cases the dynamics of flux tube jumping between these configurations cannot be properly included in a constituent quark description.[9]

In limiting cases where one string configuration is dominant the interaction (27) might describe average static properties with the information about the locations of strings contained in the color coupling factors. However, for calculating forces between hadrons or properties of multiquark matter many string configurations of equal importance arise and any realistic wave function must describe the quantum fluctuations in which strings jump from one configuration to another. There does not seem to be any hope of describing this physics with an interaction of the type (27) or any other model (e.g. bag models) which does not explicitly introduce the dynamics and topology of flux tubes or strings.

Another view of this difficulty is seen from comparison with the Abelian case. The interaction (27) also describes the four-body system of two protons and two electrons if F_i is the electric charge of particle i and $V(|\vec{x}_i - \vec{x}_j|)$ is the Coulomb interaction. If particles 1 and 2 are protons and particles 3 and 4 are electrons, the interaction (27) between proton 1 and electron 3 is always the same Coulomb interaction independent of the positions of the other particles and of whether or not proton 1 and electron 4 are bound in a hydrogen atom. The Coulomb fields of the four particles are simply additive.

In the non-Abelian case this additivity of the two-body forces appears in the interaction (27) but not in QCD. The force between quark 1 and antiquark 3 as given by interactions (3) depends only upon the positions and color couplings of the two quarks and is independent of the positions of the other particles. But in QCD the force also depends upon how the strings are drawn to the other particles.

It is also instructive to examine the four-body system for the case of a confining potential, such as a harmonic oscillator

$$V(x_i - x_j) = V_0(x_i - x_j)^2 \quad . \tag{29a}$$

Then

$$H_I = \sum_{i>j} \sum_a F_i^a \cdot F_j^a V_0(x_i - x_j)^2 =$$

$$- \frac{1}{2} \sum_z \sum_{ij} F_i^a \cdot F_j^a V_0\left(x_i^2 + x_j^2\right) + V_0 \sum_a \sum_i F_i^a \vec{x}_i \cdot \sum_j F_j^a \vec{x}_j \quad . \tag{29b}$$

For any color singlet state (or neutral state in the Abelian case) the first term on the right hand side of Eq. (29b) vanishes, by virtue of the identity (7a) for any color singlet hadron.

For the two-proton-two-electron system discussed above, with only one value of a and F_i^a denoting the electric charge of i, the interaction (29b) is seen to depend only on the distance between the center of mass of the two protons and the center of mass of the two electrons. Moving the two protons to infinity in opposite directions does not change the interaction if the center of mass of the two is held fixed. Thus, although the interaction (29) is clearly confining for the two-body system, the confinement is lost completely in the Abelian case for the four-body system. Furthermore, there are extremely pathological long-range forces. A bound two-body system (a harmonic hydrogen atom) on the earth can be broken up by moving the other proton and electron even when they are very far away; e.g. beyond the moon.

This peculiar behavior results from the fact that the interaction (29) is confining only between a pair of particles of opposite charge; i.e. for a quark-antiquark pair in a color singlet state or a quark-quark pair in a color antitriplet state. For other charge or color configurations, the expression is not positive definite and the potential is "anticonfining" and becomes negatively infinite at large distances. The total interaction is seen from Eq. (29b) to be positive definite for color singlet states and unbounded from below for all other configurations.[8]

This discussion reveals an important difference between single-hadron and multiquark states. The interaction (27) seems reasonable for meson spectroscopy, even though its remarkable success in nonrelativistic light quark spectroscopy still remains to be explained. For multiquark spectroscopy the interaction (27) correctly describes the absence of strong forces between separated color singlet clusters[23] but is clearly inadequate for finer details like the treatment of hadron-hadron interactions,[36] multiquark bound states with nontrivial spatial dependences[31] and quark matter.[53] The dynamics of the gluonic degrees of freedom and the gluon color current must be introduced explicitly to describe screening phenomena and keep local color conservation and gauge invariance.[54]

This talk is an updated version of a previous talk (unpublished) given at the University of Washington Summer Institute on Quantum Chromodynamics and Quarks, Seattle, Washington, August 1982. It is a pleasure to thank the organizers of the Seattle Institute, Gerald A. Miller and Lawrence Wilets for the opportunity to have stimulating and clarifying discussions which significantly improved the final version of this talk. This work was supported in part by the U.S. Department of Energy under contract No. W-31-109-ENG-38.

REFERENCES

1. H. J. Lipkin, in: "Common Problems in Low and Medium Energy Nuclear Physics," B. Castel, B. Goulard, and F. C. Khanna, eds., Plenum, New York (1979).
2. R. H. Dalitz, Prog. Part. Nucl. Phys. 8:7 (1982).
3. H. J. Lipkin, Phys. Rep. 8C:173 (1973).
4. F. Close, these Proceedings.
5. C. Quigg and J. L. Rosner, Phys. Lett. 71B:153 (1977).
6. C. Quigg and J. L. Rosner, Phys. Rep. 56:167 (1979).
7. N. Isgur and G. Karl, Phys. Rev. D20:1191 (1979); L. A. Copley, N. Isgur and G. Karl, *ibid.* 20:768 (1979); N. Isgur and G. Karl, Phys. Lett. 74B:353 (1978); Phys. Rev. D18:4187 (1978); D19:2653 (1979).
8. O. W. Greenberg and H. J. Lipkin, Nucl. Phys. A179:349 (1981).
9. H. J. Lipkin, Phys. Lett. 113B:490 (1982).
10. R. S. Willey, Phys. Rev. D18:270 (1978); M. B. Gavela et al., Phys. Lett. 82B:431 (1979).
11. M. Gell-Mann, Phys. Lett. 8:214 (1964).
12. G. Zweig, in: "Symmetries in Elementary Particle Physics," Academic Press, New York (1965).
13. R. Van Royen and V. F. Weisskopf, Nuovo Cim. 50A:617 (1967).
14. H. Rubinstein, F. Scheck and R. H. Socolow, Phys. Rev. 154: 1608 (1967); J. Franklin, Phys. Rev. 172:1807 (1968); 182: 1607 (1969).
15. W. Thirring, Phys. Lett. 16:335 (1965); C. Becchi and G. Morpurgo, Phys. Rev. 140:B687 (1965); R. G. Moorhouse, Phys. Rev. Lett. 16:771 (1966).
16. G. Alexander, H. J. Lipkin and F. Scheck, Phys. Rev. Lett. 17: 412 (1966); A. N. Mitra and M. Ross, Phys. Rev. 158:1630 (1967); H. J. Lipkin, H. R. Rubinstein and H. Stern, Phys. Rev. 161:1502 (1967).
17. E. M. Levin and L. L. Frankfurt, Zh. Eksperim. i. Theor. Fiz.-Pis'ma Redakt 105 (1965).
18. H. Rubinstein and H. Stern, Phys. Lett. 21:447 (1966).
19. Ya. B. Zeldovich and A. D. Sakharov, Yad. Fiz. 4:395 (1966); Sov. J. Nucl. Phys. 4:283 (1967); A. D. Sakharov, SLAC TRANS-0191 (1980).
20. P. Federman, H. R. Rubinstein and I. Talmi, Phys. Lett. 22:203 (1966).
21. O. W. Greenberg, Phys. Rev. Lett. 13:598 (1964).
22. H. J. Lipkin, in: "Physique Nucléaire, Les Houches 1968," C. de Witt and V. Gillet, eds., Gordon and Breach, New York (1969).
23. H. J. Lipkin, Phys. Lett. 45B:267 (1973).
24. M. Gell-Mann, in: "Proceedings of the XVI International Conference on High Energy Physics," J. D. Jackson and A. Roberts, eds., (1972).

25. Y. Nambu, in: "Preludes in Theoretical Physics," A. de Shalit, H. Feshbach and L. Van Hove, eds., North-Holland, Amsterdam (1966).
26. C. Llewellyn-Smith, these Proceedings.
27. A. De Rújula, H. Georgi and S. L. Glashow, Phys. Rev. D12:147 (1975).
28. H. J. Lipkin, Phys. Lett. 74B:399 (1978).
29. H. J. Lipkin, Phys. Rev. Lett. 41:1629 (1978).
30. H. J. Lipkin, Nucl. Phys. B155:104 (1979).
31. L. Montanet, G. C. Rossi and G. Veneziano, Phys. Rep. 63C:184 (1980).
32. S. Nussinov, Phys. Rev. Lett. 34:1286 (1975); Phys. Rev. D14: 246 (1976).
33. F. E. Low, Phys. Rev. D12:163 (1980).
34. J. F. Gunion and D. E. Soper, Phys. Rev. D15:2617 (1977).
35. H. J. Lipkin, Phys. Lett. 116B:175 (1982).
36. M. Harvey, these Proceedings.
37. H. J. Lipkin, Nucl. Phys. B78:381 (1974); Phys. Lett. 56B:76 (1975); Phys. Rev. D11:1827 (1975) and D17:366 (1978).
38. S. F. Biagi et al., Nucl. Phys. B186:1 (1981).
39. H. J. Lipkin, Phys. Rev. D7:846 (1973).
40. H. J. Lipkin, in: "Particles and Fields 1975," H. J. Lubatti and P. M. Mockett, eds., University of Washington, Seattle (1976).
41. H. J. Lipkin, in: "New Fields in Hadronic Physics," J. Tran Thanh Van, ed., Université de Paris-Sud, Orsay, (1976) Vol. I.
42. J. L. Rosner, in: "High Energy Physics—1980," Loyal Durand and Lee G. Pondrum, eds., AIP, New York, (1981).
43. H. J. Lipkin, Phys. Rev. D24:1437 (1981).
44. O. Overseth, in: "Baryon 1980," N. Isgur, ed., University of Toronto, Toronto (1981).
45. L. Deck, Thesis, Rutgers University (1981).
46. H. J. Lipkin, Weizmann Institute report WIS-82/28/June-ph.
47. G. E. Brown, M. Rho and V. Vento, Phys. Lett. 97B:423 (1980); G. E. Brown, Nucl. Phys. A374:630 (1983).
48. S. Théberge and A. W. Thomas, Phys. Rev. D25:284 (1982).
49. O. W. Greenberg and J. Hietarinta, Phys. Rev. D22:993 (1980).
50. R. L. Jaffe, Phys. Rev. D15:267,281 (1977); H. J. Lipkin, Phys. Lett. 70B:113 (1977).
51. N. Isgur and H. J. Lipkin, Phys. Lett. 99B:151 (1981).
52. J. Weinstein and N. Isgur, Phys. Rev. Lett. 48:659 (1982).
53. C. W. Wong, Phys. Lett. 108B:383 (1982).
54. C. N. Yang and R. L. Mills, Phys. Rev. 96:191 (1954).

VALON MODEL FOR HADRONS AND THEIR INTERACTIONS

R.C. Hwa

Institute of Theoretical Science and Dept. of Physics
University of Oregon
Eugene, Oregon 97403

INTRODUCTION

The development of hadron physics is in some ways similar to that of atomic physics. To some, this remark may seem absurd, since we have had many years of experimental investigation of elastic scattering, resonance production, multiparticle reactions in soft and hard processes on the one hand, while on the other a good theoretical understanding of the analytic properties of scattering amplitudes, their high energy behaviors, and the quark-parton model of the hadrons. Moreover, we have more recently the availability of quantum chromodynamics (QCD) as a candidate theory for strong interaction. Surely there is nothing quite comparable in atomic physics either in experiment or in theory during the formative stage of the subject. Indeed, as soon as quantum mechanics was developed, the essence of atomic physics was understood.

Precisely because QCD has not unlocked the mystery of hadron structure, our present understanding of the hadrons as bound systems of their constituents is analogous to the understanding of the atoms before 1925. The Thomson model of the atom was superceded by the Rutherford model due to information gained from scattering experiments, and then the Bohr model came into being to explain the hydrogen spectrum. Similarly, in hadron physics we have deep inelastic scattering (DIS) experiments to confirm the point-like constituents of the nucleon and bound-state descriptions in the form of potential or bag models to explain the hadron spectra. However, these two aspects of the hadron, the bound-state and scattering properties, remain to be combined satisfactorily in a unified picture that can be applied to explaining many empirical facts that have been accumulated in the past twenty years or so. It is that picture which is the subject of discussion here.

One may generally ask what we have learned about hadron structure after these many years of experimentation on high energy collisions. A simple answer is that hadrons are made of quarks. That is like saying that a nucleus is made of nucleons, a statement which is hardly adequate in answer to a question on nuclear structure. What do we know about the wave function of the quarks in a hadron, how are they revealed in a scattering problem, and how do the quarks become hadrons if the quarks are not to be observed as free particles? In both the atomic and nuclear cases the knowledge of the Coulomb potential and the nuclear potential has been sufficient to permit feasible calculations on both the bound-state and scattering properties of those composite systems. In the hadronic case, however, the knowledge of a qqq or $q\bar{q}$ potential that is adjusted to fit the hadron spectra is not enough to explain elastic scattering, let alone multiparticle production. Clearly, a basic link is still missing, which QCD at the present state of our understanding cannot satisfactorily manufacture. The valon model is an attempt to provide that link.

In the bound-state picture a proton is regarded as a composite system of three constituent quarks and a pion that of a quark and an antiquark. In scattering a hadron is regarded as consisting of valence quarks, sea quarks and gluons; collectively they are called partons[1] and there are an infinite number of them. The two pictures are unified by the valon model in which the partons are grouped into clusters, called valons, each containing one valence quark.[2,3] Thus a valon may be identified as a dressed quark or a constituent quark when confined in a bound state, but behaves as a collection of partons in a scattering state. In the latter case a valon does not behave as one integral unit, so it is inappropriate to think of it in an unconfined situation as a constituent quark whose scattering properties are unknown anyway.

Properties of the valons have been deduced from scattering experiments. Thus it is a phenomenological model. Just as it is with potential or bag models where parameters are adjusted to fit certain states in the hadron spectra, parameters in the valon model are determined by fitting certain scattering cross sections. After the parameters are fixed and the hadron is described in terms of the valon coordinates, one can make predictions about other reactions involving that hadron.

For the space available here we can only give an outline of how the hadron structure is determined in terms of the valons and how hadronic reactions can then be described in the valon framework.

THE VALON MODEL

For definiteness, let us consider the proton whose hadronic

structure we wish to describe. As is generally accepted, it consists of two u quarks and one d quark, where by "quark" we mean hereafter current quark. A Fock state for the proton, in general, also contains quark pairs ($q\bar{q}$) and gluons (g). The valon model assumes that the proton can be well represented by a state of three valons (UUD) so that we can write

$$\langle p|uudq\bar{q}\ldots g\ldots\rangle = \sum_{UUD} \langle p|UUD\rangle\langle UUD|uudq\bar{q}\ldots g\ldots\rangle \tag{1}$$

where the sum implies integration over all phase space that the valons span. The model further assumes that in a scattering process the valons may be regarded as interacting weakly with one another, so the last factor in Eq. (1) factorizes as follows

$$\langle UUD|uudq\bar{q}\ldots g\ldots\rangle = \langle U|uq\bar{q}\ldots g\ldots\rangle\langle U|uq\bar{q}\ldots g\ldots\rangle\langle D|dq\bar{q}\ldots g\ldots\rangle. \tag{2}$$

The quark distribution in a valon, U say, is symbolically

$$P_q^U(z) = \sum\int\ldots\ |\langle U|q(z)\ldots\rangle|^2 \tag{3}$$

where the integral is over all variables associated with all quarks and gluons except the quark q with momentum fraction z; the sum is over all Fock states. In recognition that the quark distribution is a manifestation of QCD which is independent of flavor we distinguish just two different types:

$$K(z) = zP_u^U(z); \tag{4a}$$

$$L(z) = zP_{\bar{u}}^U(z) = zP_d^U(z) \tag{4b}$$

which refer, respectively, to the cases where the quark has the same or different flavor as that of the valon.

The valons, being fixed in number, can have an exclusive distribution

$$G_{UUD}(y_1,y_2,y_3) = |\langle p|UUD\rangle|^2 \tag{5}$$

where y_i are the momentum fractions of the valons satisfying the constraint

$$\sum_{i=1}^{3} y_i = 1\ . \tag{6}$$

In demanding Eq. (6) we have assumed that the gluons exchanged among the valons that result in binding carry negligible momenta, since the long-distance behavior (hadronic scale) is effected by small space-like momentum exchanges. Equation (6) is an important feature of the valon model; it properly expresses the phase-space constraints of the internal variables of a hadron, more realistically than what one can do with general Fock states. From the exclusive distribution one can obtain the two-valon joint distribution by integrating out the y variable of the valon not referred to, i.e.,

$$G_{v_1 v_2}(y_1, y_2) = \int dy_3\, G_{v_1 v_2 v_3}(y_1, y_2, y_3) \tag{7}$$

and the single-valon inclusive distribution $G_v(y)$ by integrating out one more. The normalization is

$$\int_0^1 dy\, G_v(y) = 1 \tag{8}$$

for every valon, including the U in the proton. Momentum conservation further demands

$$\int_0^1 dy\, y[2G_U(y) + G_D(y)] = 1 \ . \tag{9}$$

The factorization of Eq. (2) implies that the quark distribution in a proton is a convolution of G(y) and P(z):

$$q(x) = \sum_v \int_x^1 \frac{dy}{y}\, G_v(y)\, P_q^v\left(\frac{x}{y}\right) \ . \tag{10}$$

If the quark distribution is what is probed in a DIS process with a photon having virtualness Q^2, then the Q^2 dependence of $q(x,Q^2)$ is a result of $P_q^v(z)$ on the rhs of Eq. (10) becoming Q^2 dependent. $G_v(y)$ is Q^2 independent because the valons are the constituents of the bound-state problem whose wave function is independent of the high-Q^2 virtual photon which probes the internal structure of the valon. The latter is to be described by $E_q^v(z,Q^2)$ which in QCD is the evolution function that is calculable in lower order approximations.

In all the quantities discussed above we have considered only the longitudinal momenta, ignoring the details about transverse momenta, spin and color of the quarks. Thus those variables have all been either integrated or summed over. Generalization to account for the transverse coordinates is possible and has been considered in Ref. 4 for the treatment of form factors. As for the spin and color coordinates it is not meaningful to keep track of them because the model neglects the exchange of soft gluons among the valons, which leaves essentially unaltered the longitudinal momenta of the valons and quarks, but which changes the spin and color components drastically. Thus the soft gluon leakage from a valon may be regarded as providing an averaging mechanism on the spin and color indices of the valon and of the associated partons.

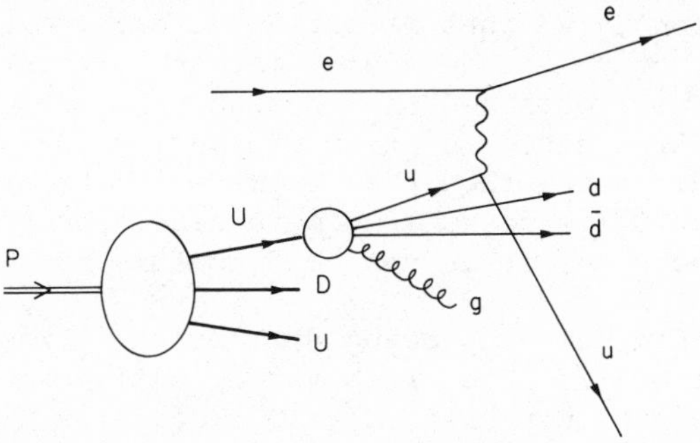

Fig. 1. Schematic diagram for deep inelastic scattering in the valon model.

If we can solve the bound-state problem from first principles, then we would determine the valon distribution $G_{UUD}(y_1, y_2, y_3)$ without free parameters. But our knowledge about confinement in QCD is not yet at that stage, so the valon distribution can only be determined phenomenologically. In the next section we discuss how that is done in connection with deep inelastic scattering.

DEEP INELASTIC SCATTERING

For deep inelastic scattering of electrons or muons by nucleons a pictorial representation of the process in the valon model is shown in Fig. 1. The structure function $F_2^N(x,Q^2)$ is given by the two-step convolution

$$F_2^N(x,Q^2) = \sum_v \int_x^1 dy\, G_v^N(y) F_2^v\left(\frac{x}{y}, Q^2\right) \tag{11}$$

where F_2^v is the structure function of valon v

$$F_2^v(z,Q^2) = \sum_j e_j^2\, E_j^v(z,Q^2) \quad . \tag{12}$$

Here, e_j is the charge of quark (or antiquark) j in units of e, and $E_j^v(z,Q^2)$ is the evolution function which is the Q^2-dependent version of $P_q^v(z)$ in Eqs. (3) and (10). Whereas $P_q^v(z)$ is not calculable, $E_j^v(z,Q^2)$ at high Q^2 can be calculated in perturbative QCD in certain approximations. Our approach is to adopt some phenomenological form for $G_v^N(y)$, use the calculated result for $F_2^v(z,Q^2)$, and apply Eq. (11) to the experimental data on $F_2^N(x,Q^2)$; from the best fit we determine the parameters governing the form of $G_v^N(y)$. The most questionable part of this procedure is the approximation for the evolution functions $E_j^v(z,Q^2)$. But before we describe the nature of the approximation, it should be stated that whatever $E_j^v(z,Q^2)$ is used, the valon distribution $G_y^N(y)$ that is extracted has the ambiguity associated with the nature of the approximation used for $E_j^V(z,Q^2)$. In that sense the $G_V^N(y)$ obtained should always be thought of as an effective distribution since an exact expression for $E_j^V(Z,Q^2)$ is nonexistent. A tacit assumption that is made in the extraction procedure is that $G_V^N(y)$ does not depend sensitively on the accuracy of $E_j^V(z,Q^2)$; otherwise, it is not very useful.

If we regard a valon as being defined at Q_V^2 where its internal structure cannot be resolved, we have, by definition, $E_j^V(z,Q_V^2) = \delta(z-1)\delta_{vj}$. As Q^2 is increased, $E_j^V(z,Q^2)$ deviates from a δ-function at z=1, signifying the possibility of finding a quark j at $z<1$ due to gluon bremsstrahlung. Although $E_j^V(z,Q^2)$ is not known exactly, its form at high Q^2 can be well approximated by the leading order result in perturbative QCD, the so-called leading-log approximation[5] (LLA). If from high Q^2 we use LLA to extrapolate $E_j^V(z,Q^2)$ down to some low Q^2 where $E_j^V(z,Q^2)_{LLA} = \delta(z-1)\delta_{vj}$ (a possibility that exists in QCD), we call that Q^2 value Q_o^2. Instead of discussing the true valon distribution at Q^2 which relies on the exact $E_j^V(z,Q^2)$ that is unknown, we consider the effective valon distribution at Q^2 which relies on the LLA of $E_j^V(z,Q^2)$ that is easy to calculate. Our tacit assumption mentioned earlier is that the effective valons at Q_o^2 do not differ drastically from the true ones at Q^2. The validity of this assumption can be checked by investigating the implications of

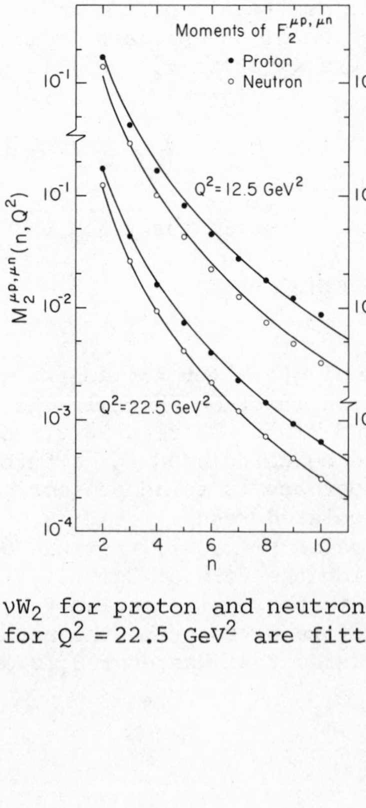

Fig. 2. Moments of νW_2 for proton and neutron data (from Ref. 7). The curves for $Q^2 = 22.5$ GeV^2 are fitted; the others calculated.

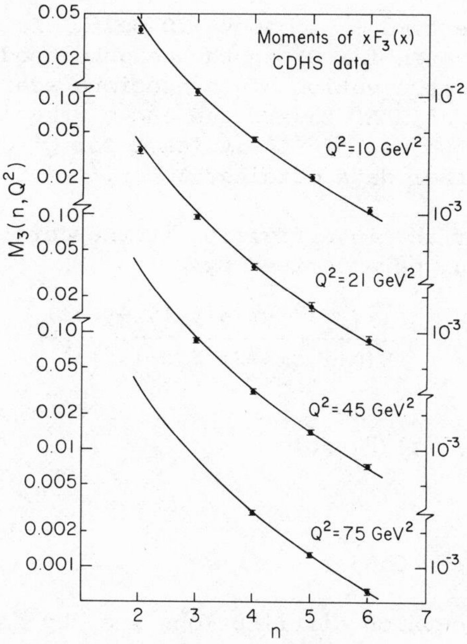

Fig. 3. Moments of xF_3 for neutrino scattering. Data are from Ref. 30. The curves are calculated using the valon parameters determined from Fig. 2 but with slightly different values of Q_0 and Λ.

the valon distributions thus obtained in such problems as low-p_T hadronic reactions and low-Q^2 form factors. As we shall describe in the following sections, they do turn out to be supportive of the model.

The problem of extracting G_{UUD} has been considered in Refs. 2 and 6; the former describes the model with flavor symmetry, while the latter takes the differences between U and D valons into account. Figure 2 shows a fit of F_2 for proton and neutron at $Q^2 = 22.5\ \text{GeV}^2$. The curves for $Q^2 = 12.5\ \text{GeV}^2$ are calculated without further adjustment of parameters, as are those shown in Fig. 3 for neutrino interactions.[6] The agreement with data is self-evident and gives support to the valon model.

The evolution parameter

$$\zeta = \ln\left[\ln(Q^2/\Lambda^2)/\ln(Q_0^2/\Lambda^2)\right] \tag{13}$$

used in the LLA is for $Q_0 = 0.8$ GeV and $\Lambda = 0.65$ GeV. Thus, for the data considered, ζ is above 2 which is rather large for a log log function. It implies that a great deal of evolution has already taken place by the time Q^2 is above 20 GeV^2. Indeed, the evolution is quite advanced even for Q^2 in the neighborhood of 2 GeV^2, thus agreeing with the observation of "precocious scaling" made in the early days of DIS at SLAC toward the end of the '60's. This partially justifies the use of LLA at least for Q^2 down to 2 GeV^2, as is supported by other data obtained later.[7]

The result of the investigation is the determination of the valon distribution, parametrized by

$$G_{UUD}(y_1,y_2,y_3) = \frac{(y_1y_2)^\alpha\ y_3^\beta\ \delta(y_1+y_2+y_3-1)}{B(\alpha+1,\alpha+\beta+2)\ B(\alpha+1,\beta+1)} \tag{14}$$

where

$$B(a,b) = \Gamma(a)\Gamma(b)/\Gamma(a+b) \quad . \tag{15}$$

We found[6]

$$\alpha = 0.65, \qquad \beta = 0.35 \tag{16}$$

so that the single-valon distributions are, by integration,

$$G_U(y) = 7.98\ y^{0.65}\ (1-y)^2 \tag{17}$$

$$G_D(y) = 6.01\ y^{0.35}\ (1-y)^{2.3} \quad . \tag{18}$$

From these distributions we can make some inference about the proton structure. In the first place they are not like δ-functions centered at $y=1/3$, which would be the case if the valons are very loosely bound and have very little momenta transferred among them. On the other hand, they are not very broadly distributed either, suggesting that the valons are not very tightly bound. Indeed, if the valons are given the constitutent quark masses of 350 MeV, the proton mass is just below the threshold for three valons, so with small binding energy we expect the spatial wave functions of the valons, not to be too overlapping. This is consistent with the momentum wave functions being not too spread out, as affirmed by Eqs. (17) and (18). Those distributions agree rather well with the solutions of the covariant-harmonic-oscillator model of the proton.[8]

From a study of the proton and neutron form factors, a subject which we shall return to in a later section, we have also learned about the transverse momentum distributions of the valons. Moreover, the valons themselves have a nontrivial size, which will be important when we consider the hadron size.

For pion structure we can infer from the production of lepton pairs that[3]

$$G_V^\pi(y) \cong 1 \quad . \tag{19}$$

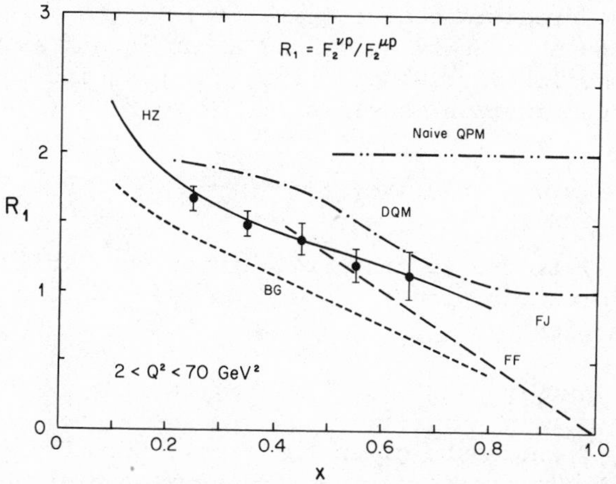

Fig. 4. Ratio of $F_2^{\nu p}$ to $F_2^{\mu p}$. The figure without the solid curve is originally given by P. Allen et al., Ref. 31. The solid curve is the prediction of the valon model without free parameters. (See Ref. 10.)

This result is in rough agreement with those from other model considerations.[9] The implication is that the pion is a tightly bound state of the valons so the momentum distribution is very spread out.

Before we leave the subject of hadron structure as inferred from hard processes, it is appropriate to mention how the knowledge of that structure enables one to calculate the various quark distributions. Starting from the valon distributions obtained, one need only use the LLA to evolve up again to high Q^2. The results are given in Ref. 6. Sharp peaks in the quark and gluon distributions are predicted at x=0 as consequences of soft gluon radiation, but they are expected to be rounded out by finite-mass effects as well as by the production of massive quarks. For a comparison with data, Fig. 4 shows the ratio of $F_2^{\nu p}$ to $F_2^{\mu p}$; the solid curve indicates the result calculated from the parametrizations determined in Ref. 6. Though there are no adjustable parameters, the agreement is nearly perfect.[10] This is yet another piece of evidence in support of the valon model.

HADRONIZATION

In addition to providing an insight into the wave functions of the constituents in a hadron, the valon model also has the virtue of specifying the probability for recombination. The recombination model has been shown to be realistic in describing the hadronization of quarks.[11] Given any process, whether it be a low-p_T hadronic reaction or the production of a quark jet, if the joint distribution $F(x_1,x_2)$ of finding a quark at x_1 and an antiquark at x_2 can be determined at some low virtualness, then the pion inclusive distribution in the recombination model is

$$\frac{x}{\sigma}\frac{d\sigma}{dx} = \int F(x_1,x_2)\ R(x_1,x_2,x)\ \frac{dx_1}{x_1}\frac{dx_2}{x_2} \tag{20}$$

where $R(x_1,x_2,x)$ is the recombination function. It is the invariant function for forming a pion at x, and is related to the valon distribution in a pion by

$$R(x_1,x_2,x) = y_1 y_2\ G^{\pi}_{VV}(y_1,y_2)\ , \qquad y_i = x_i/x \tag{21}$$

where $G^{\pi}_{VV}(y_1,y_2)$ should be given the simple form $\delta(y_1+y_2-1)$ in order that (19) can be recovered as the single-valon distribution. A number of comments must now be made to explain Eqs. (20) and (21).

In Eq. (20) we have expressed the probability for pion production as a convolution of the probabilities for a $q\bar{q}$ configuration and for recombinations. Should one not be concerned about amplitudes and phases? The answer is no because the phases of the amplitudes are averaged out by the emission of soft gluons which do not affect the probability of the quarks being at x_1 and x_2 in (20). This is the same reason why we do not have color and spin factors in (20), as we have discussed before. In a sense those soft gluons allow leakage in color, angular momentum, and phase without taking away longitudinal momentum so that any interference phenomenon not accounted for in the convolution in (20) would be averaged out to zero.

A corollary to the above observation concerns what to do about those gluons which are emitted with finite longitudinal momenta. $F(x_1,x_2)$ is the distribution for q and $\bar{q}$ after the gluon emissions (prior to hadronization) have been taken into account. But the emitted gluons must hadronize also. Since glueballs constitute only a small fraction of the hadrons produced in either hadronic collisions or quark jets, those gluons must contribute to the production of pions and other conventional hadrons. They can be taken into account by the conversion of the gluons back to $q\bar{q}$ pairs which are then to be included in the $F(x_1,x_2)$ function in (20). Such a procedure for recombination is called "saturation of the sea" and has substantial phenomenological support.[3,12-14]

The recombination mechanism expressed in Eqs. (20) and (21) has an implicit step that remains to be discussed. There appears to be an inconsistency between $F(x_1,x_2)$ being the distribution for a quark and an antiquark and $R(x_1,x_2,x)$ being the recombination function related to the valon distribution in a pion. What is implicit is a dressing process which turns the quarks to valons before recombination.[3] Such a dressing process does not change the momentum since the valon that a quark turns into comprises all the partons that the original quark has lost momentum to. Thus x_1 for the quark is also the x_1 for the recombining valon. For simplicity we have referred to the process as quark-antiquark recombination. If there is a gluon exchanged between the quarks or the valons to effect binding, it can be regarded as having been included in the recombination function, so the integrand in (20) describes the configuration before such exchanges take place.

One final remark to be made about (20) concerns directly and indirectly produced pions. Since we sum over all spin orientations of the recombining quarks, vector as well as pseudoscalar mesons can be formed. If the recombination function for vector mesons is the same as for pseudoscalars, then (20) should yield the total distribution for both π and ρ with a multiplicity of three for the latter. Since the ρ is a far more loosely bound state than a pion, its valon distribution is not expected to be as flat as (19) for a pion. Thus its recombination function would differ from (21) in the following way

$$R(x_1,x_2,x) = B^{-1}(\gamma,\gamma)\left(\frac{x_1x_2}{x^2}\right)^{\gamma}\delta\left(\frac{x_1}{x}+\frac{x_2}{x}-1\right) \tag{22}$$

where γ is greater than 1. It turns out[15] that (20) is not very sensitive to the value of γ so long as it is not smaller than 1, so the inclusive distribution for ρ is expected to be similar to that for π. A compilation[16] of data for resonance production is shown in Fig. 5, and indeed the shape of their distributions is nearly the same and roughly in agreement with the one for π. In a calculation of the pion distribution using (20) one does not require the $q\bar{q}$ to be in an antiparallel state, so the result effectively includes the formation and decay of resonances, most notably the ρ. That is, the calculated result is for both directly and indirectly produced pions. It, however, only approximates the realistic situation because the ρ produced decays in a way which, when viewed in the c.m. frame of hadron collision, favors low rather than high x, that being merely a reflection of Lorentz transformation.[17] But on the average the result from (20) can and should be compared to the all-inclusive pion production data.

With (20) and (21) as the basis for pionization of partons, the only task remaining, as well as being the most important part of the problem on particle production is the calculation of $F(x_1,x_2)$. A precise way of doing that is for the problem of quark jets produced at high Q^2, where LLA in QCD can be used. This problem has been treated in great detail in Refs. 13 and 14. The results compare very well with the data obtained in e^+e^- annihilation.

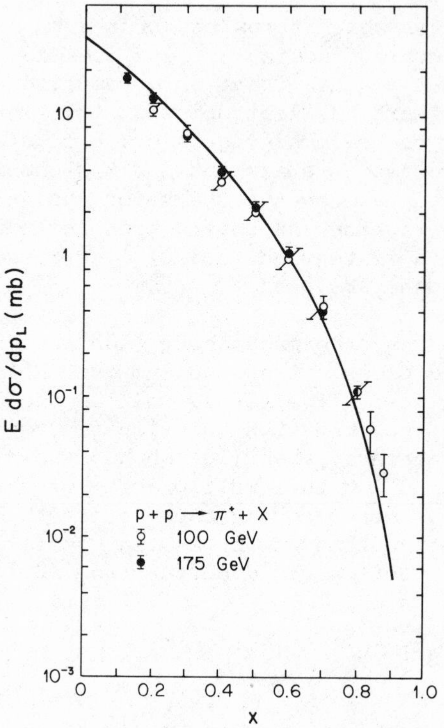

Fig. 5. Inclusive π^+ production cross section vs. x. (Theoretical curve from Ref. 16.)

LOW-p_T INCLUSIVE REACTIONS

At high energies the dominant process in hadronic collisions is the production of particles at low p_T. The interpretation of such processes in the framework of hadron constituents is therefore of paramount importance to an understanding of hadron structure and their interactions. Although Regge analysis has been applied with some success, it is an approach that does not shed much light on the structure of hadrons. Hard processes, such as large-p_T reactions, reveal the point-like nature of the hadron constituents, but a satisfactory theoretical description of the processes involves so many complications (e.g. primordial k_T, gluon bremsstrahlung, higher order QCD terms, trigger bias, etc.) that a clear interpretation of the data is still lacking. Hard processes are important for testing the dynamics of quark and gluon interactions but ineffectual for probing hadron structure. Soft processes at low-p_T, on the other hand, are free of the complications mentioned above. Because of short-range correlations, partons that are very far apart in rapidity do not interact effectively. Hence, in a hadron-hadron collision only the partons at small x of one hadron "see" the other hadron.

The quarks that have $x \gtrsim 0.2$ go through the interaction region essentially unchanged in their momentum distribution. Consequently, the hadronization of those quarks retains a great deal of information about the initial quark distribution. For that reason low-p_T reactions are well suited to learning about hadron structure. In an intersecting storage ring experiment one may therefore figuratively say that what one can see through the side windows is a murky mess but the forward windows offer clear views of the initial hadrons.

The valon-recombination model provides a framework for calculating the inclusive distribution of low-p_T reactions. The valon model specifies the quark and antiquark distributions while the recombination model describes the hadronization. The former provides more than what can be obtained from DIS, which can give quark or antiquark distributions, but not their joint distribution. The exclusive valon distribution given in (14) satisfies momentum conservation and can therefore specify a more reliable $q\bar{q}$ joint distribution. The only unknowns are the parton distributions in a valon appropriate for low-p_T reactions. They have been denoted by K(z) and L(z) in (4) for favored and unfavored distributions. They can be determined in two ways. One is by fitting the data on νW_2 at the lowest Q^2 values available,[3] which are in the range 1 to 3 GeV^2. The other is to use the valon-recombination model to fit low-p_T data, using K(z) and L(z) as adjustable inputs.[18] The two approaches are oppositely directed. The former makes possible a prediction of low-p_T inclusive distributions without free parameters, while from the latter one can calculate νW_2 and infer the effective Q^2 value relevant to low-p_T hadronic reactions.[19] One does not expect the results from the two methods to be the same, since very different assumptions are involved.

The first method was applied to the early SLAC data using flavor-independent valon distributions; the result for K(z) was parametrized[3] for the moments of the nonsinglet component of K(z), which, upon inverting, yield[18] $K_{NS}(z) = 1.2z^{1.1}(1-z)^{0.16}$. With saturated sea the parametrization for L(z) turned out to be $L(z) = 0.41(1-z)^{3.5}$. With those functions determined, the valon-recombination model is completely specified. The details of how the pion inclusive distribution can be calculated in that model are described in Ref. 3. Here we just give the result which is shown in Fig. 6; evidently, it agrees very well with data which range over three order of magnitudes. In Ref. 19 the data for $K^+ + p \rightarrow \pi^- + X$ at 70 GeV/c with the pions in the proton fragmentation region are used to determine K(z) and L(z), which then become the inputs to predict the π^+ distribution in the same experiment. The results are satisfactory. When νW_2 is calculated, the corresponding value of Q^2 is estimated to be around 1 GeV^2. The calculation was done, however, using the valon distribution derived from a study of the form factors. We quote here the results[20] for K(z) and L(z) corresponding to the valon distribution (14) and (16) from DIS: $K_{NS}(z) = K(z) - L(z) = 0.41\, z^{0.5}(1-z)^{-0.27}$ and $L(z) = 0.61(1-z)^{5.13}$.

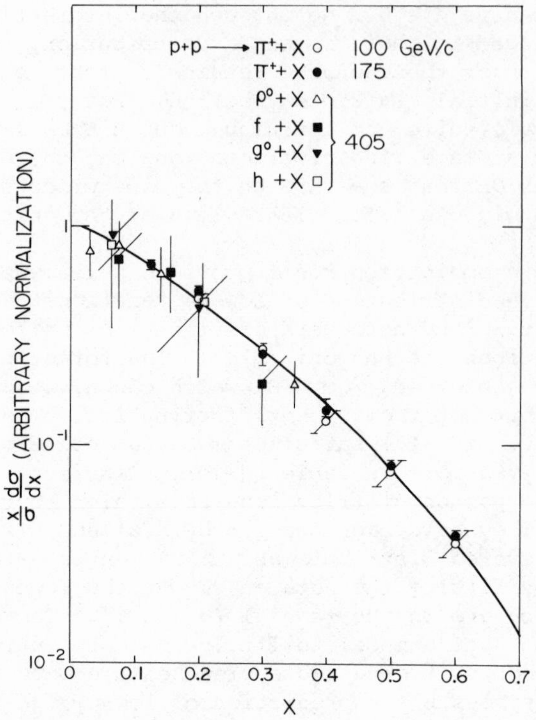

Fig. 6. Inclusive distribution as predicted by the valon-recombination model. (Ref. 3.)

Considering the differences in the input data and the methods used to extract them, these parametrizations should not be regarded as grossly different from the ones obtained from the SLAC data, especially if only the first few moments are compared.

The significance of soft processes at low-p_T is demonstrated by the utility of the reaction $K^+ + p \to K^n + X$ (where $K^n = K^0$ or $\bar{K}^0$) to determine the structure of the kaon via the valon model. The quark distributions in the kaon have been found[21] to be as shown in Fig. 7. The fact that the strange valence quark has a larger fraction of the kaon momentum than the non-strange valence quark is a reflection of the difference in the masses attributed to the valons. Since the structure of unstable hadrons cannot be probed in DIS, and can only marginally be studied in large-p_T or Drell-Yan processes, low-p_T reactions offer a unique way to learn about the internal structure of such particles. The construction of facilities with hyperon beams, for example, is therefore highly desirable for that purpose.

The valon-recombination model, though successful in giving the

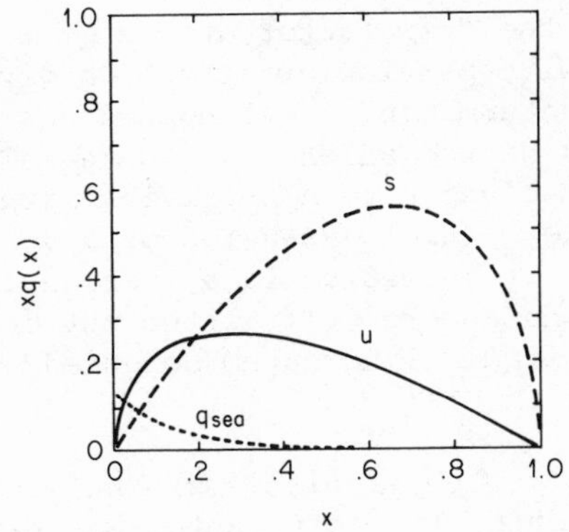

Fig. 7. Quark distributions in a kaon as determined from low-p_T reactions. (See Ref. 21.)

x distribution over the bulk of the x range, is not expected to be accurate in the large x limit because of coherence effects which must operate near kinematical boundaries (and which are not accounted for in the model). In that limit the triple-Regge analysis seems to work rather well. However, the extension of that analysis to intermediate values of x has no theoretical justification, and any agreement with data can only be regarded as fortuitous. Thus the valon model and Regge analysis complement each other in their respective domains of validity. It would be interesting to extend the former to include considerations about the transverse momentum distributions so that predictions on the t dependence of the reactions can also be made. It is a direction that remains to be developed.

PROBLEMS IN HADRON PHYSICS

In the remainder of this short review we outline the applications of the valon model to various problems in hadron physics. Generally speaking, the model is relevant to any problem in which the hadron internal wave function is needed. Due to limitation in space our discussion will be brief. Details are to be found in the references cited.

Form Factors

The valon model is particularly useful for a description of the electomagnetic form factor of a hadron at low Q^2. Unlike the behavior at high Q^2 which reflects the short-distance properties of quark interactions, at low Q^2 (say $Q^2 < 1$ GeV^2) the valon distribution in the hadron is of crucial importance. Whether the quark distribution in a valon is also important at low Q^2 has undergone a change in recent months, at least in the mind of this author.

The problem of the form factor at low Q^2 was first formulated in the valon model with particular attention given to the transverse coordinates.[4] In determining the longitudinal and transverse momentum distributions of the valons, an essential assumption made is that the valons themselves have non-trivial structure only for $Q^2 >$ 1 GeV^2. That is, the spatial extension of a valon is small compared to the hadron size. It seemed to be a reasonable working hypothesis, and the form factor data were well fitted but only with a valon distribution whose longitudinal momentum dependence differs somewhat from that determined in DIS.[4]

Subsequently, a new physical constraint was put on the problem when the consideration of the form factor was extended to the time-like region. It is well known that both the nucleon and pion form factors have poles at the masses of the vector mesons. It would be highly contrived for the valon distributions in the nucleon and pion to be so arranged that they have just the right poles at the same position while having very different space-like properties. The more natural requirement would be for the valons themselves to have nontrivial form factors which possess poles at the vector meson masses. The residual Q^2 dependence of the hadronic form factors is then to be ascribed to the different valon distributions in the various hadrons. In that way the universality of the hadron behavior in the time-like region is automatically taken care of.[22]

Without entering into the details, we mention here the obvious conclusion, namely: the valons have a size characterized by ρ exchange. Thus, valons are not small by hadronic standards. Indeed, a proton may be regarded as a dense-pack system of three valons, as may all the other hadrons be similarly regarded, except for the uncommon pion whose valon wave functions significantly overlap.[23]

Regge Behavior of Elastic Amplitudes

For non-Pomeron Regge exchange the dual diagrams[24] have been very useful in keeping track of what quantum numbers can be exchanged. In treating the dual lines as valons, one can easily see the qualitative connection between hadron-hadron scattering amplitude and valon-valon scattering amplitude. This can indeed be done quantitatively in the valon model.[22,25] Thus the study of Regge behavior of hadron scattering is reduced to an investigation of the high energy behavior of valon-antivalon annihilation amplitude, which in turn is related to the small z behavior of the non-singlet quark distribution, $F_V(z)$, of a valon.

In the ladder approximation for F_V an integral equation can be written with a kernel corresponding to the exchange of a gluon. The difficulty is in determining what that kernel is in QCD for soft processes. With the recognition that no exact method exists, an inference for the kernel is made in Ref. 26 on the bases of a phenomenological observation and perturbative QCD. In that way the moments, M(n) of $F_V(z)$ are then related to the non-singlet) anomalous dimension d_n^{NS}. The solution of the integral equation trivially indicates the possibility of a pole in M(n), the position of which is limited to the range between n=0 and n=1, where $d_n^{NS} = -\infty$ and 0, respectively. Phenomenology in the valon model provides a specific way to determine that pole position which turns out to be $\alpha = 0.40 \pm 0.04$. From the amplitude analysis relating hadron-hadron scattering to valon-antivalon annihilation it can be established that the value of α is just the position of the leading Regge pole that can be exchanged at zero momentum transfer. Thus we have connected the Regge intercept with some aspect of QCD using the valon model and some phenomenology.[26] Note that this is an attempt to obtain the Regge intercept as an output rather than to use it as an input to specify, say, the nature of the quark distribution at x=0.

Thus Regge poles are a property of valon-valon interaction independent of the host hadrons. The residues of the poles in hadronic scattering amplitudes, of course, depend on the hadron wave functions in terms of the valons. In this way we see once again how valons connect bound-state and scattering properties.

Radiation Length of Fast Quarks in Nuclear Matter

If the pp total cross section is taken to be roughly 40 mb, then one may assign 13 mb to each valon and regard it as the total cross section between a valon and a proton. To what extent does that give us a hint of how far a quark will traverse nuclear matter? None at all. The reason is that a valon contains slow partons which are mainly responsible for the total cross section. A fast quark does not interact effectively with a proton because the partons in the latter are separated by a large distance in rapidity from the incident quark. It is for that reason that we could calculate in the previous section the inclusive distribution of produced pions in the projectile fragmentation region. The target independence of the cross section is, however, only approximate, for if it were exact, then a fast quark would have infinite radiation length in nuclear matter. The fact that there is A dependence indicates not only the failure of exact factorization, but also the possibility of extracting from the experimental data the nature of nuclear attenuation.

The data on processes $p + A \to h + X$, where h is any detected hadron in the projectile fragmentation region, are generally presented in terms of the parameter $\alpha(x)$ that appears in the equation

$$x\,\frac{d\sigma}{dx} = \left(x\,\frac{d\sigma}{dx}\right)_0 A^{\alpha(x)} \tag{23}$$

where $(x d\sigma/dx)_0$ is the inclusive distribution if the target is a proton. Equation (23) does not give an outstanding fit of the data, but $\alpha(x)$ does give a rather convenient, albeit imperfect, summary of

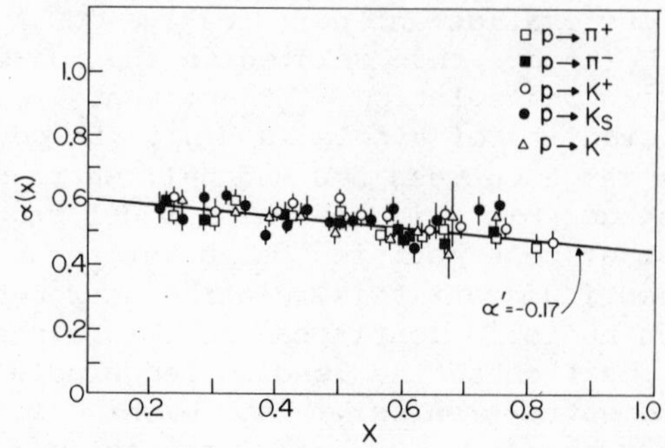

Fig. 8. A compilation of the data on $\alpha(x)$. (From Ref. 27.)

the data. It is shown in Fig. 8. The fact that it is less than 2/3 is a manifestation of nuclear attenuation of valons.[27] The fact that the slope α' is negative reveals the nature of how the valons are absorbed; in particular, the value of α' can be related to the radiation length of the fast quarks in the valons.[18]

Without entering into the details here, it is possible to see qualitatively why α' is negative. In traversing a nucleus each quark radiates gluons and slows down to a smaller value of x. The distribution of quarks that emerge from a nucleus becomes more steeply falling in x as A increases, similar to how Q(x) would behave in accordance to scaling violation in QCD.[28] This steepening of the quark distribution results in a corresponding steepening of the inclusive distribution of the produced hadron in a manner that can be calculated in the valon-recombination model, as described in the previous section. The consequence is that $\alpha(x)$ decreases with increasing x. Using α' as a phenomenological input then facilitates the derivation of the radiation length of a fast quark, which turns out to be about 50 fm.[18]

Taking a nucleon to be roughly 1 fm in size, the result implies that a fast quark loses only about 2% of its momentum in a hadron-nucleon collision. That is why factorization is a good approximation. But for a nuclear target, the 2% losses accumulate and factorization becomes invalid. What is learned here is important to any careful study of h-A and A-A′ collision processes in the fragmentation regions.

Quark-Gluon Plasma in Heavy Ion Collisions

So far it has been suggested that photons, lepton pairs, and kaons emitted from the interaction region in a heavy-ion collision should be detected and examined as possible signals for a phase transition. But why not the pions, the commonest of the hadrons?

It has been said, though not with great certainty, that hadronization of the light quarks probably occurs after a thermal system has cooled down so that the characteristics of a hot, dense quark-gluon plasma may not be revealed in the pion distribution. While the argument may have some intuitive basis in the thermodynamics of the problem which is far from being well understood, it is hard to provide a microscopic description of that picture in terms of hadron constituents. It has been shown in the previous section that the recombination model is successful in relating quark distributions in a hadron to the pion inclusive distributions at low p_T in the fragmentation region. By virtue of such a model for hadronization one can turn the problem around and infer the quark distributions from some experimentally measured pion distribution. This is just what is needed for the diagnostics of the phase transition in heavy-ion collisions. One of the most important quantities to learn about a many-body system is the momentum distribution of its constituents. The pion inclusive cross section can give us that for quarks in the fragmentation region of a nucleus. The question then is whether the quark distribution so obtained describes the plasma phase if a phase transition occurs. Theoretically there is no firm reason why quarks and antiquarks in the plasma phase cannot recombine to form pions. As the two colliding nuclei just complete their passage through each other, the compression phase is at its peak, while the proper time of interaction of the constituents has already been more than a few fermis long—longer than what is usually regarded as necessary for hadronization to occur in pp collision. Thus it is not unreasonable to expect recombination to take place at that time. The corresponding hadrons detected would then be in the projectile or target fragmentation regions. There are reasons to believe that in the deconfined phase the momentum distribution of the valence quarks will be narrower than that before phase transition. To observe a corresponding change in the pion distribution would be highly significant. It is suggested that experiments be proposed to trigger on events with high multiplicities which are expected to accompany central collisions that are optimal for the formation of quark-gluon plasma.

CONCLUDING REMARKS

What has been discussed above is just the skeleton of a phenomenological model that has proven to be useful in describing the structure of hadrons and their interactions. A major omission is on the subject of proton fragmentation as a result of deep inelastic probing. It is a problem that involves both hard and soft processes, ideally suited for the valon model to treat.[29] What has not been treated is the p_T dependence of produced hadrons in low-p_T reactions. The establishment of a connection with triple Regge analysis would be of interest.

The model clearly needs a formulation that can put it on firmer footing. There are two parts to the problem, one being the connection between valons and hadrons, the other being how the valons interact in collision. The former is the bound-state problem that is not likely to have a satisfactory solution until the confinement problem is solved. What we have obtained is perhaps the best one can expect for the moment. The latter is the scattering problem, but reduced to the valon level. How the valons scatter in an elastic process could, at least in principle, be studied by summing Feynman diagrams but with special attention paid to the infrared problem. Such coherent processes are known to be very complex. But our knowledge is scanty even for incoherent processes. How does one derive short-range correlations in QCD? Without an understanding of that it would be hard to derive in any rigorous sense the quark distribution in the fragmentation region after a collision has taken place and before hadronization occurs.

Hadron physics of the type that we have been concerned with here is no longer in the mainstream of particle physics because it is too complicated. Recently, nuclear physicists have picked up interest in the subject, thereby proving yet another time their fortitude to work on difficult problems. So long as it is not feasible to have quark beams and quark detectors to isolate quark interaction processes, hadron physics will always be a part of particle physics. With that view there is hope that there will be increasing interaction among particle and nuclear physicists.

ACKNOWLEDGMENT

I wish to thank the organizers of the NATO Advanced Study Institute for inviting me to give this lecture. I am grateful to all my collaborators without whose participation this work could not attain its present form. The work is supported, in part, by the U.S. Department of Energy under Contract Number DE-AT06-76ER70004.

REFERENCES

1. R. P. Feynman, "Photon-Hadron Interactions," W. A. Benjamin, New York (1972).
2. R. C. Hwa, *Phys. Rev.* D22:759 (1980).
3. R. C. Hwa, *Phys. Rev.* D22:1593 (1980).
4. R. C. Hwa and C. S. Lam, *Phys. Rev.* D26:2338 (1982).
5. See, for example, A. J. Buras, *Rev. Mod. Phys.* 52:199 (1980).
6. R. C. Hwa and M. S. Zahir, *Phys. Rev.* D23:2539 (1981).
7. D. W. Duke and R. G. Roberts, *Nucl. Phys.* B166:243 (1980).
8. P. E. Hussar, *Phys. Rev.* D23:2781 (1981).
9. A. De Rújula and F. Martin, *Phys. Rev.* D22:1787 (1980).
10. R. C. Hwa and M. S. Zahir, *Phys. Rev.* D25:2455 (1982).
11. K. P. Das and R. C. Hwa, *Phys. Lett.* 68B:459 (1977).

12. D. W. Duke and F. E. Taylor, Phys. Rev. D17:1788 (1978).
13. V. Chang and R. C. Hwa, Phys. Rev. Lett. 44:439 (1979); Phys. Rev. D23:728 (1981).
14. L. M. Jones, K. E. Lassila, U. Sukhatme and D. Willen, Phys. Rev. D23:717 (1981).
15. E. Takasugi, X. Tata, C. B. Chiu and R. Kaul, Phys. Rev. D20: 211 (1979); E. Takasugi and X. Tata, Phys. Rev. D21:1838 (1980).
16. Q. B. Xie, Phys. Rev. D26:2261 (1982).
17. R. G. Roberts, R. C. Hwa and S. Matsuda, J. Phys. G5:1043 (1979). This paper which considers resonance production and decay, treated as separate contributions in addition to what follows from Eq. (20), and therefore double-counted.
18. R. C. Hwa, in: "Partons in Soft-Hadronic Processes," R. T. Van de Walle, ed., World Scientific, Singapore (1981).
19. L. Gatignon et al., Zeit. f. Physik 16:229 (1983).
20. L. Gatignon, (private communication).
21. L. Gatignon et al., Phys. Lett. 115B:329 (1982).
22. R. C. Hwa, to be published in "Proc. of 13th Int'l Symposium on Multiparticle Dynamics," (Volendam, 1982).
23. R. C. Hwa (to be published).
24. H. Harai, Phys. Rev. Lett. 22:562 (1969); J. Rosner, *ibid.* 22: 689 (1969).
25. R. C. Hwa, in: "Proc. of 12th Int'l Symposium on Multiparticle Dynamics," W. D. Shephard and V. P. Kenney, eds., World Scientific, Singapore, (1982).
26. R. C. Hwa, Phys. Rev. Lett. 50:305 (1983).
27. A. Dar and F. Takagi, Phys. Rev. Lett. 44:768 (1980).
28. G. Altarelli and G. Parisi, Nucl. Phys. B126:298 (1977).
29. R. C. Hwa and M. S. Zahir, Univ. of Oregon Report OITS-204.
30. J. de Groot et al., Z. Phys. C1:143 (1979).
31. P. Allen et al., Phys. Lett. 103B:71 (1981).

MULTI-QUARK STATES AND POTENTIAL MODELS

M. Harvey

Atomic Energy of Canada Ltd.
Chalk River Nuclear Laboratories
Chalk River, Ontario K0J 1J0

INTRODUCTION

My task in these lectures is to discuss multihadron states in the quark model with special reference to the two-nucleon systems. We shall see during the course of these lectures that we are confronted by two main problems. The first is a suitable description of the dynamics of quarks and the second is the symmetries involved in multiquark (greater than three) states. I emphasize the two-nucleon systems (deuteron, NN-phase shifts) as a goal for multiquark studies because they are really the simplest systems for which we have experimental data.

Quantum chromodynamics[1-3] (QCD) now exists as a viable theory for the strong interactions but at the moment, it is untenable for normal (i.e. non-strange, non-charm) light, hadronic systems (nucleons, pions) which are of particular relevance to nuclear physics. We are forced then to a phenomenological approach "inspired" by QCD.

Two basic approaches have been adopted to describe single hadrons: the bag models[3-13] and the potential models.[14-22] The bag models are perhaps the closest to simulating the physics of hadronic systems treating the quarks as relativistic particles moving in a confined region of space—the bag. An overall fitting of hadron ground state properties is given with reasonable accuracy for such simple models. Predictions are made for magnetic moments, axial vector coupling constants, ground state energies, but the excitation spectra is considered beyond the scope of the simple models involving as they do deformed bags. The bag boundary is considered as a phase transition between the inside, weak coupling regime of interacting quarks and gluons, and the outside, strong coupling regime. The model does not really define the boundary which makes the application of such a model to dibaryon systems difficult. What happens when two baryons overlap?

The potential models have been applied to the energies and decays of excited baryon states and have successfully correlated such data. The models are largely based on non-relativistic kinematics and so must be considered as a mapping of the real dynamics onto a model space. In principle however the potential models describe the interaction between all quarks—whether they be in the same hadron or not—and so it is with these models that most of the later studies of dinucleon systems have been devoted.

These lectures will be directed to the application of the potential model to dinucleon systems mainly because it provides a useful framework in which to discuss the intricacies and the problems. My approach will be that of a critic rather than a strong advocate. I shall emphasize those aspects of the study that I feel have the greatest probability of remaining relevant in the future (e.g. the symmetries) whilst pointing out the inadequacies of the potential approach (e.g. the non-relativistic limitations, infinite energy of color states).

Any discussion of multihadron systems in the quark model should first start with a review of the quark-model predictions for single hadron states—this being the raison-d'être for the quark model. The first part of these lectures will be devoted to normal three-quark and quark-antiquark (u and d quark) systems in potential models. I emphasize the symmetries here partly for its own interest in the particular applications, but also as examples of the more complicated symmetry structures of the later six-quark (dinucleon) considerations to be discussed in the second section.

The third section will be devoted to a review of the generator coordinate and resonating group methods which have been used to describe nucleon-nucleon collisions in the quark model taking into account all aspects of the Fermi statistics and other symmetries.

Although there are large active communities studying the structure of nuclei and the structure of hadrons, there are relatively few people involved with the relationship between the two disciplines. What is the relevance of QCD and the concept of a quark-parton to nuclear physics? The two disciplines cannot forever, however, develop independent of each other as, for example, has happened between atomic and nuclear physics. In this latter case there is some excuse based on the different dimensions (10^{-8} cm compared to 10^{-13} cm) arising from the different dominant interactions (electromagnetic and strong). Between particle and nuclear physics there are no such excuses. The nucleon size is of the same order of magnitude as the nucleus and the properties of both seem to be governed by the same strong interaction. The challenge of the coming decade is to either relate the two fields or at least define criteria by which they can be pursued independently. At the very least I hope these lectures will crystallize some of the problems.

POTENTIAL MODELS FOR QUARK STATES

Quark Hamiltonians

The aim of potential models is to correlate data on excitation energies of hadrons with data on multihadronic states, e.g. the deuteron and NN-scattering cross sections. The Hamiltonian has the usual form

$$H = T + V \tag{1}$$

where T is the kinetic energy and V the potential.

The kinetic energy is often assumed to have the non-relativistic form

$$T = \sum_i \frac{1}{2m_i} P_i^2 + \sum_i m_i c^2 \tag{2}$$

although there are strong arguments why this cannot be valid. The root mean-squared momentum of a particle confined within a volume of dimension 1 fm is $\langle P_i^2 c^2\rangle^{1/2} \sim \sqrt{3}\hbar c \sim 342$ MeV. Thus for the non-relativistic treatment to be valid the quark masses $m_i c^2 >> 342$ MeV. Actually, as we shall see, we are forced to take for the u and d quark $m_i c^2 \approx 1/3\ M_N c^2 \approx 330$ MeV (where M_N is the nucleon mass) which hardly justifies the approximation. Nevertheless the approximation is useful because, using it, we can separate out the center-of-mass kinetic energy from the internal

$$T = \frac{1}{2M} P^2 + \frac{1}{2M} \sum_{i<j} \left(\frac{P_i}{m_i} - \frac{P_j}{m_j}\right)^2 m_i m_j + Mc^2 \tag{3}$$

where

$$P = \sum_i P_i \quad \text{and} \quad M = \sum_i m_i \quad . \tag{4}$$

Note that the mass associated with the kinetic energy of the center-of-mass is the sum of the masses of the constituents—ignoring their interactions. In order that the total mass of nucleons be correct for the later discussion of NN scattering we are forced then to take, in the case of u and d quarks, $m_i \approx M_N/3$.

Attempts have been made by Stanley and Robson[20,21] to

"relativize" the kinetic term in the form

$$T = \sum_i \left(m_i^2 c^4 + P_i^2 c^2\right)^{1/2} \tag{5}$$

and remarkable fits to hadron spectroscopy have been shown. However, it is equally difficult to justify such a form (or the subsequent addition of the potential) and it suffers from the drawback of not being separable in relative and center-of-mass momenta—a crucial necessity for the discussion of multihadron states.

The structure of the potential is inspired by QCD.[1-3,23] The current theory of strong interactions is that quarks possess a new type of charge-property, called color, arising from a fundamental SU(3) gauge symmetry. The three, of SU(3), implies that there are now three different color-charges—unlike the one electric charge arising from the U(1) gauge symmetry of QED. The interaction among quarks is carried by the SU(3) gauge boson called the gluon. By analogy with QED the fundamental field equations for the quarks and gluons are (using the summation convention throughout)

$$[i\,\gamma_\mu \partial_\mu + g\,\gamma_\mu G_\mu^\alpha\,\hat{\lambda}^\alpha]q = 0 \tag{6}$$

$$\Box\, G_\mu^\alpha = g\,\overline{q}\,\gamma_\mu \hat{\lambda}^\alpha q + g f_{\beta\gamma}^\alpha\, F_{\mu\nu}^\beta\, G_\nu^\gamma \quad . \tag{7}$$

Here q and G denote the quark and gluon fields respectively, γ_μ are the conventional γ-matrices; g is the coupling constant, and

$$F_{\mu\nu}^\beta = \partial_\mu G_\nu^\beta - \partial_\nu G_\mu^\beta + f_{\varepsilon\omega}^\beta\, G_\mu^\varepsilon\, G_\nu^\omega \quad .$$

The $\hat{\lambda}^\alpha$ are eight, 3×3 generating matrices for the group SU(3) [corresponding to the Pauli spin matrices for SU(2)] and

$$[\hat{\lambda}^\alpha, \hat{\lambda}^\beta] = f_{\alpha\beta}^\gamma \hat{\lambda}^\gamma \tag{9}$$

where $f_{\alpha\beta}^\gamma$ are the structure constants.

Equations (6) and (7) are very similar to those for QED except for the λ-matrices and the "extra" second term on the R.H.S. of Eq. (7) which arises from the non-Abelian character of SU(3). It is the non-Abelian character of QCD which is supposed to account for the confining property of quarks. Within the hadron, quarks are considered to interact perturbatively by the exchange of a single gluon. This interaction [arising by ignoring the extra term in Eq. (7)] leads to an effective two-body interaction of the form

$$\sum_{i<j} \hat{\lambda}_i \cdot \hat{\lambda}_j\; v_{hyp}(r_{ij}) \tag{10}$$

with $\hat{\lambda}_i \cdot \hat{\lambda}_j \equiv \hat{\lambda}_i^\alpha \hat{\lambda}_j^\alpha$.

The structure for $v_{hyp}(r_{ij})$ has been written down by De Rújula et al.[24] but, for our purposes, we shall treat it phenomenologically with $v_{hyp}(r_{ij})$ being of short range and having spin-dependent components.

As the confining aspects of QCD have not been worked out, the structure for the effective confining potential is very uncertain. We cannot be sure even that a two-body interaction is adequate— indeed many indications are that it is not. Nevertheless, most of the confining potentials that have been considered are two-body and

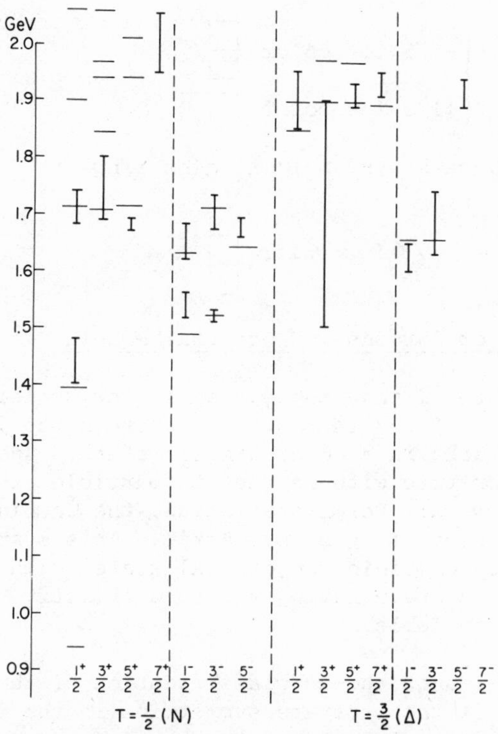

Fig. 1. Observed energy levels of the nucleon (T = 1/2) and delta (T = 3/2) states with their width marked by error bars. "The" nucleon (939 MeV) and "the" delta (1232 MeV) are stable on this scale (no error bars). The bars show the results of a least squares fitting of the strength of the spin-spin interaction and the matrix elements of the confining potential in relative s, p and d-states assuming non-relativistic kinetics (quark mass: 362 MeV) and oscillator states. A tensor potential was assumed with structure, relative to the spin-spin term, given by the one-gluon exchange potential of De Rújula et al.[24]

of the form

$$-\sum_{i<j} \hat{\lambda}_i\cdot\hat{\lambda}_j\; v_c(r_{ij}) \tag{11}$$

where $v_c(r_{ij})$ is a function that increases with distance. This is sometimes considered to have a linear[25] or logarithmic[26] form (albeit for heavy quarks) but, more conveniently, an oscillator[22,27,28]

$$v_c(r_{ij}) = Br_{ij}^2 + D \;. \tag{12}$$

The only property of the color matrices that we need to know is the expectation value of $\hat{\lambda}_i\cdot\hat{\lambda}_j$ in the two-body color states:

$$\langle\hat{\lambda}_i\cdot\hat{\lambda}_j\rangle_{qq} = \begin{cases} -2/3 \text{ for color} \\ +1/3 \text{ for color} \end{cases} \tag{13}$$

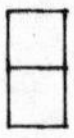

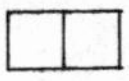 .

For a quark-antiquark system in a color singlet

$$\langle\hat{\lambda}_i\cdot\hat{\lambda}_j\rangle_{q\bar{q}} = -4/3 \text{ for color} \tag{14}$$

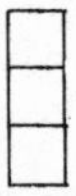 .

Quark Structure of Baryons in Potential Models

For reference, I show the states in the nucleon (T = 1/2) and delta (T = 3/2) spectra[29] in Fig. 1. According to QCD, baryons consist of three quarks in a color-singlet state. The color-singlet state is antisymmetric with respect to particle exchange and hence, to respect the overall Fermi statistics, the combined orbital, flavor and spin state must be symmetric. This symmetric state can be constructed by combining an orbital state, with a given symmetry, with a flavor/spin state having the same symmetry. The possibilities are listed in Table 1.

In Table 2 I show the subclassification of the flavor/spin states in terms of the separate symmetries in the flavor and spin.

Table 1. Symmetry Structure of Color Singlet 3-Quark States

Color	Orbital	Flavor/Spin
$[1^3]$	[3]	[3]
	[21]	[21]
	$[1^3]$	$[1^3]$

Table 2. Sub-Classification of Flavor/Spin States

Flavor/Spin	Flavor (T)	Spin (S)
[3]	[3] (3/2)	[3] (3/2)
	[21] (1/2)	[21] (1/2)
[21]	[3] (3/2)	[21] (1/2)
	[21] (1/2)	[3] (3/2)
	[21] (1/2)	[21] (1/2)
	[1^3] (-)	[21] (1/2)
[1^3]	[21] (1/2)	[21] (1/2)
	[1^3] (-)	[3] (3/2)

When there are only the two u or d flavors, the flavor symmetry is identical to the isospin T given in brackets in the table.

For the ground state of a baryon species we expect quarks to be in relative s-states. It is often convenient to think of the three quarks in independent particle s-states in some confining potential well. If this well has the oscillator structure then the independent particle states can be written

$$S(r) = \left(\frac{1}{\pi b}\right)^{3/2} \exp(-r^2/2b^2) \quad . \tag{15}$$

Defining coordinates

$$R = \sqrt{\frac{1}{3}}\,(r_1 + r_2 + r_3)$$

$$\rho = \sqrt{\frac{1}{2}}\,(r_1 - r_2)$$

$$\lambda = \sqrt{\frac{1}{6}}\,(r_1 + r_2 - 2r_3) \tag{16}$$

we can write

$$S(r_1)S(r_2)S(r_3) \equiv S(R)S(\rho)S(\lambda) \quad . \tag{17}$$

Thus the harmonic oscillator independent particle state can be written in terms of a product of states of the center-of-mass and the relative coordinates. Whichever way we consider the orbital state the internal orbital symmetry has[30,3] the Young tableau [3].

(Note that the coordinates ρ and λ belong to the [21] representation of the symmetric group with

$$\begin{cases} \rho \text{ transforming like } \begin{array}{|c|c|}\hline 1 & 3 \\ \hline 2 \\ \cline{1-1}\end{array} \\ \\ \lambda \text{ transforming like } \begin{array}{|c|c|}\hline 1 & 2 \\ \hline 3 \\ \cline{1-1}\end{array} \end{cases} . \tag{18}$$

The combination $\rho^2+\lambda^2$ which appears in the exponential of $S(\rho)S(\lambda)$ has [3] symmetry.)

With the orbital state having [3] symmetry, the flavor/spin symmetry must also be [3] (c.f. Table 1). In the case of just the two u and d flavors, this implies that the lowest states of the normal baryons have (TS) = (1/2 1/2) and (3/2 3/2) (c.f. Table 2) to be identified with the normal nucleon and delta.

The only difference between the nucleon and delta is therefore in the spin and isospin spaces. It is conventional (but possibly not entirely correct)[13] to describe the entire energy splitting between the N and Δ in terms of a spin-spin potential arising from the hyperfine term of the one gluon exchange potential (OGEP)[24]

$$\sum_{i<j} \hat{\lambda}_i\cdot\hat{\lambda}_j\ \hat{\sigma}_i\cdot\hat{\sigma}_j\ v_\sigma(r_{ij}) \tag{19}$$

with $v_\sigma(r_{ij}) \sim \delta(r_{ij})$. Actually it is probably wise to treat this potential as phenomenological as the OGEP has spin-orbit components which are too strong[18,31] as we shall discuss later. Using the property that the matrix element of a two-body operator between two n-particle antisymmetric states in n(n-1)/2 times the matrix element for any pair we can write

$$\langle\Phi|\sum_{i<j} \hat{\lambda}_i\cdot\hat{\lambda}_j\ \hat{\sigma}_i\cdot\hat{\sigma}_j\ v_\sigma(r_{ij})|\Phi\rangle = 3\langle\Phi|\hat{\lambda}_1\cdot\hat{\lambda}_2\ \hat{\sigma}_1\cdot\hat{\sigma}_2\ v_\sigma(r_{12})|\Phi\rangle \tag{20}$$

with Φ the antisymmetric nucleon or delta states:

$$\Phi = N = [1^3]_C\ [3]_O\left\{[21]_T \times [21]_S\right\}[3]_{TS} \tag{21a}$$

or

$$\Phi = \Delta = [1^3]_C\ [3]_O\left\{[3]_T \times [3]_S\right\}[3]_{TS}\ . \tag{21b}$$

Explicitly the various components in these states have the following structures

$$[3]_S = \left\{\left(\frac{1}{2}\,\frac{1}{2}\right)^1\ \frac{1}{2}\right\}^{3/2} \tag{22}$$

$$[21]_S = \begin{cases} \boxed{\begin{smallmatrix}1&2\\3&\end{smallmatrix}} = \left\{\left(\frac{1}{2}\,\frac{1}{2}\right)^1 \frac{1}{2}\right\}^{1/2} \\ \boxed{\begin{smallmatrix}1&3\\2&\end{smallmatrix}} = \left\{\left(\frac{1}{2}\,\frac{1}{2}\right)^0 \frac{1}{2}\right\}^{1/2} \end{cases} \tag{23}$$

where the superscripts refer to vector coupling and similar expressions could be written for isospin states. For the Δ

$$[3]_{TS} = [3]_T \times [3]_S \quad . \tag{24}$$

For the N

$$[3]_{TS} = \sqrt{\frac{1}{2}} \left(\boxed{\begin{smallmatrix}1&2\\3&\end{smallmatrix}}_T \; \boxed{\begin{smallmatrix}1&2\\3&\end{smallmatrix}}_S + \boxed{\begin{smallmatrix}1&3\\2&\end{smallmatrix}}_T \; \boxed{\begin{smallmatrix}1&3\\2&\end{smallmatrix}}_S \right) \quad . \tag{25}$$

In the latter case we use the usual rule that totally symmetric states can be formed by summing over the Yamanouchi symbols (numbered Young tableau) of products of two states having the *same* symmetry labels.[30]

The matrix element in Eq. (20) can now easily be evaluated knowing the two-body matrix elements for color operators in Eq. (13) and those for the spin operators

$$\langle \hat{\sigma}_1 \cdot \hat{\sigma}_2 \rangle = \begin{cases} \;\;1 & S=1 \\ -3 & S=0 \end{cases} \quad . \tag{26}$$

Thus

$$\begin{cases} \langle N | \sum_{i<j} \hat{\lambda}_i \cdot \hat{\lambda}_j \; \hat{\sigma}_i \cdot \hat{\sigma}_j \; v_\sigma(r_{ij}) | N \rangle = +2 \langle v_\sigma(r_{12}) \rangle_s \\ \langle \Delta | \sum_{i<j} \hat{\lambda}_i \cdot \hat{\lambda}_j \; \hat{\sigma}_i \cdot \hat{\sigma}_j \; v_\sigma(r_{ij}) | \Delta \rangle = -2 \langle v_\sigma(r_{12}) \rangle_s \end{cases} \tag{27}$$

where $<v_\sigma(r_{12})>_s$ denotes the matrix elements of $v_\sigma(r_{ij})$ in relative s-states. From a phenomenological point of view, this matrix element can be chosen to fit the observed N/Δ mass splitting.

Thus

$$-4 \langle v_\sigma(r_{ij}) \rangle_s = (M_\Delta - M_N)c^2 \approx 293 \text{ MeV}$$

$$\text{i.e. } \langle v_\sigma \rangle_s \approx -73.25 \text{ MeV} \quad . \tag{28}$$

Actually this fitting is only valid in a zeroth order approximation. The eigensolutions, with the potential in Eq. (19), do not necessarily have definite orbital and isospin/spin symmetries. Matrix elements exist[20,32] between the naive N structure we have been considering and states having orbital angular momentum L=0 but orbital symmetry [21]

$$N^*\left([21]_o [21]_{TS}^{1/2\ 1/2}\right) = [1^3]_c \left\{[21]_o \times [21]_{TS}^{1/2\ 1/2}\right\}_{[3]} \quad . \qquad (29)$$

Here

$$\left\{[21]_o \times [21]_{TS}^{TS}\right\}_{[3]} = \sqrt{\tfrac{1}{2}} \left(\begin{bmatrix}1&2\\3&\end{bmatrix}_o \begin{bmatrix}1&2\\3&\end{bmatrix}_{TS}^{TS} + \begin{bmatrix}1&3\\2&\end{bmatrix}_o \begin{bmatrix}1&3\\2&\end{bmatrix}_{TS}^{TS} \right) \qquad (30)$$

with

$$\begin{bmatrix}1&2\\3&\end{bmatrix}_{TS}^{1/2\ 1/2} = \sqrt{\tfrac{1}{2}} \left(\begin{bmatrix}1&2\\3&\end{bmatrix}_T \begin{bmatrix}1&2\\3&\end{bmatrix}_S - \begin{bmatrix}1&3\\2&\end{bmatrix}_T \begin{bmatrix}1&3\\2&\end{bmatrix}_S \right) \qquad (31a)$$

[i.e. the state orthogonal to that in Eq. (25)]

$$\begin{bmatrix}1&3\\2&\end{bmatrix}_{TS}^{1/2\ 1/2} = \sqrt{\tfrac{1}{2}} \left(\begin{bmatrix}1&2\\3&\end{bmatrix}_T \begin{bmatrix}1&3\\2&\end{bmatrix}_S + \begin{bmatrix}1&3\\2&\end{bmatrix}_T \begin{bmatrix}1&2\\3&\end{bmatrix}_S \right) \qquad (31b)$$

and

$$\begin{bmatrix}1&2\\3&\end{bmatrix}_o = \sqrt{\tfrac{1}{2}} \left(s'(\rho)s(\lambda) - s(\rho)s'(\lambda) \right)$$

$$= \sqrt{\tfrac{1}{2}}$$

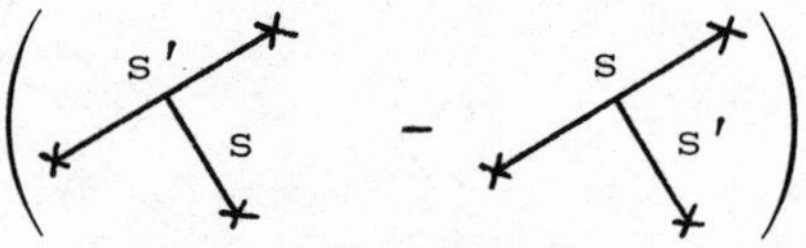

$$\begin{bmatrix}1&3\\2&\end{bmatrix}_o = \left\{p(\rho)\ p(\lambda)\right\}^0 \equiv \qquad (32)$$

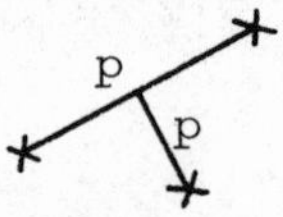

where s′ is an excited s-state orthogonal to the lowest s-state. Using Eq. (21a) and Eq. (29) with the substitutions from Eqs. (30), (31) and (32) we can deduce the value of the coupling matrix element

$$\langle N | \sum_{i<j} \hat{\lambda}_i \cdot \hat{\lambda}_j\ \hat{\sigma}_i \cdot \hat{\sigma}_j\ v_\sigma(r_{ij}) | N^* \rangle = -2 \langle s | v_\sigma(r_{ij}) | s' \rangle \ . \qquad (33)$$

If v_σ is proportional to $\delta(r_{ij})$, as is suggested from one gluon exchange, and we assume oscillator functions for s and s′ then

$$\langle s | v_\sigma | s' \rangle \approx \sqrt{\tfrac{3}{2}}\ \langle v_\sigma \rangle_s \quad . \qquad (34)$$

The absolute energy of the N* state, according to Fig. 1, is of the order of 700 MeV above the energy of the N-state. Hence, the

N-state is shifted in energy, due to the mixing, by

$$\langle N|\sum_{i<j} \hat{\lambda}_i\cdot\hat{\lambda}_j\ \hat{\sigma}_i\cdot\hat{\sigma}_j\ v_\sigma|N^*\rangle^2/700 \text{ MeV} \approx 46 \text{ MeV} \tag{35}$$

with an amplitude for admixture of the N* into the N of

$$\langle N|\sum_{i<j} \hat{\lambda}_i\cdot\hat{\lambda}_j\ \hat{\sigma}_i\cdot\hat{\sigma}_j\ v_\sigma|N^*\rangle/700 \approx 0.2 \tag{36}$$

These numbers are small enough that they can be ignored in a first run through the problem although Robson[21] has claimed that the N* admixture into the nucleon ground state is important for understanding the attraction between nucleons in a potential model.

The systematics of the energies of the negative parity $L=1$ spectra are as easy to compute as the ground state energies. In the oscillator model these states are formed from two states having either a one-quantum (p-state) excitation on the ρ-coordinate or a one-quantum excitation on the λ-coordinate. Defining

$$p(r) = \sqrt{\frac{8}{3\sqrt{\pi}}}\ \frac{r}{b}\ Y^1_m(\theta\phi)\ \exp(-r^2/2b^2) \tag{37}$$

the two states have the explicit structure

$$\phi_\rho = p(\rho)\ s(\lambda) \longrightarrow \begin{array}{|c|c|}\hline 1 & 3 \\ \hline 2 \\ \cline{1-1}\end{array}$$

and

$$\phi_\lambda = s(\rho)\ p(\lambda) \longrightarrow \begin{array}{|c|c|}\hline 1 & 2 \\ \hline 3 \\ \cline{1-1}\end{array} \tag{38}$$

Since ρ and λ transform according to $\begin{array}{|c|c|}\hline 1 & 3 \\ \hline 2 \\ \cline{1-1}\end{array}$ and $\begin{array}{|c|c|}\hline 1 & 2 \\ \hline 3 \\ \cline{1-1}\end{array}$ of the symmetric group S_3, then ϕ_ρ and ϕ_λ also transform in the same way. Negative parity states therefore have [21] orbital symmetry and hence [21] flavor/spin symmetry. From Table 2 we see that the [21] flavor/spin can be formed from (TS) = (1/2 1/2), (1/2 3/2) and (3/2 1/2) states. Vector coupling these states with orbital angular momentum $L=1$ to total angular momentum J, we find precisely this number of J^- states discovered in the spectrum at ~1.6 GeV.

Explicitly the negative parity states have the structure given in Eq. (30) except that now we interpret

$$\begin{array}{|c|c|}\hline 1 & 2 \\ \hline 3 \\ \cline{1-1}\end{array}_{\,o} \equiv \phi_\lambda \quad \text{and} \quad \begin{array}{|c|c|}\hline 1 & 3 \\ \hline 2 \\ \cline{1-1}\end{array}_{\,o} \equiv \phi_\rho \tag{39}$$

The flavor/spin state with (TS) = (1/2 1/2) is as defined in Eq. (31).

For (TS) = (1/2 3/2) we have

$$\begin{array}{|c|c|}\hline 1 & 2 \\ \hline 3 \\ \cline{1-1}\end{array}^{\,1/2\ \ 3/2}_{F/S} = \begin{array}{|c|c|}\hline 1 & 2 \\ \hline 3 \\ \cline{1-1}\end{array}_{F} \ \begin{array}{|c|c|c|}\hline 1 & 2 & 3 \\ \hline\end{array}_{S}$$

and

$$\begin{array}{|c|c|}\hline 1 & 3 \\ \hline 2 \\ \cline{1-1}\end{array}^{\,1/2\ \ 3/2}_{F/S} = \begin{array}{|c|c|}\hline 1 & 3 \\ \hline 2 \\ \cline{1-1}\end{array}_{F} \ \begin{array}{|c|c|c|}\hline 1 & 2 & 3 \\ \hline\end{array}_{S} \ . \tag{40}$$

The states with (TS) = (3/2 1/2) can be deduced from Eq. (40) by interchanging the T(F) and S labels.

For the interaction in Eq. (19) the various (TS) multiplets are degenerate with the values

$$\begin{aligned}(TS) &= \left(\tfrac{1}{2}\,\tfrac{1}{2}\right) \rightarrow \ \langle v_\sigma\rangle_p + \langle v_\sigma\rangle_s \\ &= \left(\tfrac{1}{2}\,\tfrac{3}{2}\right) \rightarrow -\langle v_\sigma\rangle_p - \langle v_\sigma\rangle_s \\ &= \left(\tfrac{3}{2}\,\tfrac{1}{2}\right) \rightarrow 3\langle v_\sigma\rangle_p - \langle v_\sigma\rangle_s \end{aligned} \tag{41}$$

where $\langle v_\sigma\rangle_s$ and $\langle v_\sigma\rangle_p$ are the expectation values of v_σ in relative s- and p-states respectively.

If we follow the inspiration from QCD that $v_\sigma \sim \delta(r_{ij})$, then $\langle v_\sigma\rangle_p \approx 0$. Thus the (TS) = (1/2 3/2) and (3/2 1/2) multiplets should be degenerate and split off from the (TS) = (1/2 1/2) by $2\langle v_\sigma\rangle_s \approx 146$ MeV. This is precisely what is seen from the experimental spectrum (see Fig. 1) and a further indication that the QCD-inspiration has some validity.

Actually the function v_σ cannot exactly be a δ-function since the energy of states like the naive nucleon ground state N and the (TS) = (1/2 1/2) negative parity states would then be unbounded from below. We can however assume v_σ to be a gaussian with small width and derive the above conclusion.

Isgur and Karl[18] have shown that the small deviations of the energy levels from that given by a spin-spin potential can be accounted for by a tensor potential whose strength, relative to the spin-spin potential, is that given by OGEP. Unfortunately, the OGEP also has spin-orbit terms which would destroy this nice understanding we have with experiment. Recently however, Schwesinger and Fiebeg[31] have shown that the spin-orbit strength in OGEP can be reduced if the gluon is considered to be confined within the hadron. For this reason then we consider the short range components of the quark potential to consist of more than one gluon exchange and hence feel free to take a phenomenological description.

The energy difference between the negative parity states and the ground state is a consequence of the confinement. Without the spin-spin potential this excitation energy is ~515 MeV. For an oscillator confinement this corresponds to $\hbar\omega = \hbar^2/mb^2$. With the mass of a quark $m = M_N/3$ we deduce $b \approx 0.46$ fm. This is a very small value for the oscillator parameter. The r.m.s. radius for the nucleon ground state (correcting for center-of-mass fluctuations) is equivalent to b. Thus there is a basic incompatibility between the observed nucleon size (~0.8 fm) and the excitation spectrum. Already we see then that other aspects of the problem must enter the calculation (e.g. a pion cloud) which effectively increases the charge radius.[7-13]

Detailed fits to hadron spectra have been given by Robson and Stanley[20,21] in a pseudo-relativistic model [see Eq. (5)] with a linear confinement potential and phenomenological one-gluon exchange

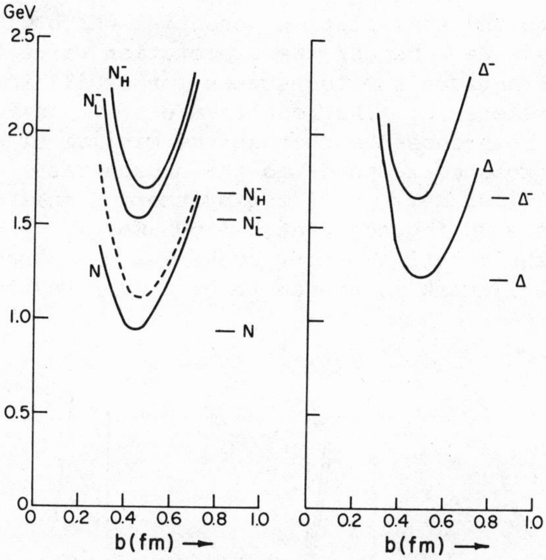

Fig. 2. A variational calculation of nucleon (N) and delta (Δ) states assuming oscillator single-quark states with parameter b, an r^2 confining potential, a spin-spin potential with gaussian radial dependence and non-relativistic kinematics with quark masses of 355 MeV (u and d quarks) and 586 MeV (s-quarks). Parameters of the potentials and absolute constants are determined from a least squares fitting of the energy minimum of the curves to the experimental spectrum shown on right. [N.B. a tensor force has been assumed to reduce some experimental levels to degeneracy for this comparison e.g. $N_{\overline{L}}(N_{\overline{H}})$ refer to the lower (higher) degenerate negative parity levels—see text.]

terms. They show that a consistent fit to both baryon and meson spectra can be achieved with the same potential parameters if the color factors are changed in going from quark-quark interaction to quark-antiquark as in Eq. (13) and (14). It is also necessary to introduce annihilation terms for the isoscalar mesons. One can understand such fits with a variational model that involves little computation. Consider the Hamiltonian

$$H = C_H + \sum_i (m_i - m)c^2 + \sum_i \frac{1}{2m_i} p_i^2 - \frac{1}{2M} P^2 + K \sum_{i<j} \hat{\lambda}_i \cdot \hat{\lambda}_j \, r_{ij}^2 + \alpha m^2 \sum_{i<j} \hat{\lambda}_i \cdot \hat{\lambda}_j \, \frac{\hat{\sigma}_i \cdot \hat{\sigma}_j}{m_i m_j} v_\sigma(r_{ij}) + V_A \qquad (42)$$

with

$$v_\sigma(r_{ij}) = \exp(-r_{ij}^2/\Lambda^2); \qquad r_{ij} = r_i - r_j \ .$$

Here V_A refers to the annihilation potential which will only act for isoscalar mesons. We consider the expectation value of H in 3-quark baryon and quark-antiquark meson states for oscillator wave functions having parameter b. The results are shown in Figs. 2 and 3. Parameters have been chosen such that the minimum of each curve for the separate hadrons corresponds to the hadron mass. Essentially the α-parameter (1193 MeV) is fitted by the Δ-N mass difference (+293 MeV) and the spring constant K (-605 MeV fm^{-2}) by the excitation energy of the negative parity states as discussed above. The mass of the normal quark is chosen to be m = 355 MeV and the range

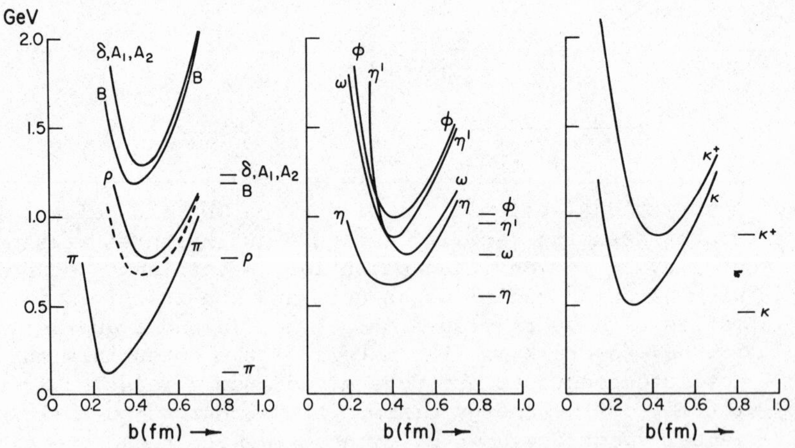

Fig. 3. A variational calculation for some $q\bar{q}$ meson states, as in Fig. 2.

of the gaussian $\Lambda = 0.2$ fm. Different constants C_H were chosen for the baryons ($C_B = -417$ MeV) and mesons ($C_M = -206$ MeV). The strange quark mass is essentially determined from the mass differences between the normal mesons and K-mesons. With no further parameter adjustment one can get the curves for the T=1 and 1/2 mesons (curves on the left and right in Fig. 3) and the baryons (Fig. 2). [We have assumed the splitting of the δ, A_1 and A_2 mesons to arise from a tensor force and show in Fig. 3 only the mean position. Similar mean positions are given for the higher (H) and lower (L) set of nucleon negative parity levels and the delta negative parity level in Fig. 2.] Note that the very low mass of the pion is achieved because of the near instability of the Hamiltonian arising from the spin-spin term. If this term had had a δ-function radial dependence the π-curve would not have had a minimum—nor would the N or $N_{\overline{L}}$. It is only the assumption of a gaussian radial dependence (considered by Stanley and Robson[20] as the finite size for the quark) that allows a minimum to occur. Because of its light mass (2 quarks instead of 3) the minimum for the pion is much more sensitive to the Λ parameter than the N (or $N_{\overline{L}}$). This great sensitivity is perhaps the signal, within the potential model, that the pion is a special object e.g. Goldstone boson.

The interactions leading to the baryon and isovector mesons can be considered (respectively) to be of the form

where the  refers to the effective interaction perhaps with many gluon exchanges. If this were the only type of interaction present then the isoscalar meson spectrum should mirror that of the isovector. States with the same spin would have the same energy e.g. $M_\eta \equiv M_\pi$ etc. That this is not the case indicates that other effects are operative. It is generally assumed that annihilation terms are responsible for the difference. These can schematically be considered of the form

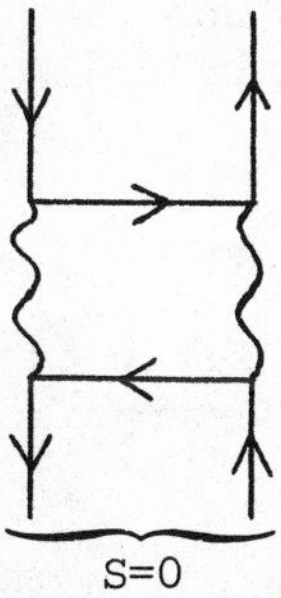

and

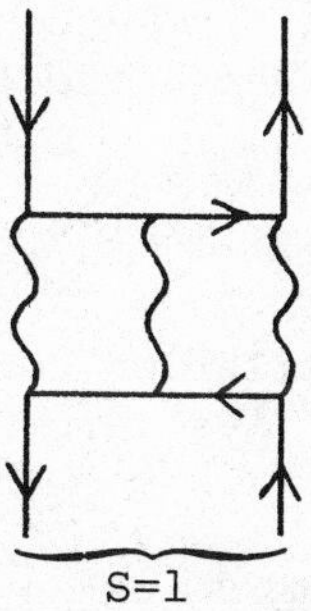

S=0 S=1

In analogy to positronium, we can consider the annihilation contributions for S=0 quark-antiquark pairs (involving two and perhaps higher even numbers of gluon exchanges to be different from the S=1 pairs (involving three and perhaps higher odd numbers of gluon exchanges). Clearly, since gluons do not have an isospin, these diagrams can only operate within $q\bar{q}$ states having T=0. Moreover, these terms will couple the $u\bar{u}$ to itself with the same amplitude as it couples to $d\bar{d}$, $s\bar{s}$, $c\bar{c}$ etc. Schematically, the annihilation diagrams have amplitudes within $q\bar{q}$ states as shown by the following matrix:

	$u\bar{u}$	$d\bar{d}$	$s\bar{s}$	...
$u\bar{u}$	a	a	a	...
$d\bar{d}$	a	a	a	...
$s\bar{s}$	a	a	a	...
⋮	⋮	⋮	⋮	

Because of the near equality of masses of the u and d, leading to equivalent interaction in the u-d sector, this annihilation matrix can be rewritten

	T = 1 $\sqrt{\frac{1}{2}}\,(u\bar{u}-d\bar{d})$	T = 1 $\sqrt{\frac{1}{2}}\,(u\bar{u}+d\bar{d})$	$s\bar{s}$	$c\bar{c}$	...
$\sqrt{\frac{1}{2}}\,(u\bar{u}-d\bar{d})$	0	0	0	0	...
$\sqrt{\frac{1}{2}}\,(u\bar{u}+d\bar{d})$	0	2a	$\sqrt{2}a$	$\sqrt{2}a$	...
$s\bar{s}$	0	$\sqrt{2}a$	a	a	...
$c\bar{c}$	0	$\sqrt{2}a$	a	a	...
⋮	⋮	⋮	⋮	⋮	

The annihilation can be considered to take place at a point (δ-function radial dependence) or over a finite region (gaussian) i.e.

$$V_A = \beta_S \frac{m^2}{m_i m_j} \exp(-r^2/\Lambda^2)$$

(Since the annihilation contributions to the energies effectively come in as repulsive forces, the δ-function assumption here causes no problem.)

Choosing the "best" values for the strengths β_S($\beta_0 = 6.5$ MeV, $\beta_1 = 0.3$ MeV) and adding the annihilation terms to the Hamiltonian in Eq. (42) yields the curves for the isoscalar mesons in Fig. 3.

Table 3. Symmetry Structure of Colored 3-Quark States

Class	Color	Orbital	Flavor/Spin	Sign at ∞
I	[21]	[3]	[21]	+
II		[21]	[3]	-
III		[21]	[21]	+
IV		[21]	[1^3]	+
V		[1^3]	[21]	+
VI	[3]	[3]	[1^3]	-
VII		[21]	[21]	-
VIII		[1^3]	[3]	-

Qualitatively the fitting is pretty good and we believe the essential physics that is in the detailed calculations of Stanley and Robson has been simulated by the model.

Non-Singlet Color States

The QCD ansatz is that only color-singlet hadrons are to be found in nature. It is tempting then to completely ignore the possibility of non-singlet, color hadron states. When we consider multiquark states however, we find we cannot ignore them.

Eight classes of three-quark, non-singlet, color states can be identified—five with color octet and three with color decuplet. (See Table 3.) The problem with such states, is that they are not all bounded from below with the simple confining potentials that people have been using in the literature.[33,34] Consider for example the class II color-octet states which have the structure

$$\mathrm{II} = \sqrt{\frac{1}{2}}\left(\begin{array}{|c|c|}\hline 1&2\\\hline 3\\\cline{1-1}\end{array}_{\,c}\;\begin{array}{|c|c|}\hline 1&3\\\hline 2\\\cline{1-1}\end{array}_{\,o} - \begin{array}{|c|c|}\hline 1&3\\\hline 2\\\cline{1-1}\end{array}_{\,c}\;\begin{array}{|c|c|}\hline 1&2\\\hline 3\\\cline{1-1}\end{array}_{\,o}\right)\begin{array}{|c|c|c|}\hline 1&2&3\\\hline\end{array}_{\,F/S} \quad . \tag{43}$$

With a two-body potential $V_c = -\sum_{i<j} \hat{\lambda}_i\cdot\hat{\lambda}_j\, v_c(r_{ij})$ using to confine color singlet triplets

$$\langle \mathrm{II}|V_c|\mathrm{II}\rangle = -\frac{3}{2}\left[\frac{1}{3}\langle v_c\rangle_a - \frac{2}{3}\langle v_c\rangle_s\right] \tag{44}$$

where $\langle v_c\rangle_a$ and $\langle v_c\rangle_s$ denote two-body matrix elements in antisymmetric or symmetric color states.

Assuming two states, one located at the point R_0 and the other at R_1 and assuming two particles are in the state R_0 and one in R_1 we can write the orbital functions explicitly as

$$\begin{array}{|c|c|}\hline 1 & 2 \\ \hline 3 & \multicolumn{1}{c}{} \\ \cline{1-1} \end{array}_{\,0} = \sqrt{\frac{1}{3}}\left[\sqrt{2}\, R_0^2 R_1 - \overline{R_0 R_1 R_0}\,\right] \quad (45)$$

$$\begin{array}{|c|c|}\hline 1 & 3 \\ \hline 2 & \multicolumn{1}{c}{} \\ \cline{1-1} \end{array}_{\,0} = \widetilde{R_0 R_1} R_0 \quad (46)$$

[$\overline{ab} = \sqrt{1/2}\,(ab + ba)$ and $\widetilde{ab} = \sqrt{1/2}\,(ab - ba)$ with particle numbering being determined by the ordering]. Thus

$$<v_c>_a = \left\langle \widetilde{R_0 R_1} \middle| v_c \middle| \widetilde{R_0 R_1} \right\rangle = \left\langle R_0 R_1 \middle| V_c \middle| R_0 R_1 - R_1 R_0 \right\rangle \quad (47)$$

and

$$<v_c>_s = \frac{2}{3}\left\langle R_0^2 \middle| v_c \middle| R_0^2 \right\rangle + \frac{1}{3}\left\langle \overline{R_0 R_1} \middle| v_c \middle| \overline{R_0 R_1} \right\rangle \quad (48a)$$

$$= \frac{2}{3}\left\langle R_0^2 \middle| v_c \middle| R_0^2 \right\rangle + \frac{1}{3}\left\langle R_0 R_1 \middle| v_c \middle| R_0 R_1 + R_1 R_2 \right\rangle . \quad (48b)$$

If $R_0 \to 0$ and $R_1 \to \infty$ then

$$<v_c>_a \to \left\langle R_0 R_1 \middle| v_c \middle| R_0 R_1 \right\rangle$$

$$<v_c>_s \to \frac{1}{3}\left\langle R_0 R_1 \middle| v_c \middle| R_0 R_1 \right\rangle$$

i.e. $\left\langle II \middle| V_c \middle| II \right\rangle = -\frac{1}{6}\left\langle R_0 R_1 \middle| v_c \middle| R_0 R_1 \right\rangle \to -\infty$ as $(R_0 - R_1) \to \infty$.

Similar techniques can be used to show the large distance behavior of the other non-singlet colored triplets. (An exercise for the student.) We find that all the color decuplets are unbounded but only the class II octets (see Table 3).

SIX QUARK STATES

Symmetry Structure of Six-Quark States

It is tempting to describe the six-quark state of two nucleons simply as the antisymmetrization of the quarks between the two nucleons[27,35,36]

$$\psi_{NN} = \mathcal{A}\left(N_{R/2}\ N_{-R/2}\right) \quad (49)$$

where $N_{R/2}$ represents the naive quark model description of a nucleon in terms of three constituent quarks in s-states. Many people have considered such an approach, but I advocate building up the dinucleon state from its underlying symmetry structure.[37,38] The reason for this is that the symmetries illuminate the structure of the system better and highlight the difficulties confronting us in the use of such states for a description of dinucleon systems whether it be the deuteron or NN-scattering. With the symmetry structure well defined, one can employ the fractional-parentage technology[39-44] for calculating matrix elements of physical operators. A third reason for staying close to the symmetry structure is that quarks are still very much unknown objects—especially in low mass hadrons—but, whatever they are, the symmetry structure of collections of them is likely to remain longer in the physics literature, and henceforth to be most useful.

From the naive quark model, the orbital symmetry of the nucleon ground state is [3]. Hence the orbital symmetries of six-quark states that are expected to be most important are

$$[3] \times [3] = [6] + [51] + [42] + [33] \quad . \tag{50}$$

This is not to say that we shall always be able to ignore the other symmetries [the nucleon has other components in its wave function than orbital symmetry [3]—see Eq. (36)], but those above are enough for a first pass through the problem of describing dinucleon properties and will illuminate some of the difficulties. With the color state of each nucleon being antisymmetric, and the assumption of only three colors, we know that the color-state of the six-quark state has the partial symmetry [222]. The problem of finding flavor/spin states that will combine with the color state and a particular orbital state to form totally antisymmetric states is now not so simple as in the three-quark problem. Now the orbital and flavor/spin states have to combine to the adjoint symmetry of the [222] color symmetry (i.e. [33]). The possible flavor/spin symmetries that will combine with the set in Eq. (50) to form states of [33] symmetry are listed in Table 4. Shown also are the isospin (T) and spin (S) subclassifications.

The precise structure of the states in Table 4 can be written down with the aid of the symmetric group Clebsch-Gordan coefficients. Thus, if states in orbital, color and flavor/spin spaces are denoted by their Young symmetries and Yamanouchi labels fY, f′Y′ and f″Y″ respectively, (we use the unprimed, primed and double-primed with this meaning throughout this section) the general antisymmetric state in Table 4 can be written (to within an overall normalization factor)

$$|Af\ f'\ Bf''\rangle \approx \sum_{YY'Y''} S(fYf'Y'|f''Y'')\,|AfY\rangle_0\,|f'Y'\rangle_C\,|B\bar{\tilde{f}}''\bar{\tilde{Y}}''\rangle \quad . \tag{51}$$

Here $\bar{f}''$ denotes the dual symmetry to f″—it has the number of rows

Table 4. Classification of Six-Quark, Antisymmetric States with [222] Color Symmetry

Orbital	SU(4)(TS)	SU(2)(T) × SU(2)(S)
[6]	[33]*	(01) (10) (12) (21) (03) (30)
[51]	[42]*	(00) (02) (20) $(11)^2$ (12) (21) (13) (31) (22)
	[321]	(01) (10) (02) (20) $(11)^2$ (12) (21)
[42]	[51]*	(01) (10) (12) (21) (23) (32) (11) (22)
	[33]*	(01) (10) (12) (21) (03) (30)
	[411]	(01) (10) (11) (12) (21) (22)
	[2211]	(01) (10)
	[321]	(01) (10) (02) (20) $(11)^2$ (12) (21)
[33]	[6]*	(00) (11) (22) (33)
	[42]*	(00) (02) (20) $(11)^2$ (12) (21) (13) (31) (22)
	[222]	(00) (11)
	[3111]	(00) (11)

*The asterisks pick out the isospin-spin SU(4) symmetries which contain dibaryon states in the cluster model and the right hand column lists the (TS) subclassifications with respect to SU(2)(T) × SU(2)(S).

and columns equal to the number of columns and rows of f″. The tilde (~) denotes that the symmetry is defined with the opposite phase convention to the standard Young-Yamanouchi following the general practice in nuclear physics.[40,41] The letters A and B denote a shorthand notation describing the configurations in orbital and F/S space. The symmetric group Clebsch-Gordon coefficients S(...| ...) can be computed using a technique given by Hamermesh.[43] If we rewrite the Y-label as Y = (pqy) where the nth and $n\text{-}1^{th}$ particles are in rows p and q, and the remaining particles have a distribution in the Young tableau given by y, then

$$S\Big(f(pqy)\ f'(p'q'y')\,\Big|\,f''(p''q''y'')\Big)$$

$$= K(fp\ f'p'|f''p'')\ S\Big(f_p(qy)\ f'_{p'}(q'y')\,\Big|\,f''_{p''}(q''y'')\Big) \qquad (52a)$$

$$= K_2\Big(f(pq)\ f'(p'q')\,\Big|\,f''(p''q'')\Big)\ S\Big(f_{pq}y\ f'_{p'q'}y'\,\Big|\,f''_{p''q''}y''\Big). \qquad (52b)$$

Here f_p denotes the symmetry for S_{N-1} derived from the symmetry f of S_N by the removal of a square in the pth row. Equation (52a) is a relationship between the C.G. coefficients of S_N with those of S_{N-1} via a matrix K(...|...). Hamermesh shows that the K-coefficient can be computed by recurrence relations. Equation (52b) takes the relationship one step further, relating the C.G. of S_N to those in

S_{N-2} where the K_2 are products of two K's relating S_{N-1} to S_N and S_{N-2} to S_{N-1}.

In Eq. (52) we have assumed the representation to be defined by the standard Young-Yamanouchi scheme where the last pair of particles do not necessarily have definite symmetry. K-matrices can be defined[37] for which the last pair do have definite symmetry

$$\overline{K}\Big(f[pq]\ f'[p'q']\,\big|\,f''[p''q'']\Big) = \Big(\gamma^{f}_{pq} + \delta^{f}_{pq}\,\hat{P}_{pq}\Big)\Big(\gamma^{f'}_{p'q'} + \delta^{f'}_{p'q'}\,\hat{P}_{p'q'}\Big)\Big(\gamma^{f''}_{p''q''} + \delta^{f''}_{p''q''}\,\hat{P}_{p''q''}\Big) \times \times K_2\Big(f(pq)\ f'(p'q')\,\big|\,f''(p''q'')\Big) \tag{53}$$

where $\hat{P}_{pq}$ denote the interchange of p and q. In Eq. (53) the γ and δ coefficients are the transformation coefficients between the standard scheme $|f_{pqy}\rangle$ and the diagonalized Young-Yamanouchi-Rutherford representation in which the last pair have definite symmetry $|f[pq]y\rangle$:

$$|f[pq]y\rangle = \gamma^{f}_{pq}\,|f(pqy)\rangle + \delta^{f}_{pq}\,|f(qpy)\rangle \ . \tag{54}$$

With the aid of the K-matrices and the two-body cfp's in orbital, color and flavor/spin spaces, the general state in Eq. (51) can be written in terms of an antisymmetric state ψ_{N-2} in the first N-2 particles and a state ϕ_2 of the "last pair"

$$\psi(f_o f'_c f''_{F/S}) = N(f)\sum \sqrt{\eta(f''_{p''q''})/\eta(f'')}\ \overline{K}\Big(f[pq]\,f'[p'q']\,\big|\,f''[p''q'']\Big) \times \times \Big(f_{pq}\alpha:[\beta]\,\big|\big\}\,Af[pq]\Big)\Big(f'_{p'q'}\alpha':[\beta']\,\big|\big\}\,f'[p'q']\Big) \times \times \Big(f''_{p''q''}\alpha'':[\beta'']\,\big|\big\}\,Bf''[p''q'']\Big)\psi_{N-2}\Big(\alpha f_{pq}\alpha' f'_{p'q'}\alpha'' f''_{p''q''}\Big) \times \times\ \phi_2\Big([\beta]_o[\beta']_c[\beta'']_{F/S}\Big) \tag{55}$$

Here η is the dimension of the given Young tableau in the symmetric group, $(\ldots|\}\ldots)$ is the cfp for the various spaces (orbital, color, flavor/spin). The (N-2) particle and 2-particle configurations in the various spaces we note by α and β respectively. The N(f) describes a normalization coefficient. Summation is over all allowed values of pq, p'q', p''q'' and structures $\alpha\beta$, $\alpha'\beta'$, $\alpha''\beta''$.

Writing Eq. (55) in the schematic form

$$\psi_N(A) = \sum_{\alpha\beta} P_{\alpha\beta}\ \ \psi_{N-2}(\alpha)\ \phi_2(\beta) \tag{56}$$

where $P_{\alpha\beta}$ denotes the product of K and cfp coefficients etc., then matrix elements of two-body operators have the form

$$\langle \psi_N(A) | \sum_{i<j} V_{ij} | \psi_N(A) \rangle = \frac{N(N-1)}{2} \sum_{\substack{\alpha\beta \\ \bar{\alpha}\bar{\beta}}} P_{\alpha\beta} P_{\bar{\alpha}\bar{\beta}} \langle \psi_{N-2}(\bar{\alpha}) | \psi_{N-2}(\alpha) \rangle \times$$
$$\times \langle \phi_2(\bar{\beta}) | V_{N\,N-1} | \phi_2(\beta) \rangle . \quad (57)$$

(Of course similar expressions can be written for one-body, three-body operators etc.)

Identifying Physical States

Let us consider six-quark states with $T=0$ and $S=1$ i.e. having the deuteron quantum numbers. From Table 4 we see that there are seven such states. The question now is to identify the combinations of the states in this symmetric basis that correspond to physical states of two nucleons, or deltas etc. Clearly only one state corresponds to two nucleons and another to two Δ's. The remaining five are formed from the colored 3-quark states of class I (see Table 3) and are known as hidden-color states. Specifically their structures are

$$\left(I^{3/2\ 1/2}\ I^{3/2\ 1/2}\right)^{01}, \left(I^{1/2\ 3/2}\ I^{1/2\ 3/2}\right)^{01}, \left(I^{1/2\ 1/2}\ I^{1/2\ 1/2}\right)^{01},$$
$$\left(I^{1/2\ 3/2}\ I^{1/2\ 1/2} \pm I^{1/2\ 1/2}\ I^{1/2\ 3/2}\right) \quad (58)$$

where the superscripts refer to the isospin and spin labels. Because the class I color-octet states are bounded from below (see Table 3) we are not plagued here by these problems.

Not all the symmetry states are involved in the structures of the clusters defined above. Thus, since nucleons and Δ's both have [3] flavor/spin symmetry, two N's and two Δ's can only have a symmetry contained within [3] × [3] [see Eq. (50)]. These "allowed" symmetry states have been labelled by an asterix in Table 4. Thus N^2 and Δ^2 states are described in terms of three symmetry states. The two-body fractional parentage reduction can tell us what the combinations of these symmetry states should be to describe two N's or Δ's. Consider for example the structure of the state with orbital symmetry [6] and flavor/spin symmetry {33}. Let the orbital state have the configuration $S_+^3\ S_-^3$ meaning that three quarks are in S-orbits (S_+) centered at +R/2 and the remaining three in S-orbits (S_-) centered at -R/2. The state of [6] symmetry has the two-body cfp expansion

$$(S_+^3 S_-^3)[6] = \sqrt{\frac{1}{5}} \left\{ (S_+^3 S_-)[4]\ S_-^2 + (S_+ S_-^3)[4]\ S_+^2 + \right.$$
$$\left. + \sqrt{3}\ (S_+^2 S_-^2)[4]\ \overline{S_+ S_-} \right\} \quad (59)$$

where we have split off the explicit structure of the symmetric last pair and the symmetric ([4]) first four. The antisymmetric color/flavor/spin state has the cfp structure

$$[222]_C\{33\}_{F/S} = \sqrt{\frac{1}{5}}\left\{\sqrt{2}\ \begin{array}{|c|c|}\hline \ & \ \\ \hline \ & \ \\ \hline 5 & 6 \\ \hline\end{array}_{\,C}\ \begin{array}{|c|c|c|}\hline \ & \ & 5 \\ \hline \ & \ & 6 \\ \hline\end{array}_{\,F/S} + \sqrt{3}\ \begin{array}{|c|c|}\hline \ & \ \\ \hline \ & 5 \\ \hline \ & 6 \\ \hline\end{array}\ \begin{array}{|c|c|c|}\hline \ & \ & \ \\ \hline \ & 5 & 6 \\ \hline\end{array}\right\} \tag{60}$$

with a normalized summation over the remaining four-particle Yamanouchi labels assumed. In Eq. (60) the F/S state is defined with the adjoint phase convention as is usual in nuclear physics. These expressions tell us that the probability of picking a pair of particles from $(S_+^3S_-^3)_{[6]}$ which are in one cluster (i.e. S_-^2 or S_+^2) is 1/5. Further, the probability that this pair is color *symmetric* is 2/5. Clearly, for an N^2 state, if two quarks are in a single N, then they cannot have color symmetry. Hence the symmetry state $[6]_0\{33\}_{F/S}$ has undesirable color components.

It is found, from the two-body cfp expansion of the state $[42]_0\{33\}_{F/S}$, that it also has similar undesirable color components but that a combination of both can be found that eliminates them.

Table 5. Transformation Coefficients between the Physical Basis States (with CC denoting a Hidden-Color State) shown on the left of each Block, and the Symmetry Basis States whose Structure is identified above each Block by the Orbital Symmetry [f] and Isospin-Spin Symmetry {f″}; each Block is for the designated (TS) Pairs

T = 1 S = 0 T = 0 S = 1	[6]{33}	[42]{33}	[42]{51}
N^2	$\sqrt{\frac{1}{9}}$	$\sqrt{\frac{4}{9}}$	$-\sqrt{\frac{4}{9}}$
Δ^2	$\sqrt{\frac{4}{45}}$	$\sqrt{\frac{16}{45}}$	$\sqrt{\frac{25}{45}}$
CC	$\sqrt{\frac{4}{5}}$	$-\sqrt{\frac{1}{5}}$	0
T = 0 S = 0	**[51]{42}**	**[33]{42}**	**[33]{6}**
N^2	$-\sqrt{\frac{5}{45}}$	$-\sqrt{\frac{4}{45}}$	$-\sqrt{\frac{36}{45}}$
Δ^2	$-\sqrt{\frac{20}{45}}$	$-\sqrt{\frac{16}{45}}$	$\sqrt{\frac{9}{45}}$
CC	$-\sqrt{\frac{4}{9}}$	$+\sqrt{\frac{5}{9}}$	0

The orthogonal combination of course retains the components and is therefore a combination of the hidden-color states in Eq. (58).

We can make the same kind of analysis on the flavor and spin symmetries. For the Δ^2 state with T=0 S=1 for example, not only does a pair of quarks in a Δ have color antisymmetry but it must also have T=1 and S=1. The color-allowed combination of the $[6]_0\{33\}_{F/S}$ and $[42]_0\{33\}_{F/S}$ does not have the required isospin and spin characteristics for the Δ^2 state, but it can be combined with the symmetry state $[42]_0\{51\}_{F/S}$ such that it does. The orthogonal combination then corresponds to N^2—a good check on the arithmetic. The transformations between the symmetry structure and the cluster structure are shown in Table 5 and 6 for six-quark states with isospin-spin labels (TS) = (01),(10),(00) and (11). For other (TS) labels see Ref. 37. The classification of 9- and 12-quark states is given in Ref. 45, 46 and the classification of multiquark states involving also strange quarks is in Ref. 47. In Tables 5,6 the hidden-color states are defined to be orthogonal to the N^2 and Δ^2 and $N\Delta$ states but having the same symmetry components. They are, in fact, linear combinations of the states in Eq. (58). Recently Chen et al.[48] have given the symmetry structure for all the cluster states in

Table 6. Transformation Coefficients between the Physical Basis States (with CC denoting a Hidden-Color State) shown on the left of each Block, and the Symmetry Basis States whose Structure is identified above each Block by the Orbital Symmetry [f] and Isospin-Spin Symmetry {f″}; each Block is for the designated (TS) Pairs

T=1 S=1	$[51]\{42\}_1$	$[51]\{42\}_2$	$[33]\{42\}_1$	$[33]\{42\}_2$	$[33]\{6\}$	$[42]\{51\}$
N^2	$-\sqrt{\frac{245}{1053}}$	$-\sqrt{\frac{80}{1053}}$	$-\sqrt{\frac{196}{1053}}$	$-\sqrt{\frac{64}{1053}}$	$-\sqrt{\frac{468}{1053}}$	0
Δ^2	$+\sqrt{\frac{20}{1053}}$	$-\sqrt{\frac{500}{1053}}$	$\sqrt{\frac{16}{1053}}$	$-\sqrt{\frac{400}{1053}}$	$\sqrt{\frac{117}{1053}}$	0
$\overline{N\Delta}$	0	0	0	0	0	1
$\widetilde{N\Delta}$	$+\sqrt{\frac{320}{1053}}$	$-\sqrt{\frac{5}{1053}}$	$\sqrt{\frac{256}{1053}}$	$-\sqrt{\frac{4}{1053}}$	$-\sqrt{\frac{468}{1053}}$	0
CC_1	$-\sqrt{\frac{4}{9}}$	0	$\sqrt{\frac{5}{9}}$	0	0	0
CC_2	0	$-\sqrt{\frac{4}{9}}$	0	$\sqrt{\frac{5}{9}}$	0	0

Eq. (58). From these tables we deduce, for example, that the hidden color state (TS) = (01) has the structure

$$(CC)^{01} = \Big\{\sqrt{10}\,(I^{1/2\ 1/2}\ I^{1/2\ 1/2}) + (^{1/2\ 3/2}\ \ I^{1/2\ 3/2}) + \sqrt{10}\,(I^{1/2\ 1/2}\ I^{1/2\ 3/2} - I^{1/2\ 3/2}\ I^{1/2\ 1/2}) + \sqrt{5}\,(I^{3/2\ 1/2}\ I^{3/2\ 1/2})\Big\}/6 \quad . \tag{61}$$

Thus we see that even though the QCD ansatz would have us ignore colored triplets of quarks, they are intimately involved in a description of dibaryons through their symmetry structures. To eliminate colored states entirely means the elimination of certain symmetry combinations. This may be acceptable when clusters are well separated—but what happens when clusters merge? What principle tells us here that certain combinations of symmetries must be eliminated?

Not yet having any principle to eliminate hidden-color states forces us to accept them. For the simple problem outlined above, the inclusion of the hidden color channels I^2 presents no problems. The color-octet states are bounded from below and, when clusters are at a great distance, do not couple to the N^2 channel thereby creating a long range van der Waals force.[22,34]

[The van der Waals force arises in the naive quark model from the coupling to the class II color-octet states by a potential of the form in Eq. (11).]

If we are to accept more complicated structures for the nucleon (e.g. with [21] orbital components) then the undesirable color-octets will come in and this implies a system which is unbounded from below. The further problem of long-range color van der Waals forces is almost superfluous in comparison.

One must therefore go beyond the two-body potential ansatz for describing confinement. What this is (if a potential solution exists at all) has not been worked out. Clearly one must examine three-body forces. From a physics point of view there is every indication that at least a three-body force is needed to describe confinement where confinement is interpreted as the creation of a $q\bar{q}$ pair when two quarks are separated by too great a distance (bag radius). (In potential models the $q\bar{q}$ would be treated as the third body.) Whether consistent models can be constructed using antiquarks—or whether $q\bar{q}$ pairs can be replaced directly by mesons—is still an open question. Despite the overall inadequacies of the two-body potential model, we press ahead with its application for six-quark systems with the naive quark structure for one nucleon because, with restriction of configurations, the undesirable features of the model do not appear and we can hope thereby that such applications do indeed correctly indicate the relationship between baryon structure and features of the NN-system.

Dinucleon Configurations

Tables 5 and 6 show the symmetry structure for $(N^2)^{TS}$ states for well separated nucleons. What is the structure when two nucleons collide? Clearly the problem can only be decided by the appropriate interaction between the quarks in the two nucleon clusters as the nucleons collide. The calculation must involve at least the symmetries shown in the tables or, equivalently, the various channels (NN, ΔΔ, CC etc.) when the clusters are a great distance apart. In early approaches to this problem it was thought that some idea as to the interaction between nucleons could be obtained in the Born-Oppenheimer approximation by constructing the cluster function

$$\tilde{\psi} = \sum_{\alpha} \Phi_{\alpha}(\hat{R})\ f_{\alpha}(\hat{R}) \tag{62}$$

and identifying the nucleon potential with

$$V_{NN} = \langle\tilde{\psi}|H|\tilde{\psi}\rangle \ . \tag{63}$$

Here $\Phi_{\alpha}(\hat{R})$ represents two clusters of the channel α(NN, ΔΔ, etc.) at a distance $\hat{R}$ apart, as defined in the symmetry basis, and $f_{\alpha}(\hat{R})$ is a function such that $f_{NN}(\hat{R}) \rightarrow 1$ as $R \rightarrow \infty$ and for all other channels $f_{\alpha}(\hat{R}) \rightarrow 0(\alpha \neq NN)$. In Ref. 28 the $F_{\alpha}(\hat{R})$ were determined by diagonalizing the Hamiltonian with color-dependent potentials fitting the baryon spectrum for each separation of the clusters. It was found that at small distances the orbital function with [42] symmetries remain in the cluster function. Note that $(S_+^3\ S_-^3)_{[42]} \rightarrow S^4P^2$ as $R \rightarrow 0$. Thus the energies of the negative-parity states containing P-orbits in the baryon spectrum were very relevant to the result of this study. In other references (e.g. 22,27,35,50) the cluster function was assumed to be the ψ_{NN} in Eq. (49). For this function $\psi_{NN} \rightarrow S^6$ as $R \rightarrow 0$. This essentially implies the use of ψ^2 in Eq. (55) with the additional condition $f_{NN}(0) = \sqrt{1/9}$, $f_{\Delta\Delta}(0) = \sqrt{4/45}$ and $f_{CC}(0) = \sqrt{4/5}$ (see Tables 5,6). Depending on what assumptions one takes for the $f_{\alpha}(\hat{R})$ one computes different forms for the Born-Oppenheimer potential V_{NN}. (c.f. Fig. 4.) With the function in Eq. (49), a repulsive core can be found, but with the diagonalization technique the core disappears. Who is correct? The answer is that both approaches are wrong—although very useful to the theoretician to get a grasp of the problem. The nucleon cluster function is not located at a single point but is distributed over a region of $R \sim 0.5$ fm (if we believe the spectrum of nucleons) or $R \sim 0.8$ fm (if we take the experimental measurement of the rms charged radius). The function V_{NN} then is non-local with a non-locality spreading over a region of space $\sim 2R$. Clearly then nothing specific can be said about local NN-potentials within such a distance (except if we consider "equivalent" local potentials). Actually this difficulty

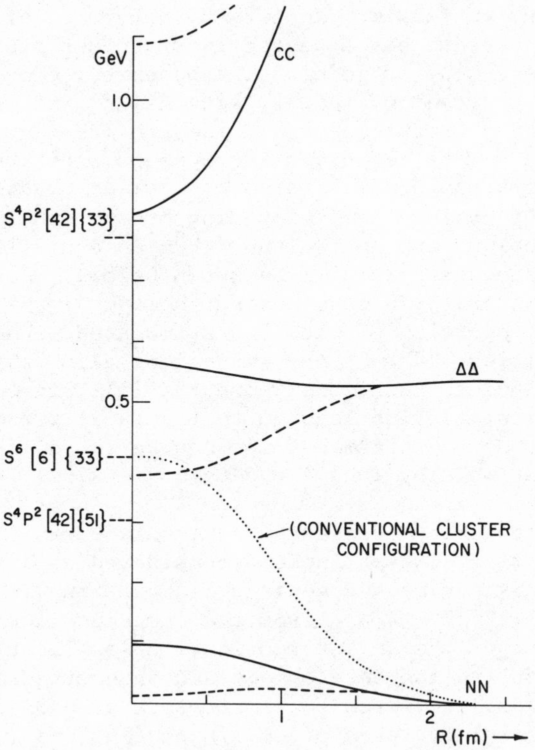

Fig. 4. Typical Born-Oppenheimer potentials for nucleons, deltas or hidden color states. The conventional cluster configuration is that in Eq. (49) with $f_{\Delta\Delta}(R) = f_{CC}(R) = 0$ for $R \to \infty$. The solid line is with configurations in Tables 5,6. The dashed lines are with configurations arising from a diagonalization of the energy matrix at each separation R of the clusters. On the left is shown the diagonal energies of various symmetry configurations. (T=1, S=0.)

should not bother us: we never measure a potential experimentally anyway. So long as we can calculate experimentally measurable quantities we should be satisfied. In the case of NN-scattering or NN-bound states we can indeed calculate phase-shifts and binding energies with the use of the resonating group or generator coordinate technique without recourse to a local potential approximation.

THE GENERATOR COORDINATE AND RESONATING GROUP TECHNIQUES

The resonating group technique[51-53] for studying clustering structure within nuclei was first proposed by Wheeler[54] in 1937 in connection with the fission property of heavy nuclei. Following the proposal the technique has appeared in structure studies in various guises e.g. the α-cluster model. The generator coordinate technique arose out of a proposal of Wheeler with Hill[55] and Griffin[56] to restore symmetry characteristics of the Hamiltonian in a wave-function ansatz. This technique too appears under various guises e.g. in the microscopic description of nuclear rotational motion.[57,58] In reality the resonating group technique and generator coordinate technique are equivalent[59,60]—an equivalence that can be proven in certain cases as will be shown below. Considerable work has been done in the 1970's to extend the techniques to describing the dynamical properties of clusters stimulated perhaps by increased computer facilities.[60-64] These extensions allow the techniques to be applied to nuclear reactions where the full microscopic properties (e.g. the Fermi statistics leading to the Pauli principle) can be taken into account. Greater details can be found in reviews and lectures, for example, by Tang[51] and LeTourneux.[53]

In these lectures I present the techniques as applied to the collision of nucleons, each nucleon considered as a cluster of three quarks. The work has been done in collaboration with J. LeTourneux (U. of Montreal).[65,66] Apart from the immediate interest of understanding aspects of the NN-interaction, the application is proving to be very useful to the development and understanding of the techniques themselves. The problem is complicated enough to exhibit many of the general features of colliding clusters (non-localities, channel coupling etc.) but simple enough that "reasonable calculations" can be performed. Greater details of this section will be given in a forthcoming paper.

The essence of the generator coordinate technique is to seek a solution of the Schrödinger equation

$$H\psi = E\psi \tag{64}$$

in the form

$$\psi = \sum_{\alpha} \int \Phi_{\alpha}(\hat{R})\, f_{\alpha}(\hat{R})\, d\hat{R} \tag{65}$$

where here the vector $\hat{R}$ is the generator coordinate.

In the two-body reaction application the "intrinsic" function $\phi_{\alpha}(\hat{R})$ represents two clusters of the lowest eigensolution of two potential wells separated by the distance $|R|$. In practice the application of the technique is only made possible by assuming each cluster is described by harmonic oscillator functions. As we review the application of the technique to the quark problem we shall see repeatedly how the harmonic oscillator assumptions facilitate the calculations, allowing the essence of (general) results to be demonstrated without enormous computational problems.

Thus, in the NN-force problem, each quark is assumed to be in an oscillator s-state shifted from the origin by $\pm R/2 = \pm(X/2, Y/2, Z/2)$

$$S_{\pm} = \pi^{-3/4}\, b^{-3/2} \exp\left\{-\left[\left(x \mp \frac{X}{2}\right)^2 + \left(y \mp \frac{Y}{2}\right)^2 + \left(z \mp \frac{Z}{2}\right)^2\right]/2b^2\right\} \tag{66}$$

The functions $\phi_{\alpha}(\hat{R})$ therefore have the orbital characteristic of $S_+{}^3S_-{}^3$. The Greek letter α represents a particular channel. As has been discussed in the previous section, a state with orbital characteristic $S_+{}^3S_-{}^3$ can have various orbital symmetries which have to be folded with the flavor/spin- and color-functions of six quarks to form totally antisymmetric states. It takes certain collections of these antisymmetric states (see Tables 5,6) to form two clusters that represent two nucleons at large distance ($R \to \infty$) in a certain isospin-spin (TS) channel. Other admixtures of the symmetries represent, asymptotically, other channels ($\Delta\Delta$, CC etc). Since we do not know, *a priori*, what the structure of the six quarks of the nucleons will be when the nucleons overlap, we must include many symmetry structures—or, equivalently, consider channel coupling: hence the sum over α in the description of the eigensolution of H in Eq. (65).

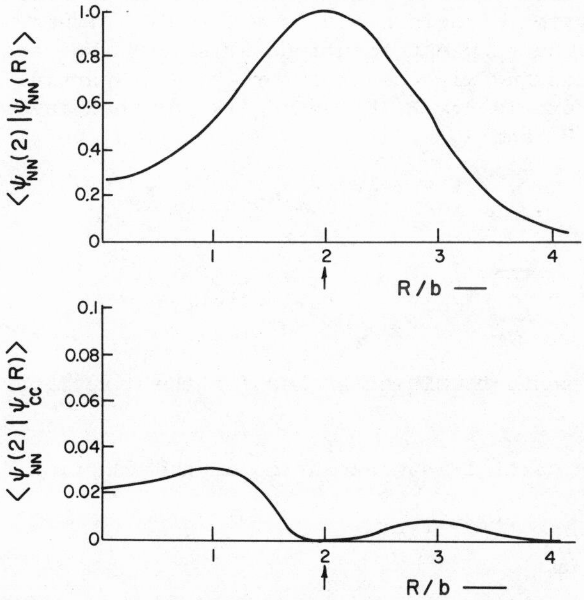

Fig. 5. Overlap matrix elements between oscillator cluster states at 2b and those at distance Rb—shown as a function of R. Here the scale is the oscillator parameter b. The cluster states NN and CC have the symmetry admixture shown in Tables 5,6 for all R.

The unknown in the ansatz of Eq. (65) is the function $f_\alpha(\hat{R})$. This is to be found by solving the Schrödinger equation. Substituting Eq. (65) into Eq. (64) multiplying on the left by $\phi_\beta(\hat{R}')$ and integrating over all particle coordinates of the intrinsic function ϕ yields the equation

$$\int [H_{\beta\alpha}(\hat{R}',\hat{R}) - EN_{\beta\alpha}(\hat{R}',\hat{R})]\ f_\alpha(\hat{R})\,d\hat{R} = 0 \tag{67}$$

where

$$H_{\beta\alpha}(\hat{R}',\hat{R}) = \langle \Phi_\beta(\hat{R}')|H|\Phi_\alpha(\hat{R})\rangle$$

$$N_{\beta\alpha}(\hat{R}',\hat{R}) = \langle \Phi_\beta(\hat{R}')|\Phi_\alpha(\hat{R})\rangle. \tag{68}$$

The Eq. (67) demonstrates how the solution for $f_\alpha(\hat{R})$ depends on the channel coupling (when $\alpha \neq \beta$) and the non-locality ($\hat{R} \neq \hat{R}'$). To demonstrate a little of both effects I show in Fig. 5 the normalization function $N_{\beta\alpha}(\hat{R}',\hat{R})$ for the case (TS) = (01) when $\alpha = \beta = NN$, and when $\alpha = NN$, $\beta = CC$ for varying $\hat{R} = Z$ ($X = Y = 0$) when $\hat{R}' = Z' = 2b$ ($X' = Y' = 0$). The various channels are here defined by the symmetry admixtures in Tables 5 and 6.

In general the solution of Eq. (67) is very complicated, involving as it does the six independent degrees of freedom of $\hat{R}$ and $\hat{R}'$. An assumption of deformed clusters would further complicate the problem. The simplest solutions arise for the case of spherical clusters interacting with central forces. The partial-wave expansions of the kernels $H_{\beta\alpha}(\hat{R}',\hat{R})$ and $N_{\beta\alpha}(\hat{R}',\hat{R})$ then involve only the angle between $\hat{R}'$ and $\hat{R}$

$$H_{\beta\alpha}(\hat{R}',\hat{R}) = \sum_{\ell m} H^{\ell}_{\beta\alpha}(R',R)\ Y^*_{\ell m}(\Omega_{R'})\ Y_{\ell m}(\Omega_R) \tag{69}$$

$$N_{\beta\alpha}(\hat{R}',\hat{R}) = \sum_{\ell m} N^{\ell}_{\beta\alpha}(R',R)\ Y^*_{\ell m}(\Omega_{R'})\ Y_{\ell m}(\Omega_R) \tag{70}$$

where Ω_R represents the Euler angles for the direction of the vector $\hat{R}$.

A similar partial-wave expansion of the function $f_\alpha(\hat{R})$ yields

$$f_\alpha(\hat{R}) = \sum_{\ell m} f^{\ell}_{\alpha}(R)\ Y_{\ell m}(\Omega_R)\ . \tag{71}$$

Substituting Eq. (67) and performing the integration over the Euler angles yields the equation

$$\sum_{\ell} \int \left[H^{\ell}_{\beta\alpha}(R',R) - EN^{\ell}_{\beta\alpha}(R',R)\right] f^{\ell}_{\alpha}(R)R^2 dR\ Y^*_{\ell m}(\Omega_{R'}) = 0\ . \tag{72}$$

A multiplication on the left by $Y_{\ell m}(\Omega_{R'})$, and integration over the angles $\Omega_{R'}$ projects from Eq. (72) equations in partial wave form:

$$\int \left[H^{\ell}_{\beta\alpha}(R',R) - EN^{\ell}_{\beta\alpha}(R',R) \right] f^{\ell}_{\alpha}(R) R^2 dR = 0 \ . \tag{73}$$

This integro-differential equation has to be solved with appropriate boundary conditions for the functions $f^{\ell}_{\alpha}(R)$ for the various channels. Note that these functions are the relative wave functions for the two effective potential wells in which the clusters of particles are defined: they do not represent the relative wave function of the clusters! The reason for this is that the center-of-mass of each cluster has a zero-point motion in each well. For oscillator states this zero point motion can be factored out exactly (another reason for using oscillator states). Thus for each cluster $S_+{}^3$ (with particles 1 2 3) and $S_-{}^3$ (with particles 4 5 6)

$$S_+{}^3 = S(\hat{\alpha}_{123})\ S(\hat{\beta}_{123})\ S\left((\hat{\rho}_{123} - \hat{R}/2)\sqrt{3}\right)$$

$$S_-{}^3 = S(\hat{\alpha}_{456})\ S(\hat{\beta}_{456})\ S\left((\hat{\rho}_{456} + \hat{R}/2)\sqrt{3}\right) \tag{74}$$

where

$$\hat{\alpha}_{123} = \sqrt{\frac{1}{2}}\ (\hat{r}_1 - \hat{r}_2)$$

$$\hat{\beta}_{123} = \sqrt{\frac{1}{6}}\ (\hat{r}_1 + \hat{r}_2 - 2\hat{r}_3)$$

$$\hat{\rho}_{123} = \sqrt{\frac{1}{3}}\ (\hat{r}_1 + \hat{r}_2 + \hat{r}_3) \tag{75}$$

and

$$S(\hat{\alpha}) = \sqrt{\frac{4}{\sqrt{\pi} b^3}}\ Y^0_0(\Omega_\alpha)\ \exp(-\hat{\alpha}^2/2b^2) \text{ etc.}$$

Then

$$S_+{}^3 S_-{}^3 = S(\hat{\alpha}_{123})\ S(\hat{\beta}_{123})\ S(\hat{\alpha}_{456})\ S(\hat{\beta}_{456})\ S(\gamma)\ S(\delta)$$

$$\equiv \chi_{123}\ \chi_{456}\ \Gamma_{rel}\ \phi_{C.M.} \tag{76}$$

Here χ_{123} and χ_{456} represent the internal structure of each cluster with

$$\Gamma_{rel} = S(\gamma) \text{ and } \gamma = \sqrt{\frac{3}{2}}\ (\hat{\rho}_{123} - \hat{\rho}_{456} - R) \equiv \sqrt{\frac{3}{2}}\ (\rho - R)$$

$$\phi_{C.M.} = S(\delta) \text{ and } \delta = \sqrt{\frac{1}{2}}\ (\hat{\rho}_{123} + \hat{\rho}_{456}) = \sqrt{\frac{1}{6}}\ (\hat{r}_1 + \hat{r}_2 + \ldots + \hat{r}_6)$$

$$= \sqrt{6}\ R_{C.M.}$$

Note that the R dependence is only contained in the function Γ_{rel}, which is also a function of the relative coordinate ρ between the centers of mass of each cluster. Explicitly we can rewrite the cluster function in Eq. (65) as

$$\Phi_\alpha(\hat{R}) = (\chi_I \chi_{II})_\alpha \, \Gamma_{rel} \, \phi_{C.M.} \tag{77}$$

where χ_I and χ_{II} are the internal states of each cluster in channel α. Defining

$$g(\rho) = \int \Gamma_{rel}(\hat{\rho}-\hat{R}) \; f(\hat{R}) dR \tag{78}$$

we can write Eq. (65) in the form

$$\psi = \sum_\alpha (\chi_I \chi_{II})_\alpha \, g_\alpha(\rho) \, \phi_{C.M.} \quad . \tag{79}$$

This is the resonating group ansatz for a solution of the Schrödinger equation. By making partial wave expansions of both the Γ-function and f-function we can write the partial waves[52]

$$g_\alpha^\ell(\rho) = \int_0^\infty \Gamma_\ell(\rho-R) \; f_\alpha^\ell(R) R^2 dR \tag{80}$$

where

$$\Gamma_\ell(\rho-R) = \frac{4\pi}{b^{-3/2}} \, i_\ell\left(\frac{\rho R}{b^2}\right) \exp\left(-\frac{(\rho^2+R^2)}{2b^2}\right) \tag{81}$$

where i_ℓ denotes a regular spherical Bessel function of imaginary argument. Note that

$$\exp(\hat{R}\cdot\hat{R}') = \sum_{\ell m} i_\ell(RR') 4\pi \; Y_{\ell m}^*(\Omega_R) \; Y_{\ell m}(\Omega_R{}') \quad . \tag{82}$$

It is the function $g_\alpha^\ell(\rho)$ that we know something about asymptotically since it represents the relative function between the centers of mass of each cluster. For the NN channel for example

$$g_{NN}^\ell(\rho) \sim [j_\ell(k\rho) - \tan(\delta_\ell) \; \eta_\ell(k\rho)]/k\rho \tag{83}$$

where $j_\ell(\eta_\ell)$ are the spherical Bessel (Neumann) functions with real arguments and δ_ℓ is the phase shift. For closed channels (like $\Delta\Delta$ if we consider NN scattering below the $\Delta\Delta$ threshold)

$$g_{\Delta\Delta}^\ell(\rho) \sim \exp(-k\rho) \quad . \tag{84}$$

For hidden-color channels the structure of the relative function depends on the form of the confinement potential. If we assume r^2 confinement then

$$g_{CC}^\ell(\rho) \sim \exp(-\rho^2/2B^2) \tag{85}$$

with B to be determined from the confinement strength and particular hidden-color channel. In practice we can take $g^{\ell}_{CC}(\rho) = 0$ for $\rho > \rho_o$ for large enough value of ρ_o. Thanks to a proof by deTakacsy[67] we know that, for channels in which the clusters do not interact asymptotically, the asymptotic values for $f^{\ell}_{\alpha}(R)$ are proportional to those for $g^{\ell}_{\alpha}(\rho)$. This allows the boundary conditions for the physical functions $g^{\ell}_{\alpha}(\rho)$ to be carried across to the generator coordinate functions $f^{\ell}_{\alpha}(\rho)$.

Several methods can be adopted to solve Eq. (67). The simplest is to assume a grid of points for $R < \overline{R}$ and hence convert Eq. (67) to the algebraic set of equations

$$\sum_{\beta}\sum_{j=1}^{n-1}\left[H^{\ell}_{\beta\alpha}(R_i,R_j) - E\,N^{\ell}_{\beta\alpha}(R_i,R_j)\right] f^{\ell}_{\alpha}(R_j)$$

$$= -\int_{\overline{R}}^{\infty}\left[H^{\ell}_{\beta\alpha}(R_i,R) - E\,N^{\ell}_{\beta\alpha}(R_i,R)\right] f^{\ell}_{\alpha}(R)R^2 dR \qquad (86)$$

where $R_n \equiv \overline{R}$ and $f^{\ell}_{\alpha}(\overline{R})$ is given by the boundary condition. For NN scattering with only the nucleon channel open this equation can be rewritten for the NN channel using Eq. (83) (with the deTakacsy theorem)

$$\sum_{\beta}\sum_{j=1}^{n-1}\left[H^{\ell}_{\beta\alpha}(R_i,R_j) - E\,N^{\ell}_{\beta\alpha}(R_i,R_j)\right] f^{\ell}_{\alpha}(R_j)$$

$$-\sum_{\beta}\int\left[H^{\ell}_{\beta\alpha}(R_i,R) - E\,N^{\ell}_{\beta\alpha}(R_i,R)\right]\eta_{\ell}(kR)RdR\,\tan(\delta_{\ell})/k$$

$$= -\int\left[H^{\ell}_{\beta\alpha}(R_i,R) - E\,N^{\ell}_{\beta\alpha}(R_i,R)\right] j_{\ell}(kR)RdR/k\ . \qquad (87)$$

For a closed channel Eq. (83) becomes

$$\sum_{\beta}\sum_{j=1}^{n-1}\left[H^{\ell}_{\beta\alpha}(R_i,R_j) - E\,N^{\ell}_{\beta\alpha}(R_i,R_j)\right] f^{\ell}_{\alpha}(R_j) = 0\ . \qquad (88)$$

The set of linear equations in Eq. (87) and (88) have to be solved for the (n-1) unknown $f^{\ell}_{\alpha}(R_j)$ (for each channel α) and the tangent of the phase shift tan (δ_{ℓ}). Alternative methods of solution have been proposed which involve the expansion of $f^{\ell}_{\alpha}(R)$ in terms of a complete set.[52] This is not as easy as it sounds because these amplitudes are highly singular. Methods for carrying out this procedure have been worked out however and can be used as well to give satisfactory answers.[51-53]

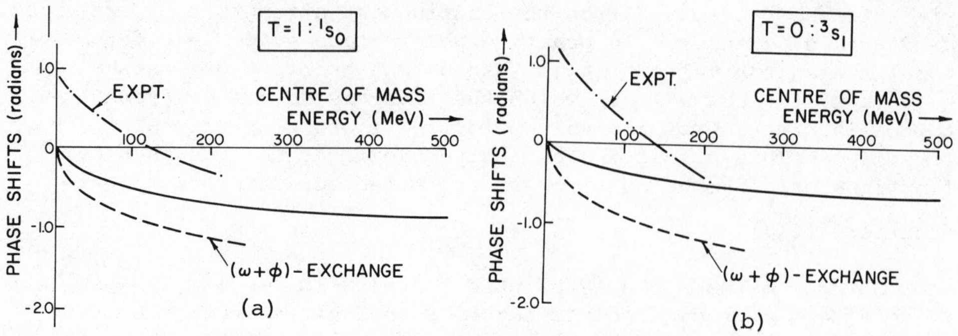

Fig. 6. Typical NN phase shifts (solid line) that emerge from using a color dependent potential between quarks fitted to the baryon spectrum. Experimental phase shifts taken from the Livermore analysis.[73] The $(\omega + \phi)$ exchange phase shift is from Ref. 75.

Whatever method is used to solve for the generator coordinate or resonating group amplitudes is independent of the physics. Similar methods can be used for nucleus-nucleus collisions (with each nucleus having constituent nucleons)[59-64,68-70] or for nucleon-nucleon collision (with each nucleon having constituent quarks). The physics is contained in the structure of the kernels $H_{\beta\alpha}(R',R)$ and $N_{\beta\alpha}(R,R')$ as defined by Eq. (68). Assuming two-body interactions these amplitudes can be computed by expanding each intrinsic cluster state in terms of the two-body fractional-parentage coefficients. Matrix elements of the six-quark states then involve products of two-body matrix elements and overlaps of the remaining four-body functions [c.f. Eq. (57)]. Note that the overlap of two functions is zero unless they have the same symmetry. Through Eq. (57), using the cluster antisymmetric states, we take into

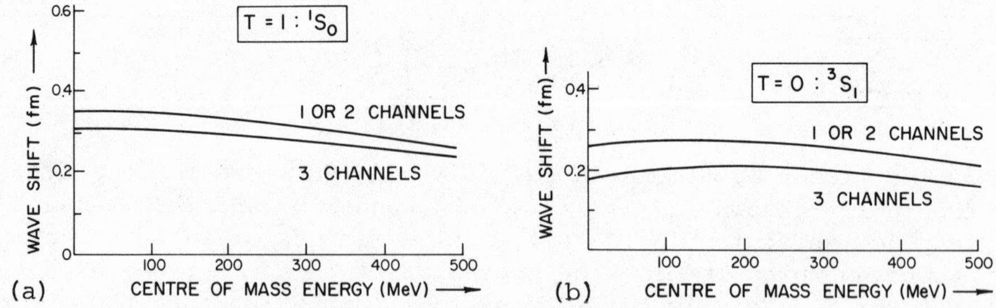

Fig. 7. Wave shifts extracted from the NN phase shifts of the quark potential model. The 1 or 2 channels refer to NN and ΔΔ channels as defined by Oka and Yazaki.[36]

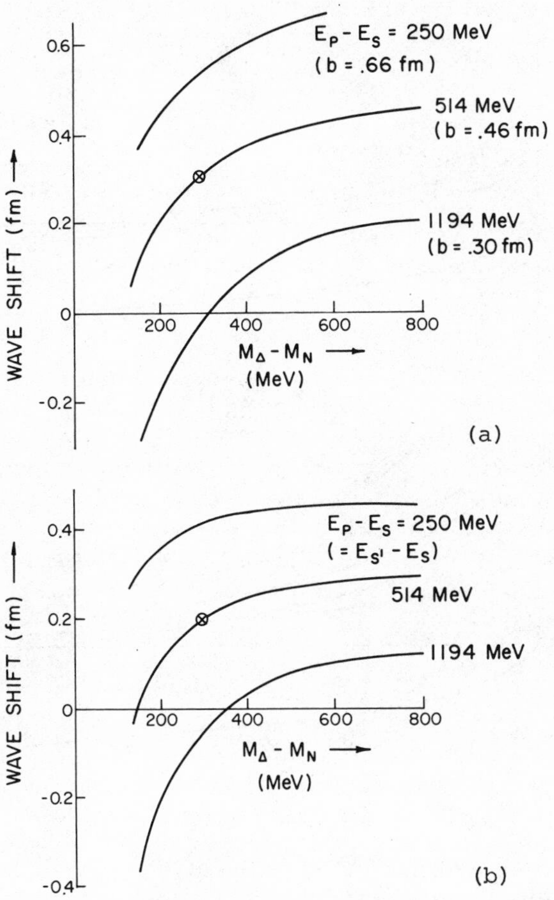

Fig. 8. Dependence of the wave shift in the quark potential model with changes in mass differences between the Δ (1232) and N and the mean excitation energy of the negative parity states. At the top (bottom) for the $^1S_0$ ($^3S_0$) channel.

account the direct and exchange terms of the interacting particles and the possible exchange of particles not involved with the interaction process.

Both the resonating group and generator coordinate techniques have been worked out for color-dependent potentials constructed to fit the N and Δ spectra. Typical results for S-wave phase shifts are shown in Fig. 6 as derived by LeTourneux and myself. Similar results are given by the Tübingen group.[71] What comes out of these calculations are negative phase shifts indicating that the color-dependent potentials lead to effective repulsion between nucleons. The same information can be plotted as the wave shifts ($-\delta/k$) as

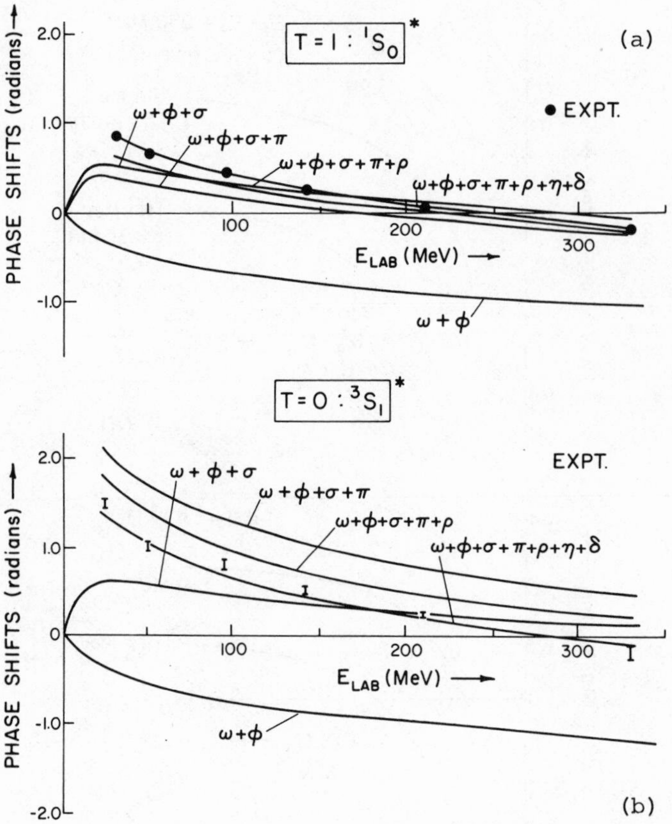

Fig. 9. Analysis of the fit to the experimental phase shifts[73] by the Bonn group.[75,76]

in Fig. 7. Here we see that the wave shift is roughly constant with energy—as if the scattering were from hard cores. The radius of the core depends on the spin i.e. ~0.3 fm (~ 0.2 fm) for S=0 (S=1). We have begun analysis to see how sensitive the results are to the input. For example Fig. 8 shows the wave shift at 100 MeV obtained by varying the Δ-N mass difference for different assumed energies for the negative parity levels. The realistic values are as marked. Clearly we have great sensitivity indicating a correlation between the precise systematics of the baryon spectrum and the occurrence of a hard core between nucleons.

Conventional nuclear forces[72] describe the interaction between nucleons as the exchange of mesons of different masses. In Fig. 9 I show the fit to the experiment[73] by the Bonn group[74,75] and an analysis[76] that was done to see the phase shifts from various components of their force. Clearly repulsive effects are being generated by $\omega + \phi$ exchanges. These mesons, being heavy, are being exchanged when the nucleons actually must have considerable overlap. The $\omega + \phi$ exchanges, at least, should not therefore be considered as an adequate physical description of the interaction between nucleons at short range. How one can successfully meld the quark picture of what is happening at short distances with the more conventional meson exchanges at larger distances is a challenge facing us in the 80's. I have tried in these lectures to touch upon some of the correlations between energy spectra and NN-phase shift that one might expect. I have pointed out the inadequacies of potential models with simple two-body interactions. But I have throughout stressed the symmetries, and techniques for dealing with them which I hope will provide a useful foundation for future studies.

REFERENCES

1. E. S. Abers and B. W. Lee, Phys. Rep. 9C:1 (1973).
2. W. Marciano and H. Pagels, Phys. Rep. 36C:137 (1978).
3. F. E. Close, "An Introduction to Quarks and Partons," Academic Press, London (1979).
4. A. Chodos, R. L. Jaffe, K. Johnson, C. B. Thorn and V. F. Weisskopf, Phys. Rev. D9:3471 (1974).
5. T. DeGrand, R. L. Jaffe, K. Johnson and J. Kiskis, Phys. Rev. D12:2060 (1975).
6. P. Hasenfratz and J. Kuti, Phys. Rep. 40:76 (1978).
7. A. W. Thomas, in: "Advances in Nuclear Physics," J. W. Negele and E. W. Vogt, eds., Plenum Press, New York (to be published).
8. S. Théberge, A. W. Thomas and G. A. Miller, Phys. Rev. D22:2838 (1980); D23:2106E (1981).
9. A. W. Thomas, S. Théberge and G. A. Miller, Phys. Rev. D24:216 (1981).
10. A. W. Thomas, J. Phys. G7:1283 (1981); Nucl. Phys. A354:51 (1981).
11. G. A. Miller, S. Théberge and A. W. Thomas, Comm. Nucl. Part. Phys. 10:101 (1981).
12. G. E. Brown, M. Rho, Phys. Lett. 82B:177 (1979).
13. M. Rho, in: "Progress in Particle and Nuclear Physics," Vol. 8, D. Wilkinson, ed., Pergamon Press, Oxford (1981).
14. O. W. Greenberg, Phys. Rev. Lett. 13:598 (1964).
15. G. Karl and E. Obryk, Nucl. Phys. B8:609 (1969).
16. R. R. Horgan, J. Phys. G2:625 (1976).
17. M. Jones, R. H. Dalitz and R. R. Horgan, Nucl. Phys. B129:45 (1977).
18. N. Isgur and G. Karl, Phys. Rev. D18:4187 (1978); D19:2653 (1979).
19. R. Koniuk and N. Isgur, Phys. Rev. D21:1868 (1980).
20. D. P. Stanley and D. Robson, Phys. Rev. D21:3180 (1980); Phys. Rev. Lett. 45:235 (1980).
21. D. Robson, in: "Prog. in Particle and Nuclear Physics," Vol. 8, D. Wilkinson, ed., Pergamon Press, Oxford (1981).

22. C. W. Wong, in: "Prog. in Particle and Nuclear Physics," Vol. 8, D. Wilkinson, ed., Pergamon Press, Oxford (1981).
23. Y. Nambu, in: "Preludes in Theoretical Physics," A. DeShalit, H. Feshbach and L. Van Hove, eds., North-Holland, Amsterdam (1966).
24. A. De Rújula, H. Georgi and S. L. Glashow, Phys. Rev. D12:147 (1975).
25. E. Eichten and K. Gottfried, Phys. Lett. 66B:286 (1977).
26. C. Quigg and J. Rosner, Phys. Lett. 71B:153 (1977).
27. D. Liberman, Phys. Rev. D16:1542 (1977).
28. M. Harvey, Nucl. Phys. A352:326 (1981).
29. Particle Data Group, Rev. Mod. Phys. 52:S1 (1980).
30. For an elementary introduction to symmetries in fermion physics see M. Harvey, AECL Report No. 7166 (1980).
31. H. R. Fiebig and B. Schwesinger, "Analysis of Splitting Mechanism in the Low Lying Negative Parity Baryons," Stony Brook Report (1982); B. Schwesinger, "Construction of One Gluon and One Pion Exchange Forces in the Chiral Bag," Stony Brook Report (1982).
32. F. E. Close and R. R. Horgan, Nucl. Phys. B164:413 (1980); B185:323 (1981).
33. D. P. Stanley and D. Robson, Phys. Rev. D23:2776 (1981).
34. O. W. Greenberg and H. J. Lipkin, Nucl. Phys. A370:349 (1981).
35. G. C. Warke and R. Shanker, Phys. Rev. C21:2643 (1980).
36. M.Oka and K. Yazaki, Prog. Theor. Phys. 66:556 (1981).
37. M. Harvey, Nucl. Phys. A352:301 (1981).
38. I. T. Obukhovsky, Yu. F. Smirnov and Yu. M. Tehuvil'sky, J. Phys. A15:7 (1982).
39. G. Racah, Phys. Rev. 61:186 (1942); 62:438 (1942); 63:367 (1943); 76:1352 (1949).
40. H. A. Jahn and H. van Wieringen, Proc. Roy. Soc. A209:502 (1951).
41. J. P. Elliott, J. Hope and H. A. Jahn, Phil. Trans. of Roy. Soc. London 246:241 (1953).
42. D. E. Rutherford, "Substitutional Analysis," Edinburgh Univ. Press, Edinburgh, (1948).
43. M. Hamermesh, "Group Theory," Addison-Wesley, Reading, Mass., (1964).
44. T. Yamanouchi, Proc. Phys.-Math. Soc. Japan 19:436 (1937).
45. Y. Suzuki, K. T. Hecht and H. Toki, in: "Symposium on Group Theory," Cocoyoc, Mexico (1982). See also M. Namiki, K. Okano and N. Oshimo, Waseda Univ. Report (1981).
46. V. G. Neudatchin, T. Obukhovsky, Yu. F. Smirnov and E. V. Tkalya, Moscow Univ. Preprint (1981).
47. R. L. Jaffe, Phys. Rev. D15:267,281 (1977).
48. J-Q. Chen, Y-J. Shi, D. H. Feng and M. Vallieres, Drexel Univ. Report (1982).
49. K. F. Liu, Univ. Kentucky Report (1981).
50. C. deTar, Phys. Rev. D16:323 (1978).
51. Y. C. Tang, Phys. Rep. 47:167 (1978), in: "Lecture Notes in

Physics," T. T. S. Kuo and S. S. M. Wong, eds., Springer-Verlag, Berlin (to be published).
52. M. Onsi and J. LeTourneux, Can. J. Phys. 58:612 (1980).
53. J. LeTourneux, Univ. of Montreal Report (1982).
54. J. A. Wheeler, Phys. Rev. 52:1083,1107 (1937).
55. D. L. Hill and J. A. Wheeler, Phys. Rev. 89:1102 (1953).
56. J. J. Griffin and J. A. Wheeler, Phys. Rev. 108:311 (1957).
57. J. P. Elliott, Proc. Roy. Soc. A245:128,562 (1958).
58. M. Harvey, in: "Advances in Nuclear Physics," Vol. 1, M. Baranger and E. Vogt, eds., Plenum Press, New York (1968).
59. K. Wildermuth and W. McClure, "Cluster Representations in Nuclei," Springer-Verlag, Berlin (1966).
60. K. Wildermuth and Y. C. Tang, "A Unified Theory of the Nucleus," Viewig, Braunschweig (1977).
61. D. M. Brink, in: "Proc. Int. Conf. on Clustering Phenomenon in Nuclei," IAEA, Vienna (1969).
62. D. R. Thompson and Y. C. Tang, Phys. Rev. C12:1432 (1975; C13:2597 (1976).
63. Y. C. Tang, Fizika (Supl. 3) 9:91 (1977).
64. D. R. Thompson, in: "Clustering Aspects of Nuclear Structure and Nuclear Reactions," W. T. H. van Oers, J. P. Svenne, J. S. C. McKee and W. R. Falk, eds., Amer. Inst. Phys., New York (1978).
65. M. Harvey, in: "Int. Symp. on Clustering Phenomena in Nuclei," P. Kramers and R. Schultheiss, eds., Werkhefte der Universität, Tübingen (1981).
66. M. Harvey, lecture at the Workshop on Quarks in Nuclei, Univ. South Carolina (1982).
67. N. B. deTakacsy, Phys. Rev. C5:1883 (1972).
68. The Resonating Group, Phys. Lett. B43:165 (1973).
69. H. Friedrich, H. Hüsken and A. Weiguny, Nucl. Phys. A220:125 (1974).
70. R. Beck, J. Borysowitz, D. M. Brink and M. V. Mihailović, Nucl. Phys. A224:45 (1975).
71. A. Faessler, F. Fernandez, G. Lübeck and K. Shimizu, Univ. Tübingen Report (1981).
72. K. Erkelenz, Phys. Rep. 13C:191 (1974).
73. M. McGregor, R. Arndt and R. Wright, Phys. Rev. 182:1714 (1969).
74. K. Holinde and R. Machleidt, Nucl. Phys. A247:495 (1975); A256:479 (1976).
75. K. Holinde, Phys. Rep. 68:121 (1981).

NUCLEAR CHROMODYNAMICS:

IMPLICATIONS OF QCD FOR NUCLEAR PHYSICS

S.J. Brodsky

Stanford Linear Accelerator Center
Stanford University
Stanford, CA 94305

INTRODUCTION

In these lectures we will discuss the synthesis of quantum chromodynamics with nuclear physics: nuclear chromodynamics.[1] In quantum chromodynamics, QCD, the fundamental degrees of freedom of hadrons and nuclei are the quanta of quark and gluon gauge fields which obey an exact internal SU(3) color symmetry.[2] If QCD is correct, then by extension it must also be the fundamental theory of nuclear forces and nuclear physics. Thus one of the most interesting questions in nuclear physics is the transition between the conventional meson-nucleon degrees of freedom to the quark and gluon degrees of freedom of QCD. As we probe to shorter distances then the meson-nucleon degrees of freedom must break down; thus we will be interested in new nuclear phenomena, new physics intrinsic to composite nucleons and mesons, and new phenomena outside the range of traditional nuclear physics.

As discussed by Chris Llewellyn Smith and Francis Close in this volume, there is now impressive experimental evidence that QCD is a viable theory of hadronic phenomena.[3] The general structure of QCD meshes remarkably well with the facts of the hadronic world, especially quark-based spectroscopy, current algebra, the approximate point-like structure of large momentum transfer lepton-hadron reaction, and the logarithmic violation of scale invariance in deep inelastic reactions. QCD has been particularly successful in predicting the features of electron-positron and gamma-gamma annihilation into hadrons, including the magnitude and scaling of the total cross sections, the production of hadronic jets with patterns conforming to elementary quark and gluon subprocesses, and phenomena associated with the production of heavy quarks. All of the empirical results are consistent with the basic postulates of QCD, that the charge and weak currents within hadrons are carried by spin-1/2 quarks and that the strength of the quark-gluon couplings become weak at short distances; i.e. asymptotic freedom.[4] Despite these many successes, it should be emphasized that strictly quantitative tests of QCD are still lacking in most part because of the difficulties in unambiguously separating perturbative and non-perturbative effects.[5]

The grand goal of nuclear chromodynamics would be to actually derive the nuclear force from the fundamental QCD degrees of freedom and interactions. The difficulty is that the nucleon-nucleon interaction in QCD is a remnant of the color forces and is even more complex than calculating the molecular force between neutral atoms. The basic ingredients are evidently quark interchange, which is related at long distances to pion and meson exchange, and multiple gluon exchange, which despite the zero mass[6] of the perturbative gluon, must still have inverse range shorter than the mass of the lowest lying gluonium bound state. Recent calculations based on bag and potential models suggest that many basic features of the nuclear force can be understood from the underlying QCD substructure. (The lectures by Harvey in this volume include a review of the progress in this field.) In these lectures we will give a direct proof from QCD that the nucleon-nucleon potential must be repulsive at short distances.[7]

The most fruitful area for immediate progress in nuclear physics is to make use of the asymptotic freedom property of QCD, which allows perturbative calculations of short distance processes, especially exclusive processes.[7,8] We will particularly emphasize the use of asymptotic freedom and light-cone quantization to derive exact results for nuclear amplitudes at short distances.[9] This includes the nucleon form factors at large momentum transfer,[8,10] meson photoproduction amplitudes, deuteron photo- and electro-disintegration;[11] and most important, the form factors of nuclei at large momentum transfer.[7,9] These nuclear reactions provide a new testing ground for QCD which are interesting from the point of view of both nuclear and elementary particle physics. Beyond this, QCD provides rigorous asymptotic constraints on nuclear amplitudes which are non-trivial constraints for constructing a complete theory of the nuclear reactions.[7] It is especially interesting to note that QCD predicts that large momentum transfer nuclear amplitudes are dominated by hidden color degrees of freedom—degrees of freedom which cannot be represented by the conventional nucleon and meson degrees of freedom of nuclear physics.

QCD could be the exact theory of the strong interactions in the same sense that QED is the theory of electromagnetic interactions. The theory is a renormalizable non-Abelian gauge theory with unbroken SU(3) color symmetry. In addition to the $g_s\overline{\psi}\gamma_\mu\psi A^\mu$ Dirac coupling of

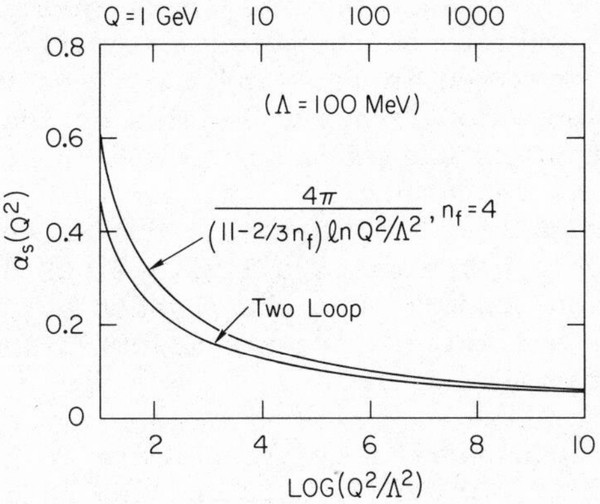

Fig. 1. The QCD coupling constant $\alpha_s(Q^2)$ to one and two-loop accuracy. For simplicity we choose $n_f = 4$.

the color octet gluon to the color triplet and anti-triplet q and $\bar{q}$, there are also non-Abelian 3-point ggg and 4-point gggg couplings of the gluons. The crucial distinction is in the net sign of the vacuum polarization: in QED, the effective coupling constant $\alpha(Q^2)$ increases with the momentum transfer; in QCD the added gluon-self couplings cause the effective strength $\alpha_s(Q^2)$ of the quark and gluon interactions to decrease logarithmically at large momentum transfer, as summarized by the asymptotic expression ($Q^2 \gg \Lambda^2_{QCD}$) (see Fig. 1)

$$\alpha_s(Q^2) = \frac{4\pi}{(11 - 2/3\ n_f)\log(Q^2/\Lambda^2_{QCD})} \quad . \tag{1}$$

The parameter Λ_{QCD} sets the basic mass scale for QCD, and n_f is the number of types or "flavors" of quarks with effective mass much smaller than Q. The fact that QCD interactions tend to zero at large momentum transfer $Q^2 \gg \Lambda^2_{QCD}$ ("asymptotic freedom") allows the systematic *perturbative* calculation of hadronic interactions at short distances.

Current phenomenological fits as well as lattice gauge theory suggest that $\Lambda_{QCD} (= \Lambda_{\overline{MS}})$ is not very large, probably less than 200 or 300 MeV. This suggests that perturbative QCD calculations become relevant at hadronic distances or nucleon-nucleon separations of a fermi or less. Since their scales are so similar, it is very difficult to argue that nuclear physics can be studied in isolation from QCD. In a sense, nuclear physics is in the regime of maximal QCD complexity. Nevertheless, there are many areas of nuclear phenomenology where we can make definite progress and find interesting variations from the conventional wisdom of nuclear physics. QCD is certainly not restricted to the high momentum transfer high energy domain of particle physics.

One reason why QCD has apparently made relatively little impact in much of nuclear physics has been the assumption that nuclear interactions can be analyzed in terms of an effective meson-nucleon field theory, in isolation from the details of the short-distance quark and gluon substructure of hadrons. In these lectures I will argue that this view is untenable. In fact, there is no correspondence principle which yields traditional nuclear physics in terms of local meson and nucleon fields as a limit of QCD dynamics. The distinctions between standard nuclear physics dynamics and QCD at nuclear dimensions are interesting and illuminating for both particle and nuclear physics. For example:

i) Meson and nucleon degrees of freedom are insufficient to describe nuclei if QCD is correct. Hidden color configurations can appear both as Fock components of ground state nuclei and as multi-quark excited nuclear states. The hidden color components mixed in the ground state wave function must contribute to the basic properties of nuclei—for example their magnetic and quadrupole moments, charge distribution, etc. More important, these hidden color components dominate nuclear amplitudes at small internucleon separation and control the normalization of the nuclear form factors at large momentum transfer.[7,12]

ii) There are many difficulties with conventional effective local Lagrangian meson-nucleon field theories. For example, if the nucleon is coupled to an isovector rho-meson then unitarity is violated in the tree graph (Born) approximation. Since such theories are not renormalizable, they have no predictive content in higher orders. In order to construct a renormalizable theory of vector mesons carrying a non-Abelian isovector charge, one would have to go to the full apparatus of non-Abelian gauge theory including trilinear and quartic vector meson couplings and a spontaneous symmetry breaking mechanism in order to provide non-zero meson masses. As we will discuss in the penultimate section the necessary criteria for a hadron or composite system to be effectively parametrized as a local field requires that its form factor at time-like momentum $Q^2 = 4M^2$ be close to 1.

iii) Although meson-exchange current contributions can be identified diagrammatically in QCD, their fall-off at large momentum transfer due to off-shell effects is much stronger than that indicated by local meson-nucleon field theory. See the penultimate section. More generally it is difficult to argue that local Lagrangian field theories for mesons and nucleons have any more predictive power than what is already contained in a non-relativistic description.

iv) Once one gives up the local Lagrangian description of meson and nucleon fields then there are a number of other consequences of conventional treatments which must be modified. For example, since quarks are the ultimate carriers of electromagnetic current in QCD the identification of nucleon-antinucleon pair production terms in the analysis of electromagnetic currents in nuclear amplitudes cannot be justified. Furthermore, the usual form of the Dirac equation for relativistic nuclei is invalid beyond first Born approximation.

v) Once one understands interactions from the standpoint of the quark and gluon substructure, then one sees that QCD rules out some standard assumptions of nuclear physics. For example, the usual impulse approximation formula for the elastic form factors of nuclei

$$F_A(Q^2) = F_N(Q^2)\ F_A^{body}(Q^2) \tag{2}$$

which is conventionally used to separate nucleon size effects from nuclear dynamics is *incorrect* in QCD because of off-shell and recoil effects. In addition, there is no calculationally tractable description of relativistic wave functions at equal time because of the inevitable presence of graphs in which particles are created from the vacuum rather than just from the wave function itself.

Despite these complications and criticisms of standard methods, QCD suggests a new nuclear physics phenomenology which we will outline in these lectures.

An essential part of our analysis is the use of quantization at equal $\tau = t + z$, rather than equal time.[8,13,14] The light-cone quantization method provides an elegant but tractable representation of relativistic bound state wave functions in a form which is a natural generalization of the usual Schrödinger non-relativistic many-body theory. Applied to QCD, this formalism allows a systematic separation of non-perturbative and perturbative effects in hadronic matrix elements and the derivation of useful factorization theorems.

In terms of their QCD Fock state description, the hadrons, including nuclei, are (color singlet) composites of quark and gluon quanta; e.g.,

$$|\pi^+\rangle = a^{\pi}_{(2)}\,|u\bar{d}\rangle + a^{\pi}_{(3)}\,|u\bar{d}g\rangle + \ldots$$

$$|p\rangle = a^{p}_{(3)}\,|uud\rangle + a^{p}_{(4)}\,|uudg\rangle + \ldots$$

$$|d\rangle = a^{d}_{(6)}\,|uud\ ddu\rangle + \ldots \tag{3}$$

This is in analogy to the Fock state expansion

$$|\text{positronium}\rangle = a_{(2)}\,|e^+e^-\rangle + a_{(3)}\,|e^+e^-\gamma\rangle \tag{4}$$

in QED. The states are specified at equal "time" $\tau = t + z$ on the light-cone. The total 3-momentum, $\vec{k}_\perp$ and $k^+ \equiv k^0 + k^z$, is then conserved ($\Sigma_i \vec{k}_{\perp i} = 0$, $\Sigma_i k_i^+ = p^+$), and the momentum-space wave function for each n-particle Fock state component of a hadron with total momentum p^μ is a function (spin-labels are suppressed) $\Psi_n = \Psi_n(x_i, \vec{k}_{\perp i})$, where $x_i = k_i^+/p^+$, $\Sigma_i^n \vec{k}_{\perp i} = 0$, and $\Sigma_i^n x_i = 1$. The states are off the "energy" shell ($k^- \equiv k^0 - k^z$)

$$p^- - \sum_{i=1}^{n} k_i^- = \frac{1}{p^+}\left[M^2 - \sum_{i=1}^{n}\left(\frac{\vec{k}_\perp^2 + m^2}{x}\right)_i\right] < 0 \quad . \tag{5}$$

This "light-front" formalism greatly simplifies relativistic bound state calculations since it is very similar to the ordinary non-relativistic many-particle theory; essentially all theorems proved in the Schrödinger theory hold for the relativistic theory at equal time on the light-cone. The "relativistic kinetic energy" $(\vec{k}_\perp^2 + m^2)/x$ plays the role of the non-relativistic kinetic energy $\vec{k}^2/2m$. Although the results are independent of the choice of Lorentz frame, the variable $x = k^+/p^+$ can be conveniently interpreted as the longitudinal momentum fraction k^z/p^z in an "infinite momentum" frame where $p^z \to \infty$. A summary of the calculational rules in light-cone perturbation theory, and examples of its use are given in the third section and the Appendix. The coupled equations of motion $H_{LC}\psi = M^2\psi$ for the Fock state wave functions in QCD and the required regularization of the infrared and ultraviolet regions is discussed in Refs. 8 and 14.

Light-cone quantization has considerable utility in nuclear physics[3,13] since it allows an exact representation of relativistic wave functions in terms of an orthonormal Fock state basis. Since the current is diagonal in this basis, all form factors, charge radii, magnetic moments[15] etc., have exact expressions in terms of the ψ_n. Furthermore, the structure functions $G(x,Q)$ (and more general multiparticle distributions) which control large momentum transfer (leading and higher twist) inclusive reactions, and the distribution amplitudes $\phi(x,Q)$ which control large momentum transfer exclusive reactions (and "directly-coupled" inclusive reactions) are each specific, basic measures of the ψ_n. Other physical quantities such as decay amplitudes provide rigorous sum rule or local constraints on the form of the valence components of meson and baryon wave functions.[16] In addition, in light-cone perturbation theory, the perturbative vacuum is also an eigenstate of the total QCD Hamiltonian on the light-cone; perturbative calculations are enormously simplified by the absence of vacuum to pair production amplitudes. In the non-relativistic limit, light-cone perturbation theory and ordinary time-ordered perturbation theory are equivalent.

An additional useful tool for analyzing relativistic nuclear dynamics is the use of the reduced amplitude formalism.[9,11] This ansatz consistently and covariantly removes the effects of nucleon compositeness from nuclear amplitudes and thus allows the study of nuclear dynamics directly. We shall also show how the use of asymptotic freedom and QCD perturbation theory provides rigorous boundary conditions for nuclear amplitudes at short distance. All of this suggests the possibility that fully analytic nuclear amplitudes can be constructed which at low energies fit the standard electromagnetic and chiral boundary conditions and low energy theorems of traditional nuclear physics while at the same time satisfying the scaling laws and anomalous dimension structure predicted by QCD at high momentum transfer.

AN OVERVIEW OF NUCLEAR CHROMODYNAMICS

In this section we will give a survey of the new perspectives for nuclear physics implied by QCD, especially new phenomena outside of the range of traditional nuclear physics. We will also briefly discuss areas in which the nucleus provides a tool for testing QCD dynamics, e.g. hidden color dibaryon states, quark and gluon propagation through nuclear matter and the space-time development of hadronic fits. Throughout these studies one is interested in signals for quark and gluon degrees of freedom in nuclei and deviations from elementary nucleon additivity. After this overview, we give more detailed technical discussions in subsequent sections and the Appendix.

Exclusive Processes in QCD

One of the most important developments in hadron physics of the last few years has been the demonstration that the predictions of perturbative QCD can be extended to the whole domain of large momentum transfer exclusive reactions,[8,9,17,18,19] including the form factors $F(Q^2)$ of hadrons as measured in large momentum transfer electron scattering reactions $eA \to eB$ $(Q^2 = -t)$, and fixed angle scattering processes $d\sigma/dt$ $(A+B \to C+D)$ for $s >> M^2$ with $z = \cos\theta_{c.m.}$ fixed. In general A, B, C, and D can be mesons, baryons, photons, and nuclei!

As an example, let us briefly consider the calculation of the nucleon form factor.[8,10] Only the "valence" $|qqq\rangle$ Fock state needs to be considered to leading order in $1/Q$ since (in a physical gauge such as $A^+ = 0$) any additional quark or gluon forced to absorb large momentum transfer (proportional to Q) yields a power-law suppression M/Q to the form factor. Further, because of the spin-1, helicity-conserving couplings of the gauge gluon, overall hadronic helicity is conserved,[20] $h_I = h_F$, again to leading order in $1/Q$. Thus QCD predicts the suppression of the Pauli form factor: $F_2(Q^2)/F_1(Q^2) \sim O(M^2/Q^2)$.

The calculation of the nuclear form factor thus reduces to a

3-body problem. The helicity-conserving form factor $G_M(Q^2)$ can be written to leading order in $1/Q$ in the factorized form:

$$G_M(Q^2) = \int_0^1 [dx] \int_0^1 [dy] \phi^\dagger(y_i,\tilde{Q}) T_H(x,y,Q) \phi(x_i,\tilde{Q}) \quad , \tag{6}$$

with $[dx] = dx_1 dx_2 dx_3 \delta(1-x_1-x_2-x_3)$. Here $T_H(x,y,Q)$ is the probability amplitude for scattering three quarks collinear with p to the final directions p+q, as illustrated in Fig. 2. To leading order in $\alpha_s(Q^2)$,

$$T_H(x,y,Q) = \left[\frac{\alpha_s(Q^2)}{Q^2}\right]^2 t(x,y) \tag{7}$$

where $t(x,y)$ is a rational function of the light-cone longitudinal momentum fractions x_i and y_i. The function $\phi(x_i,Q)$ gives the probability amplitude for finding the valence quarks in the nucleon with "light-cone" momentum fractions $x_i = k_i^+/p^+$ at small relative distance $b_\perp \sim O(1/Q)$:

$$\phi(x_i,Q) \propto \int^{k_{\perp i}<Q} [d^2k_\perp] \psi_{qqq}(x_i,\vec{k}_{\perp i}) \quad . \tag{8}$$

The "distribution amplitude" $\phi(x_i,Q)$ is the fundamental wave function

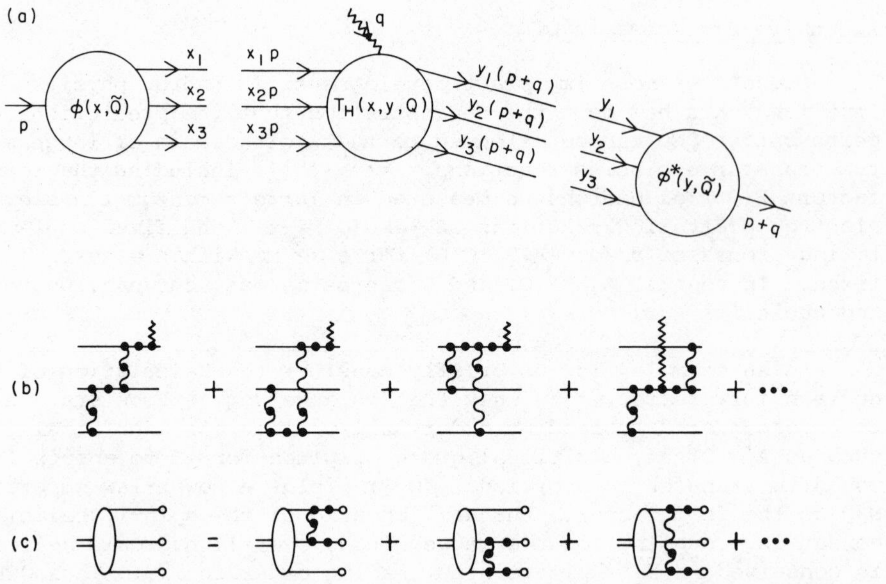

Fig. 2. (a) The general factorized structure of the nucleon form factor at large Q^2 in QCD. (b) Leading contributions to the hard-scattering amplitude T_H. (c) Bound-state equation for the baryon 3-quark wave function of large momenta.

which controls high-momentum transfer exclusive reactions in QCD;[8] it is the analogue of the wave function at the origin in the non-relativistic theory. The definition of ψ_{qqq} is discussed in detail in the following section. The (logarithmic) Q-dependence of ϕ can be completely determined by the operator product expansion at short distances[17] and the renormalization group, or by "evolution equations" computed from perturbative quark-quark scattering kernels at large momentum transfer.[8,21] The high momentum tail of the wave function for each hadron is thus controlled by QCD perturbation theory.

The QCD prediction for nucleon form factors at large Q^2 to leading order in $\alpha_s(Q^2)$ compiled from (6) and (7) then takes the form[8]

$$G_M(Q^2) = \frac{\alpha_s^2(Q^2)}{Q^4} \sum_{n,m} b_{nm} \left(\log \frac{Q^2}{\Lambda^2}\right)^{-\gamma_n^B-\gamma_m^B} \left[1 + 0\left(\alpha_s(Q^2),\ \frac{M^2}{Q^2}\right)\right]. \quad (9)$$

We can also obtain results for ratios of various baryon and baryon isobar transition form factors assuming isospin or SU(3)-flavor symmetry for the basic wave function structure. Results for the neutral weak and charged weak form factors assuming standard SU(2) × U(1) symmetry are given in Ref. 10. The γ_n^B anomalous dimensions are

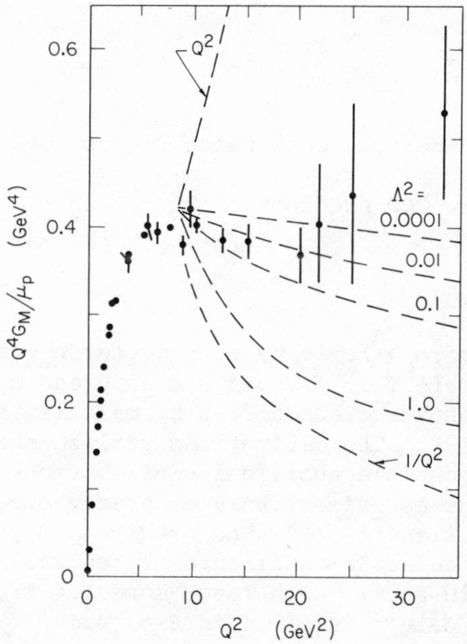

Fig. 3. QCD predictions for $Q^4G_M^p(Q^2)$ for various scale parameters Λ_s (in GeV^2). The data are from Ref. 3 and references therein.

computable positive constants determined by the evolution equation for the nucleon distribution amplitude ϕ. Since $\alpha_s(Q^2)$ is slowly varying at large Q^2, the most important dynamical behavior is the Q^{-4} power-law dependence[22] of $G_M(Q^2)$ which reflects the basic scale invariance of quark and gluon interactions, and the fact that the minimal Fock state of the nucleon contains three quarks—both non-trivial features of QCD. The prediction (9) for $G_M^p(Q^2)$ agrees with data[23] for $3 \lesssim Q^2 \lesssim 25$ GeV^2 provided $\Lambda_{QCD} \lesssim 300$ MeV, as shown in Fig. 3. A detailed discussion is given in the next section and Ref. 8.

The power-law behavior of the QCD predictions for exclusive processes at large momentum transfer can be roughly summarized by simple counting and helicity rules. To leading order in $1/Q$: (i) Total hadron helicity is conserved.[20] In particular this implies that weak and electromagnetic form factors are helicity-conserving, and independent of total spin. The dominant form factor corresponds to $h_I = h_F = 0$ or $h_I = h_F = \pm 1/2$. (ii) Dimensional counting[22] predicts the power-law scaling of fixed-angle scattering processes:

$$\frac{d\sigma}{dt}(AB \to CD) \sim \frac{1}{s^{n-2}} f(\theta_{c.m.}) \ , \tag{10}$$

where n = total number of constituent fields in A, B, C, and D, and the power-law fall-off of helicity conserving form factors:

$$F_H(Q^2) \sim \frac{1}{(Q^2)^{n_H-1}} \ , \tag{11}$$

where n_H is the number of constituent fields in H.

In particular, QCD predicts[7,9,22]

$$F_d(Q^2) \sim \left(\frac{1}{Q^2}\right)^5 \tag{12}$$

for the helicity zero → helicity zero deuteron elastic form factor. The helicity non-zero form factors are suppressed by powers of $1/Q$. The form factor results are modified by calculable logarithm corrections, as in Eq. (9). The derivations utilize the fact that Sudakov factors suppress possible anomalous contributions from end-point integration regions and effectively or nearly suppress contributions of pinch singularities.[8,19,24] The predictions for the power-law behavior reflect the scale-invariance of renormalizable interactions and appear to be in accord with large momentum transfer experiments. A comparison with data[25] is given in Fig. 4.

Applications to Nuclear Amplitudes

The deuteron's Fock state structure is much richer in QCD than

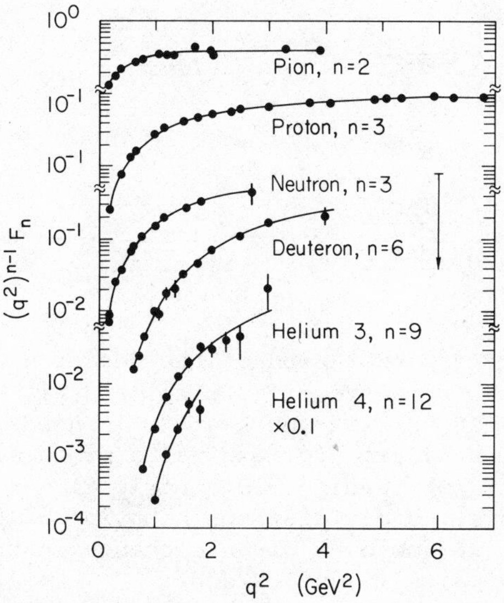

Fig. 4. Comparison of the dimensional counting rule $(Q^2)^{n-1} F(Q^2) \to$ const. $(Q^2 >> M^2)$ with data. See Ref. 25.

it would be in a theory in which the only degrees of freedom are hadrons. Restricting ourselves to the six-quark valence state, (see Fig. 5) we can readily generate states like[26]

$$|d\rangle_6 = a|(uud)_{1C}(ddu)_{1C}\rangle + b|(uud)_{8C}(ddu)_{8C}\rangle$$

$$+ c|(uuu)_{1C}(ddd)_{1C}\rangle + d|(uuu)_{8C}(ddd)_{8C}\rangle \quad . \tag{13}$$

The first component corresponds to the usual physical n-p structure of the deuteron. The second component corresponds to "hidden color" or "color polarized" configurations where the three-quark clusters are in color-octets, but the overall state is a color-singlet. The last two components are the corresponding isobar configurations. If we suppose that at low relative momentum the deuteron is dominated by the n-p configuration, then quark-quark scattering via single gluon exchange generates the color polarized states (b) and (d) at high $k_\perp$; i.e., there must be mixing with color-polarized states in the deuteron wave function at short distances. The existence of hidden color Fock state components in the nuclear wave function implies that the standard nucleon and meson degrees of freedom are not sufficient to describe nuclei. The mixing of the ground state of a nucleus with the extra hidden color states will evidently lower its energy and thus influence the nuclear magnetic moment, charge radius, and other properties. We expect that the hidden color components will be most significant in large momentum transfer nuclear processes and reactions such as the parity-violating terms in the photo-

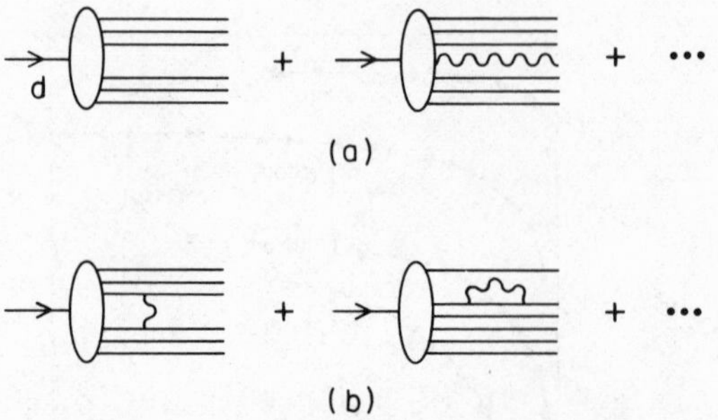

Fig. 5. (a) Fock state expansion of the deuteron wave function at equal $\tau = t + z$ on the light-cone, and light-cone gauge $A^+ = 0$. Only the six-quark state is needed to calculate large momentum transfer exclusive processes involving the deuteron. (b) Leading QCD contribution to the high momentum transfer, short-distance behavior of the deuteron wave function. A sum over gluon exchange contributions between all quark pairs is understood.

disintegration of the deuteron, which are sensitive to the structure of the nuclear wave function at short distances. Conversely, the new QCD degrees of freedom should also imply the existence of excited nuclear states which are predominantly of hidden color. These states may have narrow width if they are below the pion decay threshold. The six-quark excitation of the deuteron could possibly be found by a careful search for anomalous resonant structure in $\gamma d \to \gamma d$

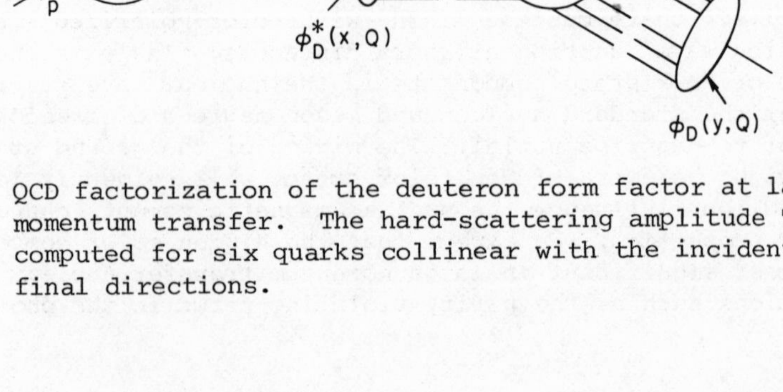

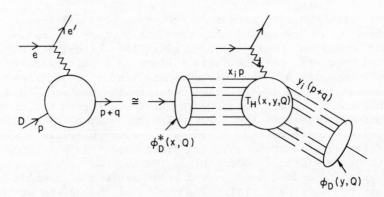

Fig. 6. QCD factorization of the deuteron form factor at large momentum transfer. The hard-scattering amplitude T_H is computed for six quarks collinear with the incident and final directions.

scattering at large angles. It is also important that detailed searches for these states be conducted in inelastic electron scattering and tagged photon nuclear target experiments. Channels such as $\gamma^* d \to pp\pi^-$ could be especially sensitive to dibaryon states dominated by hidden color configurations, since the state is not required to decay back to the n-p state.

In analogy with the nucleon form factor calculation, the QCD prediction for the leading helicity-zero deuteron form factor has the form (see Fig. 6)

$$F_d(Q^2) \sim \left[\frac{\alpha_s(Q^2)}{Q^2}\right]^5 \sum_{n,m=0} d_{nm} \cdot \left[\ln \frac{Q^2}{\Lambda^2}\right]^{-\gamma_n^d - \gamma_m^d} \tag{14}$$

where the first factor is computed from the sum of hard-scattering $6q + \gamma^* \to 6q$ diagrams. The anomalous dimensions can be calculated from evolution equations for the six-quark components of the deuteron

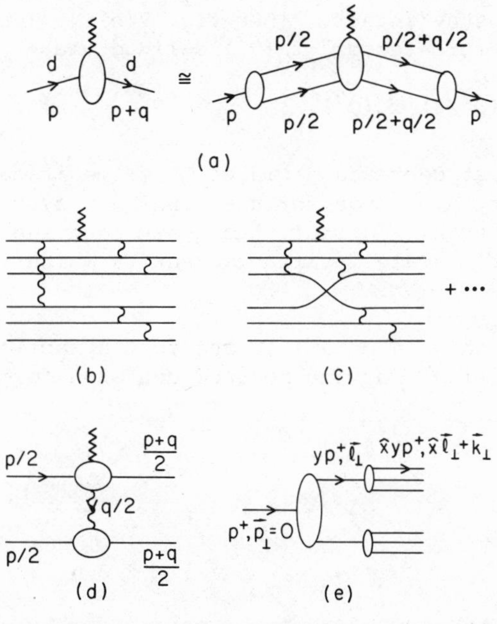

Fig. 7. (a) Representation of the deuteron form factor in the weak binding limit. (b) Typical contributions to T_H for the deuteron form factor. (c) Organization of T_H into the product of an effective nucleon form factor at $\hat{t} = Q^2/4$ and a photon-gluon nucleon Compton amplitude at $\hat{t} = Q^2/4$, consistent with the reduced nuclear amplitude formalism. (d) Structure of the deuteron wave function in the weak n-p binding limit.

wave function at short distances.[7] Results for the leading anomalous dimension are discussed in the fourth section.

For a general nucleus, the asymptotic power behavior is[9] $F_A(Q^2) \sim (Q^2)^{1-3A}$, reflecting the fact that one must pay a penalty of $\alpha_s(Q^2)/Q^2$ to move each quark constituent from p to p+q. The fact that the momentum transfer must be partitioned among the constituents implies that the truly asymptotic regime increases with the nucleon number A.

Reduced Nuclear Amplitudes

In order to make more detailed and experimentally accessible predictions we can define "reduced" nuclear amplitudes[9,11,26] which effectively remove, in a covariant fashion, the effects of nucleon composite structure. Consider elastic electron-deuteron scattering in a general Lorentz frame. The deuteron form factor $F_d(Q^2)$ is the probability amplitude for the nucleus to stay intact after absorbing momentum transfer Q. Clearly $F_d(Q^2)$ must fall at least as fast as $F_p(Q^2/4)F_n(Q^2/4)$ since each nucleon must change momentum from $\sim 1/2p$ to $\sim 1/2(p+q)$ and stay intact. [See Fig. 7(a).] Thus we should consider a "reduced form factor" $f_d(Q^2)$ defined via

$$F_d(Q^2) \equiv F_N^2(Q^2/4)\; f_d(Q^2) \quad . \tag{15}$$

Clearly $f_d(Q^2)$ must decrease at large Q^2 since it can be identified as the probability amplitude for the final n-p system to remain a ground state deuteron. In fact, the quark counting rule predicts $f_d(Q^2) \sim 1/Q^2$ for a scale invariant quantity which is the slowest fall-off possible.

As we shall show in detail in the fourth section, the exact asymptotic prediction[7] for the reduced deuteron form factor is

$$f_d(Q^2) \propto \frac{1}{Q^2}\left(\log Q^2/\Lambda_s{}^2\right)^{-\Gamma_d - I} \tag{16}$$

where

$$\Gamma_d = 2/5\; C_F/\beta \;, \quad C_F = 4/3 \;, \quad \beta = 11 - 2/3\; n_f \;.$$

Comparisons with the available large Q^2 data[27] are shown in Fig. 8. Remarkably, the QCD prediction,[16] seems to be accurate from Q^2 below 1 GeV^2 out to the limits of the experimental data. [Note also that the parameter-free prediction $Q^2 f_d(Q^2) \to$ const. was given before the large Q^2 data was taken, and that the ratio of experiment to theory is plotted on linear paper!]

In general, we can define reduced nuclear form factors[9]

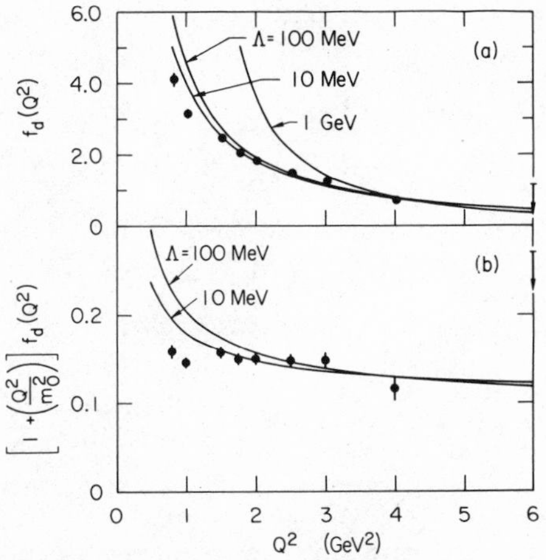

Fig. 8. (a) Comparison of the asymptotic QCD prediction $f_d(Q^2) \propto 1/Q^2(\ln Q^2/\Lambda^2)^{-1-2C_F/5\beta}$ with final data of Ref. 27 for the reduced deuteron form factor, where $F_p(Q^2) = F_n(Q^2) = (1 + Q^2/0.71)^{-2}$. The normalization is fixed at the $Q^2 = 4$ GeV^2 data point. (b) Comparison of the prediction $(1 + Q^2/m_0^2) f_d(Q^2) \propto (\ln Q^2/\Lambda^2)^{-1-2C_F/5\beta}$ with the above data. The value $m_0^2 = 0.28$ GeV^2 is used (see Ref. 9).

$$f_A(Q^2) \equiv \frac{F_A(Q^2)}{[F_N(Q^2/A^2)]^A} \quad . \tag{17}$$

QCD then predicts the power-behavior $f_A(Q^2) \sim (Q^2)^{1-A}$ (as if the nucleons were elementary). Comparisons with the data for d, ^{3}He and ^{4}He are shown in Fig. 9. We emphasize that the usual impulse approximation formula [Eq. (2)] which is supposed to remove the effects of the struck nucleon's structure is *invalid* in QCD. [A nucleon with momentum 1/2 p^μ which absorbs momentum transfer q^μ becomes far off-shell and spacelike $(1/2\ p+q)^2 \sim 1/2\ q^2$, so that the total $\gamma^* + N \to N^*$ off-shell Q^2-dependence is not given by $F_N(Q^2)$, see Fig. 10(a).] The same dynamics which controls the nucleon form factor also controls the dynamics which transfers momentum to the other constituents in the nucleus. The definition (17) of the reduced form factor $f_A(Q^2)$ takes into account the correct partitioning of the nuclear momentum, and thus, to first approximation, represents the nuclear form factor in the limit of point-like nucleon constituents.

In general it is of interest to see whether a consistent parameterization of nuclear amplitudes can be obtained if for each nuclear scattering process, reduced amplitudes are defined by dividing out the nucleon form factors at the correct partitioned momentum.

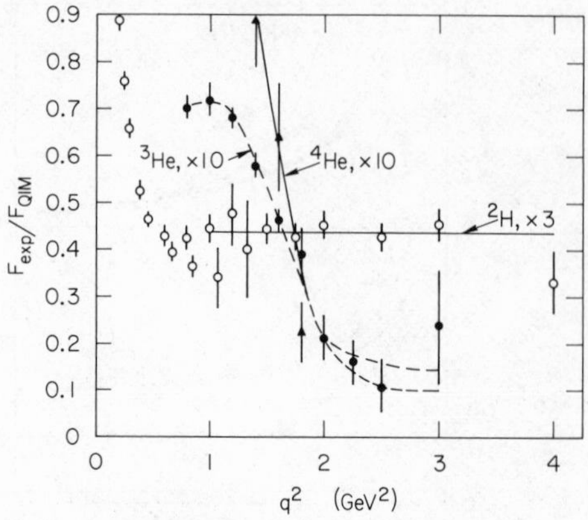

Fig. 9. Ratio of experiment to the reduced form factor predictions for d, ^3He, ^4He. The data are from Ref. 27.

For example, we can define the reduced deuteron photodisintegration amplitude:[11]

$$m_{\gamma d \to np} \equiv \frac{M_{\gamma d \to np}}{F_N(\hat{t})F_N(\hat{u})} \tag{18}$$

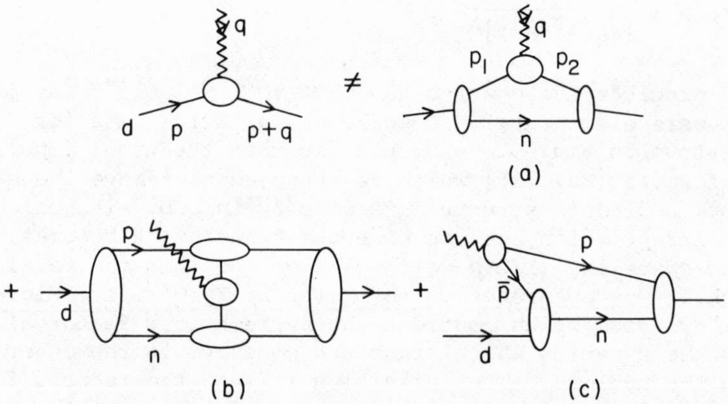

Fig. 10. Critique of the standard nuclear physics approach to the deuteron form factor at large Q^2. (a) The usual on-shell nucleon form factor $F_N(Q^2) = F_N(Q^2, p_1^2 = M_N^2, p_2^2 = M_N^2)$ does not appear. (b) Individual meson exchange current contributions are strongly suppressed at high Q^2 because of off-shell form factors predicted by QCD. (c) The nucleon-pair contribution is strongly suppressed due to nucleon compositeness.

where $\hat{t}$ and $\hat{u}$ are computed from the invariant momentum transfers to the nucleon (neglecting the deuteron binding). At large momentum transfer $m_{\gamma d \to np}$ should obey the same scaling law at fixed $\theta_{c.m.}$ as the quark amplitude $M_{\gamma\rho^+ \to u\bar{d}}$:

$$m_{\gamma d \to np} \sim \frac{f(\theta_{c.m.})}{\sqrt{s}} \tag{19}$$

with the leading helicity amplitude corresponding to hadron-helicity conservation $\lambda_d = \lambda_n + \lambda_p$. The q^2 dependence of the $M_{\gamma^* d \to np}$ amplitude (for transverse virtual photon polarization) is negligible if $|q^2| \ll \hat{t},\hat{u}$. At low momentum transfer the scaling form of the reduced amplitude can serve as a model for the background to a dibaryon resonance contribution. A more detailed discussion of these predictions and comparison with experiment is given in the fifth section.

QCD and The Nucleon—Nucleon Interaction

The asymptotic-freedom property of QCD implies that the nuclear force at short distances can be computed directly in terms of perturbative QCD hard-scattering diagrams. The basic prediction[22] for the nucleon-nucleon amplitude at large momentum transfer is (modulo small corrections)

$$M_{NN \to NN} \sim \left(\frac{1}{Q^2}\right)^4 f(\theta_{c.m.}) \tag{20}$$

where $Q^2 = -t$ is the square of the momentum transfer. The corresponding fixed angle scaling behavior,[22]

$$\frac{d\sigma}{dt}(pp \to pp) \sim \frac{1}{s^{10}} F(\theta_{c.m.}) \tag{21}$$

is consistent with the high momentum transfer data.[28] (See Fig. 11.) To actually compute the angular distribution of the N-N amplitude is a formidable task since even at the Born level there are of the order of 3×10^6 connected Feynman diagrams in which five gluons interact with six quarks; in addition, Sudakov-suppressed pinch singularities must be carefully analyzed.[29] Considerable phenomenological progress has, however, been made simply by assuming that the dominant diagrams involve quark interchange.[30] This ansatz seems to yield a good approximation to the observed large-angle baryon and meson angular distributions and charge correlations, as well as the observed crossing behavior between amplitudes such as $pp \to pp$ and $p\bar{p} \to p\bar{p}$.[30] Applications to spin correlations are discussed in Refs. 31 and 32.

The constraints of asymptotic QCD behavior, especially its power-law scaling and helicity selection rules, have only begun to

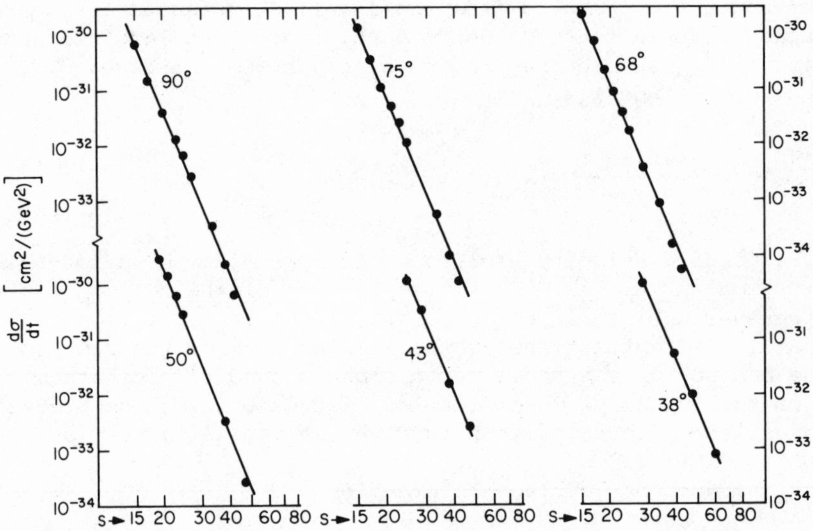

Fig. 11. Differential cross section for $pp \rightarrow pp$ scattering at wide center of mass angles. The straight lines correspond to the predicted power-law fall-off of $1/s^{10}$. The data compilation is from Ref. 28.

be exploited. For example, dispersion relations and superconvergence relations for the nuclear-nuclear helicity amplitudes should yield sum rules and constraints on hadronic couplings and their spectra. One could try to enforce a form of duality which equates the $q\text{-}\bar{q}\text{-}g$ exchange amplitudes with the sum over meson-exchange degrees of freedom. Again, this cannot be strictly correct since the existence of hidden color configurations—whether mixed with ordinary nuclear states or appearing as resonance excitations—implies that duality in terms of the low-lying hadrons cannot be a true identity: The nucleus cannot be described in QCD solely in terms of nucleon or isobar color-singlet clusters; hidden-color Fock state components are required for a complete description of nuclear matter, especially in the region of strong nucleon overlap.[7]

The perturbative structure of QCD at short distances can also be used to determine the far off-shell behavior of hadronic and nuclear wave functions and their momentum distributions. For example, the x near 1 behavior of particle distributions in the bound state (at the kinematical end of the Fermi distribution where one constituent has nearly all of the available longitudinal momentum) requires the far off-shell dependence of the wave function.[33] Modulo logarithms, the leading power behavior of perturbative QCD contributions to inclusive distributions at the edge of phase source is given by the "spectator rule"[34] $[x_a = (k_a^0 + k_a^z)/(p^0 + p^z)]$

$$\left.\frac{dN_{a/A}}{dx_a}\right|_{x_a \to 1} = C_{a/A}\left(1-x\right)^{2n_s-1} \tag{22}$$

where n_s is the number of spectator constituents in the bound state forced to carry small light-cone momentum fractions. The rule holds for the case where the helicities of a and A are identical; otherwise there is additional power-law suppression.[8,14] Examples of the spectator rule are $dN/dx \sim (1-x)^3$ for q/p, $(1-x)^{15}$ for $q/{}^3He$ and $(1-x)^{11}$ for $p/{}^3He$. These rules can be tested in forward inclusive reactions for particles produced with large longitudinal momenta, and in deep inelastic lepton scattering on hadrons and nuclei.[9,35] In general, the impulse approximation implies[9]

$$\frac{d\sigma}{dQ^2 dx}(eA \to eX) = \sum_a \left(\frac{d\sigma}{dQ^2}\right)_{ea\to ea} \frac{dN_{a/A}}{dx} , \tag{23}$$

representing the sum of incoherent contributions each of which correspond to scattering on one quark or clusters of quarks in the nuclear or hadronic target. Further discussions, applications, and tests can be found in Refs. 9,25,26,35. The transverse momentum distributions $dN_{a/A}/dk_\perp^2$ can also be predicted from the perturbative QCD processes which control the high momentum tail of the bound-state wave function.[8]

Nuclei as Probes of Particle Physics Dynamics

Thus far we have discussed several applications of nuclear chromodynamics specific to the dynamics and structure of nuclei. Conversely, there are numerous examples where a nuclear target can be used as a tool to probe particular aspects of particle physics. We will only mention a few applications here.

Parity violation in hadronic or nuclear processes. The exchange of a weak W or Z boson between the quarks of a hadron or nucleus leads to a high momentum component in the Fock state wave function

$$\Delta\psi(x,\vec{k}_\perp) \sim \frac{\alpha}{k_\perp^2 + M_W^2}\,\phi(x,M_W^2) \quad . \tag{24}$$

The interference of these amplitudes with normal QCD contributions lead
$\gamma d \to np$ and total hadronic cross sections.

The nucleus as a color filter.[36] One can study a new class of diffractive dissociative jet production processes in nuclei which isolate the valence component of meson wave functions. The fact that the wave function of a hadron is a superposition of (infrared and ultraviolet finite) Fock amplitudes of fixed particle number but varying spatial and spin structure leads to the prediction of a novel effect in QCD. We first note that the existence of the decay amplitude $\pi \to \mu\nu$ requires a finite probability amplitude for the pion to exist as a quark and diquark at zero transverse separation:

$$\psi(x,\vec{r}_\perp = 0) = \sqrt{4\pi}\,\sqrt{n_C}\; x(1-x)\, f_\pi \tag{25}$$

where f_π is the given decay constant and $n_C = 3$. (See following section.) In a QCD-based picture of the total hadron-hadron cross section, the components of a color singlet wave function with small transverse separation interact only weakly with the color field, and thus can pass freely through a hadronic target while the other components interact strongly. A large nuclear target will thus act as a filter removing from the beam all but the short-range components of the projectile wave function. The associated cross section for diffractive production of the inelastic states described by the short range components is then equal to the elastic scattering cross section of the projectile on the target multiplied by the probability that sufficiently small transverse separation configurations are present in the wave function. In the case of the pion interacting in a nucleus one computes the cross section

$$\left.\frac{d^3\sigma}{dx\, d^2r_\perp}\right|_{r_\perp^2 \sim 0} \cong \sigma_{e\ell}^{\pi A}\, 12\pi\, f_\pi^2\, x^2(1-x)^2 \tag{26}$$

corresponding to the production of two jets just outside the nuclear volume. The x distribution corresponds to $d\sigma/d\cos\theta \sim \sin^2\theta$ for the jet angular distribution in the $q\bar{q}$ center of mass. By taking into account the absorption of hadrons in the nucleus at $\vec{r}_\perp \neq 0$ one can also compute the $k_\perp$ distribution of the jets and the mass spectrum of the diffractive hadron system. Details are given in Ref. 36.

One can also use the A dependence of the nuclear cross section to separate central and diffractive mechanisms for heavy flavor production (open charm, etc.).

<u>Nuclear corrections to inclusive QCD reactions</u>. A basic tool of QCD phenomenology is the factorization ansatz, which allows inclusive cross sections at large momentum transfer to be computed directly in terms of the structure functions and fragmentation functions measured in deep inelastic lepton scattering. For example, the Drell-Yan cross section[37] for massive lepton pair production in $\bar{p}$-nucleus collisions is written as

$$\frac{d^2\sigma}{dQ^2 dx_L}(pA \to \ell\bar{\ell}X) = \sum_q e_q^2 \int_0^1 dx_a \int_0^1 dx_b\, G_{q/A}(x_a,Q)\, G_{\bar{q}/p}(x_b,Q)\, d\hat{\sigma} \tag{27}$$

where

$$d\hat{\sigma} = \frac{1}{3}\frac{4\pi\alpha^2}{3Q^2}\,\delta(x_a x_b s - Q^2)\;\delta(x_L - x_a + x_b)\left[1 + 0\left(\frac{\alpha_s(Q^2)}{\pi}\right)\right] \qquad (28)$$

is the $q\overline{q} \to \ell\overline{\ell}$ subprocess cross section, and the distributions $G_{q/A}$ and $G_{\overline{q}/\overline{p}} = G_{q/p}$ are measured in deep inelastic lepton-nucleus and lepton-nucleon scattering. The factor of 1/3 results from assuming that the annihilating q and $\overline{q}$ have uncorrelated colors. Derivations[38] of this result in QCD perturbation theory are based on the observation that the logarithmic dependence of the distributions G(x,Q) arising from collinear gluon radiation is process independent and that, at least to leading twist (i.e., leading order in $1/Q^2$), all infrared divergences cancel. The distributions can also be related to the Fock state QCD wave functions for each hadron defined at equal time on the light-cone. In principle, one can compute the radiative corrections to the subprocess cross section $d\sigma$ as an expansion in powers of $\alpha_s(Q^2)$ (although one may need to sum certain large corrections to all orders). Given Eq. (27), one can predict the normalization, scaling and logarithmic evolution, and nucleon number dependence for the cross section. In particular, one predicts $d\sigma/dQ^2\ (pA \to \ell\overline{\ell}X) \cong A^1 d\sigma/dQ^2\ (\overline{p}N \to \ell\overline{\ell}X)$ in the region of large Q^2 where deep inelastic lepton scattering is approximately additive in the nucleon number.

Although these results are familiar, they are nevertheless puzzling from a physical standpoint. Intuitively one expects that the incident $\overline{p}$ should be strongly affected by hadronic interactions upon passage through the nucleus—i.e., its Fock state and color structure should be profoundly disturbed by upstream elastic and inelastic collisions. One expects (A-dependent) modifications of the $k_\perp$ and x_L distributions of the quarks in the beam and target, accompanied by inelastic hadronic radiation. Certainly in the case of a macroscopic target, where one observes radiation losses, multiple scattering, and secondary beam production, Eq. (27) cannot be correct.

Recently these questions have been analyzed in QCD perturbation theory[39-42] including initial and final state corrections through two loop order.[40,43] Remarkably, the results are completely consistent at high energies with the factorization ansatz (27). Although the multiple scattering of the initial state particles does broaden the quark and lepton-pair transverse momentum distribution and induces hadron radiation in the central region, the integrated cross section $d^2\sigma/dQ^2dx_L$ is unaffected at high energies. A central feature of the analysis is the demonstration to all orders in perturbation theory[39,43] that hard collinear radiation is not induced by multiple scattering interactions in a target—provided that the particle momentum is large compared to a scale set by the length of the target. This effect is related to the formation zone analysis of Landau and Pomeranchuk[44] which shows that radiation from a classical current propagating between fixed target centers is limited at

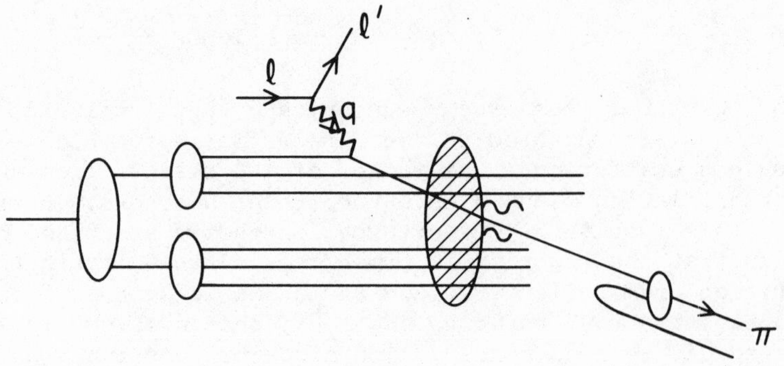

Fig. 12. Propagation of quark jets through a nuclear target.

high energies. In addition, one finds, at least through two-loop order, that initial-state color correlations which occur in individual diagrams and which could effect the 1/3 factor in the Drell-Yan cross section, cancel at high energies.[40,43] The possibility of factorization is a rmarkable attribute of gauge theory, its Ward identities and analyticity. The experimental study of the onset of the formation zone region as a function of energy and nucleon number A is of course very interesting.

In the case of particle production at large transverse momentum, the initial state interactions increase the transverse momentum fluctuations of the incident quark and gluon constituents, yielding a nuclear enhancement for the production cross section. Similar effects broaden the transverse momentum of the final state quark in deep inelastic lepton-nucleus scattering.

Propagation of quark and gluon jets in nuclear targets. In the conventional parton model picture based on the impulse approximation, the multiplicity of hadrons produced in deep inelastic lepton scattering on a nuclear target is expected to be identical to that on a single nucleon, since only one nucleon is "wounded" at large momentum transfer. (See Fig. 12.) In fact, the soft gluons radiated by the scattered quark jet in the deep inelastic process can interact in the nuclear target and produce extra associated multiplicity in the target-fragmentation and central rapidity regions. As shown in Ref. 39 only fast quanta are prevented in QCD from interacting inelastically in a nuclear target. The study of the initial and final interactions of the hadrons and jets in nuclear targets, specifically the modification of longitudinal and transverse momentum distributions, can provide important insights into the nature of QCD dynamics. Also, since binding interactions are suppressed during the transit through a nucleus of a high energy system, it really is correct to describe deep inelastic and high energy constituent processes in terms of quark and gluon propagation rather than hadron propagation; the formation of the high energy hadrons in an inclusive reaction occurs outside the nuclear volume. In the case of high momentum transfer exclusive reactions, the fact that only the minimal valence Fock state with quarks at small relative separation $b_\perp \sim 1/Q_\perp$ interacts, implies that initial and final state interactions in the nuclear target in reactions such as $eA \to ep(A-1)$, $pA \to pp(A-1)$ are suppressed as $1/Q^2$.[45]

Nuclear shadowing. A basic question in nuclear chromodynamics is the nature of shadowing and nucleon-additivity of the nuclear structure function. In Ref. 46 it is argued that initial state absorptive corrections to the virtual photoabsorption cross section vanish as $1/Q^2$ for $Q^2 > C/x_{Bj}$. In addition to the above considerations, simple additivity of the nuclear structure functions will be violated by the fact that the nuclear Fock state spectrum is more complex than that of the individual nucleon. For example, the nuclear binding associated with meson exchange contributions leads to a modification to the sea quark and antiquark distributions in the nuclear structure functions. It has also been argued that the valence wave function of a nucleon can be appreciably affected by the nuclear environment.[47] The number of strange quarks in the nuclear structure function may be different than the extrapolation from a nucleon target. We also emphasize that the existence of hidden color components in the Fock state expansion of the nuclear state implies new contributions to the nuclear structure functions, particularly in the $x > 1$ far-off-shell domain.

Interest in these effects has been heightened by the recent observation by the EMC collaboration[48] of an appreciable breakdown (±15%) of simple additivity for deep inealstic muon scattering in iron versus deuteron targets. The specific identification of additivity-violating effects in the nucleus will also require a careful study of the nuclear target dependence of lepto-production channels, e.g., the reaction $eA \to eK^+X$ which is sensitive to the intrinsic strange quark composition of the nucleus, i.e., contributions not due to QCD evolution. The identification of specific $ed \to eN^*N^*$ channels in electron-deuteron scattering may be an important clue to the $\Delta\Delta$ and hidden color Fock states of the deuteron.

It is clear that one of the most valuable experimental tools for studying the synthesis of QCD and nuclear physics is electron-nucleon and electron-nuclear scattering.[49] Electromagnetic interactions provide a unique probe of hadron and nuclear structure by a (virtual) photon beam of variable energy, tuneable mass, and selectable polarization. The invariant momentum transfer Q to the scattered electron translates directly to resolution of the target at the corresponding length scale $d \sim 1/Q$, allowing one to span the domain between strong coherence ($Q^2 < 1\ fm^{-2}$) (multiquark or meson and nucleon currents) and the short distance regime ($Q^2 \gtrsim 1\ GeV^2$)

where the electromagnetic interaction is dominated by incoherent point-like electron-quark scattering subprocesses.

The focus for studying nuclear chromodynamics is phenomena sensitive to the coherent and quark-confining QCD mechanisms underlying the nuclear force. As discussed in Ref. 49, and the conclusion, an electron accelerator with an energy of approximately 4 GeV appears to be sufficient to explore the kinematic regime for inelastic scattering up to $Q^2 \cong 2$ GeV/c^2 and $W^2 = 6$ GeV^2 with a scattered electron angular range sufficient to separate longitudinal and transverse currents. Such separation is essential for unraveling the contributions of bosonic and fermionic currents. The nuclear form factors can be probed to Q^2 beyond 6 GeV^2/c^2. This energy is also sufficient to study the production of isobars, vector mesons, and strange particles in the nuclear medium, as well as to study effects specific to nuclear dynamics, e.g., deviations from single nucleon additivity (shadowing, etc.), and the short distance effects which can reflect the extra hidden color degrees of freedom of nuclei predicted by QCD. High duty factor, high intensity, and in some cases, beam polarization are necessary to study and separate specific electroproduction channels and coincidence measurements.

HADRONIC WAVE FUNCTIONS AND THEIR MEASUREMENTS IN QCD

Even though quark and gluon perturbative subprocesses are simple in QCD, the complete description of a physical hadronic process requires the consideration of many different coherent and incoherent amplitudes, as well as the effects of non-perturbative phenomena associated with the hadronic wave functions and color confinement. Despite this complexity, it is still possible to obtain predictions for many exclusive and inclusive reactions at large momentum transfer provided we make the ansatz that the effect of non-perturbative dynamics is negligible in the short-distance and far-off-shell domain. (This assumption appears reasonable since a linear confining potential $V \sim r$ is negligible compared to perturbative $1/r$ contributions.) For many large momentum transfer processes, such as deep inelastic lepton-hadron scattering reactions and meson form factors, one can then rigorously isolate the long-distance confinement dynamics from the short-distance quark and gluon dynamics—at least to leading order in $1/Q^2$. The main QCD dynamical behavior can thus be computed from (irreducible) quark and gluon subprocess amplitudes as a perturbative expansion in an asymptotically small coupling constant $\alpha_s(Q^2)$.

An essential part of the QCD predictions is the hadronic wave functions which determine the probability amplitudes and distributions of the quark and gluons which enter the short distance subprocesses. The hadronic wave functions provide the link between the long distance non-perturbative and short-distance perturbative physics. Eventually, one can hope to compute the wave functions from the theory, e.g., from lattice or bag models, or directly from the QCD equations of motion, as we shall outline below. Knowledge of hadronic wave functions also provides explicit connections between exclusive and inclusive processes, and allows the normalization and specification of the power law (higher twist) corrections to the leading impulse approximation results. There are a number of novel QCD phenomena associated with hadronic wave functions, including the effects of intrinsic gluons, intrinsic heavy quark Fock components, diffraction dissociation phenomena, and "direct" hadron processes where the valence Fock state of a hadron enters coherently into a short-distance quark-gluon subprocess.

Relativistic Wave Functions of Composite Systems

A central problem for both hadron and nuclear physics is how to analytically represent the relativistic wave function of a multiparticle composite system. In principle, the Bethe-Salpeter formalism (or its covariant equivalents) can be used, but in practice systematic calculations are intractable. For example,[50] an infinite number of irreducible kernels is required just to reproduce the Dirac equation for an electron in an external field starting from the Bethe-Salpeter equation for muonium with $m_\mu/m_e \to \infty$. Matrix elements of currents and the wave function normalization also require, at least formally, the derivative of an infinite sum of irreducible kernels. The relative time dependence of the covariant formalism adds uncertainties and complexities.[51] The difficulties clearly become much worse in QCD for nucleons and nuclei considered as multiquark bound states.

A more intuitive procedure would be to extend the Schrödinger wave function description of bound states to the relativistic domain by developing a relativistic many-body Fock expansion for the hadronic state. Formally this can be done by quantizing QCD at equal time, and calculating matrix elements from the time-ordered expansion of the S-matrix. A glance at Fig. 13 will indicate why such an approach is analytically intractable. For example, the calculation of a covariant Feynman diagram with n-vertices requires the calculation of n! frame-dependent time-ordered amplitudes. This is illustrated for the electron anomalous moment to order α in Fig. 13(b). Even worse, the calculation of the normalization of a bound state wave function (or the matrix element of a charge or current operator) requires the computation of contributions from all amplitudes involving particle production from the vacuum. (Note that even after normal-ordering the interaction Hamiltonian density for QED, $H_I = e : \bar{\psi}\gamma_\mu\psi A^\mu :$, contains contributions $b^+d^+a^+$ which create particles from the perturbative vacuum.)

There is, however, an alternative formalism for representing bound state amplitudes in relativistic field theories which retains

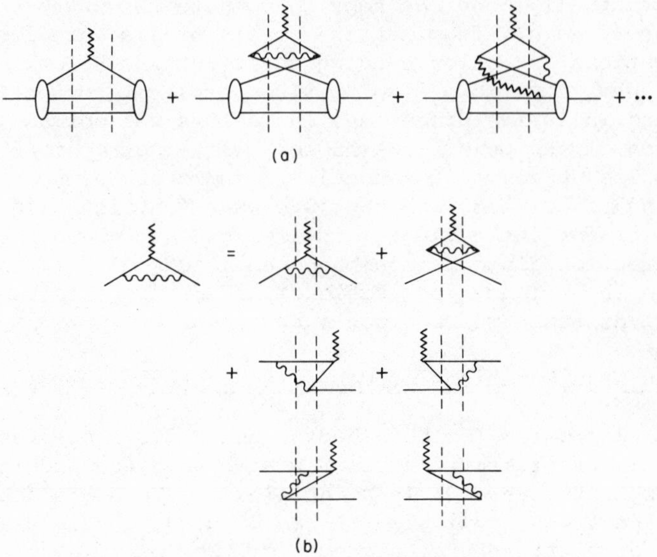

Fig. 13. (a) Calculation of the charge (or current) matrix element of a proton in ordinary time-ordered perturbation theory. (b) Calculation of the electron magnetic moment in QED to order α/π using time-ordered perturbation theory.

all of the simplicity and utility of the Schrödinger non-relativistic many-body theory. The essential step, first noted by Dirac[52] in in 1949, is to quantize the theory at equal "time" $\tau = t + z/c$ on the light-cone.[8,13] (In the non-relativistic limit $c \to \infty$, this is equivalent to equal time quantization.) One can imagine spe ifying the coordinates of a multiple quark nuclear state at equal τ as determined by an incoming light-wave directed along the z direction; on the other hand, specifying the boundary condition coordinates at equal time requires an indeterminate number of simultaneous measurements. Wave functions of composite systems quantized at equal τ can be specified in terms of a momentum space Fock state basis[8] [see Fig. 14(a)]

$$\left\{ \psi_n(\vec{k}_{\perp i}, x_i; \lambda_i) \right\} \quad \begin{matrix} n = 2,3 \ldots \\ i = 1,2 \ldots n \end{matrix} \tag{29}$$

where the λ_i are the particle helicities, the $\vec{k}_{\perp i}$ are the transverse momenta of the on-mass-shell constituents relative to the total bound state momentum P, which we assume is directed along the z-direction. The x_i are the "light-cone" longitudinal momentum fractions $x_i = (k^0 + k^3)_i/(p^0 + p^3)$. It is convenient to use the notation

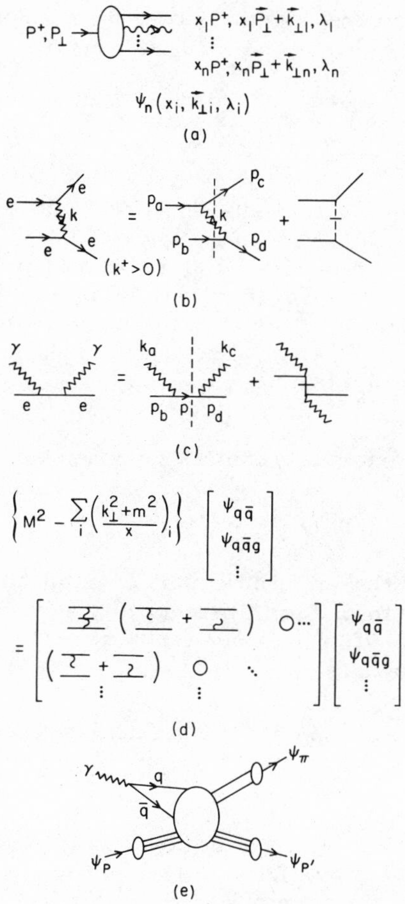

Fig. 14. (a) The n-particle Fock state amplitude defined at equal τ. The state is off the p^- light-cone energy shell [see Eq. (40)]. (b) Calculation of electron-scattering in light-cone (τ-ordered) perturbation theory in $A^+ = 0$ gauge (see Appendix). (c) Compton amplitude in LCPTh. (d) QCD equation of motion for the meson wave function. (e) Calculation of pion photoproduction using light-cone wave function. (Higher Fock state components are necessary at low momentum transfers.)

$$A^{\pm} = A^{0} + A^{3}, \quad A^2 = A^+A^- - A_{\perp}^2,$$
$$A \cdot B = \frac{1}{2}\,(A^+B^- + A^-B^+) - \vec{A}_{\perp} \cdot \vec{B}_{\perp} \;. \tag{30}$$

The mass shell condition is $k^2 = m^2$, i.e., $k^- = (k_{\perp}^2 + m^2)/k^+$. Positive particle energy $k^0 = 1/2(k^+ + k^-) > 0$ then requires $k^+ > 0$.

Transverse and light-cone momentum conservation at each vertex then requires

$$\sum_{i=1}^{n} (\vec{k}_\perp)_i = 0, \quad \sum_{i=1}^{n} x_i = 1, \quad 0 < x_i < 1 \quad . \tag{31}$$

The amplitude $\psi_n(x_i,\vec{k}_{\perp i};\lambda_i)$ then specifies the n-particle Fock amplitude off the P^- light-cone energy shell $P^- \neq \sum_{i=1}^{n} (k^-)_i$. A comparison between time-ordered and τ-ordered perturbation theory is given in Table 1. Thus at a given "time" we can define the (color singlet) basis

$$|0\rangle \, , \quad |q\bar{q}\rangle = a^{+}_{k^+,\vec{k}_\perp} \, b^{+}_{k^{+\prime},\vec{k}'_\perp} \, |0\rangle \, , \quad \ldots \tag{32}$$

The pion state, for example, can be expanded as

$$|\pi\rangle = |q\bar{q}\rangle\psi_{q\bar{q}} + |q\bar{q}g\rangle \, \psi_{q\bar{q}g} + \cdots \tag{33}$$

where $\psi_n = \langle n|\pi\rangle$ is the amplitude for finding the Fock state $|n\rangle$ in $|\pi\rangle$ at time τ. The full Fock state wave function which describes the n-particle state of a hadron with 4-momentum $P^\mu = (P^+,P^-,\vec{P}_\perp)$ and constituents with momenta

$$k^\mu = \left(k^+,k^-,\vec{k}_\perp\right) = \left(xP^+, \ \frac{\left(x\vec{P}_\perp + \vec{k}_\perp\right)^2 + m^2}{x} \ , \ x\vec{P}_\perp + \vec{k}_\perp\right) \tag{34}$$

and spin projection λ_i is

$$\psi_n = \psi_n(x_i,k_{\perp i};\lambda_i) \prod_{\text{fermions}} \frac{\varepsilon(x_iP^+,x_i\vec{P}_\perp + \vec{k}_{\perp i})_{\lambda_i}}{\sqrt{x_i}} \times$$

$$\times \prod_{\text{gluons}} \frac{\varepsilon(x_iP^+,x_i\vec{P}_\perp + \vec{k}_{\perp i})_{\lambda_i}}{\sqrt{x_i}} \quad . \tag{35}$$

Note that $\psi_n(x_i,\vec{k}_{\perp i};\lambda_i$ is independent of P^+, $\vec{P}_\perp$. The general normalization condition is

$$\sum_n \int [d^2k_\perp] \int [dx] \, |\psi_n(x_i,\vec{k}_{\perp i};\lambda_i)|^2 = 1 \tag{36}$$

where by momentum conservation

$$[d^2k_\perp] = 16\pi^3 \, \delta^2\left(\sum_{i=1} k_{\perp i}\right) \prod_{i=1} \frac{d^2k_{\perp i}}{16\pi^3} \tag{37}$$

and

Table 1. Comparison Between Time-Ordered and τ-Ordered Perturbation Theory

Equal t	Equal $\tau = t + z$
$k^0 = \sqrt{\vec{k}^2 + m^2}$ (particle mass shell)	$k^- = \frac{k_\perp^2 + m^2}{k^+}$ (particle mass shell)
$\sum \vec{k}$ conserved	$\sum \vec{k}_\perp, k^+$ conserved
$\mathcal{M}_{ab} = V_{ab} + \sum_c V_{ac} \frac{1}{\sum_a k^0 - \sum_c k^0 + i\varepsilon} V_{cb}$	$\mathcal{M}_{ab} = V_{ab} + \sum_c V_{ac} \frac{1}{\sum_a k^- - \sum_c k^- + i\varepsilon} V_{cb}$
$n!$ time-ordered contributions	$k^+ > 0$ only
Fock states $\psi_n(\vec{k}_i)$	Fock states $\psi_n(\vec{k}_{\perp i}, x_i)$
$\sum_{i=1}^{n} \vec{k}_i = \vec{P} = 0$	$x = \frac{k^+}{P^+}$, $\sum_{i=1}^{n} x_i = 1$, $\sum_{i=1}^{n} \vec{k}_{\perp i} = 0$ $(0 < x_i < 1)$
$\mathcal{E} = P^0 - \sum_{i=1}^{n} k_i^0 = M - \sum_{i=1}^{n} \sqrt{k_i^2 + m_i^2}$	$\mathcal{E} = P^+\left(P^- - \sum_{i=1}^{n} k_i^-\right) = M^2 - \sum_{i=1}^{n} \left(\frac{k_\perp^2 + m^2}{x}\right)_i$

$$[dx] = \delta\left(1 - \sum_{i=1}^{n} x_i\right) \prod_{i=1}^{n} dx_i \quad . \tag{38}$$

For the Fock state wave function in the rest system we can identify

$$x = \frac{k^o + k^3}{M} \cong \frac{m}{M} + \frac{k^3}{M} \tag{39}$$

and the off-shell light-cone energy is

$$\mathcal{E} = P^+\left[P^- - \sum_{i=1}^{n} k^-\right] = M^2 - \sum_{i=1}^{n}\left(\frac{k_\perp^2 + m^2}{x}\right)$$

$$\cong 2M\left[\mathcal{E}_{NR} - \sum_{i=1}^{n}\left(\frac{k_\perp^2 + k_3^2}{2m}\right)_i\right] \quad . \tag{40}$$

Thus, in the non-relativistic limit, the hydrogen atom wave function is

$$\psi_{1s} = \frac{C}{[k_\perp^2 + (m_e - xM)^2 + \alpha^2 m_e^2]^2} \quad . \tag{41}$$

Light-cone perturbation theory rules can be derived by either evaluating standard equal-time time-ordered perturbation theory for an observer in a fast moving Lorentz frame (the "infinite momentum" method)[13] or more directly, by quantizating at equal τ.

The rules for calculating in light-cone perturbation theory are given in detail in the Appendix. An advantage of quantizing a gauge theory at a given light-cone time τ and in $A^+=0$ gauge is the absence of ghosts or other non-physical quanta. An important feature of the light-cone rules is the presence of "instantaneous" components of the gluon propagator (analogous to the Coulomb potential) and fermion propagator (generating effective "seagull" contributions). [See Fig. 14(b),(c).] Calculations in light-cone perturbation theory are often surprisingly simple since one can usually choose Lorentz frames for the external particles such that only a few time-orderings need to be considered. All the variables have a direct physical interpretation. The formalism is also ideal for computing helicity amplitudes directly without trace projection techniques. A list of all the gluon fermion vertices which are required as gauge theory calculations is given in Tables 1 and 2 of Ref. 53.

It is straightforward to implement ultraviolet renormalization in light-cone perturbation theory. We define truncated wave functions ψ^κ and a truncated Hamiltonian H^κ such that all intermediate states with $|\mathcal{E}| > \kappa^2$ are excluded. Thus κ^{-1} is analogous to the lattice spacing in lattice field theory. Since QCD is renormalizable the effects of the neglected states are accounted for by the use of the running coupling constant $\alpha_s(\kappa^2)$ and running mass $m(\kappa^2)$, as long as κ^2 is sufficiently large compared to all physical mass thresholds. Completeness implies

$$\sum_{n,\lambda_i} \int [d^2k_\perp] \int [dx] \left|\psi_n^\kappa(x_i, k_{\perp i};\ \lambda_i)\right|^2 = 1 - 0\left(\frac{m^2}{\kappa^2}\right). \tag{42}$$

The equation of state for the meson or baryon wave function in QCD is a set of coupled multiparticle equations [see Fig. 14(d)]:[8]

$$\left[M^2 - \sum_{i=1}^{n} \left(\frac{k_\perp^2 + m^2}{x}\right)_i\right] \psi_n^\kappa = \sum_{n'} V_{nn'}^\kappa\, \psi_{n'}^\kappa, \tag{43}$$

where M^2 is the eigenvalue and $V_{nn'}$ is the set of diagonal (from instantaneous gluon and fermion exchange) and off-diagonal (from the 3 and 4 particle vertices) momentum-space matrix elements dictated by the QCD rules. Because of the κ cutoff the equations truncate at finite n,n'. In analogy to non-relativistic theory, one can imagine starting with a trial wave function for the lowest $|q\bar{q}\rangle$ or $|qqq\rangle$ valence state of a meson or baryon and iterating the equations of motion to determine the lowest eigenstate Fock state wave functions and mass M. Invariance under changes in the cutoff scale provides an important check on the consistency of the results. Note that the general solution for the hadron wave function in QCD is expected to have Fock state components with arbitrary numbers of gluons and quark-antiquark pairs.

The two-particle "valence" light-cone Fock state wave function for mesons or positronium can also be related to the Bethe-Salpeter wave function evaluated at equal τ:

$$\int \frac{dk^-}{2\pi}\, \psi_{BS}(k;p) = \frac{u(x_1,\vec{k}_\perp)}{\sqrt{x_1}}\, \frac{\bar{v}(x_2,-\vec{k}_\perp)}{\sqrt{x_2}}\, \psi(x_i,\vec{k}_\perp)$$

$$+ \text{ negative energy components,} \tag{44}$$

where ψ satisfies an exact bound state equation[8]

$$\left[M^2 - \frac{k_\perp^2 + m_1^2}{x} - \frac{k_\perp^2 + m_2^2}{x_2}\right] \psi(x_i,\vec{k}_{\perp i})$$

$$= \int_0^1 dy \int \frac{d^2\ell_\perp}{16\pi^3}\, \tilde{K}\left(x_i,\vec{k}_{\perp i};\ y_i,\vec{\ell}_{\perp i};\ M^2\right) \psi(y_i,\vec{\ell}_{\perp i})\,. \tag{44}$$

The kernel $\tilde{K}$ is computed from the sum of all two-particle-irreducible contributions to the two-particle scattering amplitude. For example, the equation of motion for the $|e^+e^-\rangle$ Fock state of positronium reduces in the non-relativistic limit to $\left(k_\perp, \ell_\perp \sim O(\alpha m),\ x = x_1 - x_2 \sim O(\alpha),\ M^2 = 4m^2 + 4m\varepsilon\right)$

$$\left\{\varepsilon - \frac{k_\perp^2 + x^2 m^2}{m}\right\} \psi(x_i, k_\perp)$$
$$= (4x_1x_2) \int_{-1}^{1} dy\ m \int \frac{d^2\ell_\perp}{(2\pi)^3} \left[\frac{-e^2}{(\vec{k}_\perp - \vec{\ell}_\perp)^2 + (x-y)^2\ m^2}\right] \psi(y_i, \ell_\perp) . \tag{46}$$

The non-relativistic solution is[8] $(\beta = \alpha m/2)$

$$\psi(x_i, k_\perp) = \sqrt{\frac{m\beta^3}{\pi}}\ \frac{64\pi\beta\ x_1 x_2}{\left[k_\perp^2 + (x_1 - x_2)^2\ m^2 + \beta^2\right]^2} \begin{cases} \dfrac{u_\uparrow \bar{v}_\downarrow - u_\downarrow \bar{v}_\uparrow}{\sqrt{2x_1x_2}} \\[2ex] \dfrac{u_\uparrow \bar{v}_\uparrow}{\sqrt{x_1 x_2}} \end{cases} \tag{47}$$

for para and ortho states respectively.

More generally, we can make an (approximate) connection between the equal-time wave function of a composite system and the light-cone wave function by equating the off-shell propagator $\mathcal{E} = M^2 - \left(\sum_{i=1}^{n} k_i\right)^2$ in the two frames:

$$\mathcal{E} = \begin{cases} M^2 - \left(\displaystyle\sum_{i=1}^{n} q^0_{(i)}\right)^2 , & \displaystyle\sum_{i=1}^{n} q_i = 0\,[\text{C.M.}] \\[2ex] M^2 - \displaystyle\sum_{i=1}^{n} \left(\frac{k_\perp^2 + m^2}{x}\right)_i , & \sum \vec{k}_{\perp i} = 0\ ,\ \sum x_i = 1\,[\text{L.C.}] \end{cases} \tag{48}$$

In addition we can identify

$$x_i = \frac{k_i^+}{p^+} \rightleftarrows \frac{(q^0 + q^3)_i}{\displaystyle\sum_{i=1}^{n} q^0_{(j)}} ,\ \vec{k}_{\perp i} \rightleftarrows \vec{q}_{\perp i} . \tag{49}$$

For a relativistic two-particle state with a wave function which is a function of the off-shell variable $\mathcal{E}$ only, we can identify[54] $(m_1 = m_2 = m,\ x = x_1 - x_2)$

$$\psi_{\text{L.C.}}\left(\frac{k_\perp^2 + m^2}{1 - x^2} - m^2\right) \rightleftarrows \psi_{\text{C.M.}}(\vec{q}^{\,2}) . \tag{50}$$

In the non-relativistic limit, this corresponds to the identification $\vec{q}_\perp = \vec{k}_\perp$, $q_3^2 = x^2m^2$.

Measures of Hadronic Wave Functions:

If we could solve the QCD equation of motion Eq. (43) for the light-cone wave functions ψ_n of a hadron then we could (in principle) calculate all of its electromagnetic properties. For example, to compute the elastic form factors $\langle p|J^\mu(0)|p+q\rangle$ of a hadron we choose the Lorentz frame[55]

$$p^\mu = (p^+,p^-,\vec{p}_\perp) = \left(p^+, \frac{M^2}{p^+}, \vec{0}_\perp\right)$$
$$q^\mu = (q^+,q^-,\vec{q}_\perp) = \left(0, \frac{2p\cdot q}{p^+}, \vec{q}_\perp\right) \tag{51}$$

where $p^2 = (p+q)^2 = M^2$ and $-q^2 = Q^2 = \vec{q}_\perp^{\,2}$. Then the only time ordering which contributes to the $\langle p|J^+|p+q\rangle$ matrix element is where the photon attaches directly to the $e_j\bar{u}_j\gamma^+u_j$ currents of the constituent quarks. The spin averaged form factor is[53,55] [see Fig. 15(a)]

$$F(Q^2) = \sum_n\sum_j e_j \int [dx]\,[d^2k_\perp] \sum_{\lambda_i} \psi_n^{*K}(x_i,\vec{k}'_{\perp i};\lambda_i)\,\psi^K(x_i,\vec{k}_{\perp i};\lambda_i) \tag{52}$$

where $\vec{k}'_{\perp j} = \vec{k}_{\perp j} + (1-x_j)\vec{q}_\perp$ for the struck quark and $\vec{k}_{\perp i} - x_i\vec{q}_\perp$ $(i \neq j)$ for the spectator quarks. (The $-x_i\vec{q}_\perp$ terms occur because the arguments $\vec{k}'_\perp$ are calculated relative to the direction of the final state hadron.) We choose $\kappa^2 \gg Q^2, M^2$. We note here the special advantage

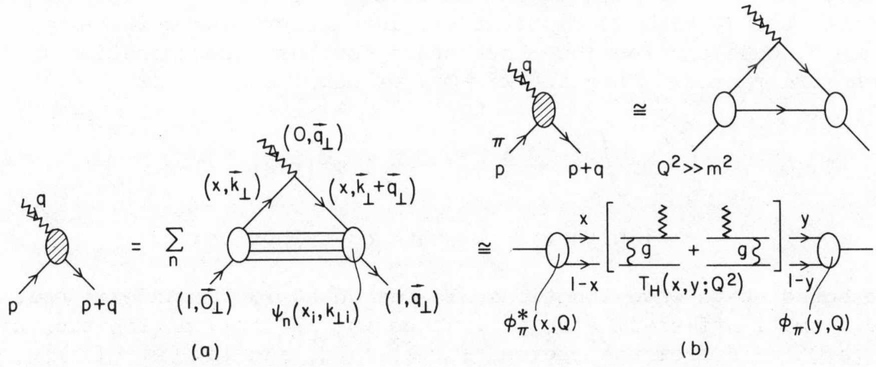

Fig. 15. (a) Calculation of current matrix elements in light-cone perturbation theory. (b) Valence Fock state contribution to the large momentum transfer meson form factor. T_H is computed for zero mass quarks q and $\bar{q}$ collinear with the meson momentum.

of light-cone perturbation theory: the current J^+ is diagonal in the Fock state basis.

Because of Eq. (42) the form factor is normalized to 1 at zero momentum transfer. We can also compute the helicity flip form factors in the same manner.[53,56] For example, the anomalous moment $a = F_2(0)$ of any spin 1/2 system can be written[56]

$$\frac{a}{M} = -\sum_j e_j \int [dx]\,[d^2k_\perp]\psi^{*\kappa}_{p\uparrow} \sum_{i\neq j}\left(\frac{\partial}{\partial k^1_i} + i\frac{\partial}{\partial k^2_i}\right)\psi^{\kappa}_{p\downarrow} \quad . \tag{53}$$

Explicit calculations of the electron anomalous moment in QED using this result are given in Ref. 56. We notice that in general all Fock states ψ^κ_n contribute to the anomalous moment of a system, although states with κ^2 much larger than the mean off-shell energy $\langle\mathcal{E}\rangle$ are not expected to be important. The general result (53) also includes the effects of the Lorentz boost of the wave function[57] from p^μ to $(p+q)^\mu$. In particular, the Wigner spin rotation contributes to $F_2(q^2)$ and the charge radius $F_1'(q^2)$ in the $q^2 \to 0$ limit and can only be neglected in the limit of non-relativistic binding $\langle\mathcal{E}\rangle \ll M^2$. This effect gives non-trivial relativistic corrections[58] to nuclear magnetic moment calculations based on simple additivity $\vec{\mu} = \langle\sum_j \vec{\mu}_j\rangle$.

Form Factors of Mesons

Results such as Eqs. (52) and (53) are formally exact but useless unless we have complete knowledge of the hadronic or nuclear wave function. However, by making use of the impulse approximation and the smallness of the QCD running coupling constant, we can calculate features of elastic and inelastic large momentum transfer processes (53) without explicit knowledge of the wave function. For example, consider the $|q\bar{q}\rangle$ Fock state component contribution to the pion form factor. Choosing $\kappa^2 = Q^2$, we have

$$F_\pi(Q^2) = \int_0^1 dx \int^Q \frac{d^2k_\perp}{16\pi^3}\,\psi^{*Q}(x,\vec{k}_\perp)\,\psi^Q\left(x,\vec{k}_\perp + (1-x)\vec{q}_\perp\right)$$

$$+ \text{ higher Fock state contributions} \quad . \tag{54}$$

The bound state wave functions are peaked at low transverse momentum, i.e., small off-shell energy . Thus the leading contribution at large Q^2 come from the regimes (a) $\vec{k}^2_\perp \ll \vec{q}^2_\perp$ and (b) $\left(\vec{k}_\perp + (1-x)\vec{q}_\perp\right)^2 \ll \vec{q}^2_\perp$. Thus

$$F^{(a)}_\pi(Q^2) \cong \int_0^1 dx\ \phi(x,Q)\ \psi^Q\left(x,(1-x)\vec{q}_\perp\right) \tag{55}$$

where

$$\phi(x,Q) \equiv \int^{Q} \frac{d^2k}{16\pi^3}\, \psi^Q(x,k_\perp) \quad . \tag{56}$$

If we simply iterate the one-gluon exchange kernel V_1 in the equation of motion for ψ, then for $q_\perp^2 >> <\ell_\perp^2>$

$$\psi^Q\left(x,(1-x)q_\perp\right) \cong \int_0^1 dy \int^Q \frac{d^2\ell_\perp}{16\pi^3} \frac{V_1\left(x,(1-x)q_\perp;y,\ell_\perp\right)\psi^Q(y,\ell_\perp)}{-q_\perp^2(1-x)/x}$$

$$\cong \int_0^1 dy \; \frac{V_1\left(x,(1-x)q_\perp;y,0_\perp\right)}{-q_\perp^2(1-x)/x}\,\phi(y,Q) \quad . \tag{57}$$

Thus we can write the gluon exchange contribution to the form factor in the form[19,53] [see Fig. 15(b)]

$$F_\pi(Q^2) = \int_0^1 dx\, dy\; \phi^*(y,Q)\; T_H(x,y;Q)\; \phi(y,Q) \tag{58}$$

where

$$T_H = \frac{16\pi\, C_F\, \alpha_s(Q^2)}{Q^2}\left[\frac{e_1}{(1-y)(1-x)} + \frac{e_2}{xy}\right] \tag{59}$$

is the "hard scattering amplitude" for scattering collinear constituents q and $\bar{q}$ from the initial to the final direction. The color factor is $C_F = (n_c^2 - 1)/2n_c = 4/3$. The "distribution amplitude" $\phi(x,Q)$ is the amplitude for finding the $|q\bar{q}>$ Fock state in the pion collinear up to the scale Q. (It is analogous to the wave function at the origin in non-relativistic calculations.) The distribution amplitude enters universally in all large momentum transfer exclusive amplitudes and is a process-independent measure of the valence quark distribution in each hadron; its (logarithmic) dependence on Q^2 can be determined directly from the operator product expansion or the light-cone or from an evolution equation, as we discuss below.

Thus the simplest estimate for the asymptotic behavior of the meson form factor is $F_\pi(Q^2) \sim \alpha_s(Q^2)/Q^2$. To see if this is correct we must examine the higher order corrections:

i) Contributions from higher particle number Fock states $|q\bar{q}g>$, $|q\bar{q}q\bar{q}>$, etc. are power-law suppressed since (in light-cone gauge) the numerator couplings cannot compensate the extra fall-off in Q^2 from the extra energy denominators.

ii) All infrared singularities and contributions from soft($\ell_\perp \to 0$) gluons cancel in color singlet matrix elements. [It is interesting to note that the quark (Sudakov) form factor falls faster at large Q^2 than $F_\pi(Q^2)$.]

iii) Vertex and vacuum polarization corrections to the T_H are higher order in $\alpha_s(Q^2)$ since we choose $\kappa^2 = Q^2$. The effective argument of α_s in T_H is $Q^2 = xyQ^2$ or $(1-x)(1-y)Q^2$ corresponding to the actual momentum transfer carried by the gluon.

iv) By definition, $\phi(x,\kappa^2)$ sums all (reducible) contributions from low momentum transfer gluon exchange in the $q\bar{q}$ wave function. Hard gluon contributions with $|\mathcal{E}| > \kappa^2$ and irreducible graphs give contributions to T_H which are higher order in $\alpha_s(Q^2)$. By analyzing the denominators in T_H one can show that the natural $\mathcal{E}$ cutoff for $\phi(x,\kappa)$ which minimizes higher order contributions is

$$\kappa^2 = Q^2 = Q^2 \min \left\{x/1-x,\ 1-x/x\right\} . \tag{60}$$

v) Although T_H is singular at $x \to 0,1$, the endpoint behavior of $\phi(x,Q^2) \sim x^\varepsilon,\ (1-x)^\varepsilon\ (\varepsilon > 0)$ is sufficient to render this region harmless.

The Meson Distribution Amplitude

The essential prediction of QCD for the pion form factor is the power-law behavior[22] $F_\pi \sim 1/Q^2$, with logarithmic corrections from the explicit powers of $\alpha_s(Q^2)$ in T_H and the Q^2 dependence of the distribution amplitudes $\phi(x,Q^2)$.

The variation of ϕ with Q^2 comes from the upper limit of the $\vec{k}_\perp$ integration (since $\psi \sim 1/k_\perp^2$) and the renormalization scale dependence:

$$\psi^Q(x,\vec{k}_\perp) = \frac{Z_2(Q)}{Z_2(Q_0)}\,\psi^{Q_0}(x,\vec{k}_\perp) \tag{61}$$

due to the vertex and self-energy insertions. Thus

$$Q^2 \frac{\partial}{\partial Q^2}\,\phi(x,Q) = \frac{Q^2}{16\pi^2}\,\psi^Q(x,\vec{q}_\perp) + \frac{d}{d\log Q^2}\log Z_2(Q^2)\ \phi(x,Q) . \tag{62}$$

To order $\alpha_s(Q^2)$ we can compute $Q^2\psi$ from one-gluon exchange [as in Eq. (57)], and $d \log Z_2(Q^2)/d \log Q^2 = \alpha_s(Q^2)\gamma_F/4\pi$. Setting $\phi(x,Q) = x(1-x)\ \tilde{\phi}(x,Q) = x_1x_2\tilde{\phi}$, we obtain[53] an "evolution equation"

$$x_1x_2Q^2 \frac{\partial}{\partial \log Q^2}\,\tilde{\phi}(x_i,Q) = \frac{\alpha_s(Q^2)}{4\pi}\int_0^1 [dy]\ V(x_i,y_i)\ \phi(y,Q) \tag{63}$$

where

$$V(x_i,y_i) = 2C_F\left\{x_1y_2\ \theta(y_1-x_1)\ \left(\delta_{h_1\bar{h}_2} + \frac{\Delta}{y_1-x_1}\right) + (1 \leftrightarrow 2)\right\} \tag{64}$$

($\delta_{h_1\bar{h}_2} = 1$ when the q and $\bar{q}$ helicities are opposite) and

$$\Delta\tilde{\phi}(y_i,Q) = \tilde{\phi}(y_i,Q) - \tilde{\phi}(x_i,Q) \quad . \tag{65}$$

The $\tilde{\phi}(x_1,Q)$ subtraction is due to the $\gamma_F\phi$ term—i.e., the infrared dependence at $y_i = x_i$ is cancelled for color singlet hadrons. Thus given the initial condition $\phi(x_i,Q_0)$, perturbation theory determines the evolution of $\phi(x,Q)$ for $Q > Q_0$. The solution to the evolution equation is

$$\phi(x_i,Q) = x_1x_2 \sum_{n=0}^{\infty} a_n(Q_0^2)\, C_n^{3/2}(x_1-x_2)\,(\log Q^2/\Lambda^2)^{-\gamma_n} \tag{66}$$

where the Gegenbauer polynomials $C_n^{3/2}$ (orthogonal on $\int [dx]x_1x_2$) are eigenfunctions of $V(x_i,y_i)$. The corresponding eigenvalues are the "non-singlet" anomalous dimensions:

$$\gamma_n = \frac{C_F}{\beta_o}\left[1 + 4\sum_{2}^{n+1}\frac{1}{k} - \frac{2\delta_{h_1\bar{h}_2}}{(n+1)(n+2)}\right] \geq 0 \quad . \tag{67}$$

These results can also be derived by using the operator product expansion for the distribution amplitude.[17] By definition

$$\phi(x,Q) = \Lambda^+ \int \frac{dz^-}{2\pi}\, e^{ixz^-/2}\, i\langle 0|\bar{\psi}(z)\psi(0)|\pi\rangle^Q \Big|_{z^+=0,\, z^2=-z_\perp^2=0(-1/Q^2)} \tag{68}$$

(Λ^+ is the positive energy spinor projection operator). The relative separation of the q and $\bar{q}$ thus approaches the light-cone $z^2 = 0$ as $Q^2 \to \infty$. Equation (67) then follows, by expanding $\psi(z)\psi(0)$ in local operators.

The coefficients a_n are determined from $\phi(x_i,Q_0)$:

$$a_n\left(\log\frac{Q^2}{\Lambda^2}\right)^{-\gamma_n} = \frac{2(2n+3)}{(2+n)(1+n)}\int_{-1}^{1} d(x_1-x_2)\, C_n^{3/2}(x_1-x_2)\, \phi(x_i,Q_0) \quad . \tag{69}$$

For $Q^2 \to \infty$, only the leading $\gamma_o = 0$ term survives

$$\lim_{Q^2\to\infty} \phi(x,Q) = a_o x_1 x_2 \tag{70}$$

where

$$\frac{a_o}{6} = \int_o^1 dx\, \phi(x,Q) = \int_o^1 dx \int^Q \frac{d^2k_\perp}{16\pi^3}\, \psi^Q(x,k_\perp) \tag{71}$$

is the meson wave function at the origin as measured in the decay $\pi \to \mu\nu$:

$$\frac{a_0}{6} = \frac{1}{2\sqrt{n_c}} f_\pi \ . \qquad (72)$$

More generally, the leptonic decay ($\rho^0 \to e^+e^-$, etc.) of each meson normalizes its distribution amplitude by the "sum rule"

$$\int_0^1 dx\ \phi_M(x,Q) = \frac{f_M}{2\sqrt{n_c}} \ , \qquad (73)$$

independent of Q. The fact that $f_\pi \neq 0$ implies that the probability of finding the $|q\bar{q}\rangle$ Fock state in the pion is non-zero. In fact all the Fock states wave functions $\psi_n^\kappa(x_i,\vec{k}_{\perp i})$ $(|\mathcal{E}| < \kappa^2)$ are well-defined, even in the infrared limit $x_i \to 0$ (since $|\mathcal{E}| \sim \langle k_\perp^2\rangle/x_i$ and $\langle k_\perp^2\rangle$ is non-zero for a state of finite radius).

The pion form factor at high Q^2 can thus be written[8,17-19,59]

$$F_\pi(Q^2) = \int_0^1 dx\ \phi^*(x,Q)\, T_H(x,y;Q)\ \phi(y,Q) \ ,$$

$$T_H = \frac{16}{3\pi}\, \frac{\alpha_s\left((1-x)(1-y)Q^2\right)}{(1-x)(1-y)Q^2} \ . \qquad (74)$$

Thus

$$F_\pi(Q^2) = \left|\sum_{n=0} a_n \log^{-\gamma_n} Q^2/\Lambda^2\right|^2 \frac{16\pi}{3}\, \frac{\alpha_s(\overline{Q}^2)}{Q^2}$$

$$\times \left[1 + O\left(\frac{\alpha_s(Q^2)}{\pi}\right) + O\left(\frac{m}{Q}\right)\right] \qquad (75)$$

where $\overline{Q}^2 \cong \langle(1-x)(1-y)\rangle Q^2$. Finally, for the asymptotic limit where only the leading anomalous dimension contributes:[60]

$$\lim_{Q^2\to\infty} F_\pi(Q^2) = 16\pi\ f_\pi^2\ \frac{\alpha_s(Q^2)}{Q^2} \ . \qquad (76)$$

The analysis of the $F_{\pi\gamma}(Q^2)$ form factor, measurable in $ee \to ee\pi^0$ reactions, proceeds in a similar manner. An interesting result is[53]

$$\alpha_s(Q^2) = \frac{F_\pi(Q^2)}{4\pi Q^2|F_{\pi\gamma}(Q^2)|^2}\left[1 + O\left(\frac{\alpha_s(Q^2)}{\pi}\right)\right] \qquad (77)$$

which provides a definition of α_s independent of the form of the distribution function ϕ_π. Higher order corrections to $F_\pi(Q^2)$ and $F_{\pi\gamma}(Q^2)$ are discussed in Ref. 59.

General Large Momentum Transfer Exclusive Processes

The meson form factor calculation which we outlined above is the prototype for the calculation of the QCD hard scattering contribution for the whole range of exclusive processes at large momentum transfer. Away from possible special points in the x_i integrations (see below) a general hadronic amplitude can be written to leading order in $1/Q^2$ as a convolution of a connected hard-scattering amplitude T_H convoluted with the meson and baryon distribution amplitudes:

$$\phi_M(x,Q) = \int^{|\mathcal{E}|<Q^2} \frac{d^2k_\perp}{16\pi^2}\, \psi^Q_{q\bar{q}}(x,\vec{k}_\perp) \quad . \tag{78}$$

and

$$\phi_B(x_i,Q) = \int^{|\mathcal{E}|<Q^2} [d^2k_\perp]\; \psi_{qqq}(x_i,\vec{k}_{\perp i}) \quad . \tag{79}$$

The hard scattering amplitude T_H is computed by replacing each external hadron line by massless valence quarks each collinear with the hadrons momentum $p_i^\mu \cong x_i\, p_H^\mu$. For example, the baryon form factor at large Q^2 has the form[53,61] (see Fig. 2)

$$G_M(Q^2) = \int [dx]\,[dy]\; \phi^*(y,\hat{Q})\; T_H(x,y;Q^2)\; \phi(s,\hat{Q}) \tag{80}$$

where T_H is the $3q+\gamma \to 3q'$ amplitude. (The optimal choice for $\hat{Q}$ is discussed in Ref. 53.) For the proton and neutron we have to leading order ($C_B = 2/3$)

$$T_p = \frac{128\pi^2\, C_B^2}{(Q^2+M_0^2)^2}\, T_1 \tag{81}$$

$$T_n = \frac{128\pi^2\, C_B^2}{3(Q^2+M_0^2)^2}\, [T_1 - T_2] \tag{82}$$

where

$$T_1 = -\frac{\alpha_s(x_3y_3Q^2)\;\alpha_s\big((1-x_1)(1-y_1)Q^2\big)}{x_3(1-x_1)^2\; y_3(1-y_1)^2} + \frac{\alpha_s(x_2y_2Q^2)\;\alpha_s\big((1-x_1)(1-v_1)Q^2\big)}{x_2(1-x_1)^2\; y_2(1-y_1)^2} - \frac{\alpha_s(x_2y_2Q^2)\;\alpha_s(x_2y_2Q^2)}{x_2x_3(1-x_3)\; y_2y_3(1-y_1)} \tag{83}$$

and

$$T_2 = \frac{-\alpha_s(x_1y_1Q^2)\;\alpha_s(x_3y_3Q^2)}{x_1x_3(1-x_1)\; y_1y_3(1-y_3)} \quad . \tag{84}$$

T_1 corresponds to the amplitude where the photon interacts with the quarks (1) and (2) which have helicity parallel to the nucleon helicity, and T_2 corresponds to the amplitude where the quark with opposite helicity is struck. The running coupling constants have arguments $\hat{Q}^2$ corresponding to the gluon momentum transfer of each diagram. Only large Q^2 behavior is predicted by the theory; we utilize the parameter M_o to represent the effect of power-law suppressed terms from mass insertions, higher Fock states, etc.

The Q^2-evolution of the baryon distribution amplitude can be derived from the operator product expansion of three quark fields or from the gluon exchange kernel, in parallel with the derivation of (63). The baryon evolution equation to leading order in α_s is[53]

$$x_1x_2x_3\left\{\frac{\partial}{\partial\zeta}\,\tilde{\phi}(x_i,Q) + \frac{3}{2}\,\frac{C_F}{\beta_o}\,\tilde{\phi}(x_i,Q)\right\} = \frac{C_B}{\beta_o}\int_o^1 [dy]V(x_i,y_i)\tilde{\phi}(y_i,Q)\ . \tag{85}$$

Here $\phi = x_1x_2x_3\phi$, $\zeta = \log(\log Q^2/\Lambda^2)$ and [see Fig. 7(b)]

$$\begin{aligned} V(x_i,y_i) &= 2x_1x_2x_3 \sum_{i\neq j} \theta(y_i-x_i)\ \delta(x_k-y_k)\ \frac{y_j}{x_j}\left(\frac{\delta_{h_i\bar{h}_j}}{x_i+x_j} + \frac{\Delta}{y_i-x_i}\right) \\ &= V(y_i,x_i)\quad . \end{aligned} \tag{86}$$

The infrared singularity at $x_i = y_i$ is cancelled because the baryon is a color singlet. The evolution equation has the general solution

$$\phi(x_i,Q) = x_1x_2x_3 \sum_{n=0}^{\infty} a_n\ \tilde{\phi}_n(x_i)\ \left(\log\frac{Q^2}{\Lambda^2}\right)^{-\gamma_n^B}\quad . \tag{87}$$

The leading (polynomial) eigensolution $\tilde{\phi}_n(x_i)$ and corresponding baryon anomalous dimensions are given in Refs. 53 and 62. Thus at large Q^2, the nucleon magnetic form factor has the form[62]

$$G_M(Q^2) \rightarrow \frac{\alpha_s^2(Q^2)}{Q^4}\sum_{n,m} b_{nm}\left(\log\frac{Q^2}{\Lambda^2}\right)^{-\gamma_n^B-\gamma_n^B}\left[1 + 0\left(\alpha_s(Q^2),\ \frac{m^2}{Q^2}\right)\right]\ . \tag{88}$$

We can also use this result to obtain results for ratios of various baryon and isobar form factors assuming isospin or SU(3)-flavor symmetry for the basic wave function structure. Results for the neutral weak and charged weak form factors assuming standard SU(2) × U(1) symmetry are given in Ref. 61.

As we see from Eq. (80), the integration over x_i and y_i has potential endpoint singularities. However, it is easily seen that

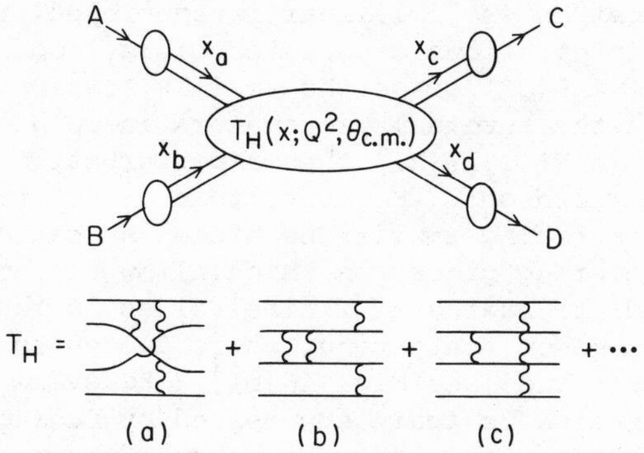

Fig. 16. QCD contributions to meson-meson scattering at large momentum transfer. Diagram (c) corresponds to a pinch singularity which is suppressed by quark form factor effects.

any anomalous contribution [e.g., from the region $x_2, x_3 \sim O(m/Q)$, $x_1 \sim 1 - O(m/Q)$] is asymptotically suppressed at large Q by a Sudakov form factor arising from the virtual correction to the $\bar{q}\gamma q$ vertex when the quark legs are near-on-shell $[p^2 \sim O(mQ)]$.[53,63] This Sudakov suppression of the endpoint region requires an all orders resummation of perturbative contributions, and thus the derivation of the baryon form factors is not as rigorous as for the meson form factor, which has no such endpoint singularity.

The most striking feature of the QCD prediction (88) is the $1/Q^4$ power-law behavior of G_M^p as G_M^n. The power-law dependence[22] reflects:

i) The essential scale-invariance of the qq scattering subprocesses within T_H.

ii) The fact that the minimal Fock state of a baryon is the 3-quark state.

In the case of hadron scattering amplitudes $A + B \to C + D$, photoproduction, Compton scattering, etc., the leading hard scattering QCD contribution at large momentum transfer $Q^2 = tu/s$ has the form[53] (helicity labels and suppressed) (see Fig. 16)

$$\mathcal{M}_{A+B\to C+D}(Q^2,\theta_{c.m.}) = \int [dx]\phi_C(x_c,\tilde{Q})\ \phi_D(x_d,\tilde{Q})\ T_H(x_i;Q^2,\theta_{c.m.}) \times \phi_A(x_a,\tilde{Q})\ \phi_B(x_b,\tilde{Q}) \quad . \tag{89}$$

The essential behavior of the amplitude is determined by T_H, computed where each hadron is replaced by its (collinear) quark constituents.

We note again that T_H is "collinear irreducible", i.e., the transverse momentum integrations of all reducible loop integration are restricted to $k_\perp^2 > 0(Q^2)$ since the small $k_\perp$ region is already contained in ϕ. If the internal propagators in T_H are all far-off-shell $0(Q^2)$ [as in Fig. 16(a)] then a perturbative expansion in $\alpha_s(Q^2)$ can be carried out. However, this is not true[60] for all hadron-hadron scattering amplitudes since one can have multiple quark-quark scattering processes which allow near-on-shell propagation in intermediate states at finite values of the x_i. The classic example is meson-meson scattering, where two pairs of quarks scatter through the same angle [see Fig. 16(c)]. However, the near-on-shell region of integration is again suppressed by Sudakov factors. (Physically this suppression occurs because the near-on-shell quarks must scatter without radiating gluons.) A model calculation by Mueller[63] for π-π scattering in QCD (using an exponentiated form of the Sudakov form factor) shows that the leading contribution comes in fact from the off-shell region $|k^2| \sim 0(Q^2)^{1-\varepsilon}$ where $\varepsilon = (2c+1)^{-1}$, $c = 8C_F/(11 - 2/3\, n_f)$ (for four flavors $\varepsilon \cong 0.281$). This region gives the contribution

$$\mathcal{M}_{\pi\pi\to\pi\pi} \sim 0(Q^2)^{-3/2-c\ell n(2c+1/2c)}$$
$$\cong (Q^2)^{-1.922} \qquad (90)$$

compared to $(Q^2)^{-2}$ from the hard scattering $|k^2| \sim 0(Q^2)$ region.

Thus, even when pinch singularities are present, the far-off-shell hard scattering quark and gluon processes dominate large momentum transfer hadron scattering amplitudes. Given this result, we can abstract some general QCD features common to all exclusive processes at large momentum transfer:

i) All of the non-perturbative bound state physics is isolated in the process-independent distribution amplitudes.

ii) The nominal power-law behavior of an exchange amplitude is $(1/Q)^{n-4}$ where n is the number of external elementary particles (quarks, gluons, leptons, photons in T_H). This immediately implies the dimensional counting rules given in (10) and (11). These power-law predictions are modified by (a) the Q^2-dependence of the factors of α_s in T_H, (b) the Q^2-evolution of the distribution amplitudes and (c) a possible small power associated with the almost complete Sudakov suppression of pinch singularities in hadron-hadron scattering. The dimensional-counting rules appear to be experimentally well-established for a wide variety of processes (see Ref. 19 and Figs. 4,8,9,11).

iii) Since the distribution amplitudes ϕ_M and ϕ_B are $L_z = 0$ angular momentum projections of the hadronic wave functions, the sum of the quark spin along the hadron's momentum equals the hadron spin:[20]

$$\sum_{i\varepsilon H} S_i^z = S_H^z \quad . \tag{91}$$

(In contrast, in inclusive reactions there are any number of non-interacting quark and gluon spectators, so that the spin of the interacting constituents is only statistically related to the hadron spin—except possibly at the edge of phase-space $x \sim 1$.) Furthermore, since all propagators in T_H are hard, the quark and hadron masses can be neglected at large Q^2 up to corrections of order $\sim m/Q$. The vector gluon interactions conserve quark helicity when all masses are neglected. Thus total quark helicity is conserved in T_H at large Q^2. Combining this with (91), we have the QCD selection rule:

$$\sum_{\text{initial}} \lambda_H = \sum_{\text{final}} \lambda_H \tag{92}$$

i.e., total hadron helicity is conserved up to corrections of order $O(m/Q)$.

Hadron helicity conservation thus applies for all large momentum transfer exclusive amplitudes involving light meson and baryons. Notice that the photon spin is not important: QCD predicts that $\gamma p \to \pi p$ is proton helicity conserving at fixed $\theta_{c.m.}$, $s \to \infty$, independent of the photon polarization. Exclusive amplitudes which involve hadrons with quarks or gluons in higher orbital angular momentum states are also suppressed by powers of the momentum transfer. An important corollary of this rule is that helicity-flip form factors are suppressed, e.g.:

$$F_{2p}(Q^2)/F_1(Q^2) \sim O(m^2/Q^2) \quad . \tag{93}$$

The helicity rule, Eq. (92), is one of the most characteristic features of QCD, being a direct consequence of the gluon's spin. A scalar or tensor gluon-quark coupling flips the quark's helicity. Thus, for such theories, helicity may or may not be conserved in any given diagram contributing to T_H, depending upon the number of interactions involved. Only for a vector theory, like QCD, can we have a helicity selection rule valid to all orders in perturbation theory.

Two-Photon Processes[65]

One of the most important applications of perturbative QCD is to the two-photon processes $d\sigma/dt\ (\gamma\gamma \to M\overline{M})$, $M = \pi, K, \rho, \omega$ at large $s = (k_1 + k_2)^2$ and fixed $\theta_{c.m.}$. These reactions, which can be studied in $e^+e^- \to e^+e^- MM$ processes, provide a particularly important laboratory for testing QCD since these "Compton" processes are, by far, the simplest calculable large-angle exclusive hadronic scattering reactions. The large-momentum-transfer scaling behavior, the helicity structure, and often even the absolute normalization can be rigorously computed for each two-photon channel.

Conversely, the angular dependence of the $\gamma\gamma \to M\bar{M}$ amplitudes can be used to determine the shape of the process-independent meson "distribution amplitudes", $\phi_M(x,Q)$, the basic short-distance wave functions which control the valence quark distributions in high momentum transfer exclusive reactions. Detailed discussions are given in Ref. 65.

The Phenomenology of Hadronic Wave Functions[16]

Thus far, most of the phenomenological tests of QCD have focused on the dynamics of quark and gluon subprocesses in inclusive high momentum transfer reactions. The Fock state wave functions $\psi_n^K(x_i, \vec{k}_{\perp i};\lambda_i)$ which determine the dynamics of hadrons in terms of their quark and gluon degrees of freedom are also of fundamental importance. If these wave functions were accurately known then an extraordinary number of phenomena, including decay amplitudes, exclusive processes, higher twist contributions to inclusive phenomena, structure functions, and low transverse momentum phenomena (such as diffractive processes, leading particle production in hadron-hadron collisions and heavy flavor hadron production) could be interrelated. Conversely, these processes can provide phenomenological constraints on the Fock state wave functions which are important for understanding the dynamics of hadrons in QCD. In addition the structure of nuclear wave functions in QCD is essential for understanding the synthesis of nuclear physics phenomenology with QCD.

The central measures of the hadron wave functions are the distribution amplitudes

$$\phi(x_i,Q) = \int^Q [d^2k_\perp]\ \psi_V^Q(x_i,\vec{k}_{\perp i}) \tag{94}$$

which control high momentum transfer form factors and exclusive processes:

$$\mathcal{M} \cong \Pi\ \phi \otimes T_H \tag{95}$$

and the quark and gluon structure functions

$$G_{q/H}(x,Q) = \sum_n \int^Q [d^2k_\perp]\ [dx]\ |\psi_n^Q(x_i,k_{\perp i})|^2 \delta(x-x_q) \tag{96}$$

which control high momentum transfer inclusive reactions

$$d\sigma \cong \Pi\ G \otimes d\hat{\sigma} \tag{97}$$

A summary of the basic properties, logarithmic evolution, and power law behavior of these quantities is given in Table 2.

The decays of the pion $\pi^- \to \mu^-\nu$ and $\pi^0 \to 2\gamma$ place two important phenomenological constraints on the construction of hadronic wave

Table 2. Comparison of Exclusive and Inclusive Cross Sections

Exclusive Amplitudes	Inclusive Cross Sections
$\mathcal{M} \sim \Pi\ \phi(x_i,Q) \otimes T_H(x_i,Q)$	$d\sigma \sim \Pi\ G(x_a,Q) \otimes d\hat{\sigma}(x_a,Q)$
$\phi(x,Q) = \int^Q [d^2k_\perp]\ \psi^Q_{val}(x,k_\perp)$	$G(x,Q) = \sum_n \int^Q [d^2k_\perp][dx]' \lvert\psi^Q_n(x,k_\perp)\rvert^2$
Measure ϕ in $\gamma\gamma \to M\bar{M}$	Measure G in $\ell p \to \ell X$
$\sum_{i\varepsilon H} \lambda_i = \lambda_H$	$\sum_{i\varepsilon H} \lambda_i \neq \lambda_H$
EVOLUTION	
$\frac{\partial\phi(x,Q)}{\partial \log Q^2} = \alpha_s \int [dy] V(x,y)\ \phi(y)$	$\frac{\partial G(x,Q)}{\partial \log Q^2} = \alpha_s \int dy\ P(x/y) G(y)$
$\lim_{Q\to\infty} \phi(x,Q) = \prod_i x_i \cdot C_{flavor}$	$\lim_{Q\to\infty} G(x,Q) = \delta(x)\ C$
POWER LAW BEHAVIOR	
$\frac{d\sigma}{dx}(A+B \to C+D) \cong \frac{1}{s^{n-2}} f(\theta_{CM})$	$\frac{d\sigma}{d^2p/E}(AB \to CX) \cong \sum \frac{(1-x_T)^{2n_s-1}}{(Q^2)^{n_{act}-2}} f(\theta_{CM})$
$n = n_A + n_B + n_C + n_D$	$n_{act} = n_a + n_b + n_c + n_d$
T_H: expansion in $\alpha_s(Q^2)$	$d\hat{\sigma}$: expansion in $\alpha_s(Q^2)$
COMPLICATIONS	
End point singularities Pinch singularities High Fock states	Multiple scales Phase-space limits on evolution Heavy quark thresholds Higher twist multiparticle processes Initial and final state interactions

functions. If one assumes a simple ansatz for the pion's non-perturbative wave function

$$\psi_{q\bar{q}}^{(\kappa)}\ (x,k_\perp) = A\ e^{-b^2k_\perp^2/x(1-x)} \tag{98}$$

[where $\kappa \sim 0(\pm\ \text{GeV})$], then one finds the interesting result that the probability $P_{q\bar{q}/\pi}$ that pion is in its valence $q\bar{q}$ Fock state is less than 1/4. Furthermore, the radius for the pion's valence state is less than 0.42 fm, significantly smaller than that measured from the slope of the pion form factor: $R_\pi^{expt} \cong 0.7$ fm. These results are consistent if the fluctuations from the higher Fock states contribute a larger size to the average pion radius.

Similarly, constraints from the nucleon form factors, the $x \to 1$ behavior of the nucleon structure functions, and the decay rule $\psi \to p\bar{p}$ also suggest that the radius of the three quark valence Fock state of the nucleon is significantly smaller than that of the total proton state. Details may be found in Refs. 1,14,16,53,54.

QCD PREDICTIONS FOR THE DEUTERON FORM FACTOR AND THE DEUTERON DISTRIBUTION AMPLITUDE

As shown in the previous section, hadronic form factors in QCD at large momentum transfer $Q^2 = (q^0)^2 - \vec{q}^{\,2}$ can be written in a factorized form where all nonperturbative effects are incorporated into process-independent distribution amplitudes $\phi_H(x_i,Q)$, computed from the equal $\tau = t+z$, six-quark Fock state valence wave function at small relative quark transverse separation $b_\perp^i \sim 0(1/Q)$. The $x_i = (k^0+k^3)_i/(p^0+p^3)$ are the light-cone longitudinal momentum fractions with $\sum_{i=1}^n x_i = 1$. In the case of the deuteron, only the six-quark Fock state needs to be considered since in a physical gauge any additional quark or gluon forced to absorb large momentum transfer yields a power law suppressed contribution to the form factor. The deuteron form factor can then be written as a convolution (see Fig. 6),

$$F_d(Q^2) = \int_0^1 [dx][dy]\phi_d^\dagger(y,Q)\ T_H^{6q+\gamma^*\to 6q}(x,y,Q)\ \phi_d(x,Q)\ , \tag{99}$$

where the hard scattering amplitude

$$T_H^{6q+\gamma^*\to 6q} = \left[\frac{\alpha_s(Q^2)}{Q^2}\right]^5\ t(x,y)\ \Big(1+0\big(\alpha_s(Q^2)\big)\Big) \tag{100}$$

gives the probability amplitude for scattering six quarks collinear with the initial to the final deuteron momentum and

$$\phi_d(x_i,Q) \propto \int^{k_{\perp i}<Q} [d^2k_\perp]\psi_{qqq\ qqq}(x_i,\vec{k}_{\perp i}) \quad , \tag{101}$$

gives the probability amplitude for finding the quarks with longitudinal momentum fractions x_i in the deuteron wave function collinear up to the scale Q. Because the coupling of the gauge gluon is helicity-conserving and the fact that $\phi_d(x_i,Q)$ is the $L_z = 0$ projection of the deuteron wave function, hadron helicity is conserved[20] and to leading order in 1/Q the dominant form factor corresponds to $\sqrt{A(Q^2)}$; i.e., $h = h' = 0$.

The distribution amplitude $\phi_d(x_i,Q)$ is the basic deuteron wave function which controls high-momentum transfer exclusive reactions in QCD; it is the analog of the wave function at the origin in non-relativistic theory. The logarithmic Q^2-dependence of ϕ_d can be determined as we show below by an evolution equation computed from perturbative quark-quark scattering kernels at large momentum transfer, or equivalently, by the operator product expansion at short distances and the renormalization group.[17]

The QCD prediction for the leading helicity-zero deuteron form factor then has the form[53,66,67]

$$F_d(Q^2) = \left[\frac{\alpha_s(Q^2)}{Q^2}\right]^5 \sum_{n,m} d_{mn} \left(\ln \frac{Q^2}{\Lambda^2}\right)^{-\gamma_n^d-\gamma_m^d} \times$$
$$\times \left[1 + O\left(\alpha_s(Q^2), \frac{m}{Q}\right)\right] \quad , \tag{102}$$

where the main dependence $[\alpha_s(Q^2)/Q^2]^5$ comes from the hard-gluon exchange amplitude T_H [see Fig. 7(b)]. The anomalous dimensions γ_n^d are calculated from the evolution equations for $\phi_d(x_i,Q)$.

The evolution equation for six quark systems in which the constituents have the light-cone longitudinal momentum fractions $x_i (i = 1,2,\ldots,6)$ can be obtained from a generalization of the proton (three quark) case discussed in the previous section. A nontrivial extension is the calculation of the color factor, C_d, of six-quark systems which will be briefly discussed in the next paragraph. Since in leading order only pair-wise interactions, with transverse momentum Q, occur between quarks, the evolution equation for the six-quark system becomes

$$\prod_{k=1}^{6} x_k \left[\frac{\partial}{\partial \xi} + \frac{3C_F}{\beta}\right] \tilde{\phi}(x_i,Q) = -\frac{C_d}{\beta}\int_0^1 [dy]V(x_i,y_i)\tilde{\phi}(y_i,Q) \tag{103}$$

where the factor 3 in the square bracket comes from the renormalization of the six quark field. In Eq. (103) we have defined

$$\phi(x_i,Q) = \prod_{k=1}^{6} x_k \tilde{\phi}(x_i,Q) \quad . \tag{104}$$

The evolution is in the variable

$$\xi(Q^2) = \frac{\beta}{4\pi} \int_{Q_0^2}^{Q^2} \frac{dk^2}{k^2}\,\alpha_s(k^2) \sim \ln\left(\frac{\ln(Q^2/\Lambda^2)}{\ln(Q_0^2/\Lambda^2)}\right) \ , \tag{105}$$

where

$$[dy] = \delta\left(1 - \sum_{i=1}^{6} y_i\right) \prod_{i=1}^{6} dy_i \ , \tag{106}$$

$C_F = (n_c^2 - 1)/2n_c = 4/3$, $\beta = 11 - (2/3)n_f$, and n_f is the effective number of flavors. Also, by summing over interactions between quark pairs $\{i,j\}$ due to exchange of a single gluon, $V(x_i,y_i)$ is given by

$$V(x_i,y_i) = 2 \prod_{k=1}^{6} x_k \sum_{i\neq j} \theta(y_i - x_i) \prod_{\ell\neq i,j}^{6} \delta(x_\ell - y_\ell)$$
$$\times \frac{y_j}{x_j}\left(\frac{\delta_{h_i\bar{h}_j}}{x_i+x_j} + \frac{\Delta}{y_i-x_i}\right)$$
$$= V(y_i,x_i) \tag{107}$$

where $\delta_{h_i\bar{h}_j} = 1(0)$ when the constituents' $\{i,j\}$ helicities are anti-parallel (parallel). The infrared singularity at $x_i = y_i$ is cancelled by the factor

$$\Delta\tilde{\phi}(y_i,Q) = \tilde{\phi}(y_i,Q) - \tilde{\phi}(x_i,Q) \tag{108}$$

since the deuteron is a color singlet.

The six-quark bound states have five independent color components ($3\times3\times3\times3\times3\times3 \supset 1+1+1+1+1$). It can be shown in general that the color factor C_d is given by

$$C_d = \frac{1}{5} S^{\alpha}_{ijklmn} \left(\frac{\lambda_\alpha}{2}\right)^{i}_{i'} \left(\frac{\lambda_\alpha}{2}\right)^{j}_{j'} S_{\alpha}^{i'j'klmn} \ , \tag{109}$$

where $\lambda_\alpha(\alpha = 1,2,\ldots,8)$ are Gell-Mann matrices in $SU(3)^c$ group and $S^{\alpha}_{ijklmn}(\alpha = 1,2,\ldots,5)$ are the five independent color singlet representations. Here we shall focus on results for the leading contribution to the distribution amplitude and form factor at large Q. Since the leading eigensolution to the evolution equation (103) turns out to be completely symmetric in its orbital dependence, the dominant asymptotic deuteron wave function is fixed by overall anti-symmetry[68] to have spin-isospin symmetry $\{33\}_{TS}$ which is dual to its color symmetry $[222]_C$. Thus the coefficient for each c (and TS) component has equal weight:

$$\phi_{6q}\left([222]_C \times \{33\}_{TS}\right) = \frac{1}{\sqrt{5}} \sum_{\alpha=1}^{6} [222]_C^{\alpha} \{33\}_{TS}^{\alpha} (-1)^{\alpha} \quad . \tag{110}$$

Since the evolution potential is diagonal in isospin and spin, C_d is computed by the trace of the color representation. The color factor[66,68] is -2/3 for the color anti-symmetric pair $\{i,j\}$ and +1/3 for the color symmetric pair $\{i,j\}$. Since three color anti-symmetric pairs $\{i,j\}$ and two color symmetric pairs $\{i,j\}$ exist in this state, the color factor is[69]

$$C_d = \frac{1}{5}\left(-\frac{2}{3} \times 3 + \frac{1}{3} \times 2\right) = \frac{-C_F}{5} \quad . \tag{111}$$

To solve the evolution equation (100), we factorize the Q^2-dependence of $\tilde{\phi}(x_i,Q)$ as

$$\tilde{\phi}(x_i,Q) = \tilde{\phi}(x_i)e^{-\gamma\xi} = \tilde{\phi}(x_i)\left(\ln\frac{Q^2}{\Lambda^2}\right)^{-\gamma} \tag{112}$$

so that Eq. (100) becomes

$$\left(-\gamma + \frac{3C_F}{\beta}\right)\tilde{\phi}(x_i) = -\frac{C_d}{\beta}\int_0^1 [dy]\, V(x_i,y_i)\, \tilde{\phi}(y_i) \quad , \tag{113}$$

where the eigenvalues of γ will provide the anomalous dimensions γ_n. The general matrix representations of γ_n with bases $|\Pi_{i=1}^{5} x_i^{m_i} >$ are given in Ref. 66. Here we give only the leading anomalous dimension γ_o corresponding to the eigenfunction $\tilde{\phi}(x_i) = 1$. The result is

$$\gamma_o = \frac{3C_F}{\beta} + \frac{C_d}{\beta}\sum_{i\neq j}^{6} \delta_{h_i\bar{h}_j} \quad , \tag{114}$$

so that the asymptotically dominant result for the helicity zero deuteron is given by

$$\gamma_o = \frac{6}{5}\frac{C_F}{\beta} \quad . \tag{115}$$

In order to make more detailed and experimentally accessible predictions, we define the "reduced" nuclear form factor to remove the effects of nucleon compositeness:

$$f_d(Q^2) \equiv \frac{F_d(Q^2)}{F_p(Q^2/4)F_n(Q^2/4)} \quad . \tag{116}$$

The arguments for the proton and nucleon form factors (F_p and F_n) are $Q^2/4$, respectively, since in the limit of zero binding energy each nucleon must change momentum $Q/2$ due to the electromagnetic interaction. If we recall that the leading anomalous dimension of the nucleon distribution amplitude is $C_F/2\beta$, then the QCD prediction for the asymptotic Q^2-behavior of $f_d(Q^2)$ is

$$f_d(Q^2) \sim \frac{\alpha_s(Q^2)}{Q^2}\left(\ln \frac{Q^2}{\Lambda^2}\right)^{-\frac{2}{5}C_F/\beta} \tag{117}$$

where $-2C_F/5\beta = -8/145$ for $n_f = 2$.

Although the QCD prediction is for asymptotic momentum transfer, it is interesting to compare (117) directly with the available high Q^2 data[27] (see Fig. 8). In general one would expect corrections from higher twist effects (e.g., mass and $k_\perp$ smearing), higher order contributions in $\alpha_s(Q^2)$, as well as non-leading anomalous dimensions. However, the agreement of the data with simple $Q^2 f_d(Q^2) \sim$ *const* behavior for $Q^2 > 1/2$ GeV^2 implies that, unless there is a fortuitous cancellation, all of the scale-breaking effects are small, and the present QCD perturbation calculations are viable and applicable even in the nuclear physics domain. The lack of deviation from the QCD parameterization suggests that the parameter Λ in (117) is small. A comparison with a standard definition such as $\Lambda_{\overline{MS}}$ would require a calculation of next to leading effects. A more definitive check of QCD can be made by calculating the normalization of $f_d(Q^2)$ from T_H and the evolution of the deuteron wave function to short distances. It is also important to confirm experimentally that the $h = h' = 0$ form factor is indeed dominant.

We note that the deuteron wave function which contributes to the asymptotic limit of the form factor is the totally antisymmetric wave function corresponding to the orbital Young symmetry given by [6] and isospin (T) + spin (S) Young symmetry given by $\{33\}$. The deuteron state with this symmetry is related to the NN, $\Delta\Delta$, and hidden color (CC) physical bases, for both the (TS) = (01) and (10) cases, by the formula[67]

$$\psi_{[6]\{33\}} = \sqrt{\frac{1}{9}}\,\psi_{NN} + \sqrt{\frac{4}{45}}\,\psi_{\Delta\Delta} + \sqrt{\frac{4}{5}}\,\psi_{CC} \ . \tag{118}$$

Thus the physical deuteron state, which is mostly ψ_{NN} at large distance, must evolve to the $\psi_{[6]\{33\}}$ state when the six quark transverse separations $b^i_\perp \leq O(1/Q) \to 0$. Since this state is 80% hidden color, the deuteron wave function cannot be described by the traditional meson-nucleon degrees of freedom in this domain. The fact that the six-quark color singlet state inevitably evolves in QCD to a dominantly hidden-color configuration at small transverse separation also has implications for the form of the nucleon-nucleon potential, which can be considered as one interaction component in a coupled channel system. As the two nucleons approach each other, the system must do work in order to change the six-quark state to a dominantly hidden color configuration; i.e., QCD requires that the nucleon-nucleon potential must be repulsive at short distances.[12] The evolution equation (113) for the six-quark system suggests that the distance where this change occurs is in the domain where $\alpha_s(Q^2)$ most strongly varies.

FURTHER APPLICATIONS OF REDUCED NUCLEAR AMPLITUDES

One of the basic problems in the analysis of nuclear scattering amplitudes is how to consistently account for the effects of the composite structure of nucleons. In nuclear physics the traditional method of treating nucleon dynamics has been to use an effective meson-nucleon local Lagrangian field theory. However, this method is deficient for a number of reasons: (1) the wrong degrees of freedom are used, (2) neither the t^{-2} power-law fall-off of nucleon form factors nor the t^{-1} fall-off of pion form factors is naturally reproduced,[70] (3) nucleon pair terms are not correctly suppressed in intermediate states, and (4) a renormalizable (i.e., calculable) field theory of massive isovector mesons requires the full apparatus of non-Abelian gauge theories, including a spontaneous symmetry breaking mechanism. Models for nuclear scattering amplitudes based on the Born approximation and local meson-nucleon couplings have the wrong dynamical dependence in virtually every kinematical variable for composite hadrons. The inclusion of *ad hoc* form factors at each meson-nucleon or photon-nucleon vertex is clearly unsatisfactory since one must understand the off-shell dependence in each leg while retaining gauge invariance. None of these traditional methods have any real predictive power.

In principle, all nuclear scattering amplitudes could be calculated from quantum chromodynamics (QCD) in terms of the basic quark and gluon degrees of freedom. The method for computing large momentum transfer exclusive scattering amplitudes for hadrons and nuclei, starting with a Fock state wave function expansion on the light-cone (equal $\tau = t + z$), is discussed in the third section. At large momentum transfer one can readily derive QCD predictions for the leading fixed angle power-law scaling behavior and spin structure of hadronic and nuclear scattering matrix elements. However, the explicit evaluation of the multiquark and gluon hard scattering amplitudes needed for predicting the normalization and angular dependence for a nuclear process, even at leading order in α_s, requires the consideration of millions of Feynman diagrams. Beyond leading order one must include contributions of non-valence Fock states, wave function and binding corrections, and a rapidly expanding number of radiative corrections and loop diagrams.

In this section we extend the definition of reduced nuclear amplitudes in order to produce a simple method for identifying the dynamical effects of nucleon substructure, consistent with QCD and covariance. Although this technique cannot replace a full QCD calculation, it does provide a basis for constructing models for "reduced" nuclear scattering amplitudes consistent with QCD scaling laws and gauge invariance.

The basic idea has already been introduced in the second section. In general, a form factor $F(Q^2 = -q^2)$ is the probability amplitude that the target remains intact after absorbing four-momentum q. To the extent that we can neglect its binding energy, the deuteron can be represented as two nucleons, each with an equal portion of the nuclear momentum. Therefore the deuteron form factor contains the probability that each nucleon remains intact after absorbing one-half of the momentum transfer. We thus define the "reduced" deuteron form factor[9]

$$f_d(Q^2) = \frac{F_d(Q^2)}{F_p(Q^2/4)\ F_n(Q^2/4)} \tag{119}$$

which effectively removes the fall-off of the measured form factor due to the internal degrees of freedom of the nucleons. It is defined separately for each helicity form factor.

The reduced form factor must still be a decreasing function of Q^2 since it still contains the probability that the scattered nucleons reform into the ground state deuteron. An important prediction of QCD is that, modulo logarithmic factors[66] that come from the running coupling constant and anomalous dimensions of the hadronic distribution amplitudes, the large Q^2 behavior is

$$f_d(Q^2) \sim \frac{\text{const.}}{Q^2} \quad . \tag{120}$$

Thus the reduced deuteron form factor and meson form factors (for helicity $\lambda = 0$ to $\lambda' = 0$) have the identical (monopole) scaling law. After removing the nucleon form factors, the nucleons are effectively reduced to point-like spin 1/2 fermions, so the reduced deuteron and meson form factors have the same dimensional scaling behavior, basically the slowest possible for two-particle composites. Similarly, if one defines for $A = 3$:

$$f_{^3He}(Q^2) = \frac{F_{^3He}(Q^2)}{[F_N(Q^2/9)]^3} \ , \quad (\lambda = 1/2 \text{ to } \lambda' = 1/2) \tag{121}$$

then QCD predicts that the reduced ^3He (and triton) form factor scales at large Q^2 in the same way as a nucleon form factor:

$$f_{^3He}(Q^2) \sim F_N(Q^2) \sim (1/Q^2)^2 \quad . \tag{122}$$

Comparisons of the data[27,71] with the QCD predictions are given in Figs. 8 and 9.

We can go beyond the case of nuclear form factors and define reduced nuclear scattering amplitudes in general. If we consider a generic process with amplitude $\mathcal{M}(s,t)$ that involves A ingoing and outgoing nucleons and transfers, in the zero binding limit, momentum q_i to nucleon i, then the reduced amplitude is defined as

$$m(s,t) = \mathcal{M}(s,t) \left[\prod_{i=1}^{A} F_N(\hat{t}_i = q_i^2) \right]^{-1} \quad . \tag{123}$$

For example, the reduced amplitude for the photo- (or electro-) disintegration of the deuteron would be written as

$$m_{\gamma d \to np} = \frac{\mathcal{M}_{\gamma d \to np}}{F_n(\hat{t}_n)\ F_p(\hat{t}_p)} \tag{124}$$

where

$$\hat{t}_n = \left(p_n - \frac{1}{2}\, p_d \right)^2 \quad , \tag{125a}$$

$$\hat{t}_p = \left(p_p - \frac{1}{2}\, p_d \right)^2 \quad , \tag{125b}$$

with p_n, p_p and p_d the momenta of the neutron, proton and deuteron, respectively.

The nominal fixed-angle scaling behavior of the reduced amplitude is predicted by dimensional counting rules.[22] Modulo logarithms they give

$$m \sim p_T^{4-n}\ f\left(\frac{\text{invariants}}{s}\right) \tag{126}$$

where $p_T^2 = tu/s$ is the transverse momentum and n is the number of "elementary" fields in the external state (ingoing and outgoing photons, leptons, gluons, quarks or reduced nucleons). Thus for deuteron photodisintegration the reduced amplitude scales as

$$m_{\gamma d \to np} \sim p_T^{-1}\ f(\theta_{c.m.}) \quad , \tag{127}$$

the angle $\theta_{c.m.}$ being that of the proton direction with respect to the beam direction in the c.m. frame. This is the same QCD scaling as that for $\mathcal{M}_{\gamma M \to q_1 \bar{q}_2}$; here M is a meson with constituents q_1 and $\overline{q_2}$.

We can motivate the definition of the reduced amplitude by returning to the basic definition of hadronic matrix elements in τ-ordered perturbation theory:

$$\mathcal{M} = \int \Pi[dx][d^2k_\perp]\ \psi'(x_f,k_{\perp f})\ T(x_f,x_i;k_{\perp f},k_{\perp i})\ \psi(x_i,k_{\perp i}) \quad (128)$$

where the ψ are the equal $\tau = t + z$ wave functions and T is the momentum-space quark-gluon scattering amplitude. A sum over the Fock state amplitudes and quark and gluon helicities is understood. In the zero nuclear binding energy limit the nuclear Fock state wave function reduces to the product of wave functions for collinear nucleons with the nuclear momentum partitioned among the nucleons in proportion to each nucleon mass. Thus one is evidently neglecting corrections of order $2m_N\Delta\varepsilon_{BE}/\mu^2$ where m_N is the nuclear mass, $\Delta\varepsilon_{BE}$ the nuclear binding energy and μ^2 a hadronic scale parameter, as well as contributions from higher Fock states in the nucleus, e.g. the hidden-color six-quark configurations.

At this stage of approximation one must compute the corresponding multi-nucleon scattering amplitude, e.g., the amplitude for the elastic electron-deuteron scattering process

$$e + p\left(\frac{1}{2}p\right) + n\left(\frac{1}{2}p\right) \rightarrow e' + p'\left(\frac{1}{2}p + \frac{1}{2}q\right) + n'\left(\frac{1}{2}p + \frac{1}{2}q\right). \quad (129)$$

If the momentum transfer occurs rapidly compared to the scale of hadronic binding then one can argue (as in the Chou-Yang model of elastic scattering)[72] that the probability amplitude for transferring the required momentum $\hat{t}_i$ to each nucleon is proportional to its elastic form factor. Since Sudakov effects always suppress near on-shell (long-distance) momentum transfer mechanisms from pinch singularities and endpoint regions of phase space, one can argue that large momentum transfer is always local in QCD. Thus this assumption is justified, with corrections of order μ^2/q^2. A specific diagram which explicitly exhibits the factorization intrinsic to the reduced deuteron form factor is shown in Fig. 7(b).

As an application of nuclear amplitude reduction, we consider deuteron disintegration. The reduced amplitude is defined in (124). Both the scaling behavior (127) and a model for the angular dependence are discussed below.

Some other processes that might be profitably treated with this reduction method are $pp \rightarrow d\pi^+$,[73] $pd \rightarrow {}^3H\pi^+$ and $\pi^{\pm}d \rightarrow \pi^{\pm}d$. The reduced amplitudes have the same QCD scaling behavior as the amplitude for $q\bar{q} \rightarrow M\pi$, $qqq \rightarrow B\pi$ and $\pi M \rightarrow \pi M$, respectively, where B represents a baryon. From (126) the scaling is

$$m_{pp\rightarrow d\pi^+} \sim p_T^{-2}\ f(t/s)\ ,$$

$$m_{pd\rightarrow {}^3H\pi^+} \sim p_T^{-4}\ f(t/s)\ ,$$

$$m_{\pi d\rightarrow \pi d} \sim p_T^{-4}\ f(t/s)\ . \quad (130)$$

The asymptotic scaling law (127) is a remarkably simple form. The scaling holds for the hadron helicity conserving amplitude with $\lambda_n + \lambda_p = \lambda_d$, independent of the photon helicity. Amplitudes with $\lambda_n + \lambda_p \neq \lambda_d$ should be suppressed by a power of μ^2/p_T^2. One could hope that the simple scaling

$$p_T\, m_{\gamma d \to np} \simeq \text{const.} \tag{131}$$

at fixed $\theta_{c.m.}$ will hold for $p_T^2 \geq 1\ \text{GeV}^2$ since the scaling (120) (see Fig. 8), begins in this region. In terms of the differential cross section the prediction is[69]

$$\left.\frac{d\sigma}{d\Omega_{c.m.}}\right|_{\gamma d \to np} \sim \frac{1}{s - m_d^2}\, F_p^2(\hat{t}_p)\, F_n^2(\hat{t}_n)\, \frac{1}{p_T^2}\, f^2(\theta_{c.m.}) \quad . \tag{132}$$

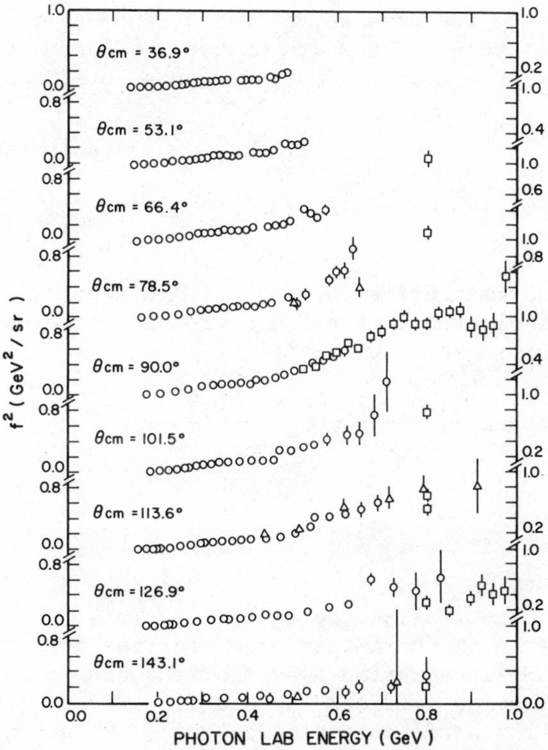

Fig. 17. Comparison of deuteron photodisintegration data with the prediction Eq. (135). The angle $\theta_{c.m.}$ is that of the proton direction with respect to the beam in the c.m. frame. The predicted scaling requires $f^2(\theta_{c.m.})$ to be independent of energy at any fixed angle. The data are from Ref. 74.

A comparison of this form with present low energy data[74] is shown in Fig. 17. The form factors were computed from the usual dipole formula

$$F_N(\hat{t}_i) = \frac{\text{const.}}{(1 - \hat{t}_i/0.71\ \text{GeV}^2)^2} \quad . \tag{133}$$

Although the results are encouraging, the available energies are too low to make a detailed check of the prediction.

We have not yet specified the form of f^2; however, it is easy to construct a model for the reduced amplitude which is gauge-invariant and has the correct helicity and scaling form. As a prototype for the reduced amplitude we shall use the amplitude for the photodisintegration of a polarized meson M into its constituent quarks q_1 and $\bar{q}_2$ (see Fig. 18), based on the lowest order QCD tree diagrams. An actual calculation of the hard-scattering amplitude for $\gamma d \to pn$ includes a coherent sum of such amplitudes with varied charge assignments and additional gluon lines attached. For the model, the charge assignments, e_1 for q_1 and $-e_2$ for $\bar{q}_2$, can be varied as parameters. The quark masses are taken to be zero. A computation of the squared amplitude summed over final spins then gives[75]

$$f^2(\theta_{c.m.}) = N \frac{(ue_1 + te_2)^2}{tu} \begin{cases} 1\ , & \text{transverse} \\ \dfrac{t^2+u^2}{4s^2}\ , & \text{longitudinal} \end{cases} \tag{134}$$

where N is a normalization constant with dimensions GeV^2/srad and "transverse" indicates an average over the two possible helicities. In the limit of $\sqrt{s} \gg m_d$ we find

$$f^2(\theta_{c.m.}) \simeq N \frac{[(2e_1 - 1) + \cos\theta_{c.m.}]^2}{1 - \cos^2\theta_{c.m.}} \times$$

$$\times \begin{cases} 1\ , & \text{transverse} \\ \dfrac{1}{8}\,(1 + \cos^2\theta_{c.m.})\ , & \text{longitudinal} \end{cases} \tag{135}$$

with the charges normalized by $e_1 - e_2 = 1$. This, when combined with (132), provides a one-parameter model for the asymptotic behavior of deuteron photodisintegration away from the beam axis. The actual angular distribution predicted by QCD from the coherent sum over the many contributing hard-scattering diagrams for six quarks is undoubtedly more complicated than that given by the above model. Nevertheless Eq. (135) should be representative of the scaling and functional dependence predicted by QCD for the reduced photodisintegration amplitude.

The simple model given in (135) makes apparent the need for data

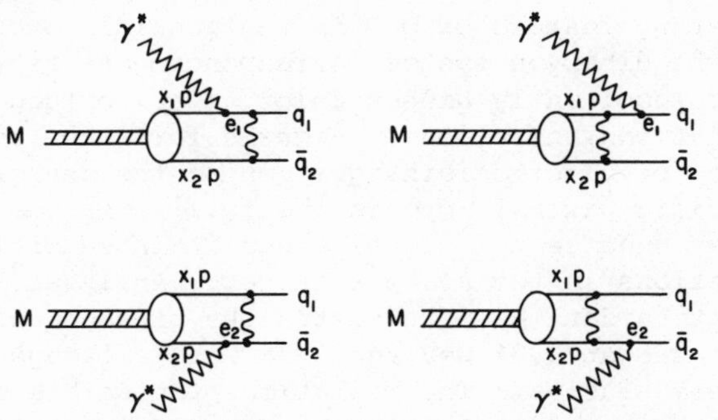

Fig. 18. Lowest order QCD diagrams for $\gamma^* M \to q_1\bar{q}_2$ where M is a bound state of q_1 and $\bar{q}_2$.

at higher energies. The points plotted in Fig. 19 were extracted by inspection from the data in Fig. 17 under the assumption that scaling had begun. The error bars reflect the range of values that would be consistent with the data. The empirical form $\sin^4\theta_{c.m.}$ fits the points fairly well but does not agree with the angular distribution predicted in (135). If the $\gamma M \to q\bar{q}$ model is a good guide, then a sign that experimental energies are approaching the true scaling limit would be that the value of $f^2(\theta_{c.m.})$ near the backward or forward direction has become large relative to the values at wider angles.

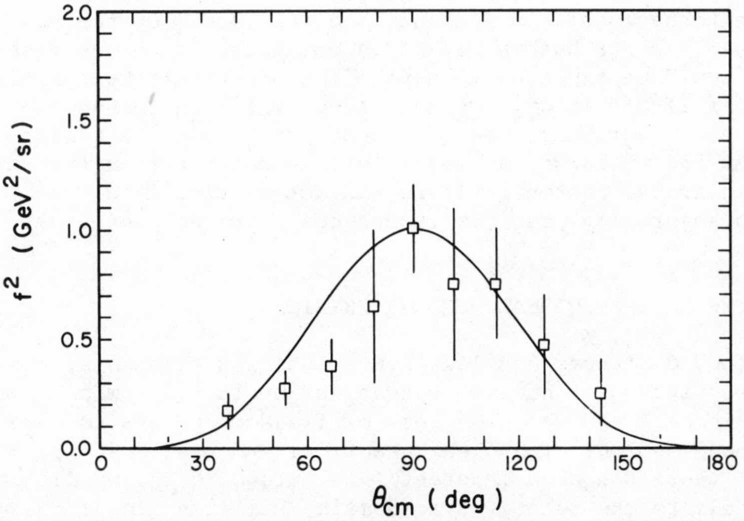

Fig. 19. Values of $f^2(\theta_{c.m.})$ extracted by inspection from the data of Ref. 74 with the assumption that scaling has begun in each data set. The solid line represents $\sin^4\theta_{c.m.}$ which was chosen empirically to summarize the extracted values.

An interesting feature of QCD is the possible occurrence of resonances in the dibaryon system corresponding to six-quark Fock states which are dominantly hidden color, i.e., orthogonal to the usual n-p and Δ-Δ configurations. Signals for such resonances could appear in photo- or electrodisintegration of the deuteron at fixed $\hat{s} = M^2$ in a specific partial wave in the full amplitude. The virtual photon probe may enhance the signal since it is sensitive to off-shell configurations in the nuclear target. Analyses[76] of deuteron photodisintegration data have suggested the presence of dibaryon resonances with masses at 2.26 GeV and 2.38 GeV, although definitive results have been elusive. The isolation of possible dibaryon contributions from the hard-scattering background is clearly interesting and important. It would be useful to have a specific model of the hard-scattering continuum since this would permit a more precise separation of the resonance and background contributions. Given the correct kinematic regime, the reduced amplitude technique leads directly to just such a model

$$m_{BG} \sim p_T^{-2} \, f\left(\frac{\text{invariants}}{s}\right) \quad . \tag{136}$$

As a model for the reduced electrodisintegration amplitude we suggest the natural extension of the model for photodisintegration, that is the electrodisintegration of a polarized meson into its constituent quarks.

In general, one would expect the dibaryon resonance and the continuum hard-scattering contributions to the electroproduction amplitude to have quite different q^2 dependence. On the one hand, the resonance contribution, if it is dominated by soft hadronic physics, would be expected to have a characteristic vector meson-dominated fall-off in q^2: $\mathcal{M}_{DB} \propto (1 - q^2/m_V^2)^{-1}$ independent of p_T^2. While on the other hand, the q^2 dependence of $\mathcal{M}_{BG}$ is minimal for $|q^2| \ll p_T^2$ and p_T^2 large, at least for the contribution from transversely polarized photons. These characteristics in q^2 should be useful in separating possible resonances from the continuum.

LIMITATIONS OF TRADITIONAL NUCLEAR PHYSICS

At long distances and low, non-relativistic momenta, the traditional description of nuclear forces and nuclear dynamics based on nucleon, isobar, and meson degrees of freedom appears to give a viable phenomenology of nuclear reactions and spectroscopy. More recently, there has been interest[77] in extending the predictions of these models to the relativistic domain, utilizing local meson-nucleon field theories in order to represent the basic nuclear dynamics, as well as using an effective Dirac equation to describe the propagation of nucleons in nuclear matter. An interesting question is whether such an approach can be derived as a "correspondence" limit of QCD, at least in the low momentum transfer ($Q^2R_p^2 << 1$) and low excitation energy domain ($M\nu << M'^2 - M^2$).

The dynamical transition from nucleon/meson to quark/gluon degrees of freedom can be understood intuitively in the case of inelastic electron scattering. According to QCD, the carriers of the electromagnetic current within hadrons and nuclei at any scale are the quark fields. At low momentum transfer, coherence between the contributing quark amplitudes together with pole dominance, can be effectively measured in terms of propagating mesons and nucleons. However, as the momentum transfer and production energy are raised, this coherence is destroyed: the electron effectively scatters incoherently on each quark, yielding Bjorken scaling and the familiar QCD logarithmic corrections due to gluon bremsstrahlung.

Nevertheless, as we have discussed in previous sections, the existence of hidden-color Fock state components in the nuclear state precludes an exact treatment of nuclear properties based on meson-nucleon-isobar degrees of freedom: these hadronic degrees of freedom do not form a complete basis in QCD. It is likely that the hidden color states give less than a few percent correction to the global properties of nuclei; nevertheless the total energy of each nuclear state is evidently lowered by the hidden-color degrees of freedom: since extra degrees of freedom lower the energy of a system it is even conceivable that the deuteron would be unbound were it not for its hidden color components!

A more serious phenomenological question is whether it is possible—even in principle—to represent composite systems such as mesons and baryons as local fields in a Lagrangian field theory, at least for sufficiently long wavelengths such that internal structure of the hadrons cannot be discerned. Here we will outline a method to construct an effective Lagrangian of this sort.[78] First, consider the ultraviolet-regulated QCD Lagrangian density $\mathcal{L}^{\kappa}_{QCD}$ defined such that all internal loops in the perturbative expansion are cut off below a given momentum scale κ. Normally κ is chosen to be much larger than all relevant physical scales. Because QCD is renormalizable, $\mathcal{L}^{\kappa}_{QCD}$ is form-invariant under changes of κ provided that the coupling constant $\alpha_s(\kappa^2)$ and quark mass parameter $m(\kappa^2)$ are appropriately defined. However, if we insist on choosing the cutoff κ to be comparable to an hadronic scale then extra (higher twist) contributions will be generated in the effective Lagrangian:

$$\mathcal{L}^{\kappa}_{QCD} = \mathcal{L}^{\kappa}_{0} + \frac{em(\kappa)}{\kappa^2}\,\overline{\psi}\sigma_{\mu\nu}\,\partial^{\mu}\,\psi A^{em}_{\nu} + e\,\frac{f_{\pi}^2}{\kappa^2}\,\phi_{\pi}\,\overleftrightarrow{\partial}^{\mu}\,\phi_{\pi}\,A^{em}_{\mu}$$
$$+ e\,\frac{f_p^2}{\kappa^2}\,\overline{\psi}_N\,\gamma^{\mu}\,\psi_N\,A^{em}_{\mu} + \frac{f_p^2\,f_{\pi}}{\kappa^6}\,\partial_{\mu}\,\overline{\psi}_N\,\gamma_5\,\gamma^{\mu}\,\psi_N\,\phi_{\pi}$$
$$+ \ldots \qquad (137)$$

where $\mathcal{L}_0^\kappa$ is the standard contribution and the "higher twist" terms of order κ^{-2}, κ^{-4}, ... are schematic representations of the quark Pauli form factor, the pion and nucleon Dirac form factors, and the π-N-N coupling. The pion and nucleon fields ϕ_π and ψ_N represent composite operators constructed and normalized from the valence Fock amplitudes and the leading interpolating quark operators. One can use Eq. (137) to estimate the effective asymptotic power law behaviors of the couplings, e.g., $F^{\text{quark}}_{\text{Pauli}} \sim 1/Q^2$, $F_\pi \sim f_\pi^2/Q^2$, $G_M \sim f_p^2/Q^4$ and the effective $\pi\, \overline{N}\, \gamma_5\, N\, F_{\pi N\overline{N}}$ coupling: $F_{\pi N\overline{N}}(Q^2) \sim M_N f_p^2 f_\pi/Q^6$. The net pion exchange amplitude for NN-NN scatterings thus falls off very rapidly at large momentum transfer $M^\pi_{NN\to NN} \sim (Q^2)^{-7}$, much faster than the leading quark interchange amplitude $M^{q\overline{q}}_{NN\to NN} \sim (Q^2)^{-4}$. Similarly, the vector exchange contributions give contributions $M^\rho_{NN\to NN} \sim (Q^2)^{-6}$.

Thus, meson exchange amplitudes and currents, even summed over their excited spectra do not contribute to the leading asymptotic behavior of the N-N scattering amplitudes or deuteron form factors, [see Fig. 10(b)], once proper account is taken of the off-shell form factors which control the meson-nucleon-nucleon vertices.

Aside from such estimates, the effective Lagrangian (137) only has utility as a rough tree graph approximation; in higher order the hadronic field terms give loop integrals highly sensitive to the ultraviolet cutoff because of their non-renormalizable character. Thus an effective meson-nucleon Lagrangian only serves to organize and catalog low energy constraints and effective couplings. It is not reliable for obtaining the actual dynamical and off-shell behavior of hadronic amplitudes due to internal quark and gluon structure.

Local Lagrangians field theories for systems which are intrinsically composite are also misleading in another respect. Consider the low-energy theorem for the forward Compton amplitude on a (spin-average) nucleon target[57]

$$\lim_{\nu\to 0} m_{\gamma p\to\gamma p'}(\nu, t=0) = -2\hat{\varepsilon}\cdot\hat{\varepsilon}'\,\frac{e^2}{M_p} \quad . \tag{138}$$

One can directly derive this result from the underlying quark currents as indicated in Fig. 20(a). On the other hand, if one assumes the nucleon is a local field, then the entire contribution to the Compton amplitude at $\nu = 0$ arises from the nucleon pair z-graph amplitude, as indicated in Fig. 20(b). Since both calculations are Lorentz and gauge invariant, both give the desired result (138). However, in actuality, the nucleon is composite and the $N\overline{N}$ pair term is strongly suppressed (see Fig. 21): each $\gamma p\overline{p}$ vertex is proportional to

$$\langle 0|J^\mu(0)|p\overline{p}\rangle \propto F_p(Q^2 = 4M_p^2) \quad ; \tag{139}$$

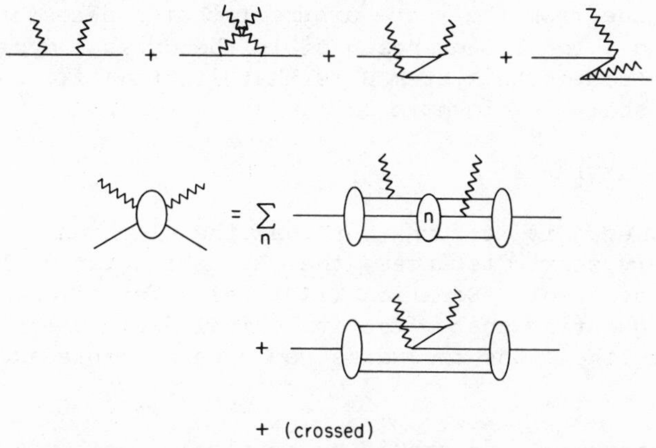

Fig. 20. (a) General calculation of the Compton amplitude for a composite target in time-ordered perturbation theory. (b) Calculation of the Compton amplitude from tree graphs in a local Lagrangian theory. Only the "z-graphs" contribute in the zero energy limit.

i.e., the timelike form factor as determined from $e^+e^- \to p\bar{p}$ near threshold. Thus, as expected physically, the $N\bar{N}$ pair contribution is highly suppressed for a composite system (even for real photons). Clearly a Lagrangian based on local nucleon fields gives an inaccurate description of the actual dynamics and cannot be trusted away from the forward scattering, low energy point. Such considerations become even more graphic if one imagines using a local field to represent a uranium nucleus.

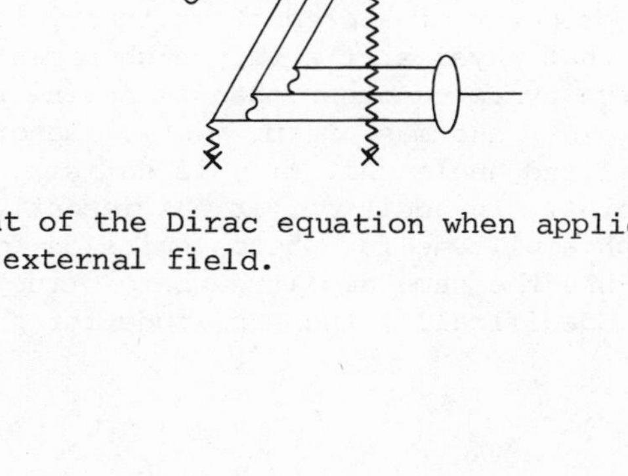

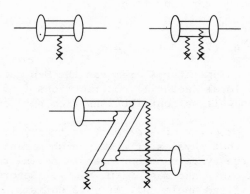

Fig. 21. Content of the Dirac equation when applied to a nucleon in an external field.

We can see from the above argument that a necessary condition for utilizing a local Lagrangian field theory as a dynamical approximation to a composite system H is that its timelike form factor at its Compton scale is close to 1:

$$F_H(q^2 \cong 4M^2) \cong 1 \quad . \tag{140}$$

For example, even if it turns out that the electron is a composite system at very short distances, the QED Lagrangian will still be a highly accurate tool. Equation (140) fails for all hadrons, save the pion, suggesting that effective chiral field theories which couple point-like pions to quarks could be a viable approximation to QCD.

More generally, one should be critical of all uses of point-like couplings involving nucleon-antinucleon pair production, e.g. in calculations of deuteron form factors, photo- and electro-distintegration. The N-$\overline{N}$ pair contributions are always suppressed by the timelike nucleon form factor. [See Fig. 10(c)]. Note: $\gamma N\overline{N}$ point-like couplings are not needed for gauge invariance, once all quark current contributions including the pointlike $q\overline{q}$ pair terms are taken into account.

We also note that a relativistic composite fermionic system, whether it is a nucleon or nucleus, does not obey the usual Dirac equation—with a momentum-independent potential—at least beyond first Born approximation. Again, the difficulty concerns intermediate states containing $N\overline{N}$ pair terms: the Dirac equation requires that $\langle p|V_{ext}|p'\rangle$ and $\langle 0|V_{ext}|p'\overline{p}\rangle$ are related by simple crossing, as in electrodynamics. In reality the pair production terms are suppressed by the timelike form factor.

It is possible that one can write an effective, approximate relativistic equation for a nucleon in an external potential of the form

$$(\vec{\alpha}\cdot\vec{p} + \beta m_N + \Lambda_+ \ V_{eff}\ \Lambda_+)\psi_N = E\psi_N \tag{141}$$

where the projection operator Λ_+ removes the N-$\overline{N}$ pair terms, and V^{eff} includes the local (seagull) contributions from $q\overline{q}$-pair intermediate states, as well as contributions from nucleon excitation.

An essential property of a predictive theory is its renormalizability, the fact that physics at a very high momentum scale $k^2 > \kappa^2$ has no effect on the dynamics other than to define the effective coupling constant $\alpha(\kappa^2)$ and mass terms $m(\kappa^2)$. Renormalizability also implies that fixed angle unitarity is satisfied at the tree-graph (no-loop) level. In addition, it has recently been shown[79] that the tree graph amplitude for photon emission for any renormalizable gauge theory has the same amplitude zero structure as classical electrodynamics. Specifically, the amplitude for photon emission during the scattering of any number of charged particles computed from tree graphs vanishes (independent of spin) in the kinematic region where the ratios $Q_i/p_i \cdot k$ for all the external charged lines are identical. This "null zone" is not restricted to soft photon momentum, and is identical to the kinematic domain for the destructive interference of the radiation associated with classical electromagnetic currents of the external charged particles. Thus the tree graph structure of gauge theories, in which each elementary charged field has zero anomalous moment ($g = 2$) is properly consistent with the classical ($\hbar = 0$) limit. On the other hand, local field theories which couple particles with non-zero anomalous moments, such as an effective meson-nucleon field theory, violate fixed angle unitarity and the above classical correspondence limit at the tree graph level. The anomalous moment of the nucleon is clearly a property of its internal quantum structure; by itself this precludes the representation of the nucleon as a local field.

The essential conflict between quark and meson-nucleon field theory is thus as a very basic level: because of Lorentz invariance a conserved charge must be carried by a local (point-like) current; there is no consistent relativistic theory where fundamental constituent nucleon fields have an extended charge structure.

CONCLUSIONS

QCD could be the theory of strong and nuclear interactions in the same sense that QED accounts for electrodynamic interactions. QCD solves many crucial problems: the meson and baryon spectra, quark statistics, the structure of the weak and electromagnetic currents of hadrons, the scale-invariance of interactions at short distances, and most likely, color (i.e., quark and gluon) confinement at large distances. Many different and diverse tests have confirmed the basic features of QCD, although the fact that the tests of quark and gluon interactions must be done within the confines of hadrons, as well as various technical difficulties, have prevented truly quantitative confirmation of the theory. The structure of the theory satisfies all prerequisites of elegance and beauty.

The synthesis of nuclear dynamics with the quark and gluon processes of quantum chromodynamics is clearly a fascinating fundamental problem in hadron physics.[80] The short distance behavior of the nucleon-nucleon interaction, determined by QCD, must join smoothly and analytically with the large distance constraints of nuclear physics.[7,81] As we have emphasized here, the fundamental mass scale of QCD is comparable with the inverse nuclear radius; it is thus difficult to argue that nuclear physics at distances below $\sim$1 fm can be studied in isolation from QCD. The meson nucleon degrees of freedom of traditional nuclear physics models become inadequate at the kinetic energy and momentum transfer scales $\gtrsim$100 MeV where nucleon substructure becomes evident.

Thus the essential question for nuclear as well as particle physics is to understand the transition between the meson-nucleon and quark-gluon degrees of freedom. There should be no illusion that this is a simple task; one is dealing with all the complexities and fascinations of QCD such as the effects of confinement and non-perturbative effects intrinsic to the non-Abelian theory. There is also fascinating physics associated with the propagation of quarks and gluons in nuclear matter and the phenomenology of hadron and nuclear wave functions. On the other hand, aside from the long-distance non-relativistic domain, models based on meson-nucleon degrees of freedom give the wrong dynamical dependence in virtually every off-shell variable.

Despite these difficulties there is reason for optimism that "nuclear chromodynamics" is a viable endeavor. For example, as we have shown in the fourth section we can use QCD to make predictions for the short distance behavior of the deuteron wave function and the deuteron form factor at large momentum transfer. The predictions give a remarkably accurate description of the scaling behavior of the available deuteron form factor data for Q^2 as low as 1 GeV^2. Asymptotically, the form factor is controlled by the 6 quark Fock state amplitude of the deuteron at small relative separation which is 80% "hidden color"; thus if QCD is correct, the traditional n-p, Δ-Δ and meson degrees of freedom can never correctly reproduce the deuteron form factor at large momentum transfers.[7,12] The QCD approach also allows the definition of "reduced" nuclear amplitudes which can be used to consistently and covariantly remove the effect of nucleon compositeness from nuclear amplitudes.[9,11]

An important feature of the QCD predictions is that they provide rigorous constraints on exclusive nuclear amplitudes which have the correct analytic, gauge-invariant, and scaling properties predicted by QCD at short distances. This suggests the construction of model amplitudes which simultaneously satisfy low energy and chiral theorems at low momentum transfer as well as the rigorous QCD constraints at high momentum transfers.[80,81] In addition, by using light cone quantization, one can obtain a consistent relativistic Fock state wave function description of hadrons and nuclei which ties on to the Schrödinger theory in the non-relativistic regime. One can also be encouraged by progress in non-perturbative methods in QCD such as lattice gauge theories, eventually these approaches should be able to deal with multi-quark source problems. The recent EMC data[48] showing breakdown of simple nucleon additivity in the nuclear structure functions also demonstrates that there is non-trivial nuclear physics even in the high momentum transfer domain.

It is essential to have direct experimental guidance in how to proceed as one develops nuclear chromodynamics. A high duty factor electron accelerator[49] with laboratory energy beyond 4 GeV is an important tool because of the simplicity of the probe and the fact that we understand the coupling of the electron to the quark current in QCD. It is also clear that:

i) One must have sufficient energy to extend electron scattering measurements from low momentum transfer to the high momentum transfer regime with sufficient production energy such that Bjorken scaling can be observed. One certainly does not want to stop at an intermediate momentum transfer domain—a regime of maximal complexity from the standpoint of both QCD and nuclear physics.

ii) One must have sufficient electron energy to separate the longitudinal and transverse currents. The σ_L/σ_T separation is essential for resolving individual dynamical mechanisms; e.g. single quark and multiple quark (meson current) contributions.

iii) One wishes to study each exclusive channel in detail in order to verify and understand the emergence of QCD scaling laws and to understand how the various channels combine together to yield effective Bjorken scaling. Helicity information is also very valuable. For example QCD predicts that at large momentum transfer, the helicity-0 to helicity-0 deuteron form factor is dominant and that for any large momentum transfer reaction, total hadronic helicity is conserved.

iv) One wishes to make a viable search for dibaryon states which are dominantly of hidden color. The argument that such resonances exist in QCD is compelling—just from counting of degrees of freedom. The calculation of the mass and width of such resonances is clearly very difficult, since the detailed dynamics is dependent on the degree of mixing with ordinary states, the availability of decay channels, etc. Since hidden color states have suppressed overlap with the usual hadron amplitudes it may be quite difficult to find such states in ordinary hadronic collisions. On the other hand, the virtual photon probe gives a hard momentum transfer to a single struck quark, and it is thus more likely to be sensitive to the short-distance hidden color components in the target wave function. Adequate electron energy is essential not only to produce dibaryon resonances but also to allow sufficient momentum transfer Q^2 to decrease backgrounds and to provide σ_L/σ_T separation.

v) One wishes to probe and parametrize the high momentum transfer dependence of the deuteron n-p and Δ-Δ components, as a clue toward a complete description of the nuclear wave function.

vi) One wishes to measure the neutron, pion, and kaon form factors.

vii) The region well beyond x=1 for deep inelastic electron-nucleus scattering is important QCD physics since the virtual quark and gluon configurations in the nuclear wave function are required to be far off shell. Understanding the detailed mechanisms which underlie this dynamics will require coincident measurements and the broadest kinematic region available for σ_L/σ_T separation. The "y-variable" approach which attributes the electron scattering to nucleon currents is likely to break down even at moderate Q^2. Coincidence measurements which can examine the importance of the nucleon component are well worth study.

viii) One wishes to study the emergence of strangeness in the nuclear state.

The fact that QCD is a viable theory for hadronic interactions implies that a fundamental description of the nuclear force is now possible. Although detailed work on the synthesis of QCD and nuclear physics is just beginning, it is clear from the structure of QCD that several traditional concepts of nuclear physics will have to be modified. These include conventional treatments of meson and baryon-pair contributions to the electromagnetic current and analyses of the nuclear form factor in terms of factorized on-shell nucleon form factors. On the other hand, the reduced nuclear form factors and scattering matrix elements discussed above give a viable prescription for the extrapolation of nuclear amplitudes to zero nucleon radius. There is thus the possibility that even the present low momentum transfer phenomenology of nuclear parameters will be significantly modified.

APPENDIX—LIGHT CONE QUANTIZATION AND PERTURBATION THEORY

In this Appendix, we outline the canonical quantization[82] of QCD in $A^+ = 0$ gauge. This proceeds in several steps. First we identify the independent dynamical degrees of freedom in the Lagrangian. The theory is quantized by defining commutation relations for these dynamical fields at a given light-cone time $\tau = t+z$ (we choose $\tau = 0$). These commutation relations lead immediately to the definition of the Fock state basis. Expressing dependent fields in terms of the independent fields, we then derive a light-cone Hamiltonian, which determines the evolution of the state space with changing τ. Finally we derive the rules for τ-ordered perturbation theory.

The major purpose of this exercise is to illustrate the origins and nature of the Fock state expansion, and of light-cone perturbation theory. We will ignore subtleties due to the large scale structure of non-Abelian gauge fields (e.g. 'instantons'), chiral symmetry breaking, and the like. Although these have a profound effect on the structure of the vacuum, the theory can still be described with a Fock state basis and some sort of effective Hamiltonian. Furthermore, the short distance interactions of the theory are unaffected by this structure, or at least this is the central ansatz of perturbative QCD.

Quantization

The Lagrangian (density) for QCD can be written

$$\mathcal{L} = -1/2 \, \mathrm{Tr}(F^{\mu\nu}F_{\mu\nu}) + \overline{\psi}(i\not{D} - m)\psi \tag{A1}$$

where $F^{\mu\nu} = \partial^{\mu}A^{\nu} - \partial^{\nu}A^{\mu} + ig[A^{\mu},A^{\nu}]$ and $iD^{\mu} = i\partial^{\mu} - gA^{\mu}$. Here the gauge field A^{μ} is a traceless 3×3 color matrix $\left(A^{\mu} \equiv \sum_a A^{a\mu}T^a,\ \mathrm{Tr}(T^aT^b) = 1/2\delta^{ab},\ [T^a,T^b] = ic^{abc}T^c, \ldots\right)$, and the quark field ψ is a color triplet spinor (for simplicity, we include only one flavor). At a given light-cone time, say $\tau=0$, the independent dynamical fields are $\psi_{\pm} \equiv \Lambda_{\pm}\psi$ and $A^i_{\perp}$ with conjugate fields $i\psi^{\dagger}_{+}$ and $\partial^{+}A^i_{\perp}$, where $\Lambda_{\pm} = \gamma^0\gamma^{\pm}/2$ are projection operators ($\Lambda_{+}\Lambda_{-} = 0$, $\Lambda^2_{\pm} = \Lambda_{\pm}$, $\Lambda_{+} + \Lambda_{-} = 1$) and $\partial^{\pm} = \partial^0 \pm \partial^3$. Using the equations of motion, the remaining fields in $\mathcal{L}$ can be expressed in terms of ψ_{+}, $A^i_{\perp}$:

$$\psi_{-} \equiv \Lambda_{-}\psi = \frac{1}{i\partial^{+}}\left[i\vec{D}_{\perp}\cdot\vec{\alpha}_{\perp} + \beta m\right]\psi_{+}$$

$$= \tilde{\psi}_{-} - \frac{1}{i\partial^{+}}\, g\, \vec{A}_{\perp}\cdot\vec{\alpha}_{\perp}\, \psi_{+} \quad ,$$

$$A^{+} = 0,$$

$$A^{-} = \frac{2}{i\partial^{+}}\, i\vec{\partial}_{\perp}\cdot\vec{A}_{\perp} + \frac{2g}{(i\partial^{+})^2}\left\{\left[i\partial^{+}A^i_{\perp},A^i_{\perp}\right] + 2\psi^{\dagger}_{+}\, T^a\, \psi_{+}\, T^a\right\}$$

$$\equiv \tilde{A}^{-} + \frac{2g}{(i\partial^{+})^2}\left\{\left[i\partial^{+}A^i_{\perp},A^i_{\perp}\right] + 2\psi^{\dagger}_{+}\, T^a\, \psi_{+}\, T^a\right\} \quad , \tag{A2}$$

with $\beta = \gamma^0$ and $\vec{\alpha}_{\perp} = \gamma^0\vec{\gamma}$.

To quantize, we expand the fields at $\tau=0$ in terms of creation and annihilation operators,

$$\psi_{+}(X) = \int_{k^{+}>0} \frac{dk^{+}\, d^2k_{\perp}}{k^{+}\, 16\pi^3} \sum_{\lambda} \Big\{ b(\underline{k},\lambda)\, u_{+}(\underline{k},\lambda)\, e^{-ik\cdot x} + d^{\dagger}(\underline{k},\lambda)\, v_{+}(\underline{k},\lambda)\, e^{ik\cdot x} \Big\}, \quad \tau = x^{+} = 0$$

$$A^i_{\perp}(X) = \int_{k^{+}>0} \frac{dk^{+}\, d^2k_{\perp}}{k^{+}\, 16\pi^3} \sum_{\lambda} \left\{ a(\underline{k},\lambda)\, \epsilon^i_{\perp}(\lambda)\, e^{-ik\cdot x} + c\cdot c\cdot \right\}, \quad \tau=x^{+}=0 \; , \tag{A3}$$

with commutation relations $\left(\underline{k} = (k^{+},\vec{k}_{\perp})\right)$:

$$\left\{b(\underline{k},\lambda),\ b^\dagger(\underline{p},\lambda')\right\} = \left\{d(\underline{k},\lambda),\ d^\dagger(\underline{p},\lambda')\right\}$$
$$= \left[a(\underline{k},\lambda),\ a^\dagger(\underline{p},\lambda')\right]$$
$$= 16\pi^3\ k^+\ \delta^3(\underline{k}-\underline{p})\,\delta_{\lambda\lambda'}\ ,$$
$$\left\{b,b\right\} = \left\{d,d\right\} = \ldots = 0\ , \tag{A4}$$

where λ is the quark or gluon helicity. These definitions imply canonical commutation relations for the fields with their conjugates $(\tau=x^+=y^+=0,\ \underline{x}=(x^-,x_\perp),\ \ldots)$:

$$\left\{\psi_+(\underline{x}),\ \psi_+^\dagger(\underline{y})\right\} = \Lambda_+\delta^3(\underline{x}-\underline{y})\ ,$$
$$\left[A^i(\underline{x}),\ \partial^+A_\perp^j(\underline{y})\right] = i\delta^{ij}\delta^3(\underline{x}-\underline{y})\ . \tag{A5}$$

As described in the third section, the creation and annihilation operators define the Fock state basis for the theory at $\tau=0$, with a vacuum $|0\rangle$ defined such that $b|0\rangle = d|0\rangle = a|0\rangle = 0$. The evolution of these states with τ is governed by the light-cone Hamiltonian, $H_{LC} = P^-$, conjugate to τ. Combining Eqs. (A1) and (A2), the Hamiltonian is readily expressed in terms of ψ_+ and $A_\perp^i$:

$$H_{LC} = H_0 + V\ ,$$

where

$$H_0 = \int d^3\underline{x}\left\{\mathrm{Tr}\left(\partial_\perp^i A_\perp^j \partial_\perp^i A_\perp^j\right) + \psi_+^\dagger\left(i\partial_\perp\cdot\alpha_\perp + \beta m\right)\frac{1}{i\partial^+}\left(i\partial_\perp\cdot\alpha_\perp + \beta m\right)\psi_+\right\}$$
$$= \sum_{\substack{\lambda \\ \text{colors}}} \int \frac{dk^+\,d^2k_\perp}{16\pi^3\,k^+}\left\{a^\dagger(\underline{k},\lambda)\,a(\underline{k},\lambda)\,\frac{k_\perp^2}{k^+} + b^\dagger(\underline{k},\lambda)\,b(\underline{k},\lambda)\ \times\right.$$
$$\left.\times\ \frac{k_\perp^2+m^2}{k^+} + d^\dagger(\underline{k},\lambda)\,d(\underline{k},\lambda)\,\frac{k_\perp^2+m^2}{k^+}\right\} + \text{constant} \tag{A6a}$$

is the free Hamiltonian and V the interaction:

$$V = \int d^3\underline{x}\left\{2g\ \mathrm{Tr}\left(i\partial^\mu\tilde{A}^\nu[\tilde{A}_\mu,\tilde{A}_\nu]\right) - \frac{g^2}{2}\,\mathrm{Tr}\left([\tilde{A}^\mu,\tilde{A}^\nu][\tilde{A}_\mu,\tilde{A}_\nu]\right)\right.$$
$$+\ g\,\bar{\tilde{\psi}}\tilde{A\!\!\!/}\tilde{\psi} + g^2\,\mathrm{Tr}\left([i\partial^+\tilde{A}^\mu,\tilde{A}_\mu]\frac{1}{(i\partial^+)^2}[i\partial^+\tilde{A}^\nu,\tilde{A}_\nu]\right)$$
$$+\ g^2\,\bar{\tilde{\psi}}\tilde{A\!\!\!/}\,\frac{\gamma^+}{2i\partial^+}\,\tilde{A\!\!\!/}\tilde{\psi} - g^2\,\bar{\tilde{\psi}}\gamma^+\left(\frac{1}{(i\partial^+)^2}[i\partial^+\tilde{A}^\mu,\tilde{A}_\mu]\right)\tilde{\psi}$$
$$\left.+\ \frac{g^2}{2}\,\bar{\psi}\gamma^+T^a\psi\,\frac{1}{(i\partial^+)^2}\,\bar{\psi}\gamma^+T^a\psi\right\}\ , \tag{A6b}$$

with $\tilde{\psi} = \tilde{\psi}_- + \psi_+$ ($\to\psi$ as $g\to 0$) and $\tilde{A}^\mu = (0,\tilde{A}^-,A_\perp^i)$ ($\to A^\mu$ as $g\to 0$). The

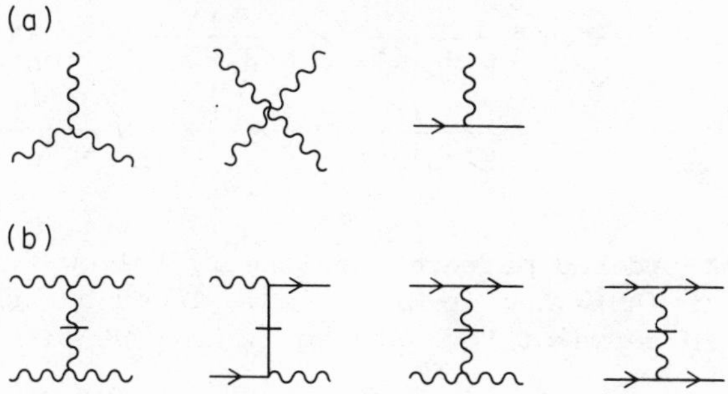

Fig. 22. (a) Basic interaction vertices in QCD. (b) "Instanteous" contributions.

Fock states are obviously eigenstates of H_0 with

$$H_0 |n:k_i^+, k_{\perp i}\rangle = \sum_i \left(\frac{k_\perp^2+m^2}{k^+}\right)_i |n:k_i^+, k_{\perp i}\rangle . \tag{A7}$$

It is equally obvious that they are not eigenstates of V, though any matrix element of V between Fock states is trivially evaluated. The first three terms in V correspond to the familiar three and four gluon vertices, and the gluon-quark vertex [Fig. 22(a)]. The remaining terms result from substitutions (A2), and represent new four-quanta interactions containing instantaneous fermion and gluon propagators [Fig. 22(b)]. All terms conserve total three-momentum $\underline{k} = (k^+, \vec{k}_\perp)$, because of the integral over $\underline{x}$ in V. Furthermore, all Fock states other than the vacuum have total $k^+ > 0$, since each individual bare quantum has $k^+ > 0$ [Eq. (A3)]. Consequently the Fock state vacuum must be an eigenstate of V and therefore an eigenstate of the full light-cone Hamiltonian.

Light-Cone Perturbation Theory

We define light-cone Green's functions to be the probability amplitudes that a state starting in Fock state $|i\rangle$ ends up in Fock state $|f\rangle$ a (light-cone) time τ later

$$\langle f|i\rangle\, G(f,i;\tau) \equiv \langle f|\, e^{-iH_{LC}\tau/2}|i\rangle$$

$$= i\int \frac{d\varepsilon}{2\pi}\, e^{-i\varepsilon\tau/2}\, G(f,i;\varepsilon)\langle f|i\rangle \; , \tag{A8}$$

where Fourier transform $G(f,i;\varepsilon)$ can be written

$$\langle f|i\rangle\, G(f,i;\varepsilon) = \left\langle f \left| \frac{1}{\varepsilon - H_{LC} + i0_+} \right| i \right\rangle$$

$$= \left\langle f \left| \frac{1}{\varepsilon - H_0 + i0_+} + \frac{1}{\varepsilon - H_0 + i0_+} V \frac{1}{\varepsilon - H_0 + i0_+} + \frac{1}{\varepsilon - H_0 + i0_+} V \frac{1}{\varepsilon - H_0 + i0_+} V \frac{1}{\varepsilon - H_0 + i0_+} + \ldots \right| i \right\rangle . \tag{A9}$$

The rules for τ-ordered perturbation theory follow immediately from the expansion in (A9) when $(\varepsilon - H_0)^{-1}$ is replaced by its spectral decomposition in terms of Fock states:

$$\frac{1}{\varepsilon - H_0 + i0_+} = \sum_{n,\lambda_i} \int \tilde{\Pi} \frac{dk_i^+ d^2k_{\perp i}}{16\pi^3 \, k_i^+} \frac{|n: \underline{k}_i, \lambda_i\rangle\langle n: \underline{k}_i, \lambda_i|}{\varepsilon - \sum_i (k^2+m^2)_i / k_i^+ + i0_+} \tag{A10}$$

where in (A9) the sum becomes a sum over all states n intermediate between two interactions. To calculate $G(f,i;\varepsilon)$ perturbatively then, all τ-ordered diagrams (i.e. all orderings of the vertices, as in Fig. 14) must be considered, the contribution from each graph computed according to the following rules:[53,83]

i) Assign a momentum k^μ to each line such that the total $k^+, k_\perp$ are conserved at each vertex, and such that $k^2 = m^2$, i.e., $k^- = (k^2+m^2)/k^+$. With fermions associate an on-shell spinor [from Eq. (A2)]

$$u(\underline{k},\lambda) = \frac{1}{\sqrt{k^+}} \left(k^+ + \beta m + \vec{\alpha}_\perp \cdot \vec{k}_\perp\right) \begin{cases} \chi(\uparrow) & \lambda = \uparrow \\ \chi(\downarrow) & \lambda = \downarrow \end{cases}$$

or

$$v(\underline{k},\lambda) = \frac{1}{\sqrt{k^+}} \left(k^+ - \beta m + \vec{\alpha}_\perp \cdot \vec{k}_\perp\right) \begin{cases} \chi(\downarrow) & \lambda = \uparrow \\ \chi(\uparrow) & \lambda = \downarrow \end{cases}$$

where $\chi(\uparrow) = 1/\sqrt{2}\,(1,0,1,0)$ and $\chi(\downarrow) = 1/\sqrt{2}\,(0,1,0,-1)^T$. For gluon lines, assign a polarization vector $\varepsilon^\mu = \left(0,\ 2\vec{\varepsilon}_\perp \cdot \vec{k}_\perp / k^+,\ \vec{\varepsilon}_\perp\right)$ where $\vec{\varepsilon}_\perp(\uparrow) = -1/\sqrt{2}\,(1,i)$ and $\vec{\varepsilon}_\perp(\downarrow) = 1/\sqrt{2}\,(1,-i)$.

ii) Include a factor $\theta(k^+)/k^+$ for each internal line.

iii) For each vertex include factors as illustrated in Fig. 23. To convert incoming into outgoing lines or vice versa replace

$$u \leftrightarrow v \qquad \bar{u} \leftrightarrow -\bar{v} \qquad \varepsilon \leftrightarrow \varepsilon^*$$

in any of these vertices.

iv) For each intermediate state there is a factor

$$\frac{1}{\varepsilon - \sum_{\rm interm} k^- + i0_+}$$

	Vertex Factor	Color Factor
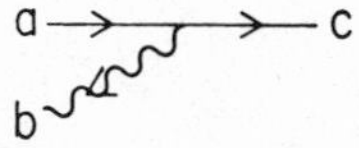	$g\,\bar{u}(c)\,\not\epsilon_b\,u(a)$	T^b
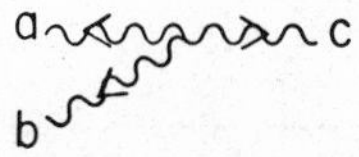	$g\left\{(p_a - p_b)\cdot\epsilon_c^*\,\epsilon_a\cdot\epsilon_b + \text{cyclic permutations}\right\}$	iC^{abc}
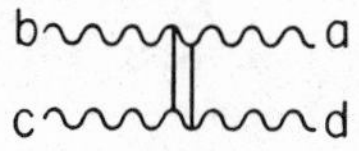	$g^2\left\{\epsilon_b\cdot\epsilon_c\,\epsilon_a^*\cdot\epsilon_d^* + \epsilon_a^*\cdot\epsilon_c\,\epsilon_b\cdot\epsilon_d^*\right\}$	$iC^{abe}\,iC^{cde}$
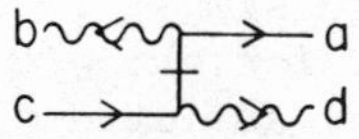	$g^2\,\bar{u}(a)\,\not\epsilon_b\,\frac{\gamma^+}{2(p_c^+ - p_d^+)}\,\not\epsilon_c^*\,u(c)$	$T^b\,T^d$
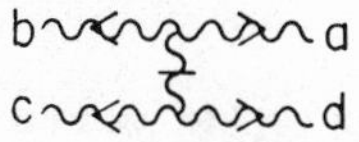	$g^2\,\epsilon_a^*\cdot\epsilon_b\,\frac{(p_a^+ - p_b^+)(p_c^+ - p_d^+)}{(p_c^+ + p_b^+)^2}\,\epsilon_d^*\cdot\epsilon_c$	$iC^{abe}\,iC^{cde}$
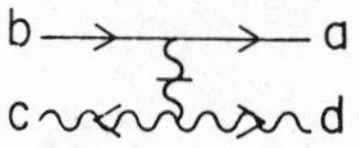	$g^2\,\bar{u}(a)\,\gamma^+u(b)\,\frac{(p_c^+ - p_d^+)}{(p_c^+ + p_d^+)^2}\,\epsilon_d^*\cdot\epsilon_c$	$iC^{cde}\,T^e$
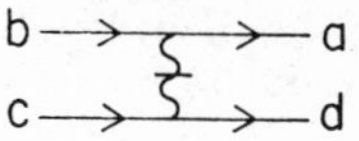	$g^2\,\frac{\bar{u}(a)\,\gamma^+\,u(b)\,\bar{u}(d)\,\gamma^+\,u(c)}{(p_c^+ - p_d^+)^2}$	$T^e\,T^e$

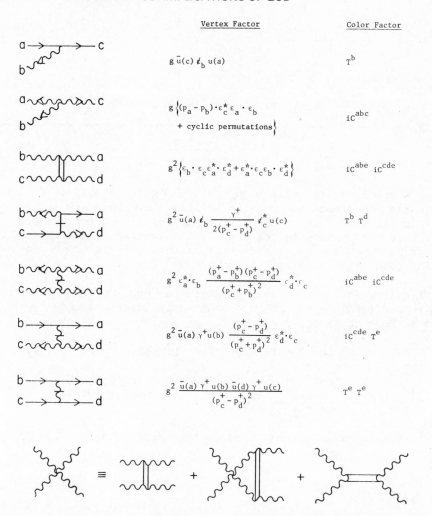

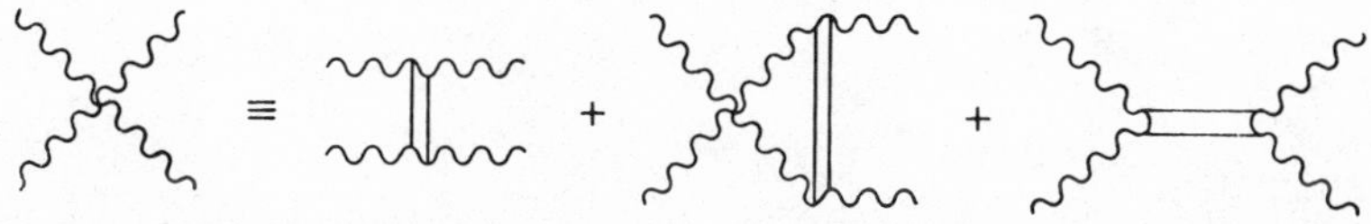

Fig. 23. Graphical rules for QCD in light-cone perturbation theory.

where ϵ is the incident P^-, and the sum is over all particles in the intermediate state.

v) Integrate $\int dk^+d^2k_\perp/16\pi^3$ over each independent k, and sum over internal helicities and colors.

vi) Include a factor -1 for each closed fermion loop, for each fermion line that both begins and ends in the initial state (i.e. $\bar{v}$ ... u), and for each diagram in which fermion lines are interchanged in either of the initial or final states.

As an illustration, the second diagram in Fig. 23 contributes

$$\frac{1}{\varepsilon - \sum\limits_{i=b,d}\left(\frac{k_\perp^2+m^2}{k^+}\right)_i} \frac{\theta(k_a^+-k_b^+)}{k_a^+ - k_b^+} \times$$

$$\times \frac{g^2 \sum\limits_\lambda \bar{u}(b)\not\epsilon^*(\underline{k}_a-\underline{k}_b,\lambda)u(a)\bar{u}(d)\not\epsilon(\underline{k}_a-\underline{k}_b,\lambda)u(c)}{\varepsilon - \sum\limits_{i=b,c}\left(\frac{k_\perp^2+m^2}{k^+}\right)_i - \frac{(k_{\perp a}-k_{\perp b})^2}{k_a^+-k_b^+}} \frac{1}{\varepsilon - \sum\limits_{i=a,c}\left(\frac{k_\perp^2+m^2}{k^+}\right)_i}$$

(times a color factor) to the $q\bar{q}\to q\bar{q}$ Green's function. (The vertices for quarks and gluons of definite helicity have very simple expressions in terms of the momenta of the particles—see for example Refs. 53,82.) These same rules apply for scattering amplitudes, but with propagators omitted for external lines, and with $\varepsilon = P^-$ of the initial (and final) states.

Finally, notice that this quantization procedure and perturbation theory (graph by graph) are manifestly invariant under a large class of Lorentz transformations:

i) boosts along the 3-direction—i.e. $p^+\to Kp^+$, $p^-\to K^{-1}p^-$, $p_\perp\to p_\perp$ for each momentum;
ii) transverse boosts—i.e. $p^+\to p^+$, $p^-\to p^- + 2p_\perp\cdot Q_\perp + p^+Q_\perp^2$, $p_\perp\to p_\perp + p^+Q_\perp$ for each momentum ($Q_\perp$ like K is dimensionless);
iii) rotations about the 3-direction.

It is these invariances which lead to the frame independence of the Fock state wave functions.

ACKNOWLEDGMENTS

These lectures are based in large measure on collaborations with a number of colleagues, including G. Peter Lepage (exclusive processes in QCD and light-cone perturbation theory); Chueng-Ryong Ji and G. Peter Lepage (nuclear amplitudes at short distances); J.R. Hiller (reduced nuclear amplitudes); Geoffrey T. Bodwin and G. Peter Lepage (initial and final state corrections to QCD processes); and Tao Huang and G. Peter Lepage (the structure of hadronic wave functions as their constraints). I also wish to thank G.E. Brown for helpful comments. My interest in applications of QCD to nuclear problems was inspired by my collaborations with Benson T. Chertok, whose experimental and theoretical work greatly motivated this field.

I am also grateful to the organizers of this productive institute, Professor David Boal, and his colleagues at Simon Fraser University and TRIUMF. This work was supported by the Department of energy, contract DE-AC03-76SF00515.

REFERENCES

1. For additional discussion of applications of QCD to Nuclear Physics and references see S. J. Brodsky, in: "New Horizons in Electromagnetic Physics," J. V. Noble and R. R. Whitney, eds., University of Virginia, Charlottesville (1983).
2. Reviews of QCD are given in: A. J. Buras, Rev. Mod. Phys. 52: 199 (1980); A. H. Mueller, Phys. Rep. 73C:237 (1981); E. Reya, Phys. Rep. 69:195 (1981); W. Marciano and H. Pagels, Phys. Rep. 36C:137 (1978); S. J. Brodsky and G. P. Lepage, Proceedings of the SLAC Summer Inst. on Particle Physics, 1979.
3. For other reviews of the phenomenology of QCD see A. J. Buras, FERMILAB-CONF-81/69-THY and A.H. Mueller, CU-TP-219 (1981), in the Proc. of the 1981 International Symposium on Lepton and Photon Interactions at High Energies, Bonn, August 1981.
4. H. D. Politzer, Phys. Rev. Lett. 30:1346 (1973); D. J. Gross and F. Wilczek, Phys. Rev. Lett. 30:123 (1973).
5. See, for example, S. J. Brodsky, SLAC-PUB-2970, presented at the XIIIth International Symposium on Multiparticle Dynamics, Volemdam (1982).
6. Non-perturbative effects can lead to an effective gluon mass in the gluon propagator, thus eliminating the problem of Van der Waals forces in QCD. See J. M. Cornwall, Phys. Rev. D26:1453 (1982).
7. S. J. Brodsky, C. R. Ji, and G. P. Lepage, SLAC report SLAC-PUB-3064.
8. G. P. Lepage and S. J. Brodsky, Phys. Rev. D22:2157 (1980) and G. P. Lepage and S. J. Brodsky, Phys. Rev. D22:2157 (1980).
9. S. J. Brodsky and B. T. Chertok, Phys. Rev. Lett. 37:269 (1976); Phys. Rev. D14:3003 (1976). S. J. Brodsky, in: "Few Body Problems in Nuclear and Particle Physics," R. J. Slobodrian, B. Cujec and K. Ramavatram, eds., Université Laval, Québec (1975).
10. S. J. Brodsky and G. P. Lepage, Phys. Rev. Lett. 43:545,1625(E) (1979). S. J. Brodsky, G. P. Lepage, S. A. A. Zaidi, Phys. Rev. D23:1152 (1981).
11. S. J. Brodsky and J. R. Hiller, SLAC report SLAC-PUB-3047.
12. Qualitative QCD-based arguments for the repulsive N-N potential at short distances are given in C. Detar, Helsinki University report HU-TFT-82-6; M. Harvey, this volume; R. L. Jaffe, Phys. Rev. Lett. 24:228 (1983). The possibility that the deuteron form factor is dominated at large momentum transfer by hidden color components is discussed in V. A. Matveev and P. Sorba, Nuovo Cim. 45A:257 (1978); Nuovo Cim. Lett. 20:435 (1977).
13. Early references to light-cone perturbation theory are P. A. M. Dirac, Rev. Mod. Phys. 21:392 (1949); S. Weinberg, Phys. Rev. 150:1313 (1966); J. D. Bjorken, J. Kogut and D. E. Soper, Phys. Rev. D3:1382 (1971); S. J. Brodsky, R. Roskies and R. Suaya, Phys. Rev. D8:4574 (1973).

14. G. P. Lepage, S. J. Brodsky, T. Huang, and P. B. Mackenzie, Cornell University report 82/522; S. J. Brodsky and G. P. Lepage, Phys. Rev. D24:1808 (1981).
15. S. J. Brodsky and S. D. Drell, Phys. Rev. D22:2236 (1981).
16. S. J. Brodsky, T. Huang, G. P. Lepage, SLAC report SLAC-PUB-2540; T. Huang, SLAC report SLAC-PUB-2580.
17. S. J. Brodsky, Y. Frishman, G. P. Lepage and C. Sachrajda, Phys. Lett. 91B:239 (1980); M. Peskin, Phys. Lett. 88B:128 (1979).
18. A. V. Efremov and A. V. Radyushkin, Rev. Nuovo Cim. 3:1 (1980); Phys. Lett. 94B:245 (1980). See also G. R. Farrar and D. R. Jackson, Phys. Rev. Lett. 43:246 (1979); V. L. Chernyak and A. R. Vhitnishii, JETP Lett. 25:11 (1977); G. Parisi, Phys. Lett. 43:246 (1979); M. K. Chase, Nucl. Phys. B167:125 (1980); E. Braaten, to be published; R. D. Field, R. Gupta, S. Otto, L. Chang, Nucl. Phys. B186:429 (1981); F. M. Dittes and A. V. Radyushkin, Dubna report JINR-E2-80-688.
19. A. Duncan and A. H. Mueller, Phys. Lett. 90B:139 (1980); Phys. Rev. D21:1636 (1980). A. H. Mueller, Ref. 1.
20. S. J. Brodsky and G. P. Lepage, Phys. Rev. D24:2848 (1980).
21. S. J. Brodsky and G. P. Lepage, in: "Quantum Chromodynamics," Wm. Frazer and F. Henyey, eds., AIP, New York (1979); Phys. Lett. 87B:359 (1979), Phys. Rev. Lett. 43:545 (1979), erratum ibid. 43:1625 (1979); S. J. Brodsky and G. P. Lepage, Ref. 1.
22. S. J. Brodsky and G. R. Farrar, Phys. Rev. Lett. 31:1153 (1973), and Phys. Rev. D11:1309 (1975); V. A. Matveev, R. M. Muradyan and A. V. Tavkheldize, Lett. Nuovo Cim. 7:719 (1973).
23. M. D. Mestayer, SLAC report 214.
24. P. V. Landshoff and D. J. Pritchard, Cambridge University report DAMTP 80/04, and references therein.
25. F. Martin et al., Phys. Rev. Lett. 38:1320 (1977); W. P. Schutz et al., Phys. Rev. Lett. 38:259 (1977); R. G. Arnold et al., Phys. Rev. Lett. 40:1429 (1978); B. T. Chertok, Phys. Lett. 41:1155 (1978); D. Day et. al., Phys. Rev. Lett. 43:1143 (1979). Summaries of the data for nucleon and nuclear form factors at large Q^2 are given in B. T. Chertok, in: Progress in Particle and Nuclear Physics, Proc. of the Int. School of Nuclear Physics, 5th Course, Erice (1978), and Proc. of the XVI Rencontre de Moriond, Les Arcs, Savoie, France, 1981.
26. S. J. Brodsky, Ref. 9. For speculations on the role of hidden color see V. A. Matveev and P. Sorba, Nuovo Cim. Lett. 20:435 (1977); A. P. Kobushkin, Kiev report ITF-77-113E; V. M. Dubovik and A. P. Kobushkin, Kiev report ITF-78-85E; Y. J. Karant, LBL report 9171; W. J. Romo and P. J. S. Watson, Phys. Lett. 88B:354 (1979).
27. R. G. Arnold et al., Phys. Rev. Lett. 35:776 (1975); R. G. Arnold, SLAC-PUB-2373 (1979).
28. P. V. Landshoff and J. C. Polkinghorne, Phys. Lett. 44B:293 (1973).
29. A. H. Mueller, Ref. 3.
30. R. Blankenbecler, S. J. Brodsky and J. F. Gunion, Phys. Rev.

D18:900 (1978), and references therein; D. Sivers, R. Blankenbecler and S. J. Brodsky, Phys. Rep. 23C:1 (1976).
31. G. Farrar, S. Gottlieb, D. Sivers and G. Thomas, Phys. Rev. D20:202 (1979); S. J. Brodsky, C. E. Carlson and H. J. Lipkin, Phys. Rev. D20:2278 (1979).
32. J. Szwed, Acta. Phys. Polonica, B14:55 (1983). See also M. Chemtob, Nuovo Cim. 71A:477 (1982), and Ref. 5.
33. For a recent discussion, see S. J. Brodsky, E. L. Berger, and G. P. Lepage, SLAC report SLAC-PUB-3027.
34. S. J. Brodsky and R. Blankenbecler, Phys. Rev. D10:2973 (1974); J. F. Gunion, Phys. Lett. 88B:150 (1979), and references therein.
35. R. Blankenbecler and I. Schmidt, Phys. Rev. D16:1318 (1977). See also Refs. 14,16,26, and L. I. Frankfurt and M. I. Strikman, Leningrad report 559 (1980); M. Chemtob, Nucl. Phys. A336:299 (1980); J. M. Namyslowski and P. Danielewicz, Warsaw report IFT/8/80; D. Kusno and M. Moravscik, Trieste report IC/80/52, and references therein.
36. G. Bertsch, S. J. Brodsky, A. S. Goldhaber and J. F. Gunion, Phys. Rev. Lett. 47:297 (1981).
37. S. D. Drell and T. M. Yan, Phys. Rev. Lett. 25:316 (1970). Possible complications in the parton model prediction due to "wee parton" exchange are discussed in this paper.
38. H. D. Politzer, Nucl. Phys. B129:301 (1977); R. K. Ellis, H. Georgi, M. Machacek, H. D. Politzer and G. C. Ross, Nucl. Phys. B152:285 (1979); S. Gupta and A. H. Mueller, Phys. Rev. D20:118 (1979). A discussion of final state interactions and possible problems with factorization for the Drell-Yan process is given in J. C. Collins and D. E. Soper, Proceedings of the Moriond Workshop, Les Arcs, France (1981).
39. G. T. Bodwin, S. J. Brodsky and G. P. Lepage, Phys. Rev. Lett. 47:1799 (1981); in: "Proceedings of the 1981 Banff Summer School on Particles and Fields," A. N. Kamal and A. Capri, eds., Plenum, New York (1983).
40. W. W. Lindsay, D. A. Ross, and C. T. Sachrajda, Southampton University report SHEP-81/82 (1982), and Phys. Lett. 117B: 105 (1982).
41. J. C. Collins, D. E. Soper, and G. Sterman, ITP-SB-82-46 (1982), and Phys. Lett. 109B:388 (1982).
42. C. A. Nelson, SUNY Binghampton report 11/25/82.
43. G. Bodwin, S. J. Brodsky, and G. P. Lepage, to be published.
44. L. Landau and I. Pomeranchuk, Dokl. Akad. Nauk SSSR 92:535 (1953).
45. This prediction has also been discussed by A. H. Mueller, Proc. of the 1982 Moriond Conference.
46. G. T. Bodwin, S. J. Brodsky, G. P. Lepage, Ref. 39.
47. C. Shakin, private communication.
48. European Muon Collaboration, J. Aubert et al., CERN report CERN-EP/83-14 (1983).
49. The Role of Electromagnetic Interactions in Nuclear Science,

A report of the DOE/NSF Nuclear Science Advisory Committee. Subcommittee on Electromagentic Interactions, P. D. Barnes, chairman, 1982. For a discussion of the design parameters for intermediate-energy, high duty factor electron beams see J. M. Laget, Proceedings of the Lund Workshop, 1982.

50. D. R. Yennie, unpublished. S. J. Brodsky, in: "Atomic Physics and Astrophysics," M. Chretien and E. Kipworth, eds., Gordon and Breach, New York (1971).

51. For practical approximations to the Bethe-Salpeter equation, see G. P. Lepage, Phys. Rev. A16:863 (1977); W. E. Caswell and G. P. Lepage, Phys. Rev. A18:810 (1978). See also F. Gross, Carnegie Mellon report (1982); Nucl. Phys. A358:215C (1981), and references therein.

52. P. A. M. Dirac, Rev. Mod. Phys. 21:392 (1949).

53. G. P. Lepage and S. J. Brodsky, Phys. Rev. D22:2157 (1980).

54. See Ref. 16. The parameterization of the hadronic wave functions presented here is preliminary; a complete discussion and final values will be given by S. J. Brodsky, T. Huang, and G. P. Lepage, to be published.

55. S. D. Drell and T. M. Yan, Phys. Rev. Lett. 24:181 (1970).

56. S. J. Brodsky and S. D. Drell, Phys. Rev. D22:2236 (1981).

57. See for example S. J. Brodsky and J. R. Primack, Ann. Phys. 52:315 (1969).

58. S. J. Brodsky and T. Huang, to be publihsed. This effect has been recently demonstrated in an explicit 3-quark nucleon model by L. S. Celenza, W. S. Pong, and C. M. Shakin, Brooklyn College report 81/101/110.

59. R. D. Field, R. Gupta, S. Otto, L. Chang, Nucl. Phys. B186:429 (1981); F. M. Dittes and A. V. Radyushkin, Dubna report JINR-E2-80-688; M. K. Chase, Ref. 18; E. Braaten (private communication.

60. G. R. Farrar and D. R. Jackson, Ref. 18.

61. See Ref. 10.

62. See Ref. 17.

63. A. Duncan and A. H. Mueller, Ref. 19.

64. P.V. Landshoff, Phys. Rev. D10:1024 (1974); P. Cvitanovic, Phys. Rev. D10:338 (1974); S. J. Brodsky and G. Farrar, Phys. Rev. D11:1309 (1975).

65. S. J. Brodsky and G. P. Lepage, Phys. Rev. D24:1808 (1981).

66. S. J. Brodsky, C. R. Ji, and G. P. Lepage, Ref. 7. See also Ref. 1.

67. As in the case of the meson form factor the "end-point" region $x_i \sim 2, \vec{k}_\perp^2$ small is power-law suppressed because of the mismatch between the struck quark and deuteron helicities. One also expects additional Sudakov form factor suppression in this region. Form more details see Ref. 3. The nominal power-law prediction $F_d(Q^2) \sim (Q^2)^{-5}$ is given in S. J. Brodsky and G. R. Farrar, Phys. Rev. Lett. 31:1153 (1975).

68. M. Harvey, Nucl. Phys. A352:301 (1981); A352:326 (1981).

69. The result $C_d = -C_F/5$ can also be obtained by requiring

cancellation of the $y_i = x_i$ singularity. This result also follows by noting that as far as color is concerned, five quarks act coherently as an effective $\overline{3}$ in the soft gluon exchange limit.

70. T. Appelquist and J. R. Primack, Phys. Rev. D1:1144 (1970).
71. W. P. Schütz et al., Phys. Rev. Lett. 38:259 (1977); F. Martin et al., Phys. Rev. Lett. 38:1320 (1977).
72. T. T. Chou and C. N. Yang, Phys. Rev. 170:1591 (1968); Phys. Rev. D19:3268 (1979). See D. E. Soper, Phys. Rev. D15:1141 (1977) for a discussion of the impact space behavior of relativistic, large momentum transfer amplitudes.
73. H. Nann et al., Phys. Lett. 88B:257 (1979). For a theoretical analysis based on a six-quark model, see G. A. Miller and L. S. Kisslinger, University of Washington report 40048-20-82.
74. H. Myers et al., Phys. Rev. 121:630 (1961); R. Ching and C. Schaerf, Phys. Rev. 141:1320 (1966); P. Dougan et al., Z. Phys. A276:55 (1976).
75. This agrees with the formula for the crossed reaction $\gamma q_1 \to Mq_2$ listed in J. A. Bagger and J. F. Gunion, Phys. Rev. D25:2287 (1982). See also R. Blankenbecler, S. J. Brodsky and J. F. Gunion, Phys. Rev. D18:900 (1978); E. L. Berger, Phys. Rev. D26:105 (1982); S. Matsuda, Kyoto University report KEK-TH 37. The origin of the amplitude zero at $ue_1 = -te_2$ is explained in S. J. Brodsky and R. W. Brown, Phys. Rev. Lett. 49:966 (1982); and R. W. Brown, K. L. Kowalski and S. J. Brodsky, FNAL report THY-82/102.
76. H. Ikeda et al., Nucl. Phys. B172:509 (1980). See also P. E. Argan et al., Phys. Rev. Lett. 46:96 (1981). A more recent measurement for $\gamma d \to pn$ by K. Baba et al., [Phys. Rev. Lett. 48:729 (1982)] is apparently not consistent with the presence of such resonances.
77. See, e.g. "New Horizons in Electromagnetic Physics," J. V. Noble and R. R. Whitney, eds., University of Virginia, Charlottesville (1983).
78. S. J. Brodsky, T. Huang and G. P. Lepage, Ref. 16.
79. S. J. Brodsky, R. W. Brown, and K. L. Kowalski, Ref. 75.
80. S. J. Brodsky and G. P. Lepage, Nucl. Phys. A353:247c (1981).
81. For a recent example of a boundary condition model incorporating quark constraints, see e.g. G. A. Miller and L. S. Kisslinger, Ref. 73.
82. From G. P. Lepage et al., Ref. 14.
83. J. B. Kogut and D. E. Soper, Phys. Rev. D1:2901 (1970), and J. D. Bjorken, J. B. Kogut, and D. E. Soper, Phys. Rev. D3:1382 (1971).

THE THERMODYNAMICS OF STRONGLY INTERACTING MATTER

H. Satz

Fakultät für Physik
Universität Bielefeld
Germany

INTRODUCTION

Changes of state are among the most familiar and yet most striking instances of collective behavior. Do they also occur in strongly interacting matter? Will a sufficient increase in density lead us out of the realm of normal nuclear matter, made up of nucleon constituents, into a new world, in which quarks and gluons form a plasma of primordial matter?

Such a transition has been discussed ever since the advent of the quark model of hadrons. Today the pursuit of these ideas, the physics of strongly interacting matter at very high density, is emerging more and more as an autonomous field of research.[1] It brings together problems and features from nuclear physics, particle physics, and statistical physics. In cosmology, it is essential for an understanding of the early stages of our universe.

The basis of this field is quantum chromodynamics; with QCD, we have today a serious and so far "uncontradicted" candidate for the theory of strong interactions. This provides us with the possibility of deriving strong interaction thermodynamics. Phenomenological models of various types have been employed for many years to discuss a possible transition from nuclear to quark matter; they always have two phases as input and obtain the transition by construction. The lattice formulation of QCD[2] now for the first time allows us to deduce the phases of strongly interacting matter and the transition behavior from one basic theory. QCD thermodynamics predicts the existence of a quark-gluon plasma at high temperature as well as that of hadronic matter at low temperature; between these regimes lies a transition region, in which—coming from low T—color is deconfined and chiral symmetry restored.

PHENOMENOLOGY

Quantum chromodynamics describes the interaction of quarks and gluons—the strong interaction. Quarks and gluons are not observed as free objects in the physical vacuum: in non-interacting form, as free particle states, we only see hadrons. The low density limit of QCD thermodynamics must therefore lead to a system of non-interacting hadrons. On the other hand, sufficiently energetic "hard" hadron-hadron or hadron-lepton scattering processes indicate independent interactions of quarks and gluons with each other or with leptons: strong interactions become weak at high energies. This novel feature, "asymptotic freedom", is in fact provided by QCD. To obtain such behavior, quarks and gluons must have an additional intrinsic degree of freedom, "color", which makes gluon-gluon interactions possible in a gauge-invariant framework. In potential language, the confining force between quarks rises linearly with increasing separation; at short distances, however, the potential becomes Coulomb-like. For the high-density limit of QCD, we therefore expect a plasma of non-interacting, "Debye-screened" quarks and gluons; here, in addition, an overall shift in energy between vacuum and plasma ground state has to be taken into account.

Let us use these limiting forms to construct a simple two-phase picture of strongly interacting matter. Our thermodynamic variables are the temperature T, the baryonic chemical potential μ, and the volume V; μ specifies the baryon number density. Here and in the following we shall for simplicity always restrict ourselves to non-strange hadrons (pions and nucleons) and quarks (u and d).

For zero baryon number (μ=0), we have at low temperatures basically a gas of pions. For a non-interacting system of massless pions, the pressure is given by

$$P_H(T,\mu = 0) = \frac{\pi^2}{90} \times 3 \times T^4 \quad , \tag{1}$$

taking into account the three charge states of the pion. The corresponding energy density is

$$\varepsilon_H(T,\mu = 0) = \frac{\pi^2}{30} \times 3 \times T^4 \quad . \tag{2}$$

At sufficiently high temperatures, we expect a plasma of non-interacting quarks and gluons, with the pressure

$$P_Q(T,\mu = 0) = \frac{\pi^2}{90} \left[2 \times 8 + \frac{7}{8} \times 2 \times 2 \times 2 \times 3 \right] T^4 - B \quad . \tag{3}$$

Here both gluons (first term in square brackets) and quarks (second term) have two spin degrees of freedom; in accord with the SU(3) gauge structure of QCD, gluons have eight and quarks three color degrees of freedom. There are, for μ=0, both quarks and antiquarks

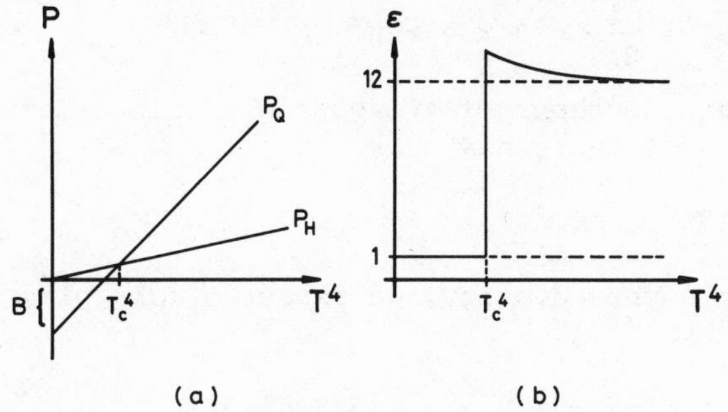

Fig. 1. Pressure (a) and energy density (b) in a two-phase ideal gas model for strongly interacting matter.

present, and we include, as mentioned, two flavors. Finally, the (positive) bag constant B accounts for the ground state shift in the plasma; to provide confinement at low density, the physical vacuum exerts a pressure on the system. The corresponding energy density becomes

$$\varepsilon_Q(T,\mu = 0) = \frac{\pi^2}{30}\left[2 \times 8 + \frac{7}{8} \times 2 \times 2 \times 2 \times 3\right] T^4 + B \quad . \tag{4}$$

To satisfy the thermodynamic requirement of minimal free energy, the system must be in the state of higher pressure. For low T, this is the pion gas, for high T, the plasma; see Fig. 1. From $P_H(T_c,0) = P_Q(T_c,0)$ we find

$$T_c = [(45/17\pi^2)B]^{1/4} \simeq 0.72\ B^{1/4} \tag{5}$$

for the transition temperature at $\mu=0$. From the description of hadronic spectra, we have 145 MeV $\leq B^{1/4} \leq$ 235 MeV; using $B^{1/4} \simeq$ 190 MeV, we obtain $T_c \simeq$ 140 MeV. The transition arrived at in this way is by construction of first order, with

$$\varepsilon_Q(T_c,0) - \varepsilon_H(T_c,0) = 4\,B \tag{6}$$

as the latent heat per unit volume. The resulting energy density is also shown in Fig. 1.

Let us now apply the same simple model to the case of cold strongly interacting matter (T=0). In the hadronic phase, we now have a completely degenerate Fermi gas of nucleons, whose pressure and energy density are given by

$$P_H(T=0,\mu) = \frac{1}{24\pi^2} \times 2 \times 2 \times \mu^4 \quad , \tag{7}$$

$$\varepsilon_H(T=0,\mu) = \frac{1}{8\pi^2} \times 2 \times 2 \times \mu^4 \quad , \tag{8}$$

with protons and neutrons of two possible spin orientations. The baryon number density is given by

$$n_B = 2\mu^3/3\pi^2 \quad . \tag{9}$$

At high baryon number density, we expect a cold plasma, whose pressure is given by

$$P_Q(T=0,\mu) = \frac{1}{24\pi^2} \times 2 \times 2 \times 3 \times \mu_Q^4 - B \tag{10}$$

including two flavor and three color degrees of freedom for the quarks; B again denotes the vacuum pressure on the system. The quark density is given by

$$n_Q = 2\mu_Q^3/\pi^2 \tag{11}$$

in terms of the quark chemical potential μ_Q; from $n_B = n_Q/3$, we obtain

$$\mu_Q = \mu \tag{12}$$

so that the subscript Q on the chemical potential in Eq. (10) can be dropped. The energy density of the plasma is

$$\varepsilon_Q(0,\mu) = \frac{1}{8\pi^2} \times 2 \times 2 \times 3 \times \mu^4 + B \quad . \tag{13}$$

We obtain the critical point by requiring minimal thermodynamic potential, which again means highest pressure. From $P_H(0,\mu_c) = P_Q(0,\mu_c)$ we get

$$\mu_c = (3\pi^2 B)^{1/4} \simeq 2.33 \; B^{1/4} \tag{14}$$

which implies

$$n_c = 2(3\pi^2)^{-1/4} \; B^{3/4} \simeq 0.85 \; B^{3/4} \tag{15}$$

for the critical density. Using as above $B^{1/4} \simeq 200$ MeV, this yields

$$n_c \simeq 5n_o \tag{16}$$

with $n_o = 0.17 \text{ fm}^{-3}$ denoting standard nuclear density. In comparison, the baryon number density inside of a nucleon is about $3n_o$, so that the value (16) seems not unreasonable. Finally, we obtain, as in Eq. (6),

$$\varepsilon_Q(0,\mu_c) - \varepsilon_H(0,\mu_c) = 4\,B \tag{17}$$

for the latent heat of the transition at T=0.

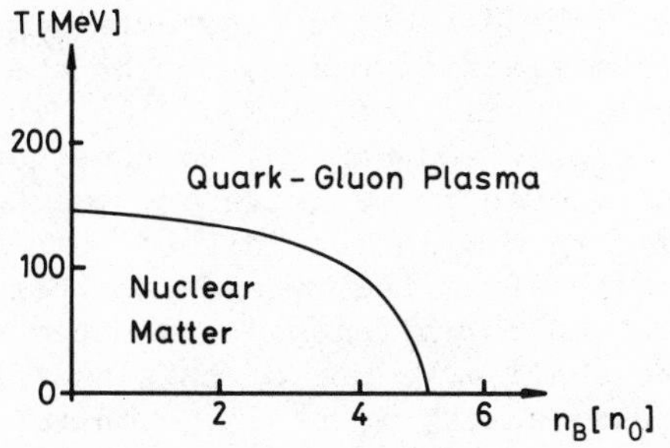

Fig. 2. Phase diagram in a two-phase ideal gas model for strongly interacting matter; n_B is a baryon number density in units of standard nuclear density n_o.

Extrapolating these results to intermediate temperatures and chemical potentials, we obtain the phase diagram shown in Fig. 2. It clearly presents only the rudiments of the picture: the inclusion of perturbative terms in the plasma would certainly modify the values of the transition parameters. Nevertheless, the basic reason for the transition would be expected to remain: we have more degrees of freedom in the plasma phase, causing it to dominate at high temperatures and pressures. Because of B, we have a higher pressure "normalization" in the hadronic phase, leading it dominate at low temperatures and pressures.

In any phenomenological model, such as the one considered here, the transition is of course obtained by construction. The basic question for QCD therefore is: can we derive both critical behavior and limiting phases from one fundamental description? In the following section, we shall see that this is indeed the case.

THE PHASE STRUCTURE OF QCD

The Lagrangian density of quantum chromodynamics is given by

$$\mathcal{L}(A,\psi,\bar{\psi}) = -\frac{1}{4}\left[\partial_\mu A_\nu^a - \partial_\nu A_\mu^a - g\, f_{bc}^a\, A_\mu^b\, A_\nu^c\right]^2 + \\ + \bar{\psi}_f^k (i\partial\!\!\!/ - g\, A\!\!\!/_a\, \lambda^a)\psi_k^f \qquad (18)$$

in terms of the gluon fields A_μ^a and the quark spinors ψ_f^k. Here the f_{abc} are the structure constants of the color gauge group, whose generators λ_a satisfy $[\lambda_a,\lambda_b] = i\, f_{ab}^c\, \lambda_c$. The gluonic color indices a,b,c run from one to eight for color SU(3), those for the quarks (k) from one to three ("red, white and blue"). As we shall here consider only u and d quarks, the flavor index f only takes on two values.

If we would set the structure constants f_{abc} equal to zero, we would recover quantum electrodynamics; it is the non-Abelian nature, with the gluon-gluon interaction in Eq. (18), which distinguishes QCD. In contrast to electrodynamics, we thus have in chromodynamics an interacting theory even if we leave out quarks altogether. The resulting Yang-Mills theory in fact already exhibits many of the essential features of the full theory and can therefore be taken as a model to introduce both formalism and evaluation techniques of QCD thermodynamics. We shall make use of this and first consider purely gluonic matter; subsequently, we shall go on to full QCD with quarks.

The theory introduced with the Lagrangian (18) is an interacting relativistic quantum field theory; so far, the only general non-perturbative way to solve such a theory, is provided by the lattice regularization approach of Wilson.[2] Together with the Monte Carlo evaluation technique pioneered by Creutz,[3] it will form the basis for our treatment of QCD thermodynamics.

The partition function for a quantum system described in terms of fields A(x) by a Hamiltonian H(A) is defined as

$$Z = \mathrm{Tr}\left\{\exp(-\beta H)\right\} , \tag{19}$$

where $\beta^{-1} = T$ is the physical temperature. The conventional lattice formulation is obtained from this in three steps, which we shall now briefly sketch for the Yang-Mills system

$$\mathcal{L}(A) = -\frac{1}{4}\left[\partial_\mu A_\nu^a - \partial_\nu A_\mu^a - g f_{bc}^a A_\mu^b A_\nu^c\right]^2 . \tag{20}$$

First, the partition function Z is rewritten in form of a path integral[4]

$$Z(\beta,V) = \int [dA] \exp\left\{\int_0^\beta d\tau \int_V d^3x\, \mathcal{L}[A(x,\tau)]\right\} , \tag{21}$$

where $\mathcal{L}[A(x,\tau)]$ is the Euclidean density, with $\tau = it$, periodic in τ. The three-dimensional integral of the Hamiltonian formulation $\left(H \sim \int d^3x\, \mathcal{H}(x)\right)$ thus becomes an asymmmetric four-dimensional integral, with the "special" dimension measuring the temperature.

In the next step, we replace the Euclidean x-τ continuum by a finite lattice, with N_σ sites and spacing a_σ in the spatial part, N_β sites and spacing a_β in the temperature direction. The integrals in the exponent of Eq. (21) now become sums, and we have $V = (N_\sigma a_\sigma)^3$, $\beta = N_\beta a_\beta$. The thermodynamic limit requires $N_\sigma \to \infty$ at fixed a_σ; the continuum limit is obtained by a_σ, $a_\beta \to 0$ with fixed $N_\beta a_\beta$, which forces also $N_\beta \to \infty$. The success of the approach rests on the (lucky) facts that already rather small lattices ($N_\sigma \sim 5 - 10$, $N_\beta \sim 3 - 5$) seem to be asymptotic, and such that scale changes (changes in lattice spacings) can be connected to changes in the coupling strength g by the renormalization group relation, indicating continuum behavior.

In the last step, we replace the gauge field "variable" $A_\mu[(x_i+x_j)/2]$ associated with the link between two adjacent sites i and j by the gauge group element

$$U_{ij} = \exp\left\{-i(x_i - x_j)^\mu A_\mu\left(\frac{x_i + x_j}{2}\right)\right\} , \tag{22}$$

where $A_\mu(x) = \lambda_a A^a_\mu(x)$. With this transformation, the partition function becomes

$$Z(\beta,V) = \int \prod_{\{\text{links}\}} dU_{ij} \exp\{-S(U)\} , \tag{23}$$

where the lattice action is, for color SU(N), given by

$$S(U) = \frac{2N}{g^2}\left\{\frac{a_\beta}{a_\sigma}\sum_{\{P_\sigma\}}\left[1 - \frac{1}{N}\,\text{Re Tr}\, U_{ij}\, U_{jk}\, U_{kl}\, U_{li}\right]\right.$$

$$\left. + \frac{a_\sigma}{a_\beta}\sum_{\{P_\beta\}}\left[1 - \frac{1}{N}\,\text{Re Tr}\, U_{ij}\, U_{jk}\, U_{kl}\, U_{li}\right]\right\} . \tag{24}$$

Here the sum $\{P_\sigma\}$ runs over all purely spacelike lattice plaquettes (ijkl), while $\{P_\beta\}$ runs over all those with two spacelike and two "temperature-like" links. If we insert Eqs. (22) and (24) in Eq. (23) and expand for small lattice spacings ($|x_i - x_j| \to 0$), then we recover in leading order the starting form (21). In Eq. (24), we have kept the color gauge group general, since the behavior of the SU(2) Yang-Mills system[5] is presently known with greater precision than that for the SU(3) system.[6] We shall therefore generally consider both; they appear to provide basically the same thermodynamics.

From Eqs. (23) and (24), the energy density

$$\varepsilon \equiv (-1/V)(\partial \ln Z/\partial\beta)_V = -(N_\sigma^3 N_\beta a_\sigma^3)^{-1}\,(\partial \ln Z/\partial a_\beta)_{a_\sigma} \tag{25}$$

is found to be[5]

$$\varepsilon \simeq 2N(N_\sigma^3 N_\beta a_\sigma^3 a_\beta g^2)^{-1}\left\{\left\langle\frac{a_\beta}{a_\sigma}\sum_{\{P_\sigma\}}\left[1 - \frac{1}{N}\,\text{Re Tr UUUU}\right]\right\rangle\right.$$

$$\left. -\left\langle\frac{a_\sigma}{a_\beta}\sum_{\{P_\beta\}}\left[1 - \frac{1}{N}\,\text{Re Tr UUUU}\right]\right\rangle\right\} \tag{26}$$

with <> denoting the usual thermodynamic average

$$\langle X\rangle \equiv \left\{\int \prod dU\, e^{-S(U)}\, X(U)\right\} / \left\{\int \prod dU\, e^{-S(U)}\right\} . \tag{27}$$

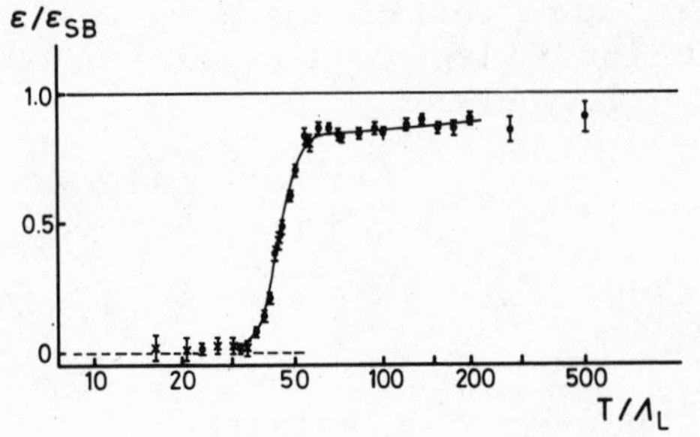

(a)

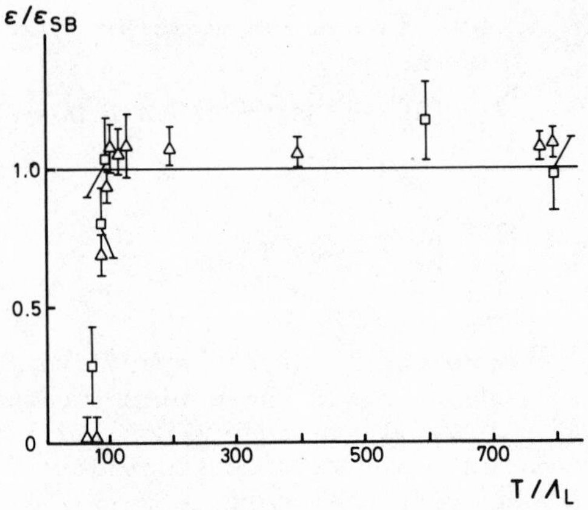

(b)

Fig. 3. Energy density of the Yang-Mills system, normalized to the ideal gas value ε_{SB}, (a) for SU(2) color group, from Ref. 5, and (b) for SU(3) color group, from Ref. 6.

Equation (26) is our starting point for the Monte Carlo evaluation of gluon thermodynamics.

The evaluation is now carried out as follows. The computer simulates an $N_\sigma^3 \times N_\beta$ lattice; for convenience we choose $a_\sigma = a_\beta = \underline{a}$. Starting from a given ordered (all U=1, "cold start") or disordered (all U random, "hot start") initial configuration, successively each link is assigned a new element U , chosen randomly with the weight exp{-S(U)}. One traverse of this procedure through the entire lattice is called one iteration. In general, it is found that five hundred or so iterations provide reasonable first indications about the behavior of the energy density (26), but for some precision one should have more. The results shown in Fig. 3 for color SU(2) are obtained with typically around three thousand iterations, after which we observe quite stable behavior; the SU(3) results are generally based on a few hundred iterations. The work was done with $N_\sigma = 7,9,10$ for $N_\beta = 2,3,4,5$; apart from expected finite lattice size effects[7] there was no striking N_σ dependence of ε, suggesting that in general the thermodynamic limit is reached. To give at least some intuitive grounds for this, note that a $10^3 \times 3$ lattice has about 12,000 link degrees of freedom.

As a result of the Monte Carlo evaluation, we obtain for a lattice of given size (N_σ, N_β) the energy density ε as function of g. In the continuum limit, g and the lattice spacing a are for color SU(N) related through

$$\underline{a}\,\Lambda_L = (11Ng^2/48\pi^2)^{-51/121} \exp\left\{-24\pi^2/11Ng^2\right\} \tag{28}$$

This relation is found by requiring a dimensional parameter Λ_L to remain constant under scale changes accompanied by corresponding changes in coupling strength. Hence once we are in the region of validity of the continuum limit, Eq. (28) gives us the connection between g and a. Since $(N_\beta a)^{-1}$ is the temperature in units of Λ_L, we then have the desired continuum form of $\varepsilon(\beta)$.

In Fig. 3, we show the resulting energy density ε as function of the temperature T, for both SU(2)[5] and SU(3).[6] We first note that at high temperatures, the results of the Monte Carlo evaluation agree quite well with the anticipated Stefan-Boltzmann form

$$\varepsilon/T^4 = \begin{cases} \pi^2/5 & SU(2) \\ 8\pi^2/15 & SU(3) \end{cases} \tag{29}$$

Let us now go to lower T, concentrating on the SU(2) case. At about $T = 50\,\Lambda_L$, ε drops sharply. The derivative of ε gives us the specific heat, shown in Fig. 4(a). At $T \simeq 43\,\Lambda_L$, it has a singularity-like peak, which signals the transition from free to bound gluons. With Λ_L taken in physical units, this gives us $T_c \simeq$ 180-200 MeV; for SU(3), we find similarly $T_c \simeq$ 160-180 MeV. How do we know that it is deconfinement which occurs here? One can study the behavior of a static $q\bar{q}$ pair immersed in a gluon system of temperature T;[8,9] the free energy F of an isolated quark then serves to define the thermal Wilson loop $\langle L\rangle = \exp\{-\beta F\}$ as order parameter. It is found that $\langle L\rangle$ is essentially zero below and non-zero above T_c [see Fig. 4(b)]. Since $\langle L\rangle = 0$ corresponds to an infinite free energy of an isolated color source, we have confinement below T_c. In accord with this, it can also be shown that for $T < T_c$ the system behaves essentially as a gas of gluonium states.[10]

All lattice results presented here were obtained with the Wilson form (24) of the action, which provides the correct continuum limit. There are, however, other lattice actions which also do this, and we may therefore ask if deconfinement, both qualitatively and

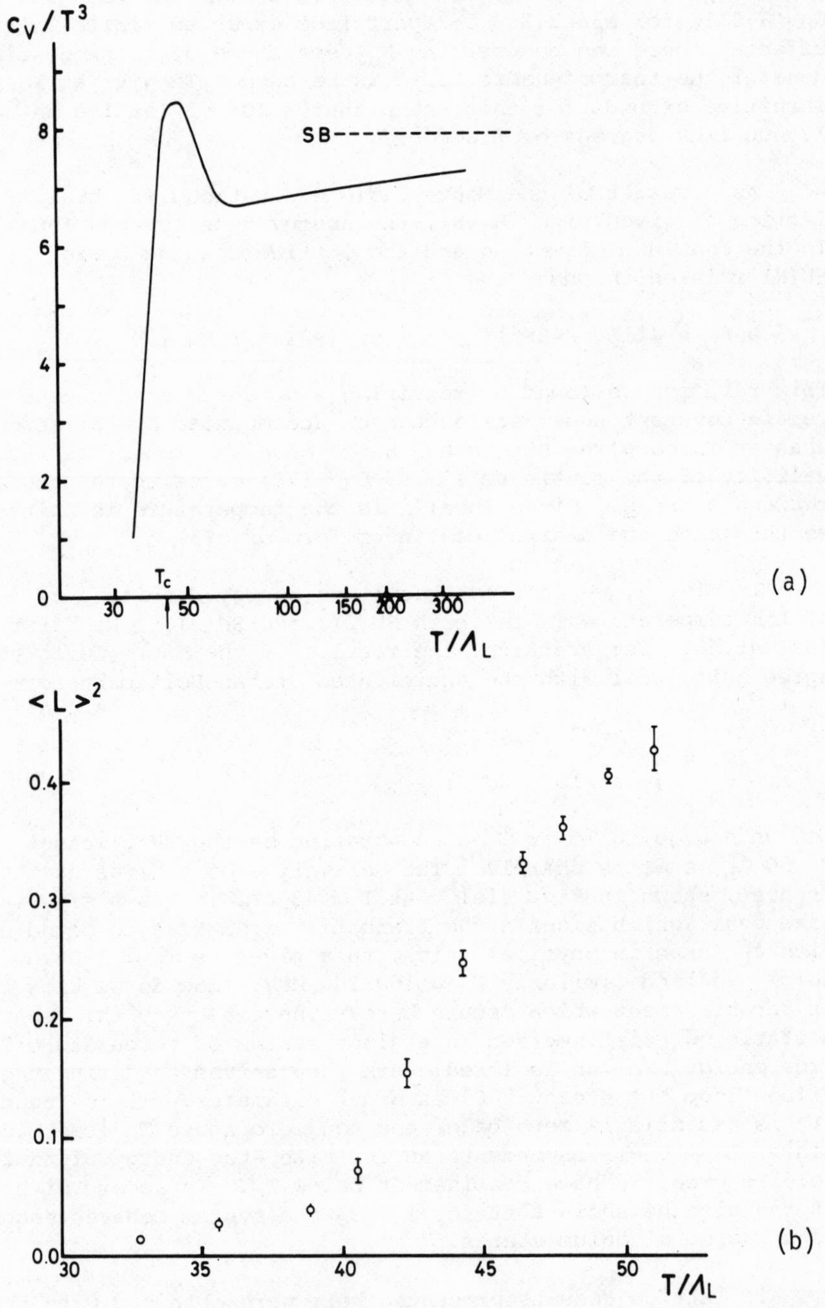

Fig. 4. Specific heat (a) and squared order parameter (b) for the SU(2) Yang-Mills system from Ref. 5.

quantitatively, is independent of the choice of action. It was recently shown that this is indeed the case.[11]

For the Yang-Mills system, we have thus seen that the lattice formulation together with Monte Carlo techniques allow us to evaluate gluon thermodynamics over the whole temperature range. The resulting behavior shows the expected two-phase nature: at low temperatures, we have a hadronic resonance gas of gluonium states; heating brings us to a deconfinement transition and beyond that to an ideal gluon gas.

We now want to extend our considerations to include quarks and antiquarks. We shall see that this brings in a basically new feature—the question of chiral symmetry restoration at high temperature. The lattice formulation encounters as a result the problem of species doubling,[2,12] and in addition the Monte Carlo evaluation becomes considerably more complex. Nevertheless, first results both on the full QCD energy density[13] and on chiral symmetry restoration[13,14] have now appeared; we shall first consider the former and then return to chiral symmetry questions.

For the full Lagrange density, the Euclidean form of the partition function on the lattice is now given by

$$Z = \int \prod_{\text{links}} dU \prod_{\text{sites}} d\psi\, d\bar{\psi}\; e^{-S^G(U) - S^F(U,\psi,\bar{\psi})} \tag{30}$$

with the dU integration to be carried out for all links, the $d\psi\, d\bar{\psi}$ integrations for all sites of the lattice. The fermion action S^F is taken in the form

$$S^F = \bar{\psi}(1 - KM)\psi \quad , \tag{31}$$

$$M_\mu = (1-\gamma_\mu)U_{nm}\,\delta_{n,m-\hat{\mu}} + (1+\gamma_\mu)U^\dagger_{mn}\,\delta_{n,m+\hat{\mu}} \quad , \tag{32}$$

while the gluon part S^G is given by Eq. (24); the coupling between quarks and gluons is given by the "hopping parameter" $K(g^2)$. The integration over the anticommuting spinor fields can be carried out[15] to give an effective boson form

$$Z = \int \prod_{\text{links}} dU\; e^{-S^G(U)} \det(1-KM) \quad . \tag{33}$$

The energy density ε is obtained from this Z; it becomes the sum $\varepsilon = \varepsilon^G + \varepsilon^F$ of a pure gluon part and a quark-gluon part[13]

$$\varepsilon^F \equiv -\xi^2 (N_\sigma^3 N_\beta a_\sigma^4 Z)^{-1} \int \prod_{\text{links}} dU\; e^{-S^G(U)} \det Q \;\times$$
$$\times \left\{ \frac{3K(g^2)}{4} \mathrm{Tr}(M_o Q^{-1}) - \frac{K(g^2)}{4} \sum_{\mu=1}^{3} \mathrm{Tr}(M_\mu Q^{-1}) \right\} \tag{34}$$

with $Q \equiv 1 - KM(U)$.

The computational problem beyond what is encountered in the pure Yang-Mills case lies in the evaluation of det Q and of Q^{-1}. We shall here use the expansion of these quantities in powers of the fermionic coupling K (hopping parameter expansion[16]), and retain in both cases only the leading term. For det Q the leading term is

$$\det Q = \det(1-KM) \simeq 1 \tag{35}$$

("quenched approximation"), while in the expansion

$$Q^{-1} = [1-KM]^{-1} = \sum_{\ell=0}^{\infty} K^{\ell}[M(U)]^{\ell} \quad , \tag{36}$$

because of gauge invariance, the first contribution to $\mathrm{Tr}(Q^{-1}M)$ arises for the shortest non-vanishing closed loop obtained from M(U). For $N_\beta = 2$ and 3, this is a thermal loop, i.e., one closed in the temperature direction; hence in that case, the first term is $\ell = N_\beta - 1$, and we obtain

$$\varepsilon^F a^4 \simeq \frac{3}{4}\,[K(g^2)]^{N_\beta}\, 2^{N_\beta+2} \, \langle L \rangle \tag{37}$$

with <L> for the expectation value of the thermal Wilson loop, and a for the lattice spacing. Comparing this with the leading term of the hopping parameter expansion for an ideal gas of massless fermions, ε^F_{SB}, we get

$$\varepsilon^F/\varepsilon^F_{SB} = [8K(g^2)]^{N_\beta} \, \langle L \rangle / N \tag{38}$$

since for the ideal gas $K = 1/8$, $\langle L \rangle = N$.

Taking $K(g^2)$ from a numerical evaluation[17] and using the Monte Carlo data[6] for <L>, we obtain for the SU(3) case the ratio $\varepsilon^F/\varepsilon^F_{SB}$ shown in Fig. 5. We note that the energy density takes on its asymptotic value for $T \gtrsim 100\ \Lambda_L$; around $T \sim 80\ \Lambda_L$ (~160 MeV), there is a sharp drop, corresponding to the onset of confinement.

For the SU(2) case, the restriction to the leading term of the hopping parameter expansion (36) has been removed;[18] including all terms up to order 50 results in the energy density shown in Fig. 5(b). We note that the qualitative features of Fig. 5(a) persist.

In Fig. 6, we show finally the overall energy density ε/T^4 for full QCD with color SU(3), obtained by combining the results for ε^F with those for the pure Yang-Mills system. We conclude that full quantum chromodynamics with fermions indeed appears to lead to the deconfinement behavior observed in the study of Yang-Mills systems alone. In particular, we note that at temperature $T \gtrsim 2T_c$ essentially all constituent degrees of freedom have been "thawed".

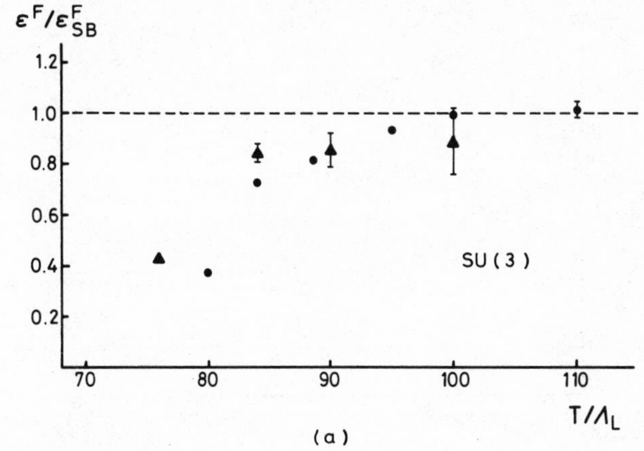

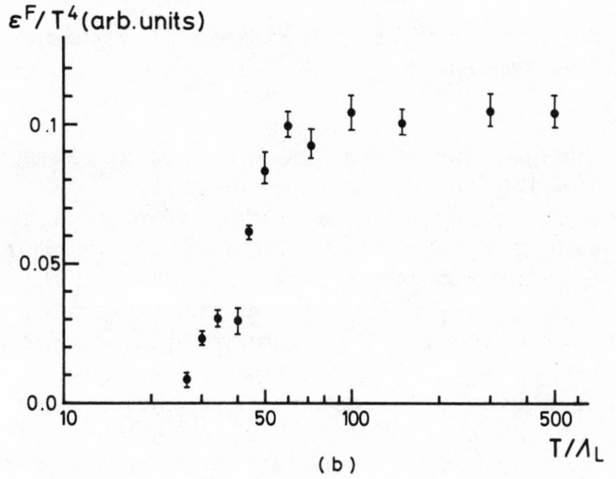

Fig. 5. (a) Energy density of the fermion sector, normalized to the ideal gas value ε^F_{SB}, for SU(3) Wilson fermions, leading term hopping parameter expansion, from Ref. 13.
(b) Energy density for SU(2) Wilson fermions, hopping parameter expansion up to order 50, from Ref. 18.

Quantum chromodynamics, for massless quarks *a priori* free of dimensional scales, contains the intrinsic potential for the spontaneous generation of two scales: one for the confinement force coupling quarks to form hadrons, and one for the chiral force binding the collective excitations to Goldstone bosons. These two lead in thermodynamics to two possible phase transitions, characterized by two critical temperatures, T_c and T_{ch}. Above T_c, the density is high enough to render confinement unimportant: hadrons dissolve into quarks and gluons. Above T_{ch}, chiral symmetry is restored, so that quarks must be massless. For T below both T_c and T_{ch}, we have a gas of massive hadrons; for T above both T_c and T_{ch}, we have a plasma of

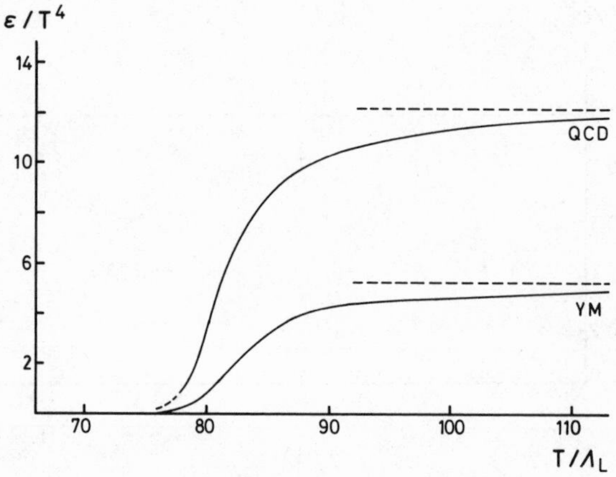

Fig. 6. Energy density of full QCD, compared to that of the SU(3) Yang-Mills system, from Ref. 13.

massless quarks and gluons. Conceptually simplest would be $T_c = T_{ch}$; the possibility $T_c > T_{ch}$ appears rather unlikely.[19] On the other hand, $T_c < T_{ch}$ would correspond to a regime of unbound massive "constituent" quarks, as they appear in the additive quark model for hadron-hadron and hadron-lepton interaction.[20] The question of deconfinement vs. chiral symmetry restoration thus confronts us with one of the most intriguing aspects of quark-gluon thermodynamics.

The fermionic action of Wilson used in the last section avoids species doubling at the cost of chiral invariance. Even an ideal gas of massless quarks in this formulation is not chirally invariant, since the expectation value $<\psi\bar{\psi}>$ is always different from zero. It has therefore been suggested[21] to use the difference between this "Stefan-Boltzmann" value and the corresponding QCD value for Wilson fermions as the physically meaningful order parameter: it would vanish when the behavior of a non-interacting system of massless fermions is reached.

In Fig. 7 we show this order parameter as calculated for color SU(2)[22] and SU(3),[13] in leading power of the hopping parameter expansion. It is non-zero up to

$$T_{ch} \simeq \begin{cases} 60\ \Lambda_L & SU(2) \\ 100\ \Lambda_L & SU(3) \end{cases} \tag{39}$$

and vanishes for higher temperatures. This suggests chiral symmetry restoration slightly above deconfinement, with

$$T_{ch}/T_c \simeq 1.3 \quad . \tag{40}$$

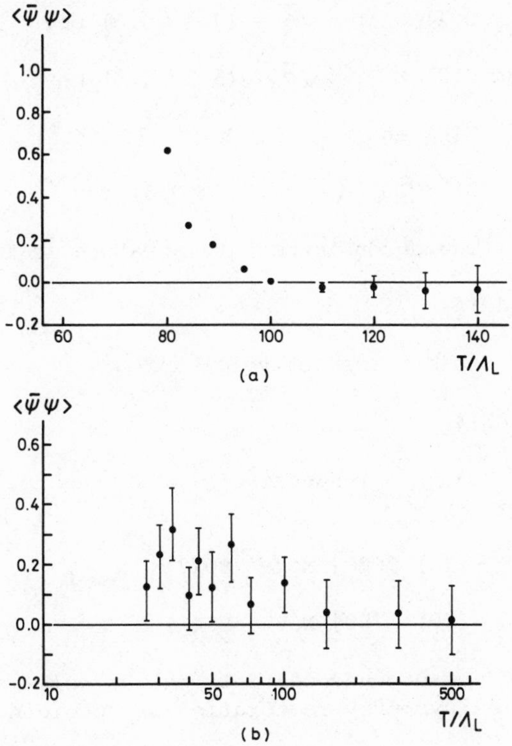

Fig. 7. Chiral symmetry order parameter, (a) for SU(3) Wilson fermions, from Ref. 13, and (b) for SU(2) Wilson fermions, from Ref. 18.

It remains open at present to what extent this will be modified by the inclusion of virtual quark loops, or if there are any significant finite lattice effects. Using for the SU(2) case a chirally invariant action with the resulting species doubling, it was found in Ref. 14 that chiral symmetry restoration occurs at

$$T_{ch} = (0.55 \pm 0.07)\sqrt{\sigma} \quad ; \tag{41}$$

this leads to similar conclusions on T_{ch}/T_c. It should be emphasized, however, that in view of possible finite size effects, both presently available calculations do not exclude the possibility $T_{ch} = T_c$.

In the lattice evaluation of QCD thermodynamics, we have calculated all physical quantities in terms of the dimensional lattice scale Λ_L. To convert Λ_L into physical units, we just have to measure one of these physical observables. String tension considerations give for the Yang-Mills systems

$$\Lambda_L = \begin{cases} (1.1 \pm 0.2) \times 10 \ \sqrt{\sigma} = (4.4 \pm 0.8)\ \text{MeV} & \text{(Ref. 22)} \\ (1.3 \pm 0.2) \times 10^{-2} \sqrt{\sigma} = (5.2 \pm 0.8)\ \text{MeV} & \text{(Ref. 23)} \end{cases} \tag{42}$$

in case of color SU(2) and

$$\Lambda_L = (5.0 \pm 1.5) \times 10^{-3} \sqrt{\sigma} = (2.0 \pm 0.6)\ \text{MeV} \qquad \text{(Ref. 24)} \tag{43}$$

for color SU(3). The deconfinement temperature is found to be

$$T_C = (38 - 43)\Lambda_L \tag{44}$$

for SU(2) (Refs. 8 and 5 respectively) and

$$T_C = (75 - 83)\Lambda_L \tag{45}$$

for SU(3) (Refs. 25 and 6 respectively). Taking the average of Eq. (42), we have

$$T_C = \begin{cases} [(170 - 210) \pm 30]\ \text{MeV} & \text{SU(2)} \\ [(150 - 170) \pm 50]\ \text{MeV} & \text{SU(3)} \end{cases} \tag{46}$$

and thus little or no dependence of T_C on the color group. The temperature for chiral symmetry restoration is accordingly given by relation (40).

From Eq. (46) and the form of Fig. 7, we can now estimate the energy density values at the two transition points. For the SU(3) Yang-Mills case, we obtain

$$\varepsilon(T_C) \simeq 200 - 300\ \text{MeV/fm}^3 \tag{47}$$

where we have assumed that the turn-over in ε occurs at about half the Stefan-Boltzmann value. This range, corresponding roughly to hadronic energy density, seems physically quite reasonable. It is not known at present if and how much it would be increased by the introduction of quarks; a shift proportional to that of the Stefan-Boltzmann limit would double the value of Eq. (47). This suggests twice standard nuclear density ($n_0 = 150$ MeV/fm^3) as lower and four times nuclear density as upper bound for the deconfinement transition. Chiral symmetry restoration, if it occurs at only slightly higher temperatures, requires considerably higher energy densities. Just a small increase beyond T_C brings up to the top of the Stefan-Boltzmann "shelf", where the energy density is above 2 GeV/fm^3.

Our basic conclusion is certainly that the lattice formulation of quantum chromodynamics appears to be an extremely fruitful approach to the thermodynamics of strongly interacting matter. It is so far the only way to describe within one theory the whole temperature range from hadronic matter to the quark-gluon plasma. It leads to deconfinement and provides first hints on chiral symmetry restoration.

We are still at the beginning. It is not really clear if $T_c \neq T_{ch}$, finite size scaling near the phase transitions has not been studied at all for $T \neq 0$, and the lattice thermodynamics of systems with non-zero baryon number has not been touched. Nevertheless, there seems to emerge today from QCD ever growing evidence for a two or three state picture of strongly interacting matter such as we have presented here.

ACKNOWLEDGMENT

It is a pleasure to thank J. Engels, R.V. Gavai, M. Gyulassy, F. Karsch and L. McLerran for stimulating discussions on various aspects of the topic.

REFERENCES

1. See e.g., "Quark Matter Formation and Heavy Ion Collisions," M. Jacob and H. Satz, eds., World Scientific Publ. Co., Singapore, (1982).
2. K. Wilson, Phys. Rev. D10:2445 (1974); in: "New Phenomena in Subnuclear Physics", A. Zichichi, ed., Plenum Press, New York (1977).
3. M. Creutz, Phys. Rev. D21:2308 (1980).
4. C. Bernard, Phys. Rev. D9:3312 (1974).
5. J. Engels, F. Karsch, I. Montvay and H. Satz, Phys. Lett. 101B: 89 (1981); Nucl. Phys. B205:545 [FS5] (1982).
6. I. Montvay and E. Pietarinen, Phys. Lett. 110B:148 (1982); Phys. Lett. 115B:151 (1982).
7. J. Engels, F. Karsch and H. Satz, Nucl. Phys. B205:239 [FS5] (1982).
8. L. D. McLerran and B. Svetitsky, Phys. Lett. 98B:195 (1981); Phys. Rev. D24:450 (1981).
9. J. Kuti, J. Polónyi and K. Szláchanyi, Phys. Lett. 98B:199 (1981).
10. J. Engels, F. Karsch, I. Montvay and H. Satz, Phys. Lett. 102B: 332 (1981).
11. R. V. Gavai, Nucl. Phys. B [FS], (in press); R. V. Gavai, F. Karsch and H. Satz, Bielefeld Report BI-TP 82/26 (1982).
12. L. Susskind, Phys. Rev. D16:3031 (1977).
13. J. Engels, F. Karsch and H. Satz, Phys. Lett. 113B:398 (1982).
14. J. Kogut, M. Stone, H. Wyld, J. Shigemitsu, S. Schenker and D. Sinclair, Phys. Rev. Lett. 48:1140 (1982).
15. T. Matthews and A. Salam, Nuovo Cim. 12:563 (1954); 2:120 (1955).

16. C. B. Lang and H. Nicolai, Nucl. Phys. B200:135 [FS4] (1982); A. Hasenfratz and P. Hasenfratz, Phys. Lett. 104B:489 (1981).
17. A. Hasenfratz, P. Hasenfratz, Z. Kunszt and C. B. Lang, Phys. Lett. 110B:289 (1982).
18. J. Engels and F. Karsch, to be published.
19. E. V. Shuryak, Phys. Lett. 107B:103 (1981); R. D. Pisarski, Phys. Lett. 110B:155 (1982).
20. H. Satz, Phys. Lett. 25B:27 (1967) and Phys. Lett. 25B:220 (1967). H. Satz, Nuovo Cim. 37A:141 (1977).
21. C. B. Lang and H. Nicolai, Nucl. Phys. B200:135 [FS4] (1982).
22. G. Bhanot and C. Rebbi, Nucl. Phys. B180:469 [FS2] (1981).
23. M. Creutz, Phys. Rev. Lett. 45:313 (1980).
24. See Ref. 23 as well as E. Pietarinen, Nucl. Phys. B190:239 [FS3] (1981).
25. K. Kajantie, C. Montonen and E. Pietarinen, Z. Phys. C9:253 (1981).

ANOMALONS, HONEY AND GLUE IN NUCLEAR COLLISIONS

M. Gyulassy

Lawrence Berkeley Laboratory
University of California
Berkeley, CA 94720

INTRODUCTION

In these lectures, I will cover three rapidly evolving areas of current research in the field of high energy nuclear collisions. Extensive reviews of additional topics can be found in Refs. 1 and 2. The topics covered here address the following questions:

i. Do novel nuclear states exist?
ii. Do nuclei flow like honey?
iii. Can a quark-gluon plasma be produced?

These questions span a wide energy range from $\sim$200 AMeV (topic ii), $\sim$2 AGeV (topic i), to $\sim$1 ATeV (topic iii). (Note AMeV $\equiv$ MeV per incident projectile nucleon, etc.) In the following section I discuss the evidence for anomalons. These are relativistic nuclear fragments that tend to interact much more frequently in matter than do familiar nuclei with the same charge. Various speculations about these objects are discussed. In particular, we ask whether anomalons could be hungry quarks with a ravenous appetite for nucleons. I also discuss the status of the search for pion field coherence in nuclear collisions. With regard to question ii, I review in the third section the in(con)clusive data showing that collective nuclear flow could take place at lower energies. This question is of interest because through the flow characteristics we hope to learn about the nuclear equation of state at high densities. Then the topic of global event analysis is discussed and cascade calculations illustrating various flow phenomena are presented. Finally, in the last section, boosting ourselves to the erg per nucleon (ATeV) energy region, the prospects of probing the quark-gluon plasma are analyzed. Nuclear stopping power and longitudinal growth are discussed.

Cosmic ray data with events showing charge particle multiplicities up to 1000 are analyzed. These data are interpreted as showing that high enough energy densities can indeed be achieved in nuclear collisions to produce an ideal quark-gluon plasma.

NOVEL NUCLEAR STATES

Anomalons

Since early cosmic ray studies, there have been recurring observations in emulsions of anomalous projectile fragments.[3-6] These fragments are anomalous because their reaction mean free path, $\lambda_a(Z)$, is much smaller than expected for nuclei with charge Z. Primary nuclei with $2 \leq Z \leq 26$ and kinetic energy between 0.2 and 2.0 AGeV are found to have a mean free path

$$\lambda(Z) = \left\{ \sum_{A \varepsilon Em} \rho_A \, \sigma_A(Z) \right\}^{-1} \tag{1}$$

where ρ_A is the density of nuclei with atomic number A in emulsions and $\sigma_A(Z)$ is the usual geometrical cross section ($r_o \approx 1.2$ fm)

$$\sigma_A(Z) \approx \pi r_o^2 \left(A^{1/3} + (2Z)^{1/3}\right)^2 \quad . \tag{2}$$

Equations (1) and (2) arise because normal nuclei behave essentially as black discs. Often it is convenient to group all charges together by parametrizing

$$\lambda(Z) = \Lambda^* Z^{-b} \quad . \tag{3}$$

Measurements of primary mean free paths[5,6] give $\Lambda^* \approx$ 25-30 cm and $b \approx 0.34 \pm 0.3$.

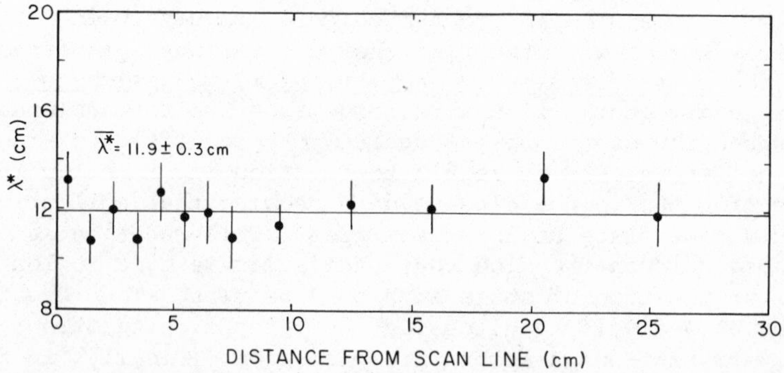

Fig. 1. Mean free path of primary ^{16}O at 2.1 AGeV in emulsion as a function of distance from the scan line.[6]

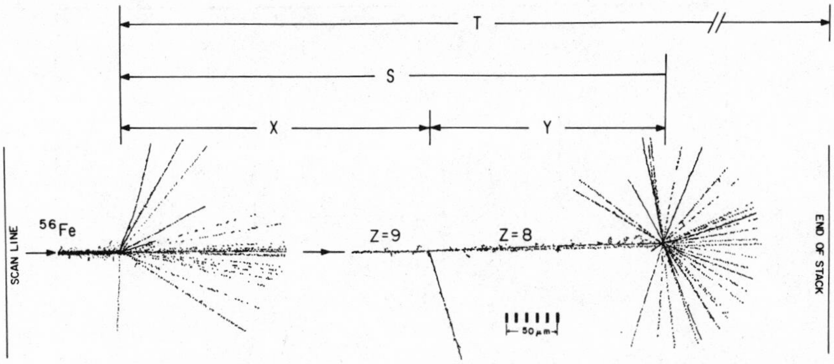

Fig. 2. Two chain event[6] produced by ^{56}Fe at 1.88 AGeV in emulsion. A secondary with Z=9 travels a distance $X = 2.6$ cm before interacting. A tertiary with $Z = 8$ then travels a distance $Y = 0.02$ cm before interacting. The mean free path for primaries with $Z \sim 8$-9 is $\lambda_{Em} \approx 10$ cm.

In practice Λ^* is determined by measuring path lengths l_i traversed by many nuclei and counting the total number of interactions according to

$$\Lambda^* = \sum_{D<l_i<D+\Delta} (l_i - D)\ Z_i^b / N(D,\Delta) \quad , \tag{4}$$

where D is an arbitrary distance into the emulsion and $N(D,\Delta)$ is the number of interaction stars observed between D and $D+\Delta$. By the definition of mean free paths, Λ^* should be independent of both D and Δ. As shown in Fig. 1, such an independence of Λ^* and D is indeed observed[6] for primary nuclei (in this case, ^{16}O).

However, the mean free paths of secondary fragments produced in nuclear collisions at ~2 AGeV do appear to be dependent on D. A "typical" multichain event seen in the emulsion is shown in Fig. 2. An incident ^{56}Fe beam collides with an Ag or Br nucleus producing among other fragments a charge Z=9 relativistic fragment. At a distance X from the primary vertex that relativistic fragment collides with another emulsion nucleus. A tertiary fragment with Z=8 is produced. This propagates a distance Y before it too interacts in the emulsion. For this event $l_i = X$ and D is the distance from the primary vertex when used in Eq. (4). On the order of 1000 events with secondary interactions have been analyzed in this way.

In Fig. 3 we see the striking result of that analysis. While $\Lambda^*(D)$ is independent of D for distances $D > 5$ cm from the primary vertex, there appears to be a significant reduction of Λ^* for $D < 5$ cm (Δ is typically ~1 cm in this analysis).

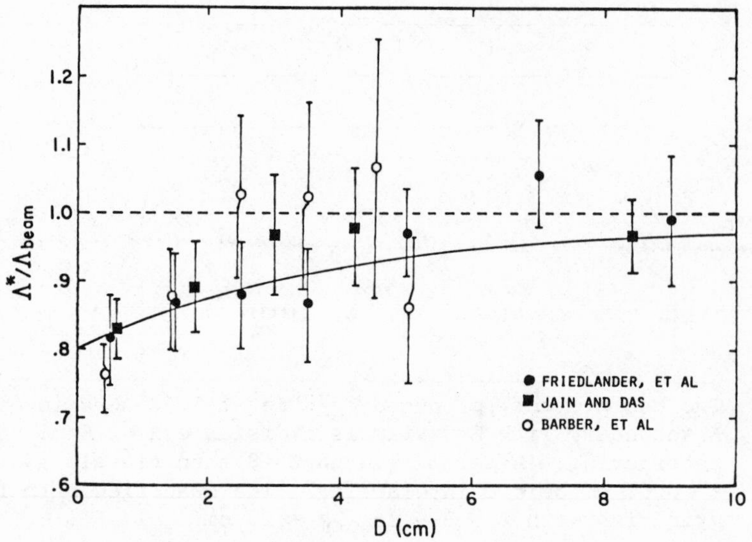

Fig. 3. The reduced mean free path Λ^*, Eq. (4), as a function of distance, D, from the primary vertex. Results from three emulsion experiments[2-4] show significant dependence of Λ^* for $D < 5$ cm.

As a parametrization to account for this dependence on D, the following model was proposed in Refs. 3, and 6: they assumed that a fraction α of all secondary fragments have an anomalously short mean free path λ_a. The solid curve in Fig. 3 was obtained by setting $\alpha = 0.06$ and $\lambda_a = 2.5$ cm. In Fig. 4 the likelihood function for different values of α and λ_a is shown. The greatest compatibility with the data is achieved for $\alpha < 0.2$ and $\lambda_a < 6$ cm. For comparison, primary nuclei with $Z < 26$ have $\lambda(Z) \gtrsim 10$ cm. Of course this parametrization is not unique. In Ref. 5 an equally good fit was obtained by assuming that 100 percent of the secondaries are anomalous but that this anomalous property decays away after $\sim 10^{-10}$ sec.

A rather mysterious property of anomalons is that global reaction characteristics, e.g., the heavy prong-multiplicity distribution $P(N_h)$, of anomalous and normal nuclear reactions do not seem to differ. In addition, there is (weak) evidence[6] that anomalons tend to produce anomalons. Thus in Fig. 2, if X is short compared to $\lambda(Z)$, then Y is also likely to be short compared to $\lambda(Z)$. This has been called the memory effect. Finally, the anomalous effect is seen convincingly only for fragments with $Z \geq 3$.

These observations taken together rule out many conventional explanations. For example, they cannot be hyper-fragments or pionic atoms because the decay of these secondaries would result in the

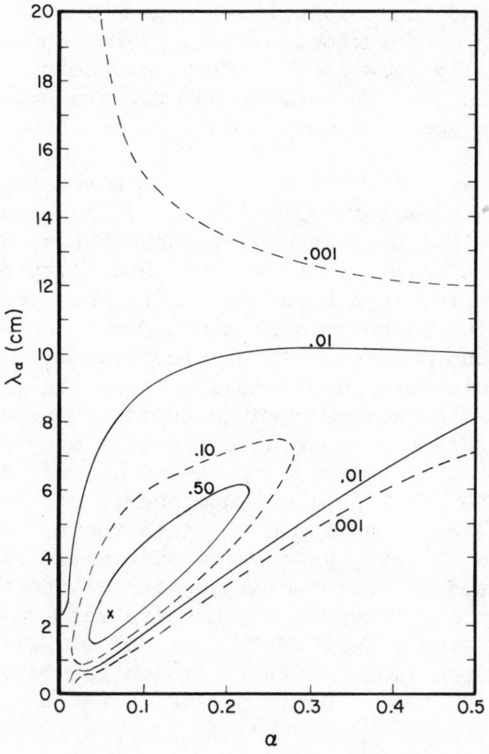

Fig. 4. Normalized likelihood contours[6] for fits to Fig. 3 assuming a fraction α of all secondaries have anomalous mean free path λ_a. The cross indicates the position of maximum likelihood.

enhancement of $P(N_h)$ for $N_h = 0,1$. Such enhancement is ruled out by the data. Nuclear molecules[7] cannot account for the observation of anomalons with $Z < 10$ nor the apparent memory effect. Neutron excess isotopes[7] cannot account for the dependence of Λ^* on D. Finally, ordinary nuclear excitations have too short a lifetime $\ll 10^{-10}$ to account for these results.

We are thus left with a puzzle. However, it is important to emphasize that only limited statistics are available and unknown systematic biases may plague emulsion data. Much higher statistics and different techniques will be needed before these anomalons can be regarded as proven. Today we can only say that this observation is the most baffling result obtained thus far at the Bevalac.

Hungry Quarks

Because anomalons are so weird, they have fueled theoretical speculations beyond the conventional boundaries of nuclear physics. Probably all such speculations are wrong, but they illustrate well how little we really know about nuclear or hadronic matter under extreme conditions of high density and temperature. In this section, I expose two wild speculations: topolons and hungry quarks.

Topolons is a generic name I give to novel topological configurations of hadronic matter. Since Wheeler, many calculations have exploited topological degrees of freedom. The main problem that cast doubt on past calculations was the stability of such configuration. More recently, it has been realized that certain classes of field theories possess nontrivial topological sectors with conserved topological quantum numbers. In particular, effective chiral Lagrangians exhibit this phenomenon.[8] Thus, the problem of stability is solved by a new conserved quantum number. These topological excitations possess rather shocking properties such as an effective fractional charge. In the model proposed in Ref. 8, the solitons predicted have the additional weird property of having high baryon *and* strangeness (>6 in magnitude). Since these states correspond to large extended objects, production of them in e^+e^- or pp collisions is suppressed by form factor and tunneling effects. However, in nuclear collisions extended objects are easy to create and a baryon number reservoir is readily available. Multiple strangeness can also be produced (more easily) through the many independent NN collisions. It is thus remotely possible that anomalons are topolons.

Further support of this conjecture can be found in the work of Boguta[9] who was able to construct chiral models consistent with normal nuclear properties. Obviously any model purporting to predict a new state of matter must be compatible with the old well-loved ones. By introducing an effective gauge theory for nuclear matter, Boguta finds that not only can normal finite nuclear properties be accounted for but also new topological excitations can be found. He calls them hadroids. Such calculations are still very preliminary but indicate that anomalons could be made compatible with nuclei within an effective chiral theory of nuclear matter.

Another exotic line of speculation revolves around the heretical view that maybe QCD is not the final word on the subject. Suppose that color is not rigorously confined! It could be that color SU(3) is spontaneously broken by some nasty Higgs mechanism.[10-13] If that is the case, then there could exist colored nuclear states or quark-nuclear complexes.

As a fun example of what such states could look like, I review the hungry quark states suggested in Ref. 10. Suppose that we add an effective interaction $L' = (\mu^2/2)A_{\mu a}A^{\mu a}$ to the QCD Lagrangian. (There is considerable debate on whether such an addition is consistent with verified QCD predictions.[11]) Thus, we suppose that gluons have a small mass μ. Consider now an MIT bag in which we place a *static* color source density ρ_a. The color potential then satisfies:

$$(\nabla^2 - \mu^2)A_a^o = -g\rho_o \quad . \tag{5}$$

Integrating Eq. (5) over the bag volume V and noting the boundary condition $\vec{n}\cdot\vec{\nabla}A_a^o = 0$ on the surface, one finds that

$$A_a^o = \frac{gQ_a}{\mu^2 V} + O(\mu^2) \tag{6}$$

where $Q_a = \int_V \rho_a$ is the total color charge. Thus, there is a field energy

$$E_F = g/2\int_V A_a^o\rho_a \approx \frac{2\pi\alpha_s}{\mu^2 V}\,C^2 \tag{7}$$

where $\alpha_s = g^2/4\pi$ and $C^2 = \Sigma Q_a^2$ is the Casimir of SU(3). ($C^2 = 16/3$, 12 for $\underset{\sim}{3}$ and $\underset{\sim}{8}$ representations.) Equation (7) shows that if $\mu \to 0$ the field energy diverges. Thus, confinement of color arises only if $\mu=0$. However, for $\mu \neq 0$ the colored state has finite energy $E = E_F + BV$, where $B \approx m_\pi^4$ is the vacuum pressure. Minimizing with respect to V, Ref. 10 finds the mass and radius of the state is

$$M = 2BV \approx 10\ \text{GeV}\left(\frac{18\ \text{MeV}}{\mu}\right) r^{1/2}$$
$$R = 2.8\ \text{fm}\left(\frac{18\ \text{MeV}}{\mu}\right)^{1/3} r^{1/6} \tag{8}$$

where $r = 3C^2/16$. Note $r=1$ for a quark-like state and $r = 9/4$ for a gluon-like state.

Thus, in this model a bare quark weighs ~10 GeV and has a radius ~3 fm. The reason it is so big is that the enormous color field pressure inside can push back the vacuum pressure B confining the quark to the bag. As shown in Ref. 10, such quark states are unlikely to have been produced in e^+e^- or hh collisions due to a large form factor suppression.

The most interesting property of these quark states is their appetite for nucleons. This is estimated by assuming that a nucleon unfortunate enough to get inside a quark bag will be dissolved into its constituents. (By definition the inside of a bag corresponds to the perturbative phase.) Adding A nucleons, a Fermi sea of up-down quarks is thereby filled up in the bag to a Fermi momentum k_F ($2k_F^3/\pi^2 = 3A/V$). The energy of this sea is

$$E_{sea} = V\,\frac{3k_F^4}{2\pi^2}\,f \tag{9}$$

where f is a fudge factor incorporating interactions in the sea: $f = 1 + 8\alpha_s/3\pi \approx 1.5$ for the MIT bag. Minimizing the total energy $E_F + BV + E_{sea}$ with respect to V gives

$$M = M_Q \frac{1+x}{\sqrt{1-x}} ,$$

$$V = \frac{M_Q}{2B} \frac{1}{\sqrt{1-x}} ,$$

$$A = \frac{M_Q}{3\sqrt{\pi}B^{1/4}} \frac{(2x/f)^{3/4}}{\sqrt{1-x}} , \tag{10}$$

where $x = E_{sea}/3BV$ is the ratio of the pressure exerted by the sea to that of the vacuum. M_Q is given by Eq. (8). Quarks will continue to eat nucleons as long as

$$\mu = \frac{\partial M}{\partial A} = 3\ (2\pi^2 Bx)^{1/4}\ f^{3/4} \leq M_N . \tag{11}$$

The value of A for which $\mu = M_N$ gives the appetite of the quark. Give the volume V, the mean free path $\lambda_Q(A)$ of this object in emulsion is given by Eq. (1) with $\sigma_A(Z)$ replaced by

$$\sigma_{AQ} \approx \pi\left(r_0 A^{1/3} + (3V/4\pi)^{1/3}\right)^2 . \tag{12}$$

In Fig. 5, the ratio of $\lambda_Q(A)$ to $\lambda(A)$ for normal nuclei with atomic number A and the energy per nucleon M_Q/A of the fat quark are shown versus A. We see that for "reasonable" fudge factors $f \sim 1.5$, up to 20 nucleons can be eaten by a quark. The weight of this satiated quark is ~2 GeV per nucleon. By fudging a little ($f \sim 1.2$) this mass can be brought down but then λ_Q/λ_A is no longer less than 1. In fact, such quark-nuclear complexes may be even *smaller* than normal nuclei.

This fun calculation illustrates that if QCD is slightly broken, then rather weird and perhaps anomalous phenomena could arise in nuclear collisions. Therefore, it is worthwhile to keep an eye out for such beasts.

The prediction of this model that is easiest to test is the occurrence of fractionally charged, high baryon number fragments. Currently, plastics give the highest charge resolution and experiments by the Price group[14] are being carried out. The preliminary results indicate *no* fractionally charged nuclei observed at a level of 10^{-3}. If this result is confirmed, then a large class of speculations can be ruled out at this energy range. Of course, integral charged topolons and glue-nuclear complexes could still be responsible for anomalons in that case.

A natural question at this point is why we have not observed

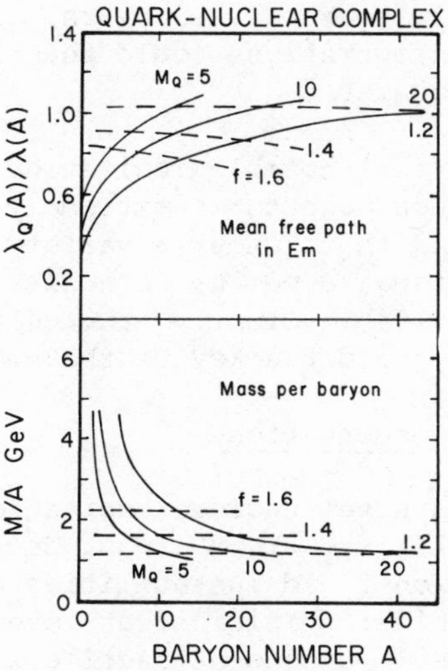

Fig. 5. The ratio of the mean free path $\lambda_Q(A)$ for quark-nuclear complexes to the normal nuclear mean free path $\lambda(A)$ in emulsions is shown on top as a function of the total number of nucleons eaten by the quark. On bottom the mass per eaten nucleon is shown. Solid curves are for different bare quark masses. Dashed curves define the appetite of a quark for different interaction fudge factors. The intersection of a dashed and solid curve gives the value of A for which $\partial M/\partial A = M_n$.

any of these novel objects in e^+e^- or pp collisions. In order to produce colored states it is necessary to pull two quarks or gluons apart by at least $2R_Q$, where $R_Q \sim 3$ fm is the radius of the quark bag. However, as we try to pull a quark and antiquarks apart, a color electric tube of area $a \simeq 4\alpha_s\pi^2\alpha'$ and electric field $E = g/a$ is formed. Numerical estimates give $gE \sim 1.8$ GeV/fm. Using the Schwinger formula[15] the probability that *no* quark-antiquark pairs are produced by that electric field in a tube of length $2R_Q$ is given by

$$P_0(2R_Q) \sim \exp\left[-\frac{2}{3}\alpha_s gER_Q^2\right] \sim \exp\left[-3(R_Q/\text{fm})^2\right] . \tag{13}$$

If $R \sim 3$ fm, then there is only a 10^{-12} probability that the tube can be stretched out far enough without creating additional pairs that neutralize the color field. It is thus far more likely that several pairs are produced in the flux tube before the tube reaches the critical size ~6 fm necessary to isolate quarks. The advantage offered by nuclei is that they are big. *If* a quark-gluon plasma can be produced in nuclear collisions (see the last section), then a quark and antiquark can separate by twice the nuclear radius without forming a flux tube. That is because the plasma state is perturbative. Once having separated by that large distance a rearrangement of field configurations could then lead to the production of isolated extended quarks.

Finally, we note that cosmological evidence for the existence or non-existence of such objects is extremely ambiguous. For example, Ref. 16 showed that, under a variety of plausible scenarios, the number of relic diquarks per nucleon lies somewhere within the narrow range 10^{-50} to 10^{10}! Clearly, the door is open to discovery. Nuclear collisions may hold the key to that door.

Last Bastion of Pion Condensation

The last topic on novel nuclear excitations that I cover here is pion field instabilities. In the past decade much interest has focused on possible pion field instabilities in nuclear matter at high densities.[17] Nuclear collisions at several hundred MeV per nucleon were recognized[18] as one attractive way to search for such phenomena. Simple estimates showed that densities up to $\rho \sim 4\rho_0$ ($\rho_0 = 0.145\ \text{fm}^{-3}$) could be easily generated. However, along with high densities high internal excitation energies must also be taken into account. In particular, the *non-equilibrium* configuration of two interpenetrating nuclear fluids should be considered. This configuration is characterized by the momentum space distribution

$$n(p) = \theta(p_F - |\vec{p}+\vec{p}^*|) + \theta(p_F - |\vec{p}-\vec{p}^*|) \quad . \tag{14}$$

We consider whether there are pion field instabilities in such a configuration. This problem is then analogous to the study of the two-stream instability in colliding plasmas.

The starting point is the calculation[18] of the pion propagator in nuclear matter with a momentum distribution given by Eq. (14). The essential point is that the pion self energy for $\omega=0$ modes is controlled by the particle-hole propagator, $\Pi_0\ (\omega,k)$, which is the Lindhard function appropriate for a zero temperature Fermi gas with Fermi momentum p_F. Recall[18] that for $\omega=0$, $\text{Re}\Pi_0(0,k) \propto p_F$ and that Π_0 is proportional to the pion self energy in one isolated Fermi sphere. With Eq. (14), each Fermi sphere is boosted and therefore the self energy is Doppler shifted[18]

$$\Pi(\omega,\vec{k}) = \Pi_0(\omega+\vec{v}^*\cdot\vec{k},\vec{k}) + \Pi_0(\omega-\vec{v}^*\cdot\vec{k},\vec{k}) \quad . \tag{15}$$

Equation (15) shows that the only standing waves ($\omega=0$) in the non-equilibrium configuration that remain standing waves in the isolated Fermi spheres are those with $\vec{v}^*\cdot\vec{k}=0$. In other words, only pion wave vectors $\vec{k}$ that are perpendicular to the beam axis can remain standing in the non-equilibrium configuration. Furthermore, for such perpendicular modes there is a large *dynamical* enhancement of the self energy

$$\Pi(0,k_\perp) = 2\Pi_0(0,k_\perp) \approx \Pi_0(0,k_\perp)\big|_{\rho=8\rho_0} \tag{16}$$

Indeed, detailed calculations[18] showed that pionic instabilities are concentrated in modes perpendicular to the beam axis. The growth rate of those modes was found to be $\gamma(k) \sim 0.1m_\pi$ for $k_\perp \sim 2m_\pi$. The large magnitude of the wave vector follows from the p-wave nature of the πNN interaction.

How would one look for such an instability? The main effect of such an instability would be the growth of a collective spin-isospin wave in the medium.[19] Because the spin-isospin current is a source of pions, any time-dependent spin-isospin wave would radiate pions. In particular, the invariant inclusive pion distribution is given by

$$\omega_k \frac{d^3n}{dk^3} = \frac{1}{2(2\pi)^3} \int d^4x d^4y \; e^{-ik(x-y)} \langle in| j_H^+(x) j_H(y) |in\rangle \tag{17}$$

where $|in\rangle$ is the state of two colliding nuclei and $j(x)$ is the divergence of the axial vector current. Thus, the pion spectra measure the on-shell Fourier transform of the spin-isospin current fluctuation. If a macroscopic collective current is generated through an instability, then Eq. (17) reduces to the well-known formula for radiation

$$\omega_k \frac{d^3n}{dk^3} = \frac{1}{2(2\pi)^3} |j(\omega_k, k)|^2 \tag{18}$$

Based on the growth rates computed in Ref. 18 and the mean field results for $j(x)$ in the presence of a pion condensate,[20] a model for the transient growth of $j(x,t)$ was constructed in Ref. 19. The predicted pion spectra are shown in Fig. 6. Although the magnitude of the pion radiation is very small, the predicted peaking of the coherent yield at large perpendicular momenta ($k_\perp \sim 2m_\pi$) makes it possible to search for this tiny signal amidst the incoherent pion yield. At high energies ($\geq$400 MeV/nucleon), ordinary incoherent $NN \rightarrow NN\pi$ processes dominate. However, below threshold the incoherent pion component is greatly suppressed, while the coherent pions due to pionic instabilities are approximately beam energy independent. This is because for perpendicular modes the self energy, Eq. (15), is independent of p^*.

New data[21] at 0.2 GeV/nucleon have provided a stringent test for coherent pions generated via pionic instabilities. As seen from Fig. 6, the incoherent background is suppressed a factor 10^5 relative to the 2 GeV/N yield at $k_\perp \sim 2m_\pi$. Unfortunately, no deviation from a pure (phase space) exponential is observed. Therefore, these data rule out the possibility of a collective spin-isospin wave in this reaction.

At least two possible explanations for the results in Fig. 6

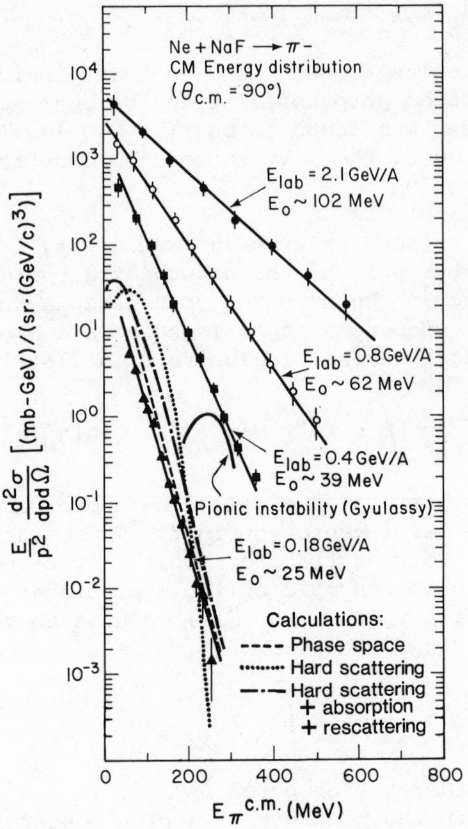

Fig. 6. Pion spectra at $\theta_{cm} = 90°$ produced in Ne + NaF 0.2-2.1 AGeV reactions.[21] An estimate[19] of the order of magnitude of the contribution due to spin-isospin current instabilities is indicated. The data at 0.2 AGeV rule out such instabilities and are consistent with simple phase space models.

can be put forward. First, if the effective Migdal parameter[17] g', representing all other interactions in the pion channel besides one pion exchange is large ($g' > 0.6$), then no instability can occur.[18] Second, Ne may be too small a system and too diffuse to be able to support a spin-isospin wave of wavelength $\lambda = 2\pi/k \sim 4.4$ fm. The second possibility is being tested currently via a similar experiment involving Xe + Xe. However, other searchers[22] for critical pion field effects tend to support the first explanation. In any case, Fig. 6 is an impressive demonstration of the high sensitivity achieved in present experiments, which now explore cross sections down to 1 μb/GeV^2 and have the ability to search for such needles in a haystack. If no positive results are observed with Xe + Xe, the last bastion of pion condensation will shift to the cores of neutron stars.[23]

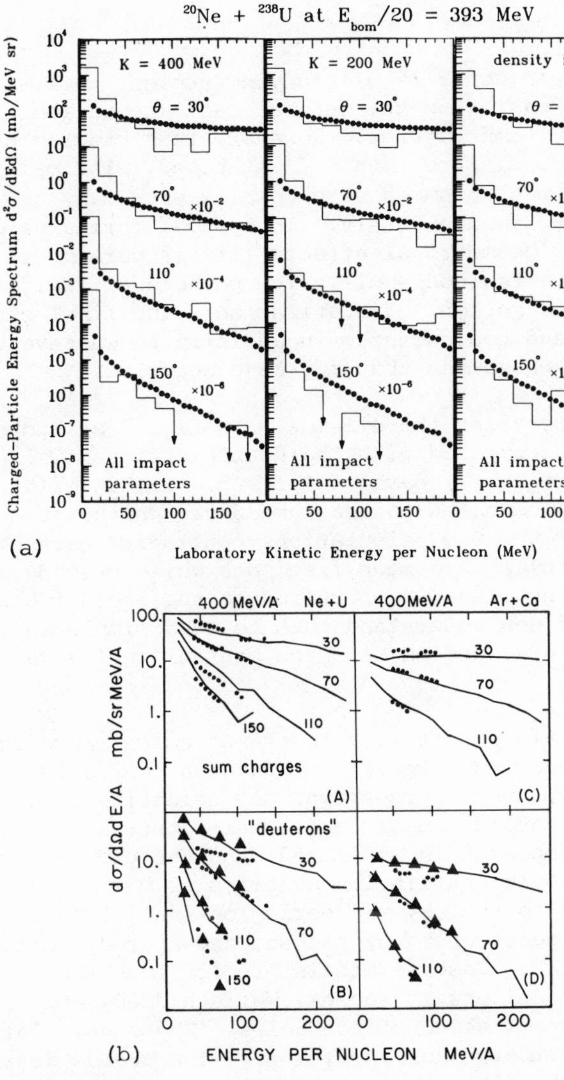

Fig. 7. (a) Comparison of charge inclusive data[25] with nonviscous hydrodynamical calculations.[24] The dots are the data and the histograms are calculations. Three different equations of state were tested. (b) Comparison of data[25] with calculations[31] using the Cugnon cascade code.[26] In parts (A), (C) charge inclusive distributions are shown. In parts (B) and (D) primordial deuteron distributions[31] are shown. The dots in (B), (D) show free deuteron spectra.[25] Triangles include d, t, ^{3}He, α.

DO NUCLEI FLOW?

In(con)clusive Data

One of the main goals[1,2] of studying nuclear collisions is to extract some information on the nuclear matter equation of state at high densities and temperatures. However, this program has turned out to be much more difficult than first hoped. The main reason is that the reaction mechanism is very complex with many competing processes going on simultaneously. In typical inclusive data we see the convolution of geometrical effects (impact parameter), finite mean free path effects (non-equilibrium phenomena), initial state interactions (Fermi motion, clusterization), and final state interactions (Coulomb and coalescence) in addition to whatever equilibrium and collective phenomena that may have occurred.

Figure 7 illustrates the dilemma. In Fig. 7(a) hydrodynamic calculations[24] are compared with the double differential charge inclusive ($\sigma_p + \sigma_d + 2\sigma_\alpha + \ldots$) data of Ref. 25. In Fig. 7(b) cascade calculations[26] are compared to the same data. Although the dynamical assumptions of the two calculations contradict each other (hydrodynamics assumes zero mean free path while cascade assumes long mean free paths), they both reproduce the inclusive data to the same accuracy. We now understand this to be a simple reflection of the fact that geometry and phase space control inclusive cross sections.[1]

In order to gain sensitivity to the dynamics, more exclusive data need to be analyzed. In Fig. 8, data on centrally triggered (high multiplicity) events are shown. By selecting a high multiplicity, the trivial convolution of impact parameters is reduced, the non-equilibrium component is minimized, and the probability of finding signatures of bulk equilibrium dynamics is maximized. The first feature to observe is that by triggering on small impact parameters, the theoretical predictions[27] of cascade and hydrodynamical models differ substantially. Cascade models predict that the inclusive yields are maximum at forward angles, while hydrodynamic models predict a sidewards emission of particles. These calculations refer to the matter flow including p,d,t,α, ... The *proton* data[28] are shown in the middle left. The great excitement arose initially because it seemed that for the first time hydrodynamic calculations were closer to the observed angular distribution than were cascade predictions. Since then the size of the experimental error bars were found to be too optimistic. Preliminary high statistics plastic ball results[29] show only a flattening at forward angles with no dip at 0°. In addition, preliminary neutron spectra[30] have caused havoc by showing a large forward *enhancement* in accord with cascade results. It was also realized[31] that copious production of composite fragments in the forward direction could cause a depletion of the unbound proton yield at zero degrees. Thus, Fig. 8 can no longer be regarded as evidence for hydrodynamical flow.

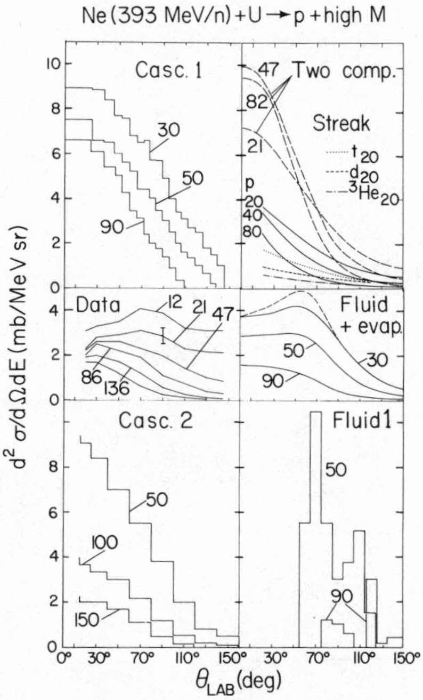

Fig. 8. Angular distribution of protons with different kinetic energies (middle left)[28] are compared to two cascade calculations, two hydrodynamic calculations and a firestreak calculation as described in Ref. 27. The reaction Ne + U at 393 MeV triggered on high multiplicity is considered.

Another observation which suggested nuclear flow phenomena is the difference between the "temperature" (slope parameter of the inclusive 90° cm yield) of protons and pions.[32] The idea was that a radial exploding fireball naturally leads to a difference on the slope parameter due to simple kinematics of superimposing collective and random motions for different mass particles. However, the recent observation[1] of an apparent kaon temperature exceeding that of protons contradicts such a model unless kaons are assumed to be decoupled from the flow. Hence, this evidence is also inconclusive.

The most recent attempt fo find evidence for compression effects was reported in Ref. 33. An outstanding problem in this field is to understand the excitation function of pions, i.e., the magnitude and dependence of the average number of pions produced versus beam energy. In Fig. 9, the measured data are compared to Cugnon cascade[26] calculations. This cascade code reproduces correctly the pion yield in proton induced reactions, yet is seen to overpredict it for nuclear reactions. One of the obvious potential deficiencies of the cascade code has been ruled out. Namely, it could be that pion absorption

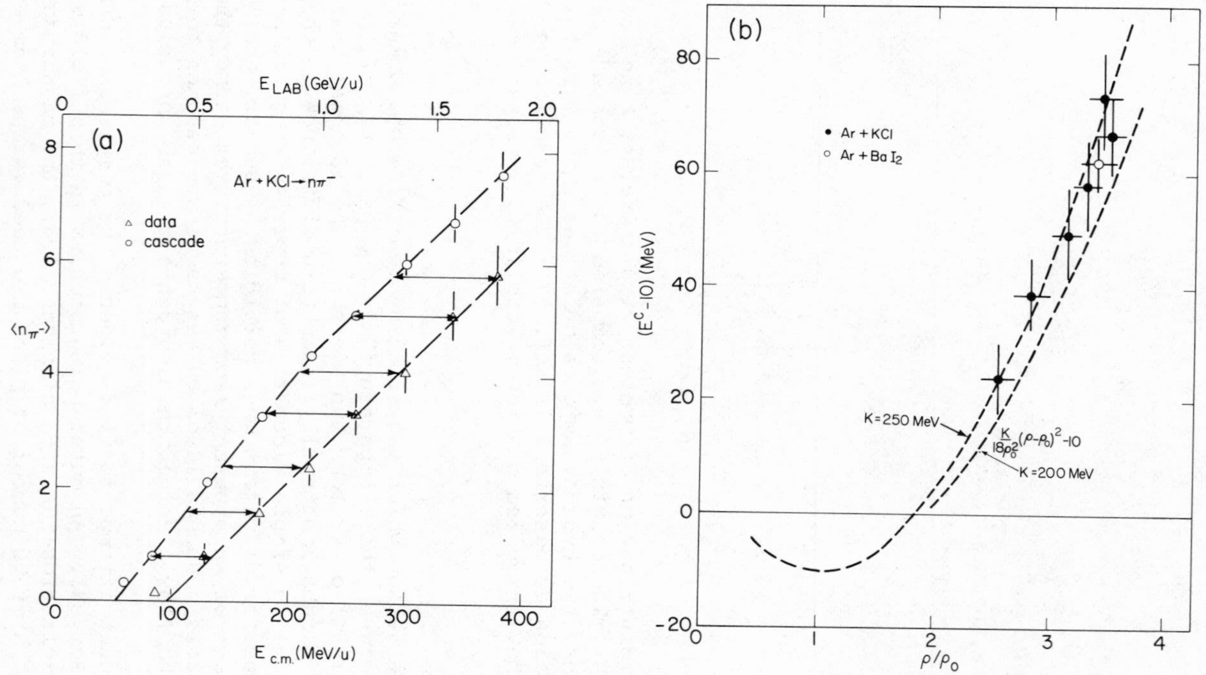

Fig. 9. (a) Mean π^- multiplicity produced in central Ar + KCl collisions versus cm kinetic energy per nucleon.[33] Open circles show predictions of Cugnon cascade.[26] Triangles are data. Horizontal arrows are used to estimate the compression energy per nucleon. (b) Compression energy deduced[33] from (a) versus baryon density as predicted by the cascade code. Dashed curves illustrate an equation of state with incompressibility constant K = 200-250 MeV.

is much stronger in dense nuclear matter than ordinary nuclear matter. However, cascade predicts that at higher energies, higher densities $\rho \sim 4\rho_0$ are achieved. Therefore, if the discrepancy were due to a poor pion absorption mechanism, then the discrepancy should be largest at the high energy region. The contrary is observed. Why then does cascade predict too many pions?

The suggestion in Ref. 33 is that due to nuclear repulsion at high densities, not as much energy is available for the production of pions as there would be if there were no potential energy. Indeed one of the main effects neglected in cascade calculations is the nuclear mean field. By assuming that the discrepancy in Fig. 9(a) is due solely to the neglected potential energy, the potential energy curve for densities up to $4\rho_0$ was obtained as shown in Fig. 9(b). This remarkable curve is the first tentative extraction of the nuclear equation of state using nuclear collision data! As such it is a landmark in this field. Note that a very sensible equation of state with compressibility $K \sim 250$ MeV is found.

However, this interpretation of the pion discrepancy is still under debate and many questions need to be answered before Fig. 9(b) can be accepted. The Cugnon cascade code does a rather good job in accounting for the inclusive charge sum and primordial deuteron spectra [see Fig. 7(b)]. Would these predictions be changed for the worse by including potential effects? If the cascade code is altered to account for the pion yield, what other observables are

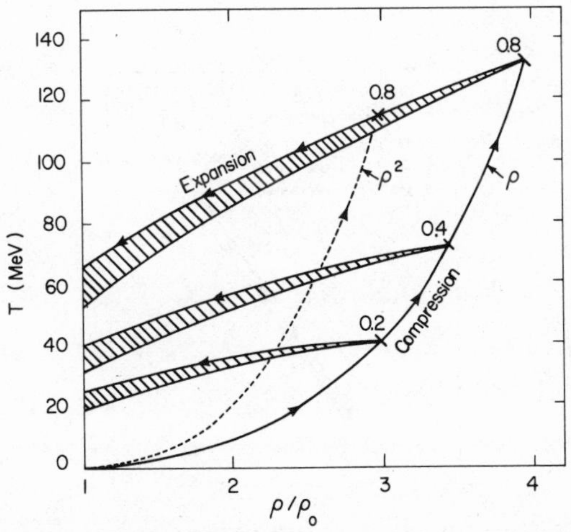

Fig. 10. The trajectories $T(\rho)$ followed in hydrodynamical compression and expansion[34] at energies 0.2, 0.4, 0.8 AGeV. Shaded area indicates temperature increase due to viscosity. A linear (ρ) and quadratic (ρ^2) dependence of the compression energy per nucleon is contrasted.

changed? Also it is known that hydrodynamical calculations of the pion multiplicity *underestimate* the pion yield by a factor of ~1/2. However, such calculations include via an assumed equation of state the compression energy. Furthermore, the pion yield is found to be insensitive[34,35] to the compressibility constant, K, assumed. This insensitivity follows because most of the entropy is produced via shock waves. However, baryon and four-momentum flux conservation (Rankine-Hugoniot equation) severely limits the magnitude of the pressure and entropy that can be achieved in shocked matter.[35] Figure 10 illustrates the problem. The dynamical path (density and temperature versus time) followed in a collision does in fact depend on the equation of state. For a compression energy rising linearly with density (ρ) an 800 MeV/nucleon collision reaches a maximum $\rho \sim 4\rho_0$ and $T \sim 140$ MeV. Subsequent isotropic expansion is indicated by the shaded area. On the other hand, for a compression energy rising quadratically (ρ^2) only $\rho \sim 3\rho_0$ is reached with $T \sim 110$ MeV. Now comes the catch. The isotropic expansion from the maximum compression point leads to nearly the same freeze-out temperature $T_f \sim 60$ MeV at $\rho_f \sim \rho_0$. Since $\langle N_\pi \rangle$ is uniquely determined by T_f and ρ_f, we conclude that $\langle N_\pi \rangle$ is insensitive to the equation of state if pions are in chemical equilibrium.

However, $\langle N_\pi \rangle$ is rather sensitive to viscosity effects.[34] Figure 11 shows the same data as in Fig. 9 together with ideal hydrodynamic (η=1) predictions. By allowing for a 20 percent increase of the entropy due to viscous heating (curve $\eta = 1.2$) rather statisfactory agreement can be found between data and calculation. Notice

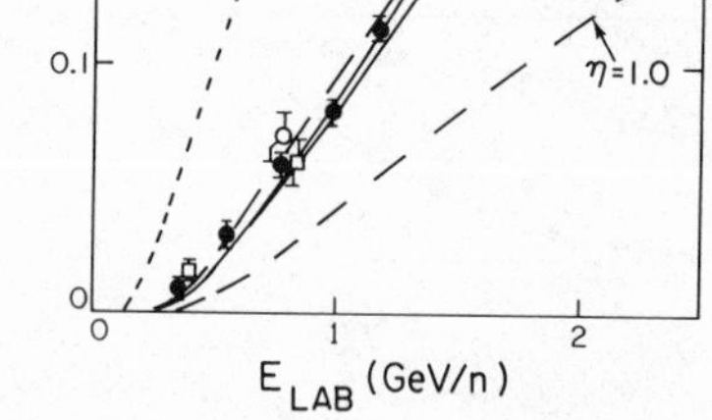

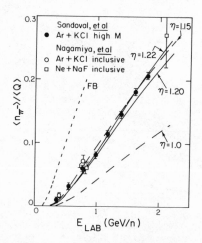

Fig. 11. Ratio of average π^- multiplicity to proton multiplicity is shown for fireball model (dotted line), non-viscous hydrodynamics ($\eta = 1.0$), and viscous hydrodynamics (η = 1.15, 1.22). Data[33] are shown by dots.

that a simple fireball model (FB) overestimates the pion yield even more than cascade. Therefore, compression effects are indeed important in reducing the entropy. However, Fig. 11 shows that viscosity effects may be crucial in getting $\langle N_\pi \rangle$ right.

We conclude that while Fig. 9 is provocative, alternate explanations of the data that are not sensitive to the compression energy can also be advanced at present. This debate is far from resolved, and it will take some time for the dust to settle.

Global Analysis

A powerful new method has been developed recently to search for compression effects directly. This involves looking for collective flow patterns on an event-by-event basis. Such global analysis requires the measurement of as many fragments as possible per event. The elaborate detector[38] called the Plastic Ball/Wall is now in operation and has the capability of measuring the mass m_ν and momenta $\vec{p}(\nu)$ of several hundred charged particles at once. In addition, streamer chamber photographs[39] provide a means to measure the momenta of all charged particles on an event-by-event basis. How can we best extract flow information from such charge exclusive data?

In high energy physics several global variables have been defined that have great intuitive appeal. These include the well-known thrust and sphericity variables

$$T = \max_{\hat{n}} \sum_{\nu} |\vec{p}(\nu) \cdot \vec{n}| / \sum_{\nu} |\vec{p}(\nu)| \tag{19}$$

$$S_{ij} = \sum_{\nu} p_i(\nu) p_j(\nu) \ . \tag{20}$$

The sphericity tensor provides the simplest three dimensional event shape parametrization. However, for nuclear collisions it is necessary to modify S_{ij} in order to take into account the copious production of nuclear fragments with $A > 1$. This is necessary because we do not want to weigh the contribution from an α-particle more than four unbound collinear nucleons in determining the flow pattern.

One way to achieve the correct weight is to calculate the kinetic flow tensor

$$F_{ij} = \sum_{\nu} p_i(\nu) p_j(\nu) / 2m_\nu \ . \tag{21}$$

Other weights,[37] e.g. $|p(\nu)|^{-1}$, can also be used. F has the advantage that for energies $\lesssim 1$ AGeV, TrF is the observed kinetic energy in the cm frame. (Of course, global analysis must be performed in the nucleus-nucleus center-of-mass.) Since F is a symmetric matrix, it is readily diagonalized

$$F = \sum_{i=1}^{3} f_i e_i e_i^{\dagger} \quad , \tag{22}$$

where f_i are the flow eigenvalues and e_i are the orthonormal eigenvectors specifying the principal axes. Therefore, F characterizes an event as an oriented ellipsoid with radii $\sqrt{f_i}$ in momentum space. The angle θ_F between the principal axis carrying the maximum kinetic energy and the beam direction is called the flow angle.

In order to display the flow characteristics we proposed the following flow diagram:[36] plot θ_F versus the kinetic flow ratio $r = f_{max}/f_{min}$. In momentum space $\sqrt{r}$ gives the aspect ratio of the ellipse in the plane of the major and minor axis of the event. The insert in Fig. 12 illustrates how an actual measured event (400 AMeV Ca + Ca) recorded by the plastic ball is represented by this flow analysis. That event corresponds to $\theta_F \simeq 19°$ and r=3.

Also shown in Fig. 12 are the results of calculations[36] using the Cugnon cascade and hydrodynamics for the reaction $^{238}U + ^{238}U$ at 400 AMeV. It is immediately obvious that there is substantial difference between the flow characteristics expected with non-viscous hydrodynamics and that expected with cascade. For such heavy sys-

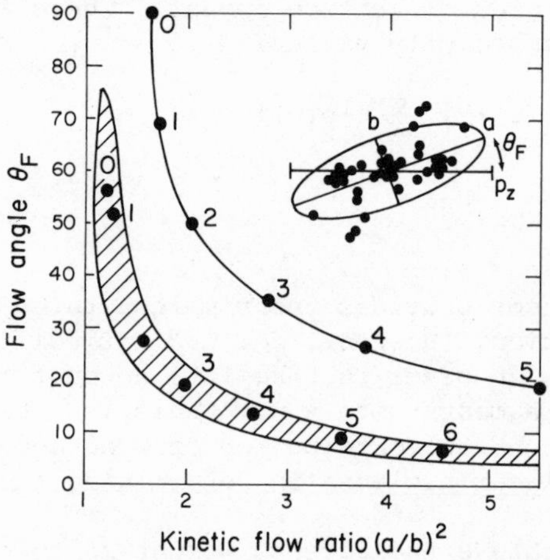

Fig. 12. Kinetic flow diagram for U+U at 400 AMeV. The flow angle is the angle between the beam axis and the principal axis carrying the largest kinetic energy. The ratio of the largest, a, to smallest, b, eigenvalues of the tensor specify the aspect ratio of the event. Numbers indicate impact parameter in units of $b_{max}/10$. The solid curve is the prediction of non-viscous hydrodynamics.[40] Shaded region is the prediction of cascade.[36] Insert shows the flow analysis of one Ca+Ca (400 MeV) event measured in the plastic ball.[38]

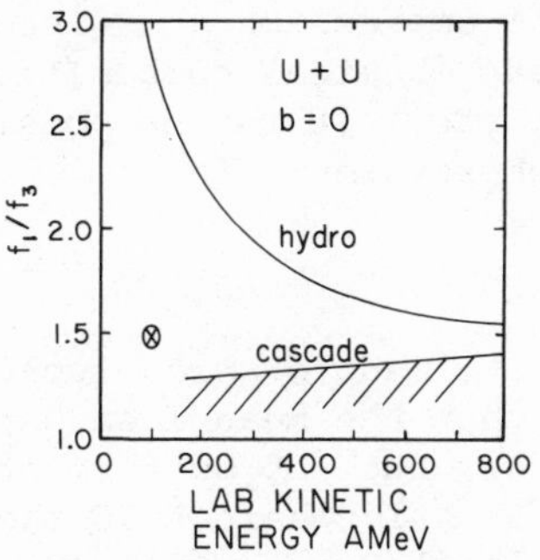

Fig. 13. Energy dependence of the kinetic flow ratio (f_{max}/f_{min}) for b=0 U+U collisions. Solid line refers to hydrodynamics,[40] while shaded region refers to cascade.[36] A density isomer leads to the x at 100 AMeV.

tems finite number fluctuations are not important. However, as shown in Ref. 36 for systems involving nuclei with $A < 100$, finite number effects lead to very large fluctuations and tend to obscure the picture. Fortunately, with the new uranium capability of the Bevalac, experiments with heavy nuclei have become possible.

Non-viscous hydrodynamics predicts much stronger sidewards flow than cascade does. Cascade dynamics leads apparently to highly viscous flow (see the following subsection). We have done calculations involving 2000 on 2000 nucleons and found no significant difference from the uranium results. The momentum distribution for b=0 collisions is close to an isotropic sphere. Non-viscous hydrodynamics predicts, on the other hand, a flattened pancake shape for b=0 events.

The first experimental global event analysis has been completed recently.[38,39] The Ca+Ca 400 AMeV reaction has been studied in most detail, but new data on Nb+Nb are now being analyzed. The preliminary Ca data show very large fluctuations as expected. However, there is some hint[39] that cascade predicts too much transparency as compared with the data. As the new Nb data become available we can expect a much clearer experimental statement on how nuclei flow.

Finally, it is important to discuss the beam energy dependence of flow phenomena.[40] In Fig. 13, the dependence of the aspect ratio on bombarding energy is shown for b=0 collisions. Observe the strong energy dependence below 400 AMeV that is predicted by hydrodynamics in contrast to the near constant isotropy predicted by cascade. Clearly global analysis is best performed in the energy range[35] 100-400 AMeV. For fun the result of a hydrodynamic calculation assuming a density isomer equation of state at 100 AMeV is also shown. The results emphasize the need for a systematic study of flow phenomena with energy. In this way can we hope to extract details of the equation of state. In practice we will always see the combined effects of compression *and* viscosity. With the combined global analysis and chemical composition studies (π multiplicities, deuteron yields, etc.) it may be possible to disentangle viscous effects from compression effects.

Nuclear Honey

Why does cascade not lead to strong collective nuclear flow? One possible answer is that the mean free path $\lambda \sim 1.7$ fm is not small enough. A necessary but not sufficient condition for hydrodynamic flow is that $\lambda/R \ll 1$, where R is the characteristic size of the system. For nuclei $\lambda/R \sim 1.4\ A^{-1/3} \approx 0.4,\ 0.2,\ 0.1$ for A = 40, 238, 2000 respectively. Thus, Ca+Ca obviously does not satisfy $\lambda/R \ll 1$. However, U+U could. In test runs[36] with A = 2000, that condition was rather well satisfied. Yet the flow was considerably weaker than predicted by *non-viscous* hydrodynamics. Is viscosity the key to the difference?

To test this idea we modified the Cugnon cascade in several ways. First we considered the effect of increasing the effective NN cross section artificially. For a fixed A, $\lambda/R = (\sigma_{eff}\,\rho R)^{-1}$ obviously decreases with increasing σ_{eff}. Second, we altered the scattering style. The conventional cascade style is to assume that nucleons propagate along straight line trajectories. At the point of closest approach, if the separation, r, between two nucleons is less than $(\sigma_{eff}/\pi)^{1/2}$, then a random momentum transfer q is given to each nucleon. The q is chosen from the experimental distribution of momentum transfers at the appropriate NN cm energy. This scattering style is referred to as stochastic because there is no correlation between $\vec{r}$ and $\vec{q}$. On the other hand, classically we expect

$$\vec{q}\cdot\vec{r} > 0 \text{ for repulsive forces} , \tag{23}$$

$$\vec{q}\cdot\vec{r} < 0 \text{ for attractive forces} , \tag{24}$$

$$\vec{q}\cdot\vec{L} = 0 \text{ for } \vec{L} = \vec{r}\times\vec{p} . \tag{25}$$

We therefore tried to impose a classical scattering style by requiring Eq. (25) to be satisfied (this insures a well defined NN scattering plane). Furthermore, we assume that if $r < r_c$ then the force is repulsive, i.e., Eq. (23) is satisfied, while for $r_c < r < (\sigma_{eff}/\pi)^{1/2}$ Eq. (24) is satisfied. Thus, our classical style corresponds to a repulsive core and an attractive longer range force. Table 1 shows the results of the calculation for U+U at 250 AMeV for impact parameter b=0. As σ_{eff} increases from σ_0 to $3\sigma_0$ the total number of NN collisions increases by a factor of 3 as expected using the stochastic style. However, the aspect ratio $a = \langle f_{max}/f_{min}\rangle$ increases only from 1.2 to 1.3. For non-viscous hydrodynamics a=1.6. (Note that there is always some "numerical" viscosity in hydrodynamics due to finite difference techniques.)

Table 1. Sensitivity of Flow Characteristics to Scattering Style.[a] Results based on Modified Cugnon Cascade Calculations[36] for $^{238}U + ^{238}U$ at $b = 0$ and 250 AMeV.

σ_{eff}	$r_c/(\sigma_{eff}/\pi)^{1/2}$	$\vec{q} \cdot \vec{L}$	n_{coll}	f_{max}/f_{min}
σ_0	stochastic		5.7	1.2
	1.0	0	4.3	1.3
$2\sigma_0$	1.0		6.6	1.7
	0.63	0	11.3	1.4
	0.50	0	18.0	1.3
	0.44	0	45.8	1.2
$3\sigma_0$	stochastic	0	16.5	1.3
	1.0	0	7.7	2.1
$5\sigma_0$	1.0	0	8.6	3.0

[a]The quantities in the table are defined as: σ_{eff} is the effective NN scattering cross section; σ_0 is the measured energy dependent total NN cross sections; r_c is the repulsive core radius; $\vec{q}\cdot\vec{L}$ is the orientation of momentum transfer to angular momentum; n_{coll} is the number of NN collisions per incident nucleon; f_{max}/f_{min} is the aspect ratio from the flow tensor, Eq. (22).

Thus, even though the number of collisions per particle, n_{coll}, increased from 5.7 to 16.5, the cascade did not reach the hydrodynamic limit. This again underscores the fact that $\lambda/R << 1$ is not sufficient for hydrodynamic flow. Now if we impose the classical repulsive style $r_c = (\sigma_{eff}/\pi)^{1/2}$, then as σ_{eff} increases from σ_0 to $3\sigma_0$, the aspect ratio increases from 1.3 to 2.1. Thus, we can make cascade flow even more than hydrodynamics by imposing a large repulsive classical scattering style. Note that by switching from stochastic to repulsive style for $\sigma_{eff} = 3\sigma_0$, the total number of NN collisions *decreases* by a factor of two while the aspect ratio *increases* by a factor of 1.6! This phenomenon is also clearly seen for $\sigma_{eff} = 2\sigma_0$ where we explored the dependence of n_{coll} and $\underline{a}$ on r_c. For purely repulsive style $f = r_c/(\sigma_{eff}/\pi)^{1/2} = 1$, $n_{coll} = 6.6$, $a = 1.7$. For $f = 0.5$, $n_{coll} = 18$ but $a = 1.3$! Note that for $f = 0.44$, n_{coll} has climbed to 46 collisions per incident nucleon but the aspect ratio has lowered to 1.2 (consistent with a sphere sampled with 238 particles). Therefore, increasing the attractive scattering component increases n_{coll} and lowers $\underline{a}$.

These results can be understood as follows: the attractive scattering style tends to bring the nucleons closer together because

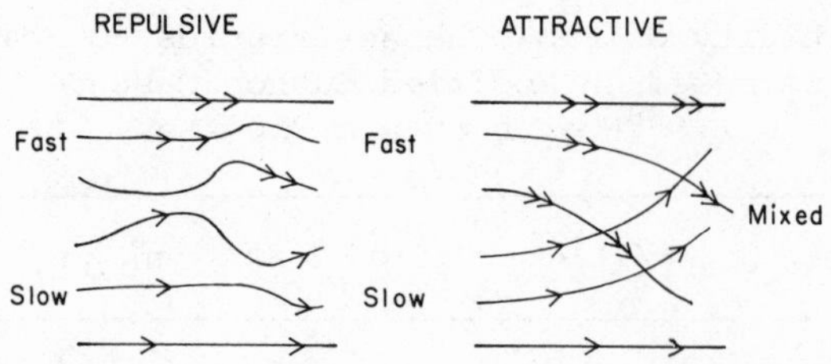

Fig. 14. Transverse flow gradients are maintained by repulsive scattering style while they are rapidly damped by attractive scattering style.

of the focusing condition, $\vec{q}\cdot\vec{r} < 0$. The higher density tends to increase n_{coll}. That the aspect ratio decreases follows because transverse momentum gradients are effectively damped if $\vec{q}\cdot\vec{r} < 0$. This is illustrated in Fig. 14. Consider a fast stream of nucleons moving parallel to a slow stream. If the scattering style is attractive, then fast nucleons will be drawn into the slow stream and vice versa. Thus, slow and fast nucleons will mix and the transverse momentum gradient is damped. This is how usual viscosity works. However, usual viscosity increases with increasing mean free path, i.e., *decreasing* σ_{eff}! The viscosity effect seen in Table 1 arises out of the dynamical enhancement of mixing due to attractive forces in a high density domain. Thus, the effect seen in Table 1 for $\sigma_{eff} > \sigma_0$ is due to dynamic viscosity. For repulsive forces, Fig. 14 shows that $\vec{q}\cdot\vec{r} > 0$ inhibits mixing of fast and slow particles and thus helps maintain flow gradients.

Finally we observe that for $\sigma_{eff} \lesssim \sigma_0$, there is virtually no sensitivity to the scattering style. In this "long" mean free path limit, dynamic viscosity is replaced by simple kinetic viscosity.

We now address the question of which scattering style is closer to reality. It appears that stochastic scattering is. There are two reasons. First, the sign of the NN force is both channel and energy dependent. Thus, the sign of $\vec{q}\cdot\vec{r}$ is likely to alternate in subsequent NN collisions as the parity of the relative wavefunction changes and as the sign phase shifts change as the relative energy is lowered. Second, and more important, there is quantum stochasticity. At these energies the relative angular momentum in NN collisions is low,[1] $\ell \lesssim 3\text{-}5$, initially and $\ell \lesssim 1\text{-}2$ later. Hence low partial waves dominate the scattering. However, for low partial waves the azimuthal dependence of the scattering wavefunction weakens. Thus, there is no scattering plane [Eq. (25)] for S and P wave scattering! The stochastic scattering style abandons Eq. (25) and choses the azimuth angle at random. Because of quantum effects this is then likely to be closer to reality.

Thus, quantum effects result in both types of processes depicted in Fig. 14. We conclude that nuclear flow is likely to be highly viscous—somewhat like honey. The global analysis of nuclear collisions with $A > 100$ will tell us soon just how sweet nuclear flow really is.

QUARK-GLUON PLASMA PRODUCTION

In this last lecture we switch from honey to glue and enter into the sticky, though exciting, subject of the formation of a quark-gluon plasma in nuclear collisions. For this purpose we consider energies $E_{lab} > 100$ AGeV. The main question we address is whether high enough energy densities can be achieved to produce a quark-gluon plasma.

Critical Parameters

One of the most striking predictions of quantum chromodynamics (QCD) is the deconfinement of hadronic matter at high energy density. This follows from the asymptotic freedom property of QCD. The best estimates for the critical energy density, ε_c, come from Monte Carlo lattice simulations of QCD. The results from two recent calculations[41,42] of the energy density ε versus temperature T in baryon free matter ($\rho_B = 0$) are shown in Fig. 15. The dots and triangles are from Ref. 41, where an approximate treatment of quarks is included. The open circles are from Ref. 42 and correspond to pure

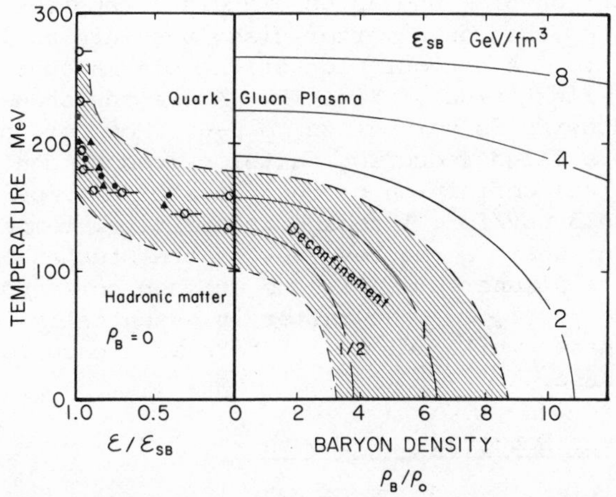

Fig. 15. Phase diagram of hadronic matter. Monte Carlo QCD data[41–43] on left indicate existence plasma transition at energy densities ~2 GeV/fm^3. Equal ε_{SB} contours, Eq. (26), versus T and ρ_B are shown on the right.

SU(3) gluon matter. On the left-hand side, the ratio of ε to that of an ideal quark-gluon plasma is plotted versus temperature for baryon density $\rho_B = 0$. The energy density of an ideal up-down-glue plasma is given by the Stephan-Boltzmann form[41-43]

$$\varepsilon_{SB}(T,\mu) = \frac{37}{30}\pi^2 T^4 + 3T^2 \mu^2 + \frac{3}{2\pi^2}\mu^4 \tag{26}$$

where μ is the chemical potential. The baryon density is given by

$$\rho_B = \frac{2}{3\pi^2}\mu^3 + \frac{2}{3}T^2 \mu \tag{27}$$

and the pressure in the plasma is simply $P_{SB} = \varepsilon_{SB}/3$.

In Fig. 15, we see that for $T > T_c \sim 200$ MeV, $\varepsilon/\varepsilon_{SB} \approx 1$, and thus QCD predicts that the state of the matter is described well as an ideal plasma. For $T < T_c$ there is a rapid departure from the Stephan-Boltzmann form as confinement sets in.

The precise nature of the deconfinement transition is still under debate, but it is likely[43] that for the SU(3) color group the transition is first order. The shaded area around the "data" points is to remind us that systematic uncertainties exist associated with the approximate treatment of quark degrees of freedom and finite lattice size corrections in present calculations and that there is uncertainty in translating the lattice cutoff Λ_L into physical units (MeV).

Based on these and other model calculations at finite baryon density,[43] the following picture of the phase diagram of hadronic matter as a function of T and ρ_B is emerging: above some critical energy density, ε_c, hadronic matter dissolves into an ideal quark-gluon plasma state. A contour plot of the plasma energy density ε_{SB} is shown on the right side of Fig. 15. Above the shaded region the actual energy density is very close to ρ_{SB}. However, below that region there is a large reduction factor caused by confinement. While the technical definition of the transition temperature[43] corresponds to $\varepsilon \sim 0.5$ GeV/fm^3, I define the critical temperature, T_c, here as the point where ε reaches ~90% of the Stephan-Boltzmann value. The critical energy density so defined corresponds to $\varepsilon = \varepsilon_{pl} \sim 2$ GeV/fm^3. For $\varepsilon \geq \varepsilon_{pl}$ the matter is essentially in a perturbative plasma phase, while below ε_{pl} there is a complicated mixed hadron-plasma phase.

Stopping Power and Longitudinal Growth

In a typical hadron-hadron collision a fraction $\eta \sim 1/2$ of the parallel momentum is lost. In terms of rapidity, y, this momentum loss corresponds to a rapidity shift[44]

$$\Delta y \approx \ln \frac{1}{1-\eta} < 1 \tag{28}$$

for both hadrons. [Recall that for a particle of mass m and momentum $(p_{\|}, p_{\perp})$, $p_{\|} = m_{\perp} \sinh y$ and $E = m_{\perp} \cosh y$ in terms of y, and $m_{\perp} = (m^2 + p_{\perp}^2)^{1/2}$.] Therefore, the rapidity of a particle after[45] $\nu \sim 0.65\, A^{0.3}$ *independent* collisions is

$$y(\nu) = y - \nu\Delta y \quad . \tag{29}$$

We say that a particle is stopped if

$$\bar{\nu} > y/\Delta y \quad . \tag{30}$$

It is important to emphasize that stopping is a frame-dependent concept. If y_L is the lab rapidity $(y_L = 2y_{cm})$, then the particle stops in the nucleon-nucleon cm frame if $y_L < 2\nu\Delta y$. In terms of lab kinetic energy, $E_L = m_N[\mathrm{ch}(y_L)-1]$, Eqs. (28)-(30) lead then to

$$E_L < \frac{0.5 \text{ GeV}}{(1-\eta)^{2\nu}} \sim 2^{2\nu-1} \tag{31}$$

as a necessary condition for a nucleon to stop in the *NN center-of-mass (i.e., midrapidity) frame*. For ^{238}U, $\nu \approx 3.4$ so most nucleons stop in a central U+U collision in the midrapidity frame if the lab kinetic energy is less than $E_L < 56$ GeV for $\eta = 1/2$. A more refined recent estimate[46] leads to a similar result. Of course Eqs. (28)-(30) cease to hold for energies above which successive collisions are not independent. We shall see explicitly that for $E > 100$ GeV this is indeed the case because of longitudinal growth.

In order to calculate the energy density, we need to estimate the compression ρ_B upon stopping. If the nuclei are thick enough to stop a nucleon in the midrapidity frame [Eq. (31)], *and* the nucleon recoil is instantaneous, then all nucleons will stop in a Lorentz contracted volume = γ_{cm}^{-1} × rest frame volume.[46] Therefore, the baryon density is at least[46]

$$\rho_B/\rho_o = 2\gamma_{cm} \approx \exp(y_L/2) \quad . \tag{32}$$

This leads to an energy density of at least

$$\varepsilon > 2\gamma_{cm}^2 M_N \rho_o \quad , \tag{33}$$

where $M_N \rho_o \approx 0.136$ GeV/fm^3. To obtain an upper bound on ρ_B consistent with baryon and four momentum conservation, we can use the Rankine-Hugoniot relation. Given an equation of state, $P = \alpha\varepsilon$, the shock compression ρ_{sh} is simply[35]

$$\rho_{sh}/\rho_o = \alpha^{-1} + (1 + \alpha^{-1})\gamma_{cm} \quad . \tag{34}$$

It is important to emphasize that Eq. (34) is independent of the shock front thickness only as long as it is smaller than the dimensions of the system. With Eq. (34) the energy density is then bounded by

$$\varepsilon < \varepsilon_{sh} = \gamma_{cm} M_N \rho_{sh} \quad . \tag{35}$$

However, ρ_B cannot increase indefinitely with γ_{cm}. There exists a characteristic proper recoil time $\tau_0 \sim (1/2\text{-}1)$ fm/c for the baryon current to change in a collision. In a frame where the nucleon has rapidity y the time required for its baryon number to stop is dilated to τ_0 ch y. Therefore, the minimum stopping distance in the midrapidity frame is $\sim\gamma_{cm}\,\tau_0$. We can also think of $\gamma_{cm}\,\tau_0$ as the minimum thickness of any shock front in the midrapidity frame. This leads to a bound on the compression

$$\rho_B/\rho_0 \lesssim gR/\tau_0\gamma_{cm} \equiv \rho_B(\tau_0)/\rho_0 \quad , \tag{36}$$

where $g \sim (1\text{-}2)$ is a geometrical factor depending on the detailed spatial distribution of $\rho_B(Z)$.

To obtain a better understanding of the origin of the finite recoil time τ_0, we need to discuss the concept of longitudinal growth.[47,48] Consider a hadron of mass M suffering a collision in which it is excited to a virtual state of energy $E^{*2} = p_0^2+M^2$. We want to know how long it takes for this virtual state to decay by emitting a particle of mass m and momentum $(p_{\|},p_{\perp})$. The final state has therefore an energy $E = \left[(p_0-p_{\|})^2 + M_{\perp}^2\right]^{1/2} + \left[p_{\|}^2 + m_{\perp}^2\right]^{1/2}$, where $m_{\perp}^2 = p_{\perp}^2 + m^2$ and $M_{\perp}^2 = p_{\perp}^2 + M^2$. The uncertainty principle states that the amplitude to emit such a particle becomes appreciable only for times[47]

$$t > t(y) \sim \frac{\hbar}{E^*-E} \xrightarrow[p_0 >> M]{} 2p_{\|}/m_{\perp}^2 = \frac{2}{m_{\perp}} \cosh y \quad . \tag{37}$$

As the rapidity of the emitted particle increases, t increases because of time dilation. We can interpret Eq. (37) as follows:[48] in the rest frame of the produced particle it takes $2/m_{\perp} \sim 1$ fm/c for the particle to come on shell. Before that time it is impossible to disentangle the wavefunction of the final particle from that of the projectile. Since the projectile is assumed highly relativistic (c=1), the position where the particle is emitted is $z(y) \sim t(y)$. A more detailed estimate of z(y) can be made by invoking the inside-outside cascade (IOC) picture of particle production.[48] In IOC particles follow classical trajectories, $z = t \times \tanh y$, but come on shell only at $t = t(y)$. For $t < t(y)$ they propagate as virtual particles with phases interlocked with the projectile. Only for $t > t(y)$ can they participate in incoherent interactions. In this IOC picture the point where a secondary particle comes on shell is thus

$$z(y) = \frac{2}{m_{\perp}} \sinh y \quad . \tag{38}$$

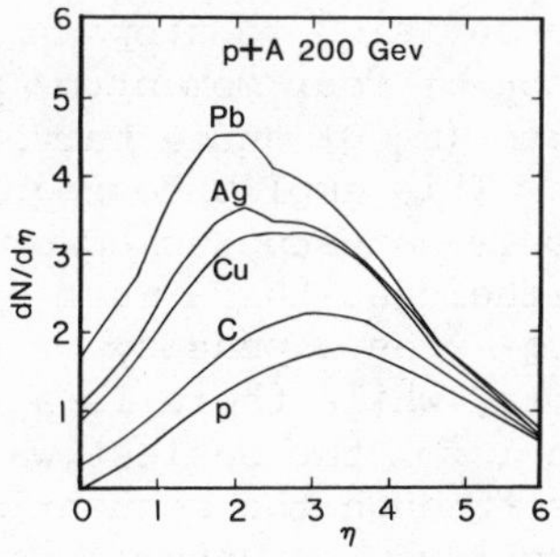

Fig. 16. Pseudo rapidity ($\eta = -\ln \tan \theta_{Lab}/2$) distributions of particles produced in p+A collisions at 200 GeV.[45]

Equations (37)-(38) imply that particles come on shell when their proper time $\tau = (t^2 - z^2)^{1/2}$ reaches $\tau = 2/m_\perp \sim 1$ fm/c, i.e., along a hyperbola in the (t,z) plane. Equations (37)-(38) specify what is meant by longitudinal growth; at very high energies the interaction region grows very rapidly along the beam direction because of the combined effects of the uncertainty principle and relativistic kinematics.

Evidence for longitudinal growth comes from hadron-nucleus data[45,49] as shown in Fig. 16. The striking feature to observe is that for large rapidity secondaries there is virtually no dependence on the target mass, A. This is a direct consequence of longitudinal growth. A pion with rapidity $y = 5$ can materialize only ~100 fm downstream from the target nucleus! The absence of cascading is particularly evident when the inelasticity $\eta(\nu)$ is computed from the data ($\eta \int dy\; dN/dy\; E(y)/E_{inc}$). We find that $\eta = 0.5$, 0.6, and 0.66 as the target changes from p, Ag, to Pb. This shows that the total energy radiated into pions increases only very slowly ($d\eta/d\nu \sim 0.07$) with ν, in complete disagreement with the naive independent scattering model, Eq. (29). On the other hand, models[49] incorporating nuclear transparency and longitudinal growth have been, on the whole, successful in accounting for high energy hadron-nucleus data. Note finally that the modification of the stopping distance proposed in Eq. (36) is consistent with the longitudinal growth of the reaction zone.

Therefore, the finite recoil time τ_0 in Eq. (36) arises from the requirement that in the rest frame of the nucleon it takes $\tau_0 \sim$ 1 fm/c to radiate a pion and that this time is dilated in any other frame. Of course, it is possible to contemplate other mechanisms besides soft pion emission for the nucleon to slow down. For example, we can imagine that two colliding nucleons with cm momentum p get excited into a very massive resonance state with $m \sim p$. However, this requires enormous momentum transfers $q \sim P$. Such amplitudes are greatly inhibited by propagators and form factors. Another generic amplitude we can envision to stop a nucleon quickly is one where the incident nucleon of four momentum $(p,\vec{p})$ decays into a nucleon with four momentum $(m_N,0)$ and a baryon number zero fireball of four momentum $\sim(p,\vec{p})$. This amplitude requires only a small momentum transfer and thus is not suppressed by the propagators of the exchanged quanta. Nevertheless, there must still appear a form factor suppression with $q\sim p$ as a penalty of stopping an extended object quickly. Therefore, while there is a small finite amplitude to stop a nucleon within 1 fm, the easiest way (greatest amplitude) for a nucleon to stop is through the sequential emission of low $p_\perp$ pions. This dominant amplitude, however, has longitudinal growth built in.[47] Therefore, on the average stopping requires a distance $\sim\gamma_{cm}\tau_o$ from the interaction point. This leads to Eq. (36).

We can now estimate at what energy the maximum baryon density is likely to be achieved. As long as the shock thickness $\gamma_{cm}\tau_o$ is smaller than $R/(2\gamma_{cm})$ in the cm frame, it pays to go to higher energies according to Eq. (34). However, for $\gamma_{cm}\tau_o > (R/2\gamma_{cm})$, shocks can no longer be maintained and Eq. (36) implies that ρ_B begins to decrease. Thus, we expect the maximum baryon density to be achieved when

$$\gamma_{cm}\tau_o = R/2\gamma_{cm} \quad . \tag{39}$$

Taking $R=14$ fm for the diameter of uranium, Eq. (36) gives $\gamma_{cm}=2.7$ and thus the kinetic energy in the lab is $E_{lab}\approx 11$ AGeV. At this point, the density may reach $\rho_B/\rho_o\sim 14$. At higher energies ρ_B decreases as $1/\gamma_{cm}$ until around $E_L\sim 50$ AGeV at which point transparency sets in. Beyond that point, nuclei pass through each other although they do get compressed to $\sim(4\text{-}5)\rho_o$ in the process.[48,44]

The energy density, in contrast to the baryon density, continues to increase with energy as we shall see in the next section.

Proper Energy Density Achieved

Because Eq. (38) gives a one-to-one correspondence between the rapidity and the production point of a particle, it is possible to compute the energy deposition per unit *length*, dE/dz, knowing the rapidity distribution dN/dy:

$$\frac{dE}{dz} = m_\perp \cosh y \,\frac{dN}{dy}\frac{dy}{dz} = \frac{m_\perp^2}{2}\frac{dN}{dy} \quad , \tag{40}$$

where $y=\mathrm{sh}^{-1}(m_\perp z/2)$. To compute the energy density, ε, we must divide dE/dz by the beam area. More precisely, we should take into account the dependence of $\varepsilon(z,x_\perp)$ on the transverse coordinate $x_\perp$. If we assume, as in most models, that $\varepsilon(z,x_\perp)$ is proportional to the number of struck nucleons along a tube at transverse coordinate $x_\perp$, then for a *central* (b=0) nuclear collision

$$\varepsilon(z,x_\perp) \approx \varepsilon_{max}(1 - x_\perp^2/R_{min}^2)^{1/2} \quad , \tag{41}$$

with

$$\varepsilon_{max} = \frac{3}{2\pi R_{min}^2} \frac{dE}{dz} \quad . \tag{42}$$

In Eqs. (41)-(42), R_{min} is the radius of the smaller nucleus. Note that $\int d^2x_\perp\ \varepsilon = dE/dz$ and that $\langle\varepsilon\rangle = 2/3\ \varepsilon_{max}$. Inserting $R = 1.18\ A^{1/3}$ and $\langle m_\perp\rangle \sim 0.3$ GeV, we obtain an estimate for the maximum energy density in the central region

$$\varepsilon_{max} \approx 0.1\ \frac{\text{GeV}}{\text{fm}^3}\ A^{-2/3}\ dN/dy \quad . \tag{43}$$

Clearly, there is at least a factor of 2 uncertainty in the conversion factor in Eq. (43). However, Eq. (43) allows us to estimate ε_{max} from measured rapidity densities.

As a first application of Eq. (43) consider pp collisions at ISR energies where $dN/dy \lesssim 3$ for $y_{cm} \approx 0$. In that case $\varepsilon_{max} \lesssim 0.3$ GeV/fm^3, which is too small to create a plasma. Even at $p\bar{p}$ collider energies[50] $dN/dy \sim 5$ is still too small on the average. The rare events with $dN/dy \sim 10$ lead to $\varepsilon \sim 1$ GeV/fm^3, but this is still below the Stephan-Boltzmann domain.

Next consider nuclear collisions. At present the only source of experimental information comes from cosmic-ray studies.[51,52] The most spectacular event observed thus far is the so-called JACEE event[52] Si + Ag at 4-5 ATeV. Over 1000 charged particles were produced with a pseudorapidity distribution shown in Fig. 17.

Note that in the central region ($\eta \sim 4$), $dn_{ch}/dy \sim 200$ is observed! This leads, assuming $\langle n_{\pi^0}\rangle = \langle n_{ch}\rangle/3$, to

$$\varepsilon_{max}(\text{JACEE}) \sim 3\ \text{GeV/fm}^3 \quad . \tag{44}$$

At this point it is important to ask whether this event is just a lucky accident. To answer this question we apply the color neutralization model of Ref. 49, which, as was mentioned before, is consistent with hadron-nucleus data. For nucleus-nucleus collisions, this predicts

$$\langle n\rangle_{AB}/\langle n\rangle_{pp} = W_P\nu_T(1+\nu_T)^{-1} + W_T\nu_P(1+\nu_P)^{-1} \tag{45}$$

where $W_P \approx A_P$ and $W_T \approx A_T[1 - (A_P/A_T)2/3]^{(3/2)}$ are the number of wounded nucleons in the projectile and target for b=0, and ν_P, ν_T are the average number of mean free paths through the projectile and target. Taking[50] $\langle n_{ch}\rangle \approx 0.88 + 0.44\ \ln s + 0.118\ \ln^2 s \approx 15$ and $W_P = 28$, $W_T = 58$, $\nu_P = 2.4$, $\nu_T = 5.0$ for b=0, Eq. (45) predicts $\langle n_{ch}\rangle_{SiAg} \approx 940$, which is close to the observed value. Thus, the JACEE event is not

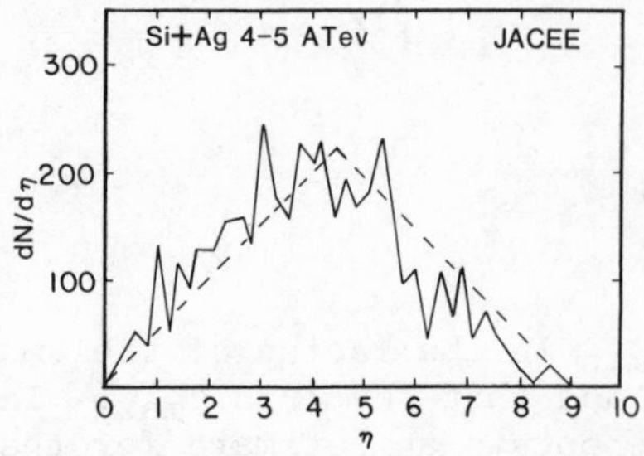

Fig. 17. Pseudo rapidity distribution[51] of Si (4-5 ATeV) + Ag → 1000 charges + X. The most spectacular nuclear collision ever recorded! Dashed triangle is to guide the eye.

unusual in this respect. Nevertheless, the achieved energy density Eq. (44) is well within the Stephan-Boltzmann domain!

A more systematic study of the energy density in the central region is shown in Fig. 18. We have included the 15 high energy cosmic-ray events tabulated in Ref. 52. In addition, the theoretical expectations, based on the color neutralization model for a

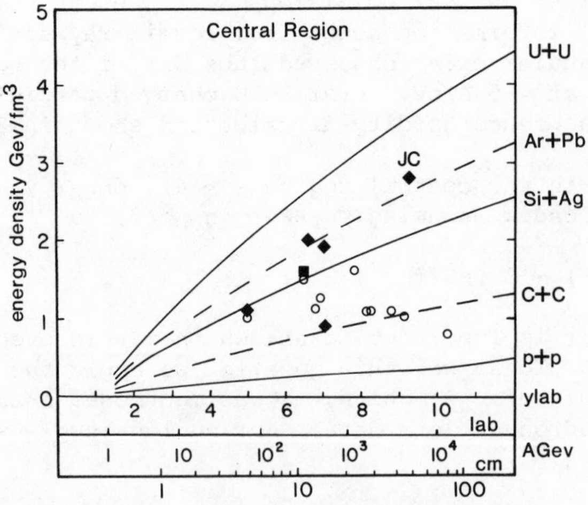

Fig. 18. Maximum energy density achieved in low baryon density regions[53] (midrapidity). Equation (43) was used to convert measured multiplicities[51,52] into proper energy densities. Diamonds correspond to Si+Ag, square to Ar+Pb, open circles to "light" (α,B,C,N) + Ag collisions. Theoretical estimates for various systems are based on Eqs. (43),(45) using tube-tube geometry as discussed in text.

variety of systems are also shown. For these estimates,[53] we have divided the transverse geometry into independent tube-tube collisions and applied Eq. (45) to each tube separately. We assumed for simplicity that $dN/dy \approx \langle n\rangle/y_{cm}$ for $y_{cm} = 0$, as appropriate for the rough triangular distributions observed[52] in nuclear collisions (see also Fig. 17). The plotted curves are for the maximum energy density at $\tau=1$ and $x_\perp = 0$ for $b=0$ in the midrapidity frame.

It is remarkable that within the factor of 2 uncertainties in the theoretical curves, the available data are consistent with expectation. We interpret Fig. 18 as experimental indication that high enough energy densities can indeed be obtained in nuclear collisions to probe the quark-gluon plasma domain. For Si+Ag the threshold for $\varepsilon_{central} > \varepsilon_{pl}$ seems to occur ~1 ATeV, while for U+U $E_{lab} \sim$ 100 AGeV seems sufficient in the midrapidity region.

Now let us return to the fragmentation regions. For $E_{lab} >$ 100 AGeV the baryons are certainly not stopped. However, compression caused by recoil and "slow" pion rescattering can lead to high energy densities.[48] An estimate for ε_{frag} can be obtained as follows: only pions with small enough relative rapidity $y_c(A)$ can rescatter within the target or projectile nuclei. Specifically, we must have $z(y) < 2R_A$ for the pion to be produced and interact within the target nucleus. From Eq. (38) this means that

$$y_c(A) = \sinh^{-1} m_\perp R_A \lesssim 3 \quad . \tag{46}$$

Therefore, the maximum energy density achieved in the fragmentation regions for $y_{lab} > 6$ is given approximately by Eq. (43) with dN/dy evaluated at $y_{cm} = y_{lab}/2 - y_c(A)$. In Fig. 19 results of calculations including nuclear recoil energies along the lines of Ref. 44 are shown. A triangular rapidity density has been assumed. These results are in accord with earlier results[48] where $\varepsilon_{frag} \sim 2$ GeV/fm^3 was obtained for U+U.

Also shown in Fig. 19 are estimates of the energy density in the stopping range. For these estimates we used the following interpolation formula

$$\varepsilon_{stop} \sim \gamma_{cm} m_n \left(\rho_{sh}^{-2} + \rho_B(\tau_0)^{-2}\right)^{-1/2} \quad . \tag{47}$$

where ρ_{sh} is given by Eq. (34) and $\rho_B(\tau_0)$ by Eq. (36). An obvious feature in Fig. 19 is that the asymptotic energy densities predicted with the modified stopping scenario [that is valid for $y_{lab} < 5$, Eq. (47)] agree within uncertainties with the estimate based on the inside outside cascade model[48] (that is valid only for $y_{lab} > 6$). Note also that the constancy of ε_{frag} with y_{lab} is expected on grounds of scaling in the fragmentation region. In contrast, the energy density in the central region, Fig. 18, continues to grow linearly with y_{lab} because dN/dy does not scale in this energy range[50] at $x_F = 0$.

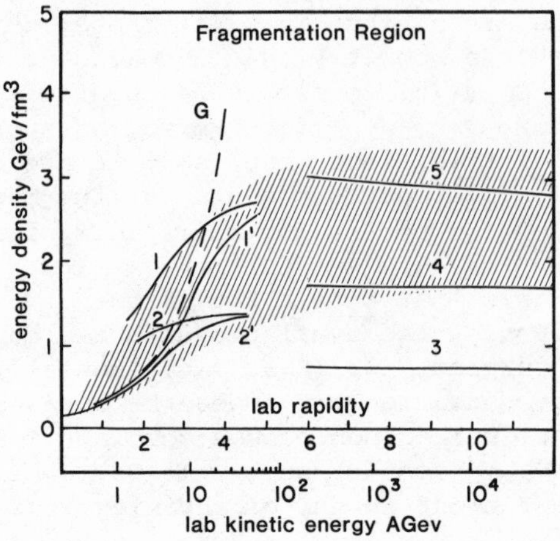

Fig. 19. Energy density achieved in high baryon density regions. Curve G illustrates Eq. (33). Shock curves Eq. (47) for Stephan-Boltzmann gas ($\alpha = 1/3$), $gR/\tau_0 = 20$, 10 are given by 1 and 2. Stiff equation of a state curves 1′, 2′ correspond to $\rho_{sh} = 2\gamma_{cm}\,\rho_0$ and $gR/\tau_0 = 20$, 10 respectively in Eq. (47). Curves 3,4,5 based on inside-outside cascade[44,48] and Eq. (46) with $m_\perp R/2 = 5,10,15$ respectively. Shaded area is best guess for central U+U collisions.

What Figs. 18, 19 show is that the domain of the quark-gluon plasma is indeed accessible via nuclear collision. They do not show, of course, what experimental signatures could result from such a plasma. Several suggestions have been put forward including strangeness abundancies,[54] dilepton yields,[55] and $\langle p_\perp \rangle$ growth.[56] We suggest a new signature: fluctuations of dN/dy on an event-by-event basis. It has been observed for some time[57] that for high energy cosmic-ray events with $E_{lab} > 10$ AGeV there are substantial fluctuations about the mean rapidity density that exceed those expected assuming Poisson statistics. In Fig. 17 there is a hint of such fluctuations in rapidity intervals $\Delta y \sim 1$. However, the most spectacular fluctuations are observed in the events discussed in Ref. 58. It is also observed that the excess dN/dy fluctuations are correlated with large $p_\perp$ gamma rays [compare Fig. 13(b) and Fig. 18 in Ref. 58]. Could these fluctuations be related to the first order phase transition from the plasma state back into the hadronic world? This speculation is fueled by a recent suggestion[59,60] that seeds for fluctuations leading to galaxy formation could arise from such a phase transition soon after the Big Bang. If the transition is indeed first order, then the plasma would not simply expand but could burn or detonate as the latent heat is converted into hadronic kinetic energy. Clearly much more thought needs to be given to the dynamics of first order phase transitions. However, it could be that we are already seeing the quark-gluon phase transition in the large fluctuations of dN/dy and the correlation of those fluctuations with high $p_{\perp}$.

ACKNOWLEDGMENT

Special thanks go to R.E. Gibbs, I. Otterlund, H.H. Heckman, and E. Friedlander for introducing me to cosmic-ray data. Helpful discussions with L. McLerran, H. Satz, J. Kapusta, P. Danielewicz, H. Stocker, E. Remler, B. Müller, J. Kuti, L. van Hove, and J. Rafelski are also gratefully acknowledged. This work was supported by the Director, Office of Energy Research, Division of Nuclear Physics of the Office of High Energy and Nuclear Physics of the U.S. Department of Energy under Contract DE-AC03-76SF00098.

REFERENCES

1. S. Nagamiya and M. Gyulassy, in: "Advances in Nuclear Physics," J. Negele, E. Vogt, eds., Plenum, New York (in press).
2. E. M. Friedlander and H. Heckmann, Lawrence Berkeley Lab report LBL-13864.
3. E. M. Friedlander et al., Phys. Rev. Lett. 45:1084 (1980).
4. P. L. Jain and G. Das, Phys. Rev. Lett. 48:305 (1982).
5. H. B. Barber et al., Phys. Rev. Lett. 48:856 (1982).
6. E. M. Friedlander et al., Lawrence Berkeley Lab report LBL-10573.
7. B. F. Bayman et al., Phys. Rev. Lett. 49:532 (1982).
8. A. P. Balachandran et al., Phys. Rev. Lett. 49:1124 (1982).
9. J. Boguta, Phys. Rev. Lett. 50:148 (1983).
10. A. De Rújula et al., Phys. Rev. D17:285 (1978); D22:227 (1980).
11. H. Georgi, Phys. Rev. D22:225 (1980).
12. R. Slansky et al., Phys. Rev. Lett. 47:887 (1981).
13. R. Saly et al., Phys. Lett. 115B:239 (1982).
14. B. Price et al., Phys. Rev. Lett. 50:566 (1983); W. Heinrich et al., Univ. of Siegen report S-I-82-15.
15. A. Casher et al., Phys. Rev. D20:179 (1979).
16. E. Kolb et al., UC Santa Barbara report NSF-ITP-81-66.
17. A. B. Migdal, Rev. Mod. Phys. 50:107 (1978).
18. M. Gyulassy and W. Greiner, Ann. Phys. 109:485 (1977).
19. M. Gyulassy, Nucl. Phys. A354:395c (1981).
20. B. Banerjee, N. K. Glendenning and M. Gyulassy, Nucl. Phys. A361:326 (1981).
21. S. Nagamiya et al., Phys. Rev. Lett. 48:1780 (1982).
22. See J. Delorme, H. Toki, W. Weise, in: "Lecture Notes in Physics 137," H. Arenhövel and A. M. Sarius, eds., Springer-Verlag, Heidelberg (1981).
23. M. Soyeur, in: "Proc. Symposium on Neutron Stars," European Physical Society, Istanbul, (1981).

24. J. R. Nix and D. Strottmann, Phys. Rev. C23:2548 (1980).
25. A. Sandoval et al., Phys. Rev. C21:1321 (1980).
26. J. Cugnon, Phys. Rev. C22:1885 (1980).
27. H. Stocker et al., Phys. Rev. Lett. 47:1807 (1981).
28. R. Stock et al., Phys. Rev. Lett. 44:1243 (1980).
29. H. Gutbrod, private communication.
30. R. Madey, private communication.
31. M. Gyulassy, K. Frankel and E. Remler, Lawrence Berkeley Laboratory report LBL-15285.
32. P. J. Siemens and J. O. Rasmussen, Phys. Rev. Lett. 42:844 (1979).
33. R. Stock et al., Phys. Rev. Lett. 49:1236 (1982); and GSI report GSI-82-31.
34. H. Stocker et al., Nucl. Phys. A387:205c (1982).
35. H. Stocker, M. Gyulassy and J. Boguta, Phys. Lett. 103B:269 (1981).
36. M. Gyulassy, K. A. Frankel and H. Stocker, Phys. Lett. 110B:185 (1982).
37. J. Cugnon, J. Knoll, C. Riedel and Y. Yariv, Phys. Lett. 109B: 167 (1982).
38. H. H. Gutbrod et al., Lawrence Berkeley Laboratory report LBL-14980.
39. H. Stroebele et al., GSI report GSI-82-32; D. Beavis et al., to be published.
40. H. Stocker et al., GSI report GSI-82-8.
41. J. Engels, F. Karsch and H. Satz, Phys. Lett. 113B:398 (1982).
42. I. Montvay and E. Pietarinen, Phys. Lett. 110B:148 (1982); 115B:151 (1982).
43. See contribution of H. Satz in these proceedings.
44. M. Gyulassy, in: "Quark Matter Formation and Heavy Ion Collisions," M. Jacob and H. Satz, eds., World Scientific Pub. Co., Singapore (1982).
45. J. E. Elias et al., Phys. Rev. D22:13 (1980).
46. A. S. Goldhaber, Nature 275:114 (1978); W. Busza and A. S. Goldhaber, Stony Brook report ITB-SB-82-22.
47. E. L. Feinberg, Phys. Rep. 5:237 (1972); J. Koplik and A. H. Mueller, Phys. Rev. D12:3638 (1975); A. S. Goldhaber, Phys. Rev. Lett. 35:748 (1975).
48. R. Anishetty, P. Koehler and L. McLerran, Phys. Rev. 22D:2793 (1980).
49. S. J. Brodsky, in: "Proceedings of 1st Workshop on Ultra-Relativistic Nuclear Collisions," Lawrence Berkeley Laboratory report LBL-8957; see also A. Bialas *ibid.*
50. UA5 Collaboration, Phys. Lett. 107B:315 (1981).
51. JACEE Collaboration, in: "Proceedings of Workshop on Very High Energy Cosmic Ray Interactions," Univ. Penn., Philadelphia (1982).
52. I. Otterlund and E. Stenlund, Physica Scripta 22:15 (1980).
53. M. Gyulassy, L. McLerran and H. Satz, to be published.
54. J. Rafelski and B. Muller, Phys. Rev. Lett. 48:1066 (1982).

55. K. Kajantie and H. I. Mietinnen, Z. Phys. C9:341 (1981).
56. L. van Hove, CERN report TH-3391.
57. J. Iwai, N. Suzuki and Y. Takahashi, Prog. Theor. Phys. 55: 1537 (1976).
58. J. Iwai et al., Nuovo Cim. 69A:295 (1982).
59. M. Crawford and D. N. Schramm, Nature 298:538 (1982).
60. J. D. Bjorken, Fermilab report Pub.-82/44-THY; K. Kajantie and L. McLerran, Univ. Helsinki report HU-TFT-82-24.

PIONS FROM AND ABOUT HEAVY IONS

J.O. Rasmussen

Lawrence Berkeley Laboratory
University of California
Berkeley, CA 94720

INTRODUCTION

When nuclei collide at energies in the center-of-mass system that exceed the rest mass energy of the pion of 140 MeV, it is energetically possible to produce a pion. I recall from a dozon years ago a flurry of excitement over the possibility that pions might be observable from heavy ion linear accelerators and cyclotrons at the modest heavy ion energies then available. I was working at the old Yale Heavy Ion Accelerator at the time, and on a visit to the University of Maryland it was suggested to me that we ought to look for pions with the 400 MeV argon ion beams at the Yale HIA. Though it sounded unlikely that all the nucleons could somehow concentrate their energy into producing a pion, it was argued that there was a coherence effect among the various nucleon-nucleon collisions that would greatly enhance pion production. The skeptics argued that if this coherence worked as advertised, the best way of making pions would be to drop an elephant into a pit—total energy well above threshold and a tremendous number of nucleons to act coherently.

We didn't actually get around to trying such long-shot experiments at Yale, either with argon ions or elephants, but ten years later I found myself very much immersed in this kind of experiment at the Berkeley BEVALAC. To this pre-history I should add that at Maryland they did succeed in observing neutral pions from ^{3}He on ^{12}C at the picobarn per steradian per MeV level.[1] Work of Schimmerling et al.,[2] at the old Princeton-Penn accelerator measured pion production by nitrogen-14 ions at 520 A MeV. That the early ideas were not so preposterous was brought home to us in a MSU-Berkeley-Tokyo collaboration when we found[3] we could still observe pions when the energy of our ^{20}Ne beam was lowered to around 100 A MeV. More recently at the CERN synchrocyclotron π^+ spectra have been measured from 86 A MeV C^{12} ions.[4] Even more surprising are the CERN experiments[5] that have measured pion production with the fused nucleus going to ground and first excited states, namely the reaction $^3He + {}^3He = {}^6Li + \pi^+$. That this unlikely process goes with a substantial (i.e. 111 ± 11 nb) cross section seems amazing and certainly provides some challenge to theorists.

PION THERMOMETRY

The various particles from high energy heavy ion collisions exhibit spectra that asymptotically approach an exponential fall-off with energy suggesting a thermal distribution. There are two main problems: one, the slopes depend on angle, and two, the apparent temperatures indicated by the different particles are not the same. We restrict ourselves here to the case of symmetric systems, where target and projectile are nearly the same charge and mass. The slopes determined at the more forward angles are not reliable indicators of any thermal equilibrium situation. In the case of protons there is a persistence of initial momentum through several collisions, since elastic scattering is forward-peaked at energies greater than 300 A MeV. In the case of pions there is a natural forward-backward peaking from the initial formation in a nucleon-nucleon collision, such distribution arising from the role of the Δ(1232) intermediate. For composite particles, like the deuteron, pick-up processes can give direct reaction components in the forward direction. Thus, we believe that the 90° c.m. spectra of various particles are the least admixed with particles from direct processes and may provide the best thermometry of the hot reaction region. A closely related method is applied by Kevin Wolf to streamer chamber π^- data, namely to use the Hagedorn approach of plotting cross invariant cross section vs. observed particle kinetic energy.

Figure 1 is taken from Nagamiya et al.,[6] and it shows proton and pion spectra from various heavy ion reactions at 0.8 A GeV. They are careful not to call the slope parameter E_0 the temperature, but it is closely related to a temperature. Figure 1 is a plot of Lorentz-invariant cross sections, whereas to get a temperature from a Boltzmann distribution, one should plot just the differential cross section without multiplying by the relativistic total energy. The effective temperatures will be somewhat lower than Nagamiya's E_0 values.

The correction to E_0 to get a temperature value is rather easy to make. Figure 2 from Nagamiya shows π^- invariant cross sections $E(d^3\sigma/d^3p)$ vs. pion kinetic energy for Ne + NaF at several different bombarding energies. Note that the following simple relations connect T and E_0:

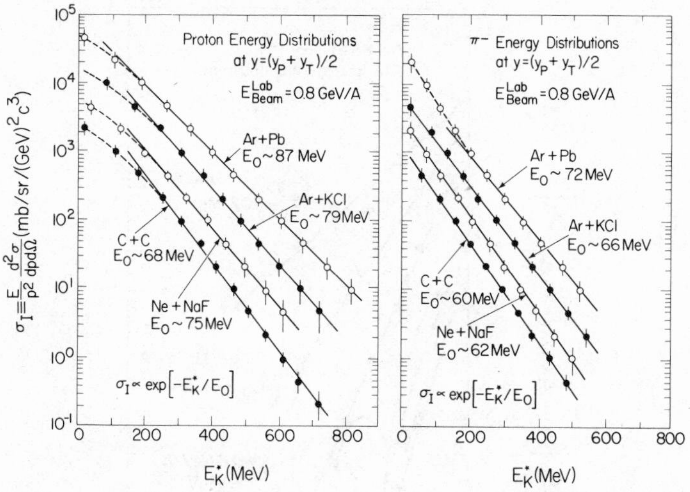

Fig. 1. Semi-logarithmic plots of Lorentz-invariant production cross sections at 90° vs. particle kinetic energy. The left figure is for protons and the right for negative pions. Various projectile-target systems are shown, all at a beam energy of 0.8 A GeV. (From Nagamiya et al.[6])

$$\sigma_{INT} = E \frac{d^3\sigma}{d^3p}$$

where E is the relativistic total energy of the pion. Taking the natural logarithm of both sides and differentiating with respect to E gives us

$$-\frac{1}{E_0} = \frac{1}{E_x} - \frac{1}{T} ,$$

where E_x is the total energy at which the slopes are to be determined. Now we see that the effective temperature T will not be so well defined as is E_0, for a plot of $d^3\sigma/d^3p$ against pion energy will be concave downward rather than the straight line forms of Nagamiya. Such behavior is physically reasonable for systems with a relatively small number of particles. We know especially from lower energy nuclear physics that the particle evaporation spectra depart from exponential behavior at high energies and indeed go to zero where all the available energy of the finite system is carried off. If we choose to evaluate the effective temperature at a kinetic energy around twice the E_0 value (total E three times greater), then the above equations tell us that T, the effective temperature is 0.75 times E_0. With this in mind we can determine temperatures from Fig. 3 of Nagamiya.

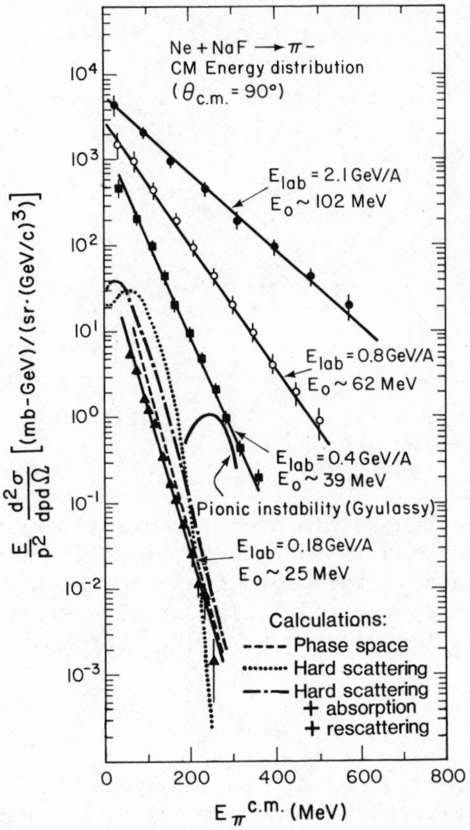

Fig. 2. Same as Fig. 1, except for the Ne + NaF system and negative pions at various beam energies. The lowest energy data are compared with various theoretical predictions. (From Nagamiya et al.[6])

The pion probe indicates temperatures smaller than given by the simple fireball model in which all the available kinetic energy of the geometrically overlapping matter (participants) is assumed to be thermalized. For example, at E_{lab} = 0.18 A GeV there is 45 MeV kinetic energy per nucleon available for a fireball model temperature of 2/3 of this, or 30 MeV. This is to be compared with a pion temperature of 0.75 times 25 MeV, or 18 MeV. Note also that the temperature indicated by the pions for a given system is always less than that indicated by the protons. Similarly, the deuterons and tritons indicate progressively higher temperatures according to Nagamiya.

In an attempt to resolve the discrepancy between proton and pion temperatures Phil Siemens and I some years ago proposed[7] a blast-wave

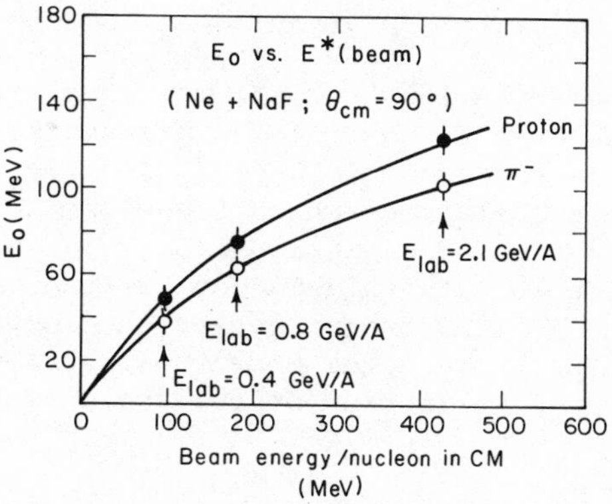

Fig. 3. Summary of slope parameters E_0 for various beam energies. (From Nagamiya et al.[6])

model. We derived expressions for the spectra from a system in which the available energy was divided between thermal and ordered motion of a spherically expanding blast wave. If the energy were about equally distributed between these forms, the difference in apparent temperatures was resolved. Today I would not want to defend the literal blast wave but rather point out that any ordered motion of the thermal sources in the perpendicular direction can give systematic differences in the limiting slopes of the various mass particles boiling off. Arguments have been given that a "bounce-off" effect occurs,[8] and this effect provides an ordered sidewise component in the thermal sources. Furthermore, cascade-code calculations of Gyulassy and Frankel[9] show a fluctuation in sideways momentum, even of head-on central collisions. Nagamiya in Ref. 10 presents a kaon temperature which does not fit into my general theme that it is ordered perpendicular motion that causes the difference in slopes. The temperature inferred from the kaon (K^+) production is even higer than that of the proton, rather than lying between proton and pion temperatures. I would merely point out that the kaons are not likely to be very good thermal probes, since their mean free paths in nuclear matter are longer than the other particles. Since they require the greatest energy for production, they may also probe just the most violent initial parts of the collision, before any thermal equilibration has set in.

Then what are these slopes and pion and proton temperatures telling us generally about the heavy ion collision processes? The simple fireball model of complete thermalization, as has certainly been pointed out before, cannot be used literally. At least for systems up to mass 40 on mass 40 at BEVALAC energies the forward momentum of nucleons is not totally degraded. One way of dealing with this incomplete degradation was outlined by C.Y. Wong.[11] He analyzes 800 MeV per nucleon $^{40}Ar + KCl$ results with an ellipsoidal Gaussian distribution for the proton spectra. He states the following:

"We observe that the longitudinal momentum distribution has a larger width as compared to that of the transverse distribution, as one expects from the cascade model.[12,13] Clearly there are substantial nucleon-nucleon collisions which transform the two-center momentum distribution into an ellipsoidal Gaussian distribution by filling in the space in between. The collisions are not numerous enough to erase the initial preference of the longitudinal momentum direction. The anisotropy in the momentum distribution indicates that complete thermal equilibrium of the whole system is not yet achieved."

Another way of dealing with the incomplete thermalization in terms of a simple model is the two-fireball model of Das Gupta and Lam.[15] The overlapping matter distributions in the collisions pass through one another, depositing part of their translational kinetic energy into thermal energy, so one has two hot sources of pions, protons, etc. instead of the single source of the simple fireball model. Radi et al.[16] adapted this model to describe the pion source function in Monte Carlo trajectory studies of pion Coulomb effects. Figure 4 is from this work, and it shows final fireball velocities and temperatures as a function of impact parameter of the collision for a $^{20}Ne + ^{20}Ne$ system. Our beam energy was E/A of 655 MeV, whereas the dashed comparison from Das Gupta was for 800 MeV. Our calculations are done for two cases, for a longitudinal momentum decay length of 2.6 fm, the theoretical value given by Sobel et al.,[17] and for double that value, which we believe to be more realistic and in line with various other evidence that collision mean free paths of nucleons in nuclei are longer than the simple first theories give. (cf. Refs. 18,19.) The one-fireball temperature shown in Fig. 4 is, of course, independent of impact parameter. The two-fireball model, as we used it with a longitudinal momentum decay length or friction coefficient, qualitatively accounts for the variation of E_0 values with the mass of the system, as shown in Fig. 1.

Knoll[20] has taken the two-fireball model a step forward in sophistication by allowing for sideways momentum of the fireballs arising from processes, such as, the "bounce-off" effect of hydrodynamics. He calls this a two-source model, and he gives the following equation (non-relativistic) for the particle distribution:

$$P(\vec{V}) \approx \exp\left[-\frac{(\vec{V}\pm\vec{V}_{coll})^2}{2\langle V^2_{thermal}\rangle}\right] .$$

Let us now recap this brief discussion of pion thermometry in heavy-ion collisions. (1) The asymptotic slope of the pion spectra at 90° c.m. gives probably the best simple measurement of temperature of the hot reaction region, since the 90° direction gives least

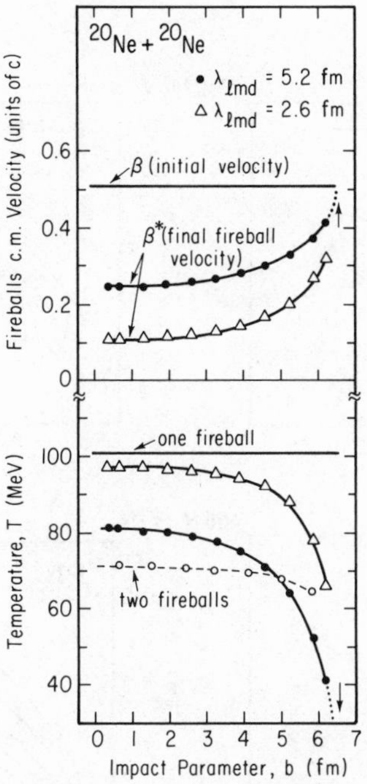

Fig. 4. Fireball velocities and temperatures in the two-fireball model for Ne + Ne at 655 MeV. The calculations are plotted against impact parameter, and two different cases are shown. The dots represent the more realistic case in which the longitudinal-momentum mean free path is double the first published estimate; the open triangles represent the case with the published longitudinal-momentum-mean free path of 2.6 fm. (From Radi et al.[16])

contamination from direct unscattered pions and since the pion's low rest mass makes it less sensitive than protons to sideways collective motions of whatever sort in the sources. (2) The temperatures so determined, as well as other evidence, suggest that for mass 20-40 symmetric collision systems at BEVALAC energies about half the available energy in the c.m. system is thermalized, with the remainder in ordered motion along the beam direction and perpendicular to it. Global analysis of streamer-chamber events offers the hope of delineating details.

HILLS AND VALLEYS IN PION SPECTRA

In the preceding section we discussed information to be gained from the smooth exponentially-falling region of highest energy pions.

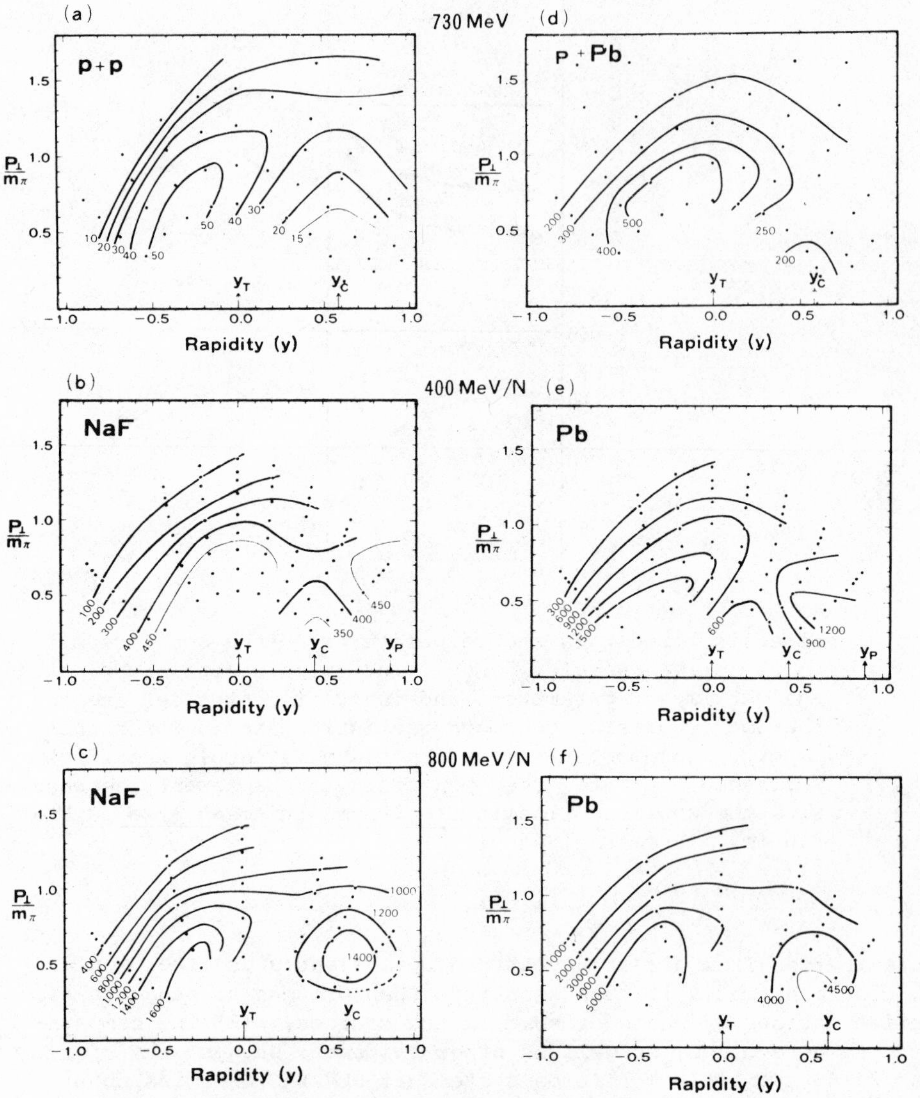

Fig. 5. Contour plots of Lorentz-invariant π^+ cross sections in the plane of rapidity y and perpendicular momentum. The numbers on the contour lines are the Lorentz-invariant cross sections for π^+ production in units of mb sr^{-1} GeV^{-1} c. The dots indicate points at which data are given. The arrows on the abscissas indicate target, nucleon-nucleon center-of-mass and projectile rapidities. (From Nakai et al.[22])

At lower pion energies a number of special features have been found in the spectra. I discussed these at some length in the Relativistic Heavy Ion Winter School in Banff last February, and the Proceedings can be consulted for details.[21] These features I referred to there as "funny hills" of the first, second, and third kind.

The first kind is a peaking that occurs at low $P_\perp$. This peaking is seen also in free nucleon-nucleon production of pions, as seen in Fig. 5(a), and is to be associated with Δ decay. These hills of the first kind are clearly seen in Ne + NaF at E/A of 0.8 GeV [Fig. 5(c)] and of 0.4 GeV [Fig. 5(b)] but are pretty well washed out in Ar + Ca at 1.05 GeV. Contour plots of the Ar + Ca data are shown in Fig. 6, where the lower plot shows data of Ref. 22 alone, and the upper plot combines it with our data.[23] The Ar + Ca system is apparently sufficiently large to be effective at scattering and thermalizing the nascent pions from nucleon-nucleon collisions.

The hills (and valleys) of the third kind are clearly Coulomb effects of the spectator fragment charges. The hills, often quite sharp and pronounced, appear in π^- spectra at or just below beam velocity. They were first reported by our MSU-Berkeley-Tokyo collaboration,[24] and reports from more comprehensive studies are more recently published by Sullivan et al.[25] Figure 7 from Ref. 21 shows from work of Murphy et al.[26] a nice illustration of the beam-velocity π^- peak. Radi et al.[27] have made a theoretical analysis of some of these π^- peaks and conclude that the Coulomb focussing is sensitive to the primary distribution of projectile fragments before proton or alpha evaporation. Furthermore, they extract a velocity dispersion

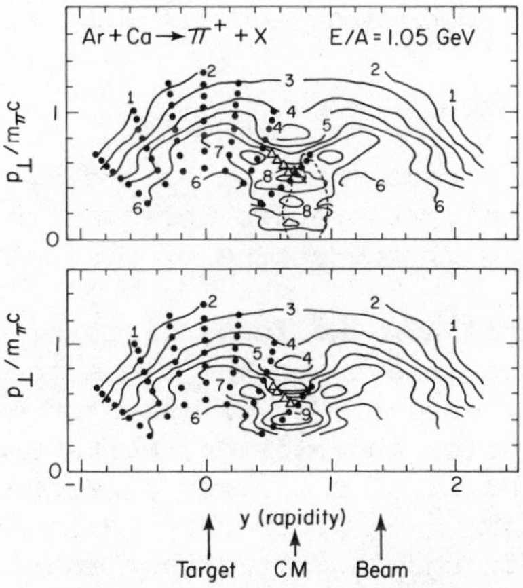

Fig. 6. Same as Fig. 5, except for the Ar + Ca system, and change of units to b sr^{-1} GeV^{-2}. The upper figure combines data of two groups, and the lower figure is for the data of one group, Wolf et al.[23]

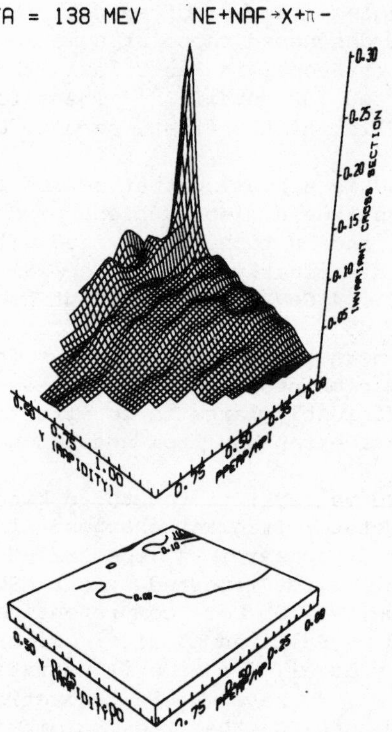

Fig. 7. Isometric and contour plots of "sub-threshold" negative pion production data. Cross section units are those of Fig. 6. Note the pronounced peak at beam velocity, and the smooth exponentially-falling behavior elsewhere. The small ripples are probably statistical fluctuations.

of the final fragments after nucleon boil-off. Corresponding to the π^- peaks at beam velocity there are holes in the π^+ spectra, caused by Coulomb defocussing by the charge of projectile fragments.

I have saved until now the funny hills of the second kind. They appear as hills or ridges in the cross sections at very low energy in the c.m., around 15 MeV or momentum of 0.4-0.5 $m_\pi c$. They can be seen in Figs. 5(c) and 5(f) and perhaps in 5(e); also Fig. 6 shows them. On the basis of the first experimental evidence from π^+ in symmetric systems[22,23] Libbrecht and Koonin[28] made a Monte Carlo trajectory study and suggested that these bumps were a Coulomb effect. Subsequent studies[29] revealed (see Fig. 8) similar low energy bumps for π^- as well as π^+, and π^+ studies showed the bump did not move out for high atomic number systems.[30] On the basis of these studies it seems that the funny hills of the second kind cannot be explained solely as some trivial Coulomb effects. We

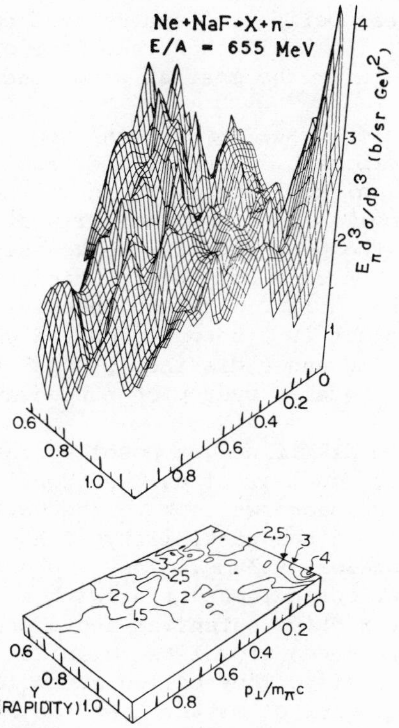

Fig. 8. Same as Fig. 7, except for the higher beam energy of E/S = 655 MeV.

should point out that they are not seen in the p+p → π^++X reaction, nor are they seen at E/A of 400 MeV. It has been suggested that they may be indicating some side-splash or collective flow effects or that spectator-shadowing could play a role. However, I currently favor the idea that these low-energy pion bumps are giving us a momentum space snapshot of some sort of pion orbiting of nuclear dimensions.

Pionic Orbits of Nuclear Size?

What is striking about the mid-rapidity bumps is that a simple application of the uncertainty relation gives a distance about equal to the nuclear radius. That is, ħ over the bump momentum of 0.5 $m_\pi c$ gives a size parameter of 2 pion Compton wave lengths, or about 2.8 fm. Of course, the bump may be shifted somewhat by Coulomb effects, so we will need in the future to accumulate data for both π^+ and π^- for systems of widely differing mass to see if the mid-rapidity bumps move inward with increasing mass, as the uncertainty principle argument would predict. Ideally we should have data somewhat sorted between peripheral and central collisions to extract the most information. We could then take Fourier transforms of the momentum spectrum to infer the spatial wave functions of the pions.

It is worthwhile to trace some of the roots of these notions about possible pion orbiting. Kitazoe and Sano[31] solved equations for thermal and chemical equilibrium among particles in the hot dense matter of heavy ion collisions, and they drew attention to a "zero-energy" component of the pion spectra. Likewise, Zimanyi, Fai, and Jakobsson[32] considered such a component, which they gave a finite width according to the nuclear size and uncertainty principle. They claim that this component is a boson-condensed component analogous to superfluid liquid ^{4}He and a distinctly separate phenomenon to the virtual pions of the Migdal-Sawyer pion condensation.[33,34]

Now it is fair to ask if we get boson condensation among identical pions in a box of nuclear size, what confines the pions to the nuclear volume? For guidance we look to the work of Ericson and Myhrer,[35] who pointed out the possibility of hadronically-bound pion orbitals within the nuclear volume. They approximated the interaction by a pion-nuclear optical potential with a Kisslinger-type velocity-dependent attractive potential and a static repulsive potential. Parameters were taken from fits to shifts and widths of pionic atom s- and p-state levels. They called attention to a peculiar feature for π^- in neutron-rich matter that the effective mass could go through a singularity and become negative, inverting the order of binding of states in a well. Mandelzweig, Gal, and Friedman[36] also studied the problem and clarified the behavior of the wave function near the singularity. They show that the singularity can divide solutions into the classes of inner and outer solutions. For the inner solution where the effective mass of the pion is negative the pion energy is lower than the potential energy, essentially a negative kinetic energy. This "Kisslinger" catastrophe with its infinity of bound states needs some qualifying remarks. The Kisslinger term goes as ∇^2, just as the kinetic energy in the Schrödinger equation, and the Kisslinger term can take on a greater magnitude than kinetic energy for attainable neutron-richness. However, the Kisslinger term is only part of an expansion of a non-local potential and should probably be regarded as insufficient for higher kinetic energies comparable to the 130 MeV pions from decay of Δ's.

I should say that the states calculated in Refs. 35 and 36 for the most part had very large widths. That is, the true absorption of pions by pairs of nucleons is so great that these broad states may be of only academic interest for pionic states of stable nuclei or mildly neutron-rich beta-unstable species. However, in heavy-ion reactions there are two possible new species of interest. First, there are the fireballs of double or higher nuclear density, where even broad states of pion orbiting might give rise to measurable bumps in the spectra. Second, there are highly neutron-rich spectator fragments about which π^- orbiting could occur. This latter situation provides the candidate for Bill McHarris' and my explanation[37] of the anomalon puzzle, to which I return shortly.

Pion Confinement in the Fireball

To get some orientation on pion orbiting conditions I have developed a computer code to solve for eigenvalues of pions in a nuclear optical potential. The methods of Ericson and Myhrer[35] have been employed except that we obtain eigenvalues first in the real part of the potential, using the WKB approximation and applying Bohr-Sommerfeld quantization conditions, with modifications of Ref. 36 where there is a singularity in the effective mass. The widths of the states are calculated perturbatively from the imaginary parts of the potential; the states of greatest interest to us will have rather small widths and thus be adequately estimated by the perturbative treatment of the imaginary phase. I have taken potential-parameter starting values from the recent work of Carr, McManus, and Stricker-Bauer,[38] though they are not very different from the pionic atom values used by Ericson and Myhrer[35] except in the isospin dependence of the static repulsive potential. We shall refer to the 25 MeV pion potential (set F) of Ref. 38 as the CMS potential.

The optical potential in the notation of Ref. 38 is as follows:

$$2\omega U = -4\pi\left[b + B - \nabla\cdot\frac{L}{1+4\frac{\pi}{3}\lambda L}\nabla\right]$$

plus the Coulomb term. Here

$$b = \bar{b}_0\rho - \varepsilon_\pi b_1\delta\rho$$

$$L = c + C$$

$$c = c_0\rho - \varepsilon_\pi c_1\delta\rho$$

$$B = B_0\rho^2 - \varepsilon_\pi B_1\rho\,\delta\rho$$

and

$$C = C_0\rho^2 - \varepsilon_\pi C_1\rho\,\delta\rho$$

where ρ is the nucleon density and $\delta\rho$ is the neutron density minus the proton density. Capitalized parameters B and C are coefficients of terms arising from true pion absorption and are dependent on the square of the nuclear density. The lower case parameters b and c denote terms arising from single nucleon scattering and are dependent linearly on the density. Isoscalar and isovector terms are distinguished by the subscripts zero and one, respectively. The Lorenz-Lorentz-Ericson-Ericson (LLEE) parameter for polarization of the medium is denoted by λ. In writing the above equations I have suppressed the radial argument r, which affects all the densities and density dependent variables. Furthermore, I have set the kinematic factors of Ref. 38 to unity for simplicity, since they differ from unity only by terms of the order of the pion-nucleon mass ratio. The quantity ω is the relativistic total energy of the pion at infinity. The quantity ε_π is +1 for π^+ and -1 for π^-.

First, with the nuclear potential turned off I have verified that the code gives about the right Bohr value for the pionic binding energy in lowest s- and p-states. Then the first calculations with the double density of a fireball made clear that the central part of the nucleus will have a negative effective pion mass for both π^+ and π^- under a broad range of neutron-to-proton ratios. The mass-singularity near the surface thus provides always a confining boundary for inner solutions. In general they are deeply bound states, but they show large widths.

ANOMALONS

From research on relativistic heavy ions has come a remarkable puzzle with which the name anomalon has been associated. When heavy ions in the GeV/nucleon energy range pass through photographic emulsions or other matter, they often undergo peripheral fragmentation reactions from which a fragment of somewhat reduced charge continues on with nearly the beam velocity. Figure 9 is a photomicrograph of a multiple fragmentation of a 1.88 GeV/nucleon ^{56}Fe ion from the Bevalac. Three fragmentation processes, giving successively charges of 24, 20, and 11, are seen before the chain is terminated by a violent central collision.[39]

The puzzle is that some of the secondary, tertiary, etc. fragments appear to exhibit anomalously short mean free paths for further reactions. Primary beam fragments exhibit normal mean free paths that can be well represented by the expression

$$\lambda_Z^* = \Lambda^* Z^{-b}$$

The coefficient Λ^* is called the "charge independent" mean free path, and for beam particles it takes the value of 30 cm with parameter b about 0.4. Now this reduced mean free path for other-than-primary fragments is plotted in Ref. 39 and in Gyulassy's talks in this school. The plot is against path length for three different experimental groups, combining experiments[40,41] with Bevalac beams of ^{56}Fe, ^{16}O, and ^{40}Ar and with cosmic rays.[42] The reduced mean free path dips to 80% of its normal value and returns to normal over a characteristic length of 2.5 cm. No special characteristics have been identified that would distinguish the reactions in the early part of

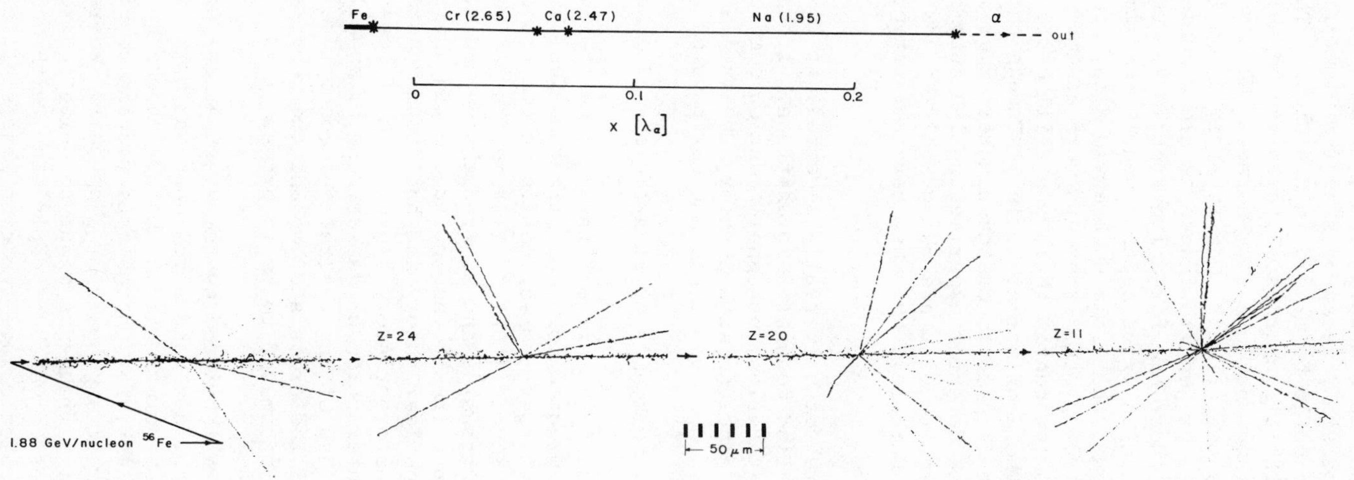

Fig. 9. Photomicrograph mosaic illustrating repeated projectile fragmentation of an incident ^{56}Fe nucleus at E/A of 1.88 GeV. The occurrence of abnormally short mean free paths of secondary, tertiary, etc. fragments in the first 2 or 3 cm of their path is called the anomalon effect.

the path from those further along the path or those of primary beams. That is, if the anomalons have a decay mode, it must be by neutrons or gamma rays, though possibly decay by a near-minimum-ionizing singly charged particle could be missed in emulsions. The anomalous behavior seems to go with all charges 3, with the behavior of charge 1 and 2 perhaps, but not clearly, normal. What must we assume to explain these deviations? One interpretation is based on the model that 6% of all projectile fragments are anomalous with a MFP of 2.5 cm. Such a short MFP for a light nucleus, such as, ^{12}C would mean it has a reaction cross section nearly ten times geometric. Nor does it help to assume that some unobserved decay process is involved, since even if 100% of the fragments were assumed anomalous with decay lengths of a few centimeters, one still would need cross sections around twice those of normal nuclei. Though this behavior is by no means rare, it is hard to design counter experiments to check it. Plastic track detector experiments with automated microscopes to eliminate scanning bias are being analyzed now by Prof. Price's group at Berkeley, but they have not yet presented results.

While anomalons are not universally accepted, it is incumbent on theorists to try to find possible explanations. I shall not try to catalog the various theories, but shall mention some of the exotic possibilities and then conclude with a theory McHarris and I are pursuing that is closer to conventional nuclear physics.

I quote from Harry Heckman's brief article[39] of June, 1982:

Theories that suggest nuclear collisions can alter the quark structure in nuclei to produce color polarization inside the nucleus, thereby giving rise to larger nuclear collision cross sections, have been put forward by Y. Karant (LBL),[43] W. Romo, and P. Watson (Carleton University, Ottawa)[44] and S. Fredriksson and M. Jandel (Stockholm).[45] J. Boguta (LBL)[46] has pursued a Lagrangian field theoretic approach in nuclear theory and has found "hadroid" solutions that exhibit the appropriately "long-range" forces required to explain the large interaction cross sections of anomalons. So far it is beyond the grasp of any theoretical concept to explain both the enhancement of the cross section and the remarkably long lifetime of anomalons, estimated to be in excess of 10^{-10}.

McHarris and I were intrigued by the speculations of van Dantzig and van der Velden[47] that a π^- and a few neutrons might form a bound system, a "pineut". The normal strong absorption channel for pionic atoms would not be present, for charge conservation requires that "true absorption" occur on a pn pair or a pp pair. We reasoned that at least 6% of projectile fragments might be quite neutron-rich, thus have neutron haloes to which π^- could bind, with quite small absorption if only the pionic wave function were small enough in the nuclear interior, where the proton density is significant.

SCHRÖDINGER EQUATION SOLUTIONS FOR PIONIC ATOMS

It is not easy to make quantitative the above speculations about exotic pionic atoms, but we have made a start along the lines pioneered by Ericson and Myhrer.[35] We have started with the pion optical potential outlined above, using parameters of Ref. 38. Their parameters have never been tested on very neutron-rich species, so the isovector terms are uncertain.

We parametrize the nuclear density as the usual Woods-Saxon form but also give the proton-to-nucleon ratio a Woods-Saxon form with separately adjusted radius, so as to set up a neutron skin for neutron-rich species. One may note that the true absorption should greatly decrease for the very deeply bound pion internal solutions of the type proposed by Ericson and Myhrer. This decrease results from the fact that the number of channels open for the dominant absorption process of ejection of two neutrons into the continuum will strongly decrease. True absorption of a negative pion converts a proton into a neutron and thus drives in the opposite direction to negative beta decay, which may be exoergic by more than 10 MeV in the neutron-rich light nuclei.

As pointed out in Ref. 36, the radius at which the effective mass singularity occurs divides the pionic solutions into essentially internal and external classes. Mandelzweig et al. give the Bohr-Sommerfeld phase values appropriate for solutions bounded by the mass singularity, and we use their prescriptions.

Use of the CMS pionic atom potential did not give us a mass singularity, even for the neutron-rich ^{34}Na test nucleus. Thus, we first softened the LLEE effect by reducing the λ parameter from 1.4 to 1.0, certainly within the range of values used in the pion optical parameters. Secondly, we reasoned that the neutron-rich nuclei would have a large neutron-density polarizability which would result in the local nuclear density, especially of neutrons, in the vicinity of an orbiting pion to be larger than normal. With these modifications the effective mass became negative throughout the nuclear interior.

We have not attempted yet to calculate the degree of excess nuclear density induced by pions. It will not be easy to calculate, but it is analogous to the effective charge problem in the nuclear shell model. Rather have we simply made calculations for three different degrees of neutron-density enhancement about the pion. Our calculations are summarized in Table 1, where the first column is the orbital angular momentum L, the second column is the fractional excess neutron density assumed (proton density was unaltered). The third column gives the location of the mass singularity in units of the Woods-Saxon half-density radius. The fourth column gives the negative eigenvalues (binding energies); for the system as a whole

Table 1. π^- Solutions Inside the Effective-Mass Singularity (^{34}Na).

L	$\Delta\rho_N$(polariz.)/ρ_{N_0}	$R_{sing.}/R_N$	$-E_\pi$(MeV)	$\langle r\rangle_{orbit}/R_N$
0	0.86	1.011	17.32	0.50
1	0.86	1.011	59.20	0.67
2	0.25	0.863	24.08	0.62
3	0.25	0.863	40.07	0.65
0	0.40	0.932	9.72	0.47
1	0.40	0.932	27.60	0.62
2	0.40	0.932	53.12	0.67
3	0.40	0.932	80.94	0.71
4	0.40	0.932	110.7	0.74
5	0.40	0.932	142.1	0.76

the binding energy should be reduced by the energy cost of local neutron density enhancement, which we have not attempted to calculate. The last column gives the average orbital radius of the solution in units of the half-density nuclear radius.

The last two solutions, L=4 and 5 for intermediate neutron density polarizability, are so deeply bound that all or most channels of true pion absorption may be closed altogether, and the lifetime could lengthen to that required for anomalons. Let us ask why the reaction cross sections of these internal pion species should be so much larger than normal nuclei. The long-range part of the nuclear force is generally assumed to come from virtual pion exchange. That is, the intermediate state is classically energy-forbidden by the rest mass energy of 140 MeV, so the characteristic range is the pion Compton wave length. Contrast this with the range of chemical binding forces in, say, the simplest molecule, the hydrogen molecule ion. The electron, real, not virtual, is the exchanged particle, and the characteristic force range is then the tunneling length of an electron with the Rydberg binding of 13.5 eV. By the same token, the Ericson-type deeply bound pions are real, not virtual, so their tunneling range outside the nucleus will be longer than the pion Compton wave length. Indeed, the more lightly bound the pion, the longer the characteristic distance it can tunnel. We are now faced with a trade-off, though. The deeply bound pion states may have the longest lifetimes, but the largest nuclear interaction cross sections will go with the more lightly bound solutions. It remains to try to calculate estimates of these properties. One also is faced with the problem of explaining the sheer abundance of anomalon production. To the extent that there is any thermal equilibration in the production of reaction fragments, we note that the masses of the deeply bound pionic species are not much greater than the corresponding ordinary nuclei.

Finally, we mention one more interesting possibility and complication. If two π^- are associated with a single nuclear fragment, it may be that extra binding will occur by virtue of admixture of "di-deltas", strongly bound combinations of two delta particles in the isospin 3 and spin 0 state.[46,47,48,49]

We must conclude that there is need for fundamental theoretical work to pin down the effects on the pion optical potential for multiple pions and the consequent di-delta admixtures. Pending such work we can only say that moderate alterations of the CMS potential of neutron-rich nuclei can give somewhat deeper bound pion s-states with lifetimes that can get into the anomalon range. The quark-bag rearrangement candidates for anomalons have at least as great difficulty explaining the properties of anomalons. Whatever the outcome of anomalon physics it continues to excite the imagination and stimulate new experiments. In particular, I am much intrigued by possibly learning something about pionic atoms with more than one pion, a field of study uniquely accessible by relativistic heavy ion research.

ACKNOWLEDGMENTS

This work was supported by the Office of Energy Research, Division of Nuclear Physics of the Office of High Energy and Nuclear Physics of the U.S. Department of Energy under Contract DE-AC03-76F00098. The ideas here presented have been shaped over time by discussions and interactions with various colleagues, particularly those of our TOSABE (Tokyo-Osaka-Berkeley) collaboration and its successor the MOBIT (Michigan State, Orsay, Berkeley, Indiana, Tokyo) collaboration. To Profs. Hafez Radi and Bill McHarris, who spent their most recent sabbaticals working in our group, I am indebted for many discussions and collaboration on theoretical calculations. Finally, I wish to express special appreciation to Arlene Spurlock of the LBL Technical Information Division for her rapid and skillful pioneering of the link-up between my Osborne-1 computer text and the large computer and advanced printers at LBL via the UNIX system.

REFERENCES

1. N. S. Wall, J. N. Craig and D. Enrow, Nucl. Phys. A214:459 (1976).
2. W. Schimmerling et al., Phys. Rev. Lett. 33:1170 (1974); K. G. Vosburgh et al., Phys. Rev. D11:1743 (1975).
3. W. Benenson et al., Phys. Rev. Lett. 43:683 (1979); errata 44:54 (1980).
4. B. Jakobsson, "Light Particle Production in ^{12}C Induced Reactions at CERN SC Energies," Univ. of Lund report LUIP (1982).
5. Y. Le Bornec et al., Phys. Rev. Lett. 47:1870 (1981).
6. S. Nagamiya et al., Phys. Rev. C24:971 (1981).

7. P. J. Siemens and J. O. Rasmussen, Phys. Rev. Lett. 42:844 (1979).
8. H. Stocker, J. A. Maruhn and W. Greiner, Phys. Rev. Lett. 44: 725 (1980).
9. M. Gyulassy, K. A. Frankel and H. Stöcker, Phys. Lett. 110B:185 (1982).
10. S. Nagamiya, Proc. Fifth High Energy Heavy Ion Study, Berkeley (May, 1981), LBL Report No. 12652, p. 141 (see Fig. 33).
11. C. Y. Wong, Proc. Fifth High Energy Heavy Ion Study, Berkeley (May, 1981), LBL Report No. 12652, p. 61.
12. Y. Yariv and Z. Fraenkel, Phys. Rev. C20:2227 (1979).
13. J. Cugnon, Phys. Rev. C22:1885 (1980).
14. S. Das Gupta, Phys. Rev. Lett. 41:1451 (1978).
15. S. Das Gupta and C. S. Lam, Phys. Rev. C21:1192 (1979).
16. H. M. A. Radi, J. O. Rasmussen, K. A. Frankel, J. P. Sullivan and H. C. Song, LBL Report No. 13768, submitted to Phys. Rev. C.
17. M. I. Sobel et al., Nucl. Phys. A251:512 (1975).
18. J. W. Negele and K. Yazaki, Phys. Rev. Lett. 47:71 (1981).
19. H. H. Heckman, D. E. Greiner, P. J. Lindstrom and H. Shwe, Phys. Rev. C17:1735 (1978).
20. J. Knoll. Proc. Fifth High Energy Heavy Ion Study, Berkeley (May, 1981), LBL Report No. 12652, p. 210.
21. J. O. Rasmussen, Proc. Relativistic Heavy Ion Winter School, Banff (Feb., 1982), also LBL Report No. 14174.
22. K. Nakai et al., Phys. Rev. C20:2210 (1979).
23. K. L. Wolf et al., Phys. Rev. Lett. 42:1448 (1979).
24. J. P. Sullivan et al., Phys. Rev. C25:1499 (1982).
25. J. P. Sullivan, Ph.D. Thesis, U. C. Berkeley, LBL Report No. 12546, unpublished.
26. D. Murphy et al., Bull. Am. Phys. Soc. 26:1148 (1981).
27. H. M. A. Radi, J. O. Rasmussen, J. P. Sullivan, K. A. Frankel O. Hashimoto, Phys. Rev. C25:1518 (1982).
28. R. G. Libbrecht and S. E. Koonin, Phys. Rev. Lett. 43:1581 (1979).
29. K. A. Frankel et al., Phys. Rev. C25:1102 (1982).
30. K. L. Wolf et al., "Pion Production and Charged-Particle Multiplicity Selection in Relativistic Nuclear Collisions," preprint, submitted to Phys. Rev. D.
31. H. Kitazoe and M. Sano, Lett. Nuovo Cim. 14:400 (1975).
32. J. Zimanyi, G. Fai and B. Jakobsson, Phys. Rev. Lett. 43:1705 (1979).
33. A. B. Migdal, Rev. Mod. Phys. 50:107 (1978).
34. R. F. Sawyer and D. J. Scalapino, Phys. Rev. D7:953 (1973).
35. T. E. O. Ericson and F. Myhrer, Phys. Lett. 74B:163 (1978).
36. V. B. Mandelzweig, A. Gal and E. Friedman, Ann. Phys. (N.Y.) 124:124 (1980).
37. W. C. McHarris and J. O. Rasmussen, "Anomalons as Pineuts Bound to Nuclear FragmentS: A Possible Explanation", LBL Report No. 14075, to be revised and resubmitted to Phys. Lett.

38. J. A. Carr, H. McManus and K. Stricker-Bauer, Phys. Rev. C25: 952 (1982).
39. H. H. Heckman, "Anomalons", LBL Report No. 14562, to be published in the Yearbook of Science and Technology, McGraw Hill, New York, NY (1982).
40. E. Friedlander et al., Phys. Rev. Lett. 45:1084 (1980).
41. P. L. Jain and G. Das, Proc. Fifth High Energy Heavy Ion Study, Berkeley, (May, 1981), LBL Report No. 12652, p. 404; Phys. Rev. Lett. 48:305 (1982).
42. H. B. Barber et al., Phys. Rev. Lett. 45:1084 (1980).
43. Y. Karant, LBL Report No. 9171 (1979).
44. W. J. Romo and P. J. Watson, Phys. Lett. 88B:354 (1979).
45. S. Fredriksson and J. Jandel, Royal Inst. of Technology, Stockholm Report TRITA-TFY-81811 (1981).
46. J. Boguta, LBL Report No. 14753, submitted to Phys. Rev. Lett. (1982) and LBL Report No. 14446, submitted to Phys. Lett. (1982).
47. R. Van Dantzig and J. M. Van der Velden, preprint NIKHEF, Amsterdam, and private communication (1981).
48. F. J. Dyson and N.-H. Xuong, Phys. Rev. Lett. 13:815 (1964).
49. T. Kamae and T. Fujita, Phys. Rev. Lett. 38:471 (1977).

NUCLEAR AND PARTICLE PHYSICS IN THE EARLY UNIVERSE

D.N. Schramm

Astronomy and Astrophysics Center
University of Chicago
Chicago, IL 60637

INTRODUCTION

In these lectures a review will be made of various events in the early universe where nuclear and elementary particle physics effects come into play. Several events will only briefly be mentioned, since they have been reviewed previously. However, some detail will be given to new results on big bang nucleosynthesis, massive neutrinos and inflationary scenarios.

The recent high level of activity at this boundary of particle physics and cosmology comes from the combination of a general acceptance of the big bang model of the universe and the fact that recent developments in particle theory point towards effects which will only be important at energies far beyond those accessible with accelerators.

The general acceptance of a hot-dense early universe (the big bang) began with the discovery of the 3 K background radiation by Penzias and Wilson.[1] However, complete acceptance didn't occur until the experiments of Richards and co-workers[2] showed that the background radiation had the appropriate thermal turnover at wavelengths of $\sim$1 mm. The existence of this radiation tells us that the universe was at one time at least $\gtrsim 10^4$ K. At $\sim 10^4$ K hydrogen would be ionized and the free electrons would easily scatter the photons. Thus, the present observed radiation is merely the last scattered thermal radiation from $\sim 10^4$ K.

We actually have confidence that the universe was a good deal hotter than this. Gamow and his co-workers predicted that there would be this thermal background on the basis of assuming that nuclear reactions occurred in the big bang. To have nuclear reactions requires that the temperature had to be greater than $\sim 10^9$ K. The verification of a temperature at least as hot as 10^{10} K comes from the fact that the ^{4}He abundance is about 25% by mass. This helium abundance comes as a natural consequence of the standard big bang if the temperature was greater than $\sim 10^{10}$ K (see Schramm and Wagoner[3] and references therein). As we will see, big bang nucleosynthesis which is based on well studied physics at temperatures and densities accessible in terrestrial laboratories is perhaps the strongest support for the standard model.

There are at present no direct observational ties to earlier times in the universe when the temperature was even higher (unless one accepts the existence of matter as an indication of grand unification decoupling at $\sim 10^{15}$ GeV). We do have some feeling that it is not totally absurd to discuss temperatures in the early universe as high as 10^{15} GeV or maybe even 10^{19} GeV. This complete lack of fear comes from the singularity theorems of Hawking, Ellis and Penrose (see Hawking and Ellis[4] and references therein). The theorems state that if the universe has a net free energy then the world lines must have come out of a singularity. Since we know the universe is roughly homogeneous and isotropic then the singularity becomes the global one which we call the big bang. We have the 3 K radiation so we know there exists positive free energy and we can extrapolate our world lines back to higher densities and temperatures.

The place where the positive energy conjecture and our extrapolation runs into trouble is at the Planck time, 10^{-43} sec after the big bang, at a temperature of 10^{19} GeV. At this temperature, gravity becomes quantized and all bets are off. It may be that at the Planck time all of space-time is a foam of mini-black holes which are forming and exploding via the Hawking process on a Planck timescale. Such instantaneous black hole formation throughout space-time would violate the positive energy condition and our extrapolation to higher densities and temperatures must break down.

However, there seems little doubt that for times $\gtrsim 1$ sec we can have complete confidence that the universe we live in is some sort of hot big bang universe. In fact, the Soviet physicist Yacob Zeldovich mentioned recently at the International Astronomical Union meeting in Patras, Greece, that we can regard the big bang model to be as well established as celestial mechanics. Within the last decade we have developed an elaborate theoretical model that agrees in detail with experiment. We are no longer discussing what the basic cosmological model should be—big bang or steady state. We are concerned now with working out the details of our big bang model. Just as in celestial mechanics one worries about the details of perihelion shifts and so on, we are now worrying about such things as how minor perturbations can lead to the formation of galaxies.

In the past, we used telescopes to look out to larger and larger regions, to try to see more and more of the 10^{90} bits of data out there. However, now that the general model is established, we need to use a different kind of telescope to understand more about the original big bang experiment. In particular, we can try to duplicate parts of the initial apparatus with high energy accelerators. We would like to use Fermilab, CERN or SLAC to duplicate some of the conditions just following the big bang and low energy nuclear accelerators to explore slightly later conditions. We are not able to observe with telescopes back to those early times, because the universe was optically opaque at times earlier than 10^5 years after the projected singularity. So with telescopes we cannot see back to the exciting very early times. We can only look back to times when the universe was not that much different than it is today.

To help us use accelerators in place of the traditional astronomical tool to explore the really early times we need to review what things we know about the original big bang. We believe that conditions were hot enough to unify all the forces. Another very interesting piece of information that we know is that this early apparatus for the big bang could not have had within it more than four types of long-lived low-mass particles (neutrinos or other "inos")—in addition to the photons and electrons which are low mass by this definition. This is a rather important prediction that comes out of the standard big bang model. We believe that we have identified at least three of these low-mass particles—the electron neutrino, the muon neutrino and the tau neutrino. If we believe the big bang predictions, it may be that we've found *all* of the low mass "inos" that exist. (This may be the first instance since Newton that astronomical observations have made specific predictions for fundamental physics. Normally the flow of information goes the other way.) We suspect that the early universe was sufficiently dense that it consisted entirely of free quarks and yet now there are no free quarks. Thus, the early universe goes through the quark-hadron phase transition which may best be studied in ultra-relativistic heavy ion collisions.

We also know that somehow the big bang managed to get rid of monopoles. At least most of them, if not the one in Palo Alto. Even if there is one in Palo Alto, there are certainly not as many in Palo Alto or any place else as there are baryons. And yet, the standard unification theories predict as many monopoles as baryons. So we have somehow to get rid of monopoles, and we do not know yet quite how to do this. One possible way is through "inflation", a new theoretical concept we will discuss in detail later on.

We also know that the event was rather special, in that it managed to smooth everything down to parts in 10^4. That is, the temperature of the cosmic background radiation appears uniform to within parts in 10^4.

The more you think about this fact, the more amazing it seems. The 3 K radiations that we get from the horizon in opposite directions was emitted about 10^5 years after the big bang. But at the time of emission, these two sources were separated by about 10^7 light-years. Thus they could not have had any causal connection with each other, to know to be exactly at 3 K to within parts in a thousand or parts in 10^4.

How did everything get to be homogeneous and isotropic? It had to be a very, very carefully laid out machine. Or we have to have some way of naturally having it occur, naturally getting everything smooth. We will see that in the past year there have been some rather interesting ideas on how one might do this smoothing, again with the idea of inflation.

If one problem of the universe is "why is it so smooth?", another problem is "why is it so bumpy?" On small scales the universe is not smooth. It is all lumped up in these bumps called galaxies, stars, people and so on, and we need to understand how these clumps formed in this very smooth background.

So on the large scale the universe is smooth, and on the small scale it is clumpy, neither of which we understand. We also know that the universe seems to be very finely tuned. In particular the universe seems to be tuned to about 50 decimal places with regard to a particular parameter which we call Ω—the ratio of the density of the universe to the critical density.

Ω controls how the density of the universe evolves with expansion. In the early life of the universe, if Ω is greater than 1 then Ω will continue to increase with a typical time scale of 10^{-43} sec. If Ω is less than 1, it will decrease with a time scale of 10^{-43}

Now, 15 billion years after the big bang, Ω is still close to 1, and 15 billion years is a large multiple of 10^{-43} sec. This means that in the big bang the parameter Ω was tuned to equal 1 to a large number of decimal places, at least 50. This result seems miraculous, unless again you can invent some way of bringing it about naturally. Some people in the field call this the flatness problem. Others refer to it as the age problem—why is the universe so old? (It's even a problem if you believe the universe is only 6,000 years old.)

There is another problem that we don't yet have a good answer for, but again the answer may be coming from work in particle physics. The problem is that more than 90% of the matter of the universe seems to be in a form that we cannot identify. This matter doesn't seem to be baryonic. What is it?

At the outset we had mentioned unified field theories, the theories that have been developed in particle physics in the past decade, giving us to my mind one of the most exciting times in physics ever.

Roughly one hundred years ago, Maxwell showed that electricity and magnetism were unified. Fifteen years ago Weinberg, Salam and Glashow[5] showed us how the weak and electromagnetic forces are really one and the same interaction. If you get above energies of 100 GeV (the mass of the W and Z bosons that carry the weak interaction), then the weak and electromagnetic forces are identical. Their spontaneously broken symmetry, which we see in our day-to-day life, is restored if you get above 100 GeV. In fact you can see this change as a phase transition. When the universe was at higher temperatures, this was a unified symmetry. As the universe cooled there was a phase transition and the symmetry was broken. Over the last decade experiments have essentially verified this theory and this gives theorists the courage to look for further unification.

As we go up to higher energies still, people like Georgi and Glashow[6] have predicted that there should be a unification of strong, weak and electromagnetic forces. From extrapolation of the physics at lower energies they predict this unification should occur at energies of the order of 10^{14} GeV. The gauge bosons of the unification are the X and Y bosons and it is at this energy you produce the monopoles we discussed before. The theories unifying these three forces are called Grand Unified Theories—or GUTs.

The hope is to bring in gravity as well and achieve "Supergrand" Unification or SuperGUTs. Theorists tend to believe that SuperGUTs should become relevant when the quantum effects of gravity become important at the order of 10^{19} GeV, the Planck energy. Although there is a lot of work on SuperGUTs, it is even less firm than the strong, weak and electromagnetic unification.

Obviously energies as high as those involved in GUTs or SuperGUTs are far beyond what we can get in terrestrial laboratories. Even when the Tevatron colliding beam machine begins to operate we will only be up to a couple of TeV. If we build the "Desertron", we are only up to tens of TeV. So we are still a long, long way from 10^{14} GeV or 10^{19} GeV. [SLAC scaled up to 10^{19} GeV would stretch from here to Alpha Centauri which would ease vacuum leak problems but would make data analysis difficult, not to mention problems with the gross national (world) product.] We have some other ways of probing these kinds of energies indirectly through things like proton decay and, if monopoles are around, monopolonium and monopole annihilation. But we have no direct ways of exploring these energies except for the one event that took place at these energies—the big bang.

So there are strong motivations for nuclear and particle theorists and cosmologists to get together now, because the early history of the universe provides our best testing ground for the grand unifying ideas.

We hope that at these high temperatures, when the unification occurs, we will find that it explains away some of our problems in cosmology.

BARYOSYNTHESIS

One of the very hopeful findings in the last few years has been the success of the Grand Unified Theories in explaining the baryon number of the universe. Up until now, we have had no idea why the universe contains about 10^{-10} baryons for each photon. Now with Grand Unification, the ratio can arise naturally. As the universe cools (at $\sim 10^{-25}$ sec), the interactions which violate both baryon number conservation and CP symmetry give rise to a slight excess of quarks over antiquarks. Later these quarks form baryons, and then the baryons and antibaryons annihilate. One is left with a small net surplus of baryons—an excess of matter over antimatter—which is exactly what we observe. By adjusting the otherwise unknown parameters of the Grand Unified Theories, the correct ratio of baryons to photons can be obtained. It is interesting that the GUT known as minimal SU(5) could not yield the observed cosmological baryon to photon ratio and this particular model has now also been found to be inconsistent with the Irvine-Michigan-Brookhaven proton decay experiments. Thus the GUT must be more complex than minimal SU(5).

Prior to Grand Unification, we had to say "In the beginning there were 10^{-10} baryons per photon." Now we can say something esthetically more pleasing "In the beginning there were GUTs!"

QUARK-HADRON PHASE TRANSITION

As the universe expands it will eventually drop below the density where the quarks are so close together that they interact with the entire sea of quarks to where the quarks only interact in color triplets or in quark anti-quark pairs. At very high densities, hadrons overlap and lose meaning as individual particles. We may then refer to the universe as being a quark-gluon fluid. It is expected that this transition occurs when the total number density of hadrons is greater than 10 times nuclear density, n_0. This will not be pursued in detail here other than to refer to arguments by Shenker, Satz, Olive and others which showed that the transition seems to be a first order phase transition.

A heavy ion collision experiment can be proposed to explore this transition. Such an experiment may in fact yield a signature for quark matter (see Olive[7] and references therein).

Schramm et al.[8] have also mentioned that small density seed perturbations may naturally occur at this transition. Current ideas on quark confinement tell us that the quark-quark color interaction is stronger the farther apart the quarks are. Therefore as quarks are removed from the quark soup to make a hadron, it may be that this lower quark density site is maximally unstable to further hadron condensation. In addition, the removing of the quarks reduces the Debye-color screening of the remaining quarks in the vicinity and thus enables those particular quarks to have longer range interactions. It thus seems that the quark-hadron phase transition may be unstable to the growth of density fluctuations. Crawford and Schramm[9] showed that such perturbation growth is possible and has led to planetary mass black holes that could be the dominant matter of the universe.

BIG BANG NUCLEOSYNTHESIS

Let us now look at the very important events at 1 sec to 3 min after the singularity when nucleosynthesis occurs everywhere in the universe. The input physics here is well verified in the laboratory so the conclusions drawn from this epoch are very firm as compared to baryosynthesis or the quark-hadron transition.

When the universe cools below 1 MeV the weak interaction freezes out (its rate becomes slow compared to the expansion rate), the ratio of neutrons to protons in the universe also freezes out. Exactly what this ratio turns out to be depends on the neutron decay half-life, which then determines what the helium abundance is. Given the measurements of the neutron half-life, the standard big bang model predicts the actual helium abundance, as we mentioned before, of one quarter of the mass of the universe in the form of ^{4}He. Moreover the observed abundances of ^{3}He, ^{2}H and ^{7}Li all agree perfectly with the predictions of the big bang theory. See Fig. 1. The agreement is really rather impressive. Each of these abundances evolves very differently in the galaxy, and yet each of them, even though they are obtained by completely different observational techniques, falls right in line with the big bang prediction. This very impressive agreement, as we noted previously, is in many ways as strong a proof for the big bang as the 3 K radiation. In fact these results can be used to predict the 3 K radiation.

One of the things that I find particularly exciting, perhaps because I was involved with it, is the prediction from the standard big bang model of the number of neutrino types, or any kind of "ino",

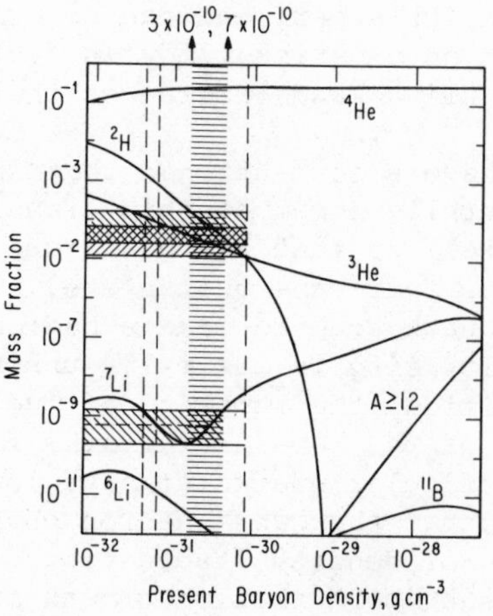

Fig. 1. Summary of Big Bang Nucleosynthesis results showing the phenomenal agreement of the 4 isotopes, ^{4}He, ^{2}H, ^{3}He, and ^{7}Li, with observed primordial abundances for $\eta \equiv n_b/n_\gamma$ of about 5×10^{-10}.

such as gravitinos and photinos. The standard big bang puts limits on the number of these weakly or semi-weakly interacting light particles.

We saw that the big bang model predicts that helium makes up approximately 25% of the mass of the universe. There is a small variation in this number, depending on how many inos there are. In particular, if there are two kinds of neutrinos, electron and muon, say, we get the lower curve of helium abundance shown in Fig. 2. We get the middle curve for three neutrinos and the upper curve for four. The uncertainty showing in the curves is due mainly to the uncertainty in the neutron half-life.

Now recall that the baryon-to-photon ratio comes out of the Grand Unified era. We can measure the actual value in a variety of ways, and we are confident we can put a lower bound on it. The consistency of big bang abundance for ^{3}He and ^{2}H with the observed values gives us a lower limit of 1.5×10^{-10} for the baryon-to-photon ratio.

We can get the upper limit from Fig. 2. With three neutrinos (if we believe e, μ and τ exist, all with low mass neutrinos) and a primordial helium abundance lower than 25 percent, the baryon-to-photon ratio can not be more than 6×10^{-10}. If we say that ν_τ does not have a low mass, then the limit would be as large as 10^{-9}.

But in any case, we get these limits on the baryon-to-photon ratio from big bang nuclear physics. Notice also from Fig. 2 that

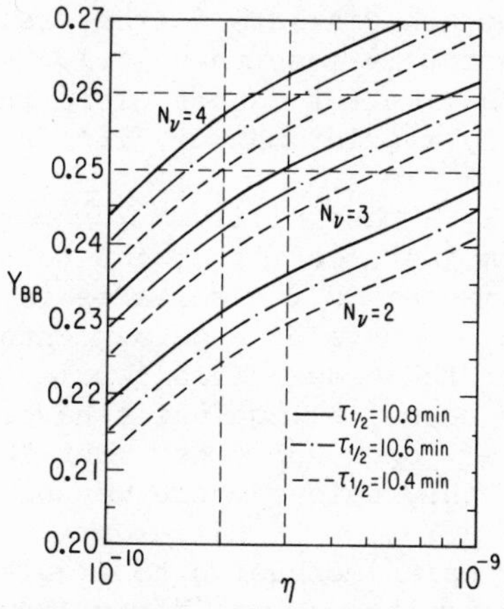

Fig. 2. A blow up of the Big Bang ^{4}He abundance versus n_b/n_γ showing the dependence on the number of neutrino flavors and the uncertainty caused by ±0.2 min in the neutron half life. Note that 3 ν's work and a 4th already requires extreme values of η and the ^{4}He mass fraction Y_{BB}.

four neutrinos would just barely satisfy the constraints. The better fit is with three, and so we may already have found all the neutrinos there are. At most, we could squeeze in one more. Now the quantum mechanical width (or equivalent lifetime) of the intermediate vector boson, the Z_0, will provide a direct test. The width is directly proportional to the number of neutrino types. When we have believable measurements of this width we will know whether or not the prediction of three neutrinos, which comes out of the standard big bang model, is correct or not.

In many ways, the width of the Z_0 will be one of the most powerful tests of the standard big bang model. It will allow us to test the model at an earlier time than any other observational test to date. Thus experiments with particle accelerators will be giving us one of the best tests of cosmology.

On the other hand cosmology is predicting a fundamental property of nature—how many fundamental particles there are. We know that each neutrino family is coupled to a quark flavor pair. The prediction of how many neutrinos there are is also a prediction of the number of quark flavor pairs. This is a surprisingly powerful statement to be coming from something that happend so long ago.

TOTAL MASS OF UNIVERSE

If we convert the upper limit on the baryon-to-photon ratio

from big bang nucleosynthesis into Ω (the fraction of the critical density), we find that Ω is less than ~ 0.1. This value implies that the density of baryonic matter cannot close the universe. So if the universe contains only regular matter, it's an open universe.

This is certainly a possibility. However, there are some other ways to obtain Ω from just looking at the galaxies as they exist. We can estimate the masses of the galaxies in a variety of ways. If we look at the rotation curve of a galaxy, standard Newtonian mechanics gives a mass of the order of 10^{11} solar masses for a galaxy such as ours. If we assume all galaxies have this mass, we obtain Ω of about 0.01—a very open universe. But this value of Ω comes from looking only at the regions where we see light being emitted.

If we apply Newtonian mechanics to a galaxy rotating around another galaxy, we calculate a mass per galaxy of $\sim 10^{12}$ solar masses, an order of magnitude larger than our previous estimate and an implied Ω of ~ 0.1. The galaxy appears to be interacting as if it has a much larger mass than the mass of the visible region. The extra mass seems to clearly be there. It is not that there is any mass missing. It is rather that the light is missing from the additional mass. We have a missing light problem. What is this mass that is not radiating, this dark matter? We can picture a halo around each galaxy of dark matter that is not radiating.

Now, if we consider larger clusters of galaxies—in which a thousand galaxies are all whirling around each other—we can do the same physics, but in terms of a statistical ensemble of the average separation distance and the average dispersion velocity. We can calculate the fraction of the mass attributed to each of the galaxies in this large cluster, and we find that the mass per galaxy approaches the order of 10^{13} solar masses. This additional order-of-magnitude increase implies an Ω almost equal to one if all galaxies really have this much mass.

Although most galaxies are not in large clusters, a growing trend in the observational data on large scales implies that an Ω of greater than 0.1 may be appropriate. In other words, galaxies may have such large haloes around them that even the orbits of a binary don't include all of the mass.

If this is the case, and Ω is actually bigger than 0.1, then we have a problem. The matter of the universe cannot be primarily baryons.

The options include massive neutrinos. Experiments are under way to check this possibility. The neutrino mass can be quite small (of order of tens of eV). Because there are so many neutrinos (roughly as many neutrinos as there are photons or 10^{10} more neutrinos than baryons), if neutrinos only had a mass of an eV they could contribute more to the mass of the universe than the baryons. If massive neutrinos are proven not to work, there are other "inos" that have been proposed—gravitinos or photinos might do for instance. However, we have recently shown that other "inos" only work well if they interact like neutrinos and have masses of tens of eV.

Currently experiments do not as yet strongly constrain the massive neutrino option. To do the job we only require that the most massive neutrino have a mass of 20 to 30 eV. The Soviet tritium endpoint experiment claims an electron neutrino mass in this range. However, even if this mass were shown to be less than an eV it could always be the tau neutrino which dominated the universe. The best way to probe that mass possibility is through tau neutrino mixing with muons and electrons but such experiments only put limits on a combination of the mass difference squared and the mixing angle. So, a stringent limit on mass differences can always be avoided if the mixing angle is sufficiently small.

Another possibility being considered is that the extra mass could come, not from a tiny elementary particle, but from a black hole. The ordinary big (solar mass) black hole would not work. Big black holes were still baryons at the time of nuclear synthesis, so their mass is included in our nucleosynthesis account.

The suggestion is to look at little black holes. On the other hand recall that Hawking[10] predicts black holes below a certain size will evaporate. So we have to look at black holes in the rather "narrow" mass range of 10^{15} to 10^{33} grams.

If this does not seem that narrow a range, consider the two standard ways in which black holes are created: (1) Stars collapsing which produces big black holes or (2) a phase transition at the Planck time, which gives you Planck mass (10^{-5} g) ones. But we have come up with another way to make black holes with masses that fall right in the middle of this mass range. At the time of the Weinberg-Salam phase transition and the quark-hadron phase transition, the amount of matter contained within the largest causally connected region was about a planetary mass. As we previously mentioned, quark-hadron transition black holes could be made that would measure a meter in diameter but have the mass of Jupiter. They would be created not by gravitational clustering, but by the interaction involved in the phase transition.

INFLATION

Finally let us discuss inflation—an exciting new idea that seems to solve a large number of problems of the early universe:

The horizon problem—why is the universe the same in different regions, even though they appear to be causally disconnected? The flatness problem—why is the universe more than 10^{-43} sec old? How was Ω able to start out so close to 1 (within 50 decimal places) that we still don't know the answer to whether the universe is open or closed today after 15 billion years? The monopole problem—where did all the monopoles go? How did we get the bumps that make galaxies?

The solution, first proposed by Alan Guth is that during the Grand Unified epoch the universe settled into a state called a false vacuum. In that state the energy density of space is very large, even in the absence of any matter. If this is the case, then the false vacuum would drive an expansion (see Fig. 3). But it drives an expansion that is exponential in time, rather than a normal power law expansion. In this very, very rapid expansion, every tiny part of space, every one of those remnants of the mini-foamy black holes that we talked about before, could inflate to a region that is as large as our universe is today. Since each tiny part of space was causally connected before the inflation began, the horizon problem disappears. To begin with everything is flat across these tiny regions, so when they inflate to the size of the universe Ω is still exactly equal to 1—the flatness problem is solved.

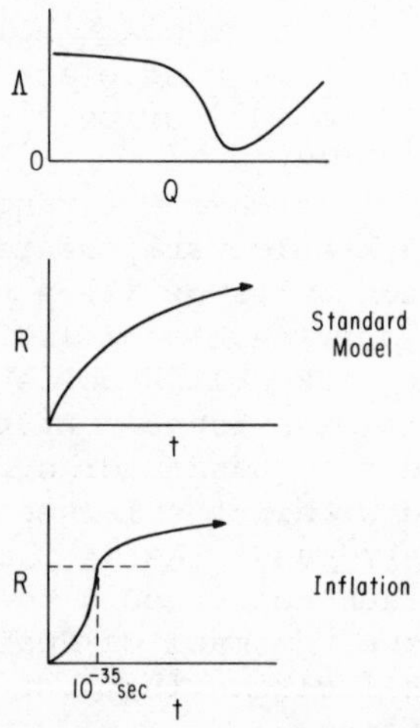

Fig. 3. These three schematics show how inflation works. The top figure shows how the vacuum energy or equivalently the cosmological constant could be non-zero at symmetry (high temperature) but drops to zero for broken symmetry. The middle figure shows the standard scale factor R vs time, t, for a Big Bang cosmology. The bottom figure shows how if Λ is large at 10^{-35} sec then there is inflation which switches to the normal expansion when Λ goes to zero.

The inflation vastly dilutes the density of monopoles, so after inflation there would only be about one monopole in the observable universe. More might be generated later on in collisions, as Michael Turner has shown. But the original monopoles we were concerned about would be inflated away to only one. This phase transition, depending on how it occurs, could have a few bumps and wiggles, because phase transitions are never totally smooth. The bumps and wiggles that occur could become our galaxies and so on.

So inflation is a very nice idea. Unfortunately it creates a new problem: how do you end inflation? Inflation seems to solve everything—gives us homogeneity, flatness, gets rid of monopoles, makes bumps that can grow into galaxies and so on. But the problem is that then every part of the universe is moving apart very fast. Now we go through the phase transition from the Grand Unified phase to the more normal universe that we live in. Normal space is a non-inflating 3-D space, in which the vacuum has no energy. As you go through this transition, you make bubbles of the new phase in the old, expanding phase. To have a phase transition go to completion, the new bubbles need to grow and gradually coalesce and take over all the space. Unfortunately they are inflating away from each other so fast that the walls of the bubbles are not expanding as fast as they are moving apart from each other, and so the phase transition never goes to completion.

Fortunately, there is a solution. The solution is that we live in a single bubble. Actually it is a little more complicated than this. You have to propose a special kind of phase transition to make a single bubble look like our universe, because normally a bubble would not contain nearly as many particles as we observe in our universe. But if we live in a single bubble, then of course we have all these other bubbles some place out there, which makes for a different philosophical idea about the universe than we have had before. (But of course, our knowable universe stays the same, it is only embedded in a far larger and richer manifold.)

Now to verify these kinds of ideas we will really need to know the nature of the Grand Unified Theory or the Super Grand Unified Theory. We will have to probe into this regime using a variety of approaches—trying to observe monopoles, proton decay, the electric dipole moment of the neutron and so on.

CONCLUSION

Consider how developments in the past few years, at this boundary of particle physics and astrophysics, have really changed our whole philosophical view on things. We have known for 500 years that the earth is not the center of the solar system. Then Harlow Shapely showed us that our Sun is not at the center of the galaxy.

It is out near the edge. Hubble and others showed us that the galaxy is not at the center of the universe. In fact, they went so far as to say there is no spatial center—all points are equivalent.

But now, if we believe Ω is larger than 0.1, we are even further removed from being the center. We are not even made out of the dominant matter of the universe. And then with inflation, we are not even the only bubble. This, I call, the extreme Copernican principle.

ACKNOWLEDGMENT

I would like to thank Alan Guth, Mike Turner, Gary Steigman, Keith Olive and Matt Crawford for their discussion and collaborative efforts. Parts of this manuscript were excerpted from a transcript of another lecture I gave at the American Institute of Physics Corporate Associates Meeting. This research was supported in part by NSF, DOE and NASA at the University of Chicago.

REFERENCES

1. A. A. Penzias and R. W. Wilson, Ap. J. 142:419 (1965).
2. D. P. Woody, J. C. Mather, N. S. Nishioka and P. L. Richards, Phys. Rev. Lett. 34:1036 (1975).
3. D. N. Schramm and R. V. Wagoner, Ann. Rev. Nucl. Sci. 27:37 (1977).
4. S. W. Hawking and G. F. R. Ellis, "The Large Scale Structure of Space-Time," Cambridge University Press (1973).
5. S. Weinberg, Rev. Mod. Phys. 46:255 (1974), and references therein.
6. H. Georgi and S. L. Glashow, Phys. Rev. Lett. 32:438 (1974), and references therein.
7. K. A. Olive, Phys. Lett. 89B:299 (1980).
8. D. N. Schramm, M. Crawford and K. A. Olive, in: "Proc. of the 1st Workshop on Ultra-Relativistic Nuclear Collisions," Lawrence Berkeley Laboratory Report, LBL 8597 (1979).
9. M. Crawford and D. N. Schramm, Nature 298:538 (1982).
10. S. Hawking, Mon. Not. R. Astr. Soc. 152:75 (1971).

THE INTERACTING BOSON MODEL

I. Talmi

Department of Physics
Weizmann Institute of Science
Rehovot, Israel

INTRODUCTION—SIMPLE SHELL MODEL CONFIGURATIONS

In order to present the interacting boson model and its relationship to other nuclear structure theories a short introduction is necessary.

Some theorists view the nucleus as a dense pea soup. Protons and neutrons, whether elementary or quark bags, are exchanging pions and heavier mesons. In a certain fraction of the nuclear wave functions some nucleons are replaced by heavier baryons, the most popular of which is the Δ resonance. Still it is quite possible that for most low energy problems a much simpler picture is sufficient. Most theories of nuclear structure make a very drastic simplification. They consider just elementary protons and neutrons interacting by two-body forces. Perhaps what is omitted is most interesting and exciting but the simplified problem is far from being simple.

The problem of A bodies ($A > 2$) strongly interacting via given forces cannot be exactly solved even in classical mechanics. *A priori*, one could have expected such systems to have very complex spectra which would not exhibit any simple regularities. In parallel, it would be difficult to think of an effective theoretical approach to such systems, apart from statistical models. Still, the experimental nuclear data exhibit remarkably simple and regular features which have been a challenge to nuclear structure theorists. Accepting the axiom that simple features could be derived from simple models, theorists developed very successful approaches to problems of nuclear structure.

The simplest and best known regularities are associated with shell structure in nuclei.[1] The experimentally observed "magic numbers" of protons and neutrons, as well as many properties of odd mass nuclei (e.g., their magnetic moments) led to the shell model. Shell model wave functions describe nucleons moving independently in a common, static, spherical potential well. Magic numbers are obtained whenever protons (or neutrons) completely fill all single nucleon j orbits in a major shell.

Shell model wave functions, especially those describing ground states of doubly magic nuclei, are very simple and easy to use. Yet it is clear that they could not be the real wave functions which are exact solutions of the nuclear many-body problem. This fact has been always stressed by nuclear many-body theorists. The very strong and very singular nucleon-nucleon interaction leads to strong short-range correlations between nucleons. Such correlations are completely missing from any shell model wave function of *independent* nucleons. To account for them a linear combination of very many shell model wave functions should be constructed. In other words, the strong short range correlations admix into any given shell model wave function many other wave functions corresponding to shell model states with very high energies. This fact is sometimes ignored for the sake of convenience but in such cases it is difficult to trust an even otherwise sophisticated calculation.

This invites clarification of the success of the shell model. To some many-body theorists shell model wave functions are just a complete and convenient set of states. The eigenstates of the nucleus are then linear combinations of such basic states. The observed regularities of the nuclear data dictate the use of rather *simple* shell model wave functions. A clear demonstration of the usefulness of the simple shell model is given in Fig. 1. Nuclear many-body theory has provided a neat way to deal with the situation. Simple shell model wave functions can be used to calculate energies only with *renormalized* interactions into which effects of short range correlations have been incorporated. Such renormalized interactions may include, in addition to two-body terms, also three-body, four-body and even higher order interactions. These could ruin the simplicity of the model. No reliable calculation has shown that these terms are much smaller than the two-body interactions. In fact, no reliable calculation has so far yielded the shell model as a good approximation and correctly computed the renormalized two-body interaction.

Other operators like those of electromagnetic moments and transitions, as well as of weak interactions, should also be renormalized if used with shell model wave functions. Also these renormalizations have not been reliably calculated. The results may depend strongly on the finite size of nuclei as opposed to the idealized system of "nuclear matter".

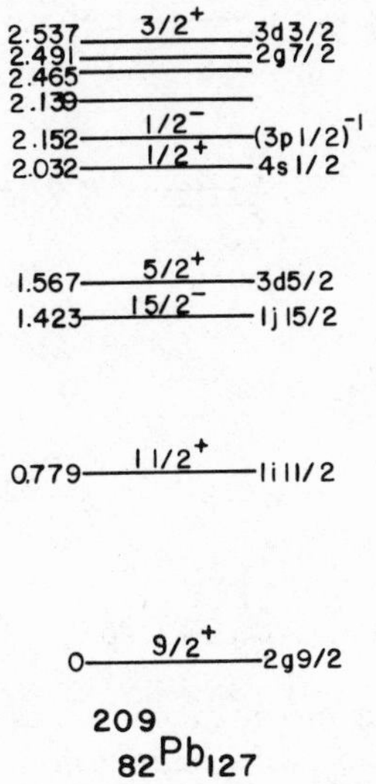

Fig. 1. Levels of ^{209}Pb. All levels up to 2 MeV excitation are due to the single valence neutron occupying the various j orbits in the major shell.

An alternative approach to nuclear structure does not depend on the results of future many-body calculations. The proof of the pudding is in the eating and a necessary test of the shell model is in its agreement with experiment. If only *effective* two-body interactions are used with simple shell model wave functions the agreement with experiment can be made in a quantitative way.[2] Matrix elements of the effective interaction are determined by fitting shell model results to the experimental energies. If the number of two-body matrix elements, taken as free parameters, is substantially smaller than the number of experimental energies, the fit is meaningful. If good agreement is thus obtained between calculated and experimental energies, the use of the model (including the adopted shell model configuration) is consistent. Very good agreement has been obtained in many cases with few nucleons outside (or missing from) closed shells. Nice examples are given in Fig. 2 and Fig. 3. This approach has also been used successfully for calculating various transitions and moments.

To demonstrate how the method works, let us consider a simple case of identical nucleons in the $2d_{5/2}$ orbit for neutrons beyond $N=50$ in Zr isotopes.[3] The ground state of ^{91}Zr has $J=5/2$ (positive parity) and is separated by 1.2 MeV from other levels. The antisymmetric states of the $(d_{5/2})^2$ configuration allowed by the Pauli principle have $J=0,2,4$. We see such a set of levels in the spectrum of ^{92}Zr (Fig. 4). If neutrons are indeed in this orbit we predict

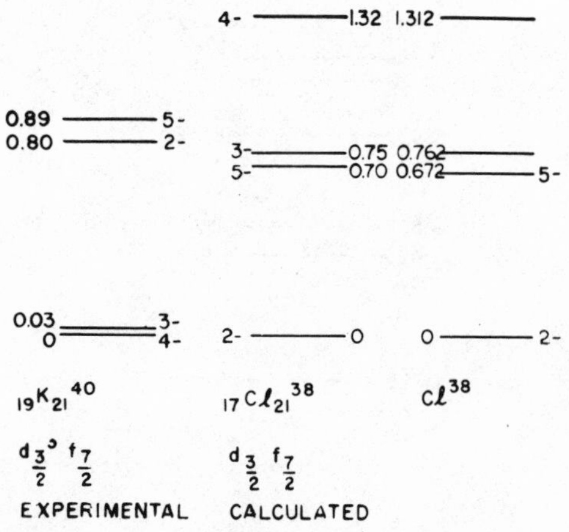

Fig. 2. Experimental and calculated levels in ^{38}Cl. Two-body matrix elements of the effective interaction determined from the $(d_{3/2})^3\ f_{7/2}$ configuration in ^{40}K were used to calculate directly the $d_{3/2}\ f_{7/2}$ levels in ^{38}Cl without extra assumptions or parameters.

such levels with the same spacings in the two hole $d_{5/2}^{-2}$ ($\equiv d_{5/2}^{4}$) configuration. This consistency check is very well satisfied by the ^{94}Zr levels. In ^{96}Zr we should have a closed $2d_{5/2}$ orbit and indeed the first excited state of that nucleus is quite high. The only level of the $(d_{5/2})^5 \equiv d_{5/2}^{-1}$ configuration has $J = 5/2$ which is the ground state of ^{95}Zr. The first excited state is at 0.95 MeV above it. The only other configuration to be considered is the $(d_{5/2})^3$ configuration in ^{93}Zr.

The allowed states of the $(d_{5/2})^3$ configuration have $J = 3/2, 5/2, 9/2$. There are several simple ways to calculate their energies in terms of the $d_{5/2}^2$ matrix elements taken from ^{92}Zr (or ^{94}Zr). The method we adopt will be useful in the following. Energies of all states of $(2d_{5/2})^n$ configurations are determined by the 3-matrix elements $V_J = \langle d_{5/2}^2\ JM|V_{12}|d_{5/2}^2\ JM\rangle$, $J = 0,2,4$, of the effective two-body interaction. Hence, the most general two-body interaction in *this shell model subspace* can be expressed as a linear combination of three independent operators. We can take for V_{12} the interaction

$$V_{12} = a + b\ 2(\vec{j}_1 \cdot \vec{j}_2) + cq_{12} \qquad (1)$$

where the parameters a, b and c should be determined by V_0, V_2 and V_4. In Eq. (1) the two-body scalar operator q_{12} is defined by

$$\langle j^2\ JM|q_{12}|j^2\ JM\rangle = [(2j+1)\delta_{J0}]\ . \qquad (2)$$

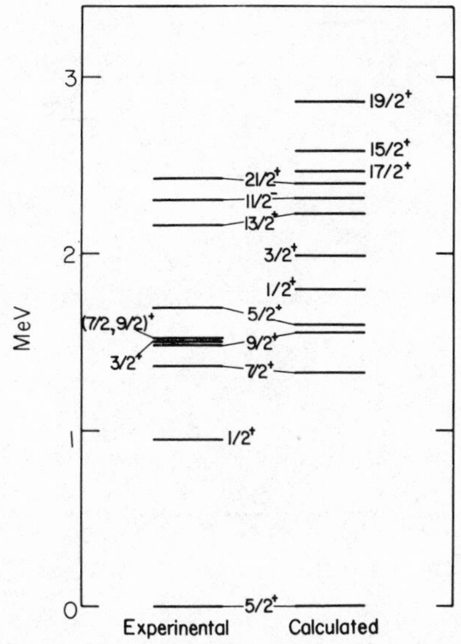

Fig. 3. Experimental and calculated levels in ^{93}Mo. Matrix elements of the effective interaction determined from neighboring nuclei were used for calculating energies which agree with experiment. Properties of the proton-neutron interaction give rise to the isomeric $21/2^+$ state ("yrast trap").

It was introduced by Racah[4] in 1943 and is known better as the pairing interaction. Using (1) and (2) and making use of the identity $(\vec{j}_1 + \vec{j}_2)^2 = j_1^2 + j_2^2 + 2\vec{j}_1 \cdot \vec{j}_2$ we obtain the V_J in terms of a,b and c.

$$V_J = a + b\left[J(J+1) - 2\,\frac{5}{2}\cdot\frac{7}{2}\right] + 6c\delta_{J0} \quad . \tag{3}$$

Solving the three equations (3) we can express a,b and c in terms of the V_J:

$$a = (5\ V_2 + 23\ V_4)/28 \tag{4a}$$

$$b = (V_4 - V_2)/28 \tag{4b}$$

$$c = (4\ V_0 - 5\ V_2 + V_4)/24 \quad . \tag{4c}$$

The parameter <u>a</u> is of little interest since it does not affect energy *spacings*. The latter are determined by b and c only. It should be stressed that the form of the interaction (1) is the most general one and is not related to any specific effective interaction. The parameters a,b,c do not have any deeper physical sense beyond (4). There

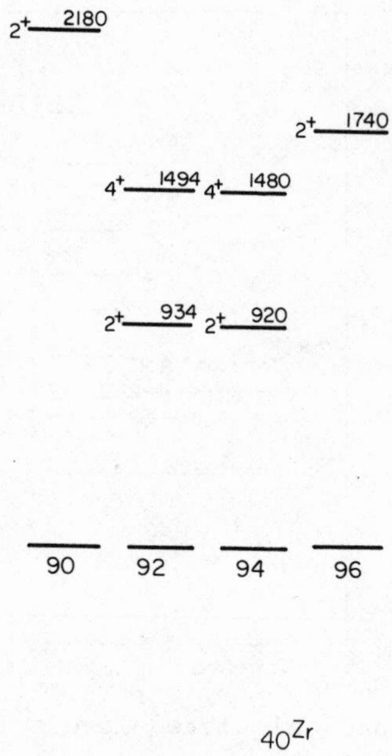

Fig. 4. Levels in even Zr isotopes. High excitation energies of $J = 2$ levels in ^{90}Zr and ^{96}Zr. indicate shell closure at $N = 50$ and closure of the neutron $2d_{5/2}$ orbit at $N = 56$.

is certainly no general dipole-dipole term in the effective interaction. The term with the parameter b in (1) could be the result of any other interaction, like the quadrupole-quadrupole one. In the j^n configuration we should consider the eigenvalues of

$$\sum_{i<k}^{n} V_{ik} = \frac{n(n-1)}{2} a + b\left[\left(\sum \vec{j}_i\right)^2 - \sum \vec{j}_i{}^2\right] + c \sum_{i<k}^{n} q_{ik} =$$

$$= \frac{n(n-1)}{2} a + b\left[J(J+1) - nj(j+1)\right] + c \sum_{i<k} q_{ik} \ . \qquad (5)$$

We will calculate the eigenvalues of the pairing term using a technique[5] introduced almost twenty years after the first calculation by Racah.

The operator Σq_{ik} measures the amount of pairing in any given state. A state of one j^2 pair with $J = 0$ can be constructed in the second quantization formalism by applying the operator

$$S_j^{+} = \frac{1}{2} \sum (-1)^{j-m} a_{jm}^{\dagger} a_{j-m}^{\dagger} \qquad (6)$$

to the vacuum state $|0\rangle$ which has no j-nucleons. The operator S_j^+ satisfies with its hermitian conjugate $S_j^- = (S_j^+)^\dagger$ the commutation relation

$$[S_j^+, S_j^-] = \sum_m a_{jm}^\dagger a_{jm} - \frac{2j+1}{2} = 2S_j^0 \qquad (7)$$

From (7) it follows that the operator Σq_{ik} is given in this formalism by $2S_j^+S_j^-$. To calculate its eigenvalues we realize that S_j^0 defined in (7) satisfies

$$[S_j^0, S_j^+] = S_j^+$$
$$[S_j^0, S_j^-] = -S_j^- \qquad (8)$$

Thus, S_j^+, S_j^- and S_j^0 form the Lie algebra of SU(2), identical with that of angular momentum. Using the well known properties of angular momentum components we obtain

$$S_j^+S_j^- = \vec{S}_j^2 - (S_j^0)^2 + S_j^0 = s(s+1) - S_j^0(S_j^0-1)$$
$$= s(s+1) - \frac{1}{4}\left(\frac{2j+1}{2} - n\right)\left(\frac{2j+5}{2} - n\right) . \qquad (9)$$

The quantum number s determines the eigenvalues of the pairing term. Racah introduced the latter to define the seniority scheme.[4] Its eigenstates are characterized by the seniority v which is loosely speaking the number of unpaired nucleons. More precisely, any state in the j^n configuration can be expressed as

$$(S_j^+)^{(n-v)/2} \, |j^v vJM\rangle$$

where

$$S_j^- |j^v vJM\rangle = 0 \ . \qquad (10)$$

This definition of v establishes the connection between it and s. Applying (9) to a state $|j^v vJM\rangle$ we obtain

$$0 = s(s+1) - \frac{1}{4}\left(\frac{2j+1}{2} - v\right)\left(\frac{2j+1}{2} - v + 2\right) .$$

Since $s > 0$ we discard the possible value $s = -1/2[(2j+1)/2 - v + 2]$ and obtain

$$s = \frac{1}{2}\left(\frac{2j+1}{2} - v\right) . \qquad (11)$$

Operating on a state with given s (or v) by S_j^+ does not change its seniority and only increases n by 2. The eigenvalues of the pairing term Σq_{ik}, given by (9) in terms of s, are expressed as functions of n and v by[6]

$$\frac{n-v}{2}(2j+3-n-v) \ . \tag{12}$$

The state with J=0 in the j^2 configuration as well as one J=0 state in all j^n configurations with even n have seniority v=0. They can all be constructed by $(S_j^{+})^n|0\rangle$. One state with J=2, J=4,..., J = 2j-1 in such configurations has v=2. A single j state has by definition seniority v=1 and this is true for one J=j state in each j^n configuration with n odd. There may be other J=j states which have higher seniorities $v > 1$. Racah introduced seniority in order to distinguish between states with the same value of J.

Coming back to the $(d_{5/2})^3$ configuration we can express (5) explicitly as

$$\frac{n(n-1)}{2}a + \left[J(J+1) - nj(j+1)\right]b + \frac{n-v}{2}(2j+3-n-v)c \ . \tag{13}$$

The state with $J = 5/2$ has seniority v=1, the others v=3. Putting n=3, $j = 5/2$ in (13) and using the values (4), energies of these states can be calculated. In Fig. 5 there is a graphical solution of the problem.[2] Good agreement is obtained for the first excited $J = 3/2$ state (the $J = 9/2$ state has not been experimentally observed). It is much closer to the ground state than the prediction of the over-simplified pairing interaction (obtained by putting b=0).

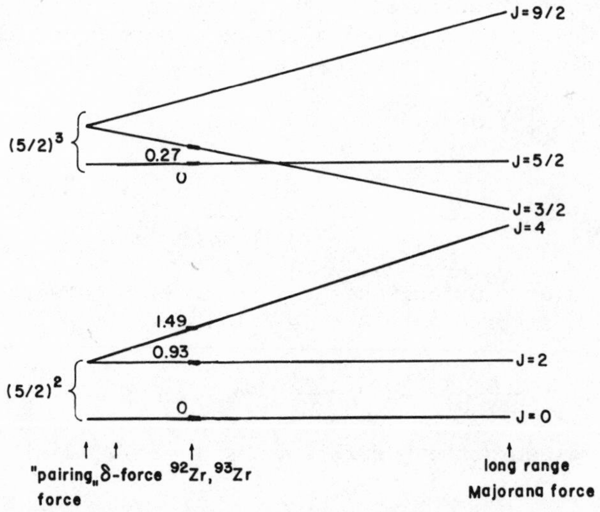

Fig. 5. Experimental and calculated levels of $(2d_{5/2})^n$ neutron configurations. The 0-2 spacing is constant and equal to the experimental one. The J=4 level is linearly interpolated between the pairing limit and the J(J+1) limit. Calculated $(2d_{5/2})^3$ level spacings are linear functions of the position of the J=4 level.

LOW-LYING ("COLLECTIVE") STATES IN SEMI-MAGIC NUCLEI

There are very simple properties of interactions between identical nucleons which are diagonal in the seniority scheme (their eigenstates have definite seniorities). The best way to derive them is to refer to the quasi-spin operators S_j^+, S_j^- and S_j^0. Using the components of angular momentum it is possible to define systems of operators that transform irreducibly among themselves under the corresponding infinitesmal rotations. Each of such sets of operators is that of the (2k+1) components of an irreducible tensor operator of rank k. Using S_j^+, S_j^- and S_j^0 it is possible to define irreducible tensor operators with respect to quasi-spin. Single nucleon operators are linear combinations of products $a^\dagger_{jm}\, a_{jm'}$, and thus can be grouped into quasi-spin irreducible tensors of ranks s=1 and s=0. Two-body Hamiltonians can be formed by products of single-nucleon operators and thus can be classified as quasi-spin tensors with ranks s = 0,1,2.

Any two-body Hamiltonian which is a quasi-spin scalar H_0 is diagonal in the seniority scheme (as a rotationally invariant Hamiltonian has vanishing matrix elements between states with different values of total angular momentum). There are, however, other terms which share this property. Products of components of the quasi-spin vector itself do not connect states with different seniorities. Since the shell model Hamiltonian conserves nucleon number, only powers of S_j^0 can be used. The restriction to two-body interactions limits such products to S_j^0 and $(S_j^0)^2$. In view of (7) we can use instead of these, terms proportional to n and n^2. Thus, the most general Hamiltonian which is diagonal in the seniority scheme has the form[7]

$$H = H_0 + \frac{1}{2}\beta n + \alpha\,\frac{n(n-1)}{2} \tag{14}$$

with arbitrary coefficients α and β.

If we apply the Hamiltonian (14) to the v=0, J=0 states of the j^n configuration (of identical nucleons) we obtain the eigenvalues

$$E(n, v = 0, J = 0) = \alpha\,\frac{n(n-1)}{2} + \beta\,\frac{n}{2} \; . \tag{15}$$

The operator H_0 does not contribute. Being a quasi-spin scalar, its eigenvalues for a given value of v (or s) are independent of n (or S_j^0) and hence

$$H_0 | j^n \; v{=}0 \; J{=}0 \rangle = H_0 | 0 \rangle = 0 \; .$$

Applying (14) to the v=1, J=j state of the j^n configuration we make use of the fact that $H|j, v=1, J=j\rangle = 0$ since (14) is a two-body interaction. This implies $(H_0 + \frac{1}{2}\beta)|j, v=1, J=j\rangle = 0$ which fixes the eigenvalues of H_0 for *any* v=1 state as $-\frac{1}{2}\beta$ (the eigenvalues of a quasi-spin scalar are independent of n). Hence, for odd values of n we obtain the eigenvalues

$$E(n,v=1,\ J=j) = \alpha\,\frac{n(n-1)}{2} + \beta\,\frac{n-1}{2}\ . \qquad (16)$$

Combining (15) and (16) and adding n equal single nucleon energies C we obtain a very simple mass formula for ground states of semi-magic nuclei which holds for *any* two-body interaction diagonal in the seniority scheme[2,6,7,8]

$$B.E.(j^n) = B.E.(n=0) + nC + \frac{n(n-1)}{2}\,\alpha + \frac{n-[1-(-1)^n]/2}{2}\,\beta\ . \qquad (17)$$

As a result of (17), nucleon separation energies should lie on two straight and parallel lines. Such a plot is presented in Fig. 6 and we see that the agreement with experiment is very good. As in all other cases the coefficient β of the pairing term is large and attractive. The values of the parameter α are rather small and always *repulsive* (it is -0.189 MeV in Fig. 6). The coefficient α and β are linear combinations of the various V_J. They are given by[2,6]

$$\alpha = \frac{2(j+1)\overline{V}_2 - V_0}{2j+1}$$

$$\beta = \frac{2(j+1)}{2j+1}\,(V_0 - \overline{V}_2) \qquad (18)$$

where $\overline{V}_2$ is the center-of-mass of the $J = 2,4,\ldots,2j-1$ levels with seniority v=2

$$\overline{V}_2 = \sum_{J>0\ \text{even}} (2J+1)V_J \Big/ \sum_{J>0\ \text{even}} (2J+1)\ . \qquad (19)$$

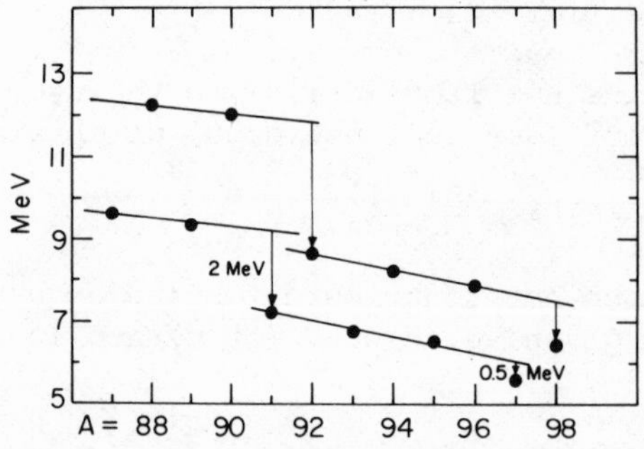

Fig. 6. Experimental single neutron separation energies of Zr isotopes. Within the $(2d_{5/2})^n$ configurations they lie on two straight and parallel lines. Note the strong drop in the separation energy beyond N = 50 and the smaller but distinct drop beyond N = 56.

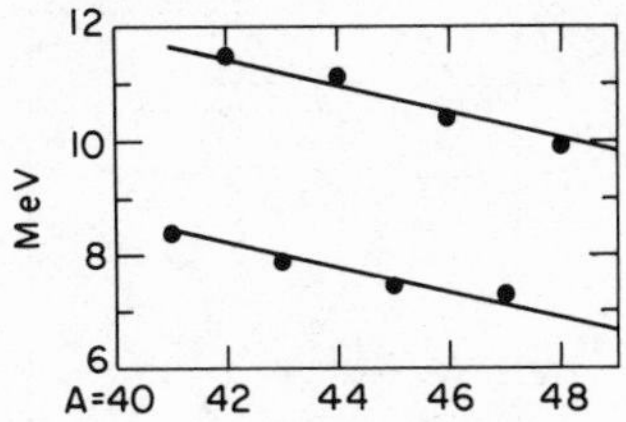

Fig. 7. Experimental single neutron separation energies of Ca isotopes. Within the $(1f_{7/2})^n$ configuration they lie on two straight and parallel lines.

The value of β given by (18) for the case discussed above can be computed from the energy differences in Fig. 4 and it is equal to 1.511 MeV. This is rather close to the value of β deduced from Fig. 6 which is 1.625 MeV. Another example of separation energies following the expression (17) is given in Fig. 7.

Another important property can be deduced directly from the form (14). *Spacings* between energy levels are determined only by H_0 and hence are independent of n.[6] Thus, for any Hamiltonian which is diagonal in the seniority scheme energy spacings between levels with given values of v of j^n configurations are independent of n. In particular, the spacings between the v=0, J=0 ground states and v=2, J = 2,4,... levels are independent of n and are equal to the spacings in the j^2 configuration.

We see an example of this feature in Fig. 4. There it was a result of symmetry between nucleons and holes. In fact, the explicit use of seniority is not required for the $2d_{5/2}$ orbit. In no $(5/2)^n$ configuration are there two states with the same value of J. Hence any two-body interaction is trivially diagonal in the seniority scheme [as explicitly demonstrated by (5) and (13)]. In $(f_{7/2})^n$ configurations (an example of which is Fig. 7) this is no longer the case but it is still true that any two-body configuration is diagonal in the seniority scheme.[6] For $j \leqslant 7/2$ seniority just provides a general and convenient scheme. To find out how good is the seniority scheme we must look at orbits with $j \geqslant 9/2$. Proton $(1g_{9/2})^n$ configurations have been thoroughly investigated. They are appreciably perturbed by $(2p_{1/2})^2$ pairs. The latter must be coupled to J=0 and hence cause equal shifts to all levels with the same seniority. The levels of even configurations shown in Fig. 8 have fairly equal spacings independent of n. This is an indication that the effective interaction in these configurations is diagonal in the seniority scheme. Matrix elements connecting states with different seniorities can be calculated in terms of energy spacings taken from Fig. 8. They turn out to be negligible in comparison with the differences between corresponding diagonal matrix elements.

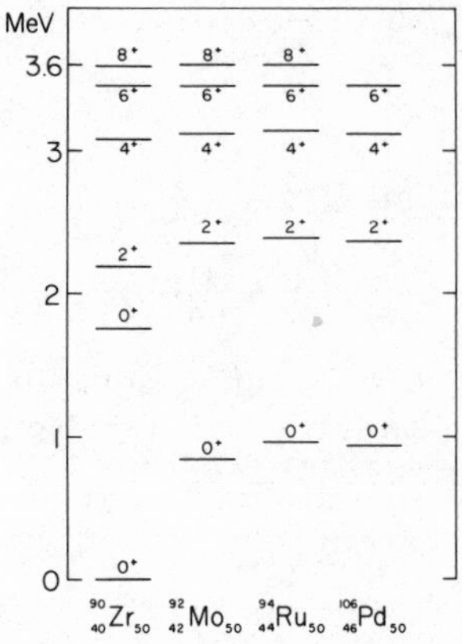

Fig. 8. Experimental level spacings of proton configurations of $N = 50$ nuclei. Spacings between $v=2$ levels with $J = 2,4,6,8$ are fairly independent of Z. They belong to $(1g_{9/2})^n$ configurations interacting with $(1g_{9/2})^{n-2}$ $(2p_{1/2})^2$ configurations. Since a $(2p_{1/2})^2$ pair has only $J=0$, only $J=0$ levels are shifted with respect to the others.

The seniority scheme provides the proper frame to describe important nuclear properties. Still, its direct application is rather limited. There are not very many nuclei in which the limitation to a single j orbit is justified. The shell model configurations in most semi-magic nuclei cannot be as simple as j^n. Several j orbits must be considered together and the determination of the effective interaction from experimental data becomes much too complicated. Still, the experimental data show remarkable regularities, similar to those in simpler cases. As we shall see below, the seniority scheme shows how to deal with these more complicated cases.

The spectrum of ^{112}Sn could serve to illustrate this point. Tin isotopes have closed proton shells ($Z = 50$) and valence neutrons occupy the orbits in the major shell $N = 50$ to $N = 82$. These are the $2d_{5/2}$, $1g_{7/2}$, $2d_{3/2}$, $3s_{1/2}$ and $1h_{11/2}$. There are altogether 160 two-body matrix elements of the effective interaction between the various orbits. This is a large number to be determined from experimental energies. There is an even bigger difficulty. These matrix elements should go into the interaction matrices constructed for each value of total spin and parity for every nucleon number. For several valence neutrons these matrices become very, very large. For example, in ^{112}Sn the 12 valence neutrons occupying the 5 valence orbits give rise to 56907 states with $J=0$ and positive parity, 267720 states with $J=2$ and positive parity and 426558 states with $J=2$ and positive parity. Even if the two-body matrix elements were given, it would have been very difficult to construct and diagonalize matrices of such high order. Moreover, most of these millions of states lie at high energies where the limitation to the valence shell is certainly invalid. Excitations into higher shells or from closed shells occur at only a few MeV above ground states. We are mainly interested in the low lying levels which exhibit simple and regular features. The actual low lying part of the spectrum is exceedingly simple (Fig. 9). It would be practically impossible to understand the origin of the simple features by staring at a computer printout of states with a myriad of components.

The only hope of dealing with such situations is to find a very drastic truncation of the giant matrices. Limiting the number of valence orbits does not seem to be a good way since tin isotopes show smooth and regular behavior throughout the major shell. Since seniority gives a good description it is worthwhile to look for a generalization of it. The Bardeen, Cooper and Schrieffer theory of superconductivity has been extensively applied to nuclei. This version, called pairing theory, is indeed a generalization of seniority but it has some severe disadvantages. First, it diagonalizes (approximately) the pairing interaction which is not a very good effective interaction (cf. Fig. 5). The second difficulty is that pairing theory does not conserve the number of particles. This is not too bad for electrons in metals but for nuclei where n may be rather small this is no longer the case. The last disadvantage of pairing theory is that it deals with the J=0 state in a very different way than with the other (e.g. the J=2) states. Interesting regularities involve excited states in semi-magic nuclei and they should be treated simultaneously with the ground states.

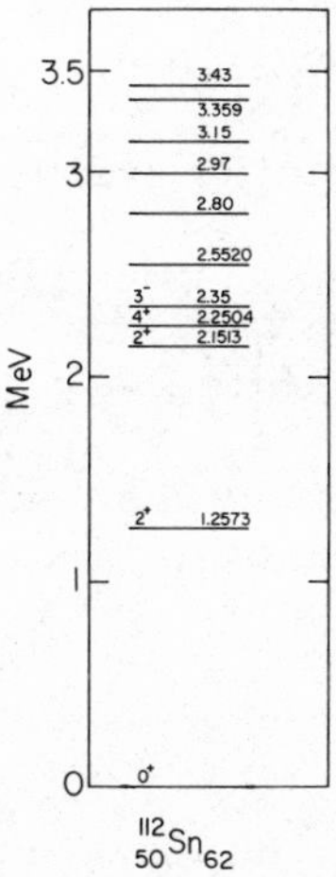

Fig. 9. Levels of ^{112}Sn. Up to 2.25 MeV there are only 3 levels in addition to the ground state.

We can generalize seniority by considering a J=0 pair creation operator which will be an extension of (6). We introduce the operator

$$S^{+} = \Sigma\, \alpha_j\, S_j^{+} \quad . \tag{20}$$

It creates a J=0 correlated pair to which each j orbit in the major shell contributes the amplitude α_j. Had all amplitudes α_j been equal, (20) would become a quasi-spin operator and would constitute, with its hermitian conjugate and one half their commutator, the SU(2) Lie algebra. These operators would be simply components of the total quasi-spin of the system. All results of seniority would hold also in this case provided $2j+1$ in the formula is replaced by $2\Omega = \Sigma(2j+1)$.[5] The situation would become very simple but it would be too simple. Had it been a good approximation it would have implied energy spacings independent of n also for odd nuclei. This is contrary to experiment as we shall see later on.

The more general operator (20) can still be used even if it is not a generator of a well known Lie algebra. As we shall soon see, its use preserves some attractive features of seniority which happen to agree with experiment. We first consider generalization of the v=0, J=0 ground states by applying successively several operators (20) to the vacuum state $|0>$ (with closed shells only). We ask under which conditions are such states eigenstates of the shell model Hamiltonian (normalized by $H|0>$), namely

$$H(S^{+})^{n}\,|0> = E_n(S^{+})^{n}\,|0> \quad . \tag{21}$$

From (21) we obtain for n=1 the condition that the pair state created by (20) be an eigenstate

$$HS^{+}|0> = [H,S^{+}]|0> = V_0 S^{+}|0> \tag{22}$$

with the eigenvalue V_0. Putting n=2 we obtain

$$H(S^{+})^{2}(0) = [H,S^{+}]S^{+}|0> + S^{+}HS^{+}|0> = 2V_0(S^{+})^{2}|0> +$$
$$+ \Big[[H,S^{+}],S^{+}\Big]|0> = E_2(S^{+})^{2}|0> \quad .$$

This will be satisfied provided the double commutator acting on the vacuum state is proportional to $(S^{+})^{2}|0>$. Since H includes single nucleon and two-body terms only, we can write this condition simply as

$$\Big[[H,S^{+}],S^{+}\Big] = \Delta(S^{+})^{2}|0> \tag{23}$$

where Δ is a number. It can be shown by induction that for any n

$$H(S^{+})|0> = n(S^{+})^{n-1}[H,S^{+}]|0> + \frac{n(n-1)}{2}(S^{+})^{n-2}\Big[[H,S^{+}],S^{+}\Big]|0>$$

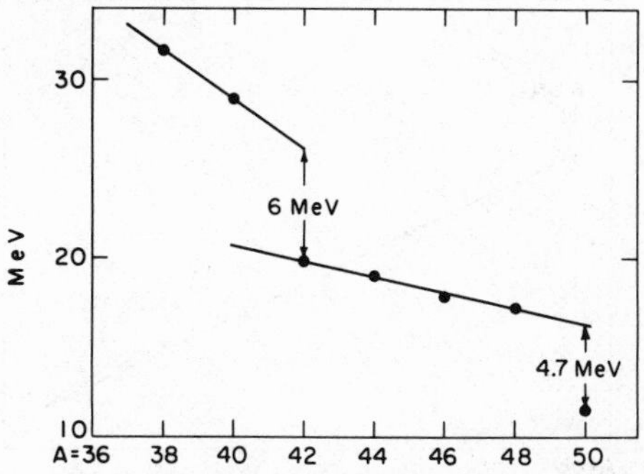

Fig. 10. Experimental neutron pair separation energies of Ca isotopes. Within the $1f_{7/2}$ orbit they lie on a straight line. Note the strong drops in the separation energy beyond the N = 20 and N = 28 magic numbers.

since higher order commutators of H with S^+ vanish. Hence, if conditions (22) and (23) are satisfied we obtain[9] for any n

$$H(S^+)^n|0> = \left(nV_0 + \frac{n(n-1)}{2}\Delta\right)(S^+)^n|0> . \quad (24)$$

The prescription (21) for the eigenstates leads directly to the form (24) of the corresponding eigenvalues. Thus, if ground states have the form (21), with generalized seniority v=0, binding energies behave exactly as in the seniority scheme. Note however, that unlike the case with equal α_j values, the single mass formula holds only for *even* nuclei. Pair separation energies of even semi-magic nuclei should lie on a straight line like the case of a single j orbit (Fig. 10). The configuration mixing due to the prescription (21) is so thorough that no sub-shell breaks occur in the pair separation energy curve like that shown in Fig. 11.

There is a price to pay for the generalization to unequal α_j values. There is no longer a complete scheme of seniority classification of all J=0 states. In fact, given two orthogonal states with the same nucleon numbers $|1>$ and $|2>$, the states $S^+|1>$ and $S^+|2>$ may both exist but need not be orthogonal. Nevertheless, the one state (21) has the nice feature of seniority v=0 and can be a good description of ground states of semi-magic nuclei.

It is still possible to construct states with J > 0 which can be assigned generalized seniority v=2. Consider the operator which creates a correlated pair with J=2 (and positive parity)[9]

$$D_M^+ = \sum_{j<j'} \beta_{jj'}(1+\delta_{jj'})^{-1/2} \sum_{\mu\mu'} (j\mu j'\mu'|jj'2M)\, a_{j\mu}^\dagger \, a_{j'\mu'}^\dagger . \quad (25)$$

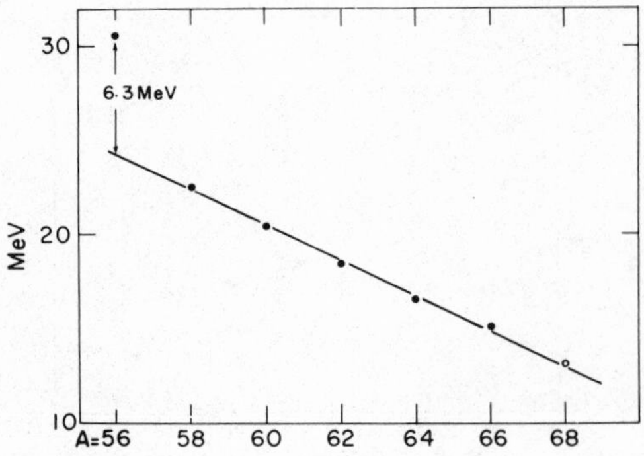

Fig. 11. Neutron pair separation energies in Ni isotopes. Both calculation and experiment yield a smooth line even though the $2p_{3/2}$, $1f_{5/2}$ and $2p_{1/2}$ orbits are strongly mixed.

The state $D^{+}_{\mu}|0>$ is analogous to the state $|j^2 v = 2, J=2>$. States with the same v=2 and higher n were then given by $(S^{+}_{j})^{\frac{n-v}{2}} |j^2 v = 2, J=2>$. In analogy, we ask for the conditions on H under which the states $(S^{+})^{n-1} D^{+}_{\mu}|0>$ will be eigenstates. The first condition, for n=1, is that the pair state be an eigenstate

$$HD^{+}_{\mu}|0> = V_2 D^{+}_{\mu}|0> \qquad (26)$$

with eigenvalue V_2. We further obtain by induction, making use of (22) and (23),

$$H(S^{+})^{n-1} D^{+}_{\mu}|0> = (V_2 + E_{n-1})(S^{+})^{n-1} D^{+}_{\mu}|0> + + (n-1)(S^{+})^{n-2} \left[[H,S^{+}],D^{+}_{\mu}\right]|0> . \qquad (27)$$

For this to be an eigenstate it is necessary and sufficient that the double commutator of H with S^{+} and D^{+} be proportional to $S^{+}D^{+}_{\mu}$. It was shown that for any case of interest[9] the proportionality factor must be equal to Δ introduced in (23). The condition can then be expressed as

$$\left[[H,S^{+}],D^{+}_{\mu}\right] = \Delta S^{+} D^{+}_{\mu} . \qquad (28)$$

Putting this value of the double commutator in (27) we obtain

$$H(S^{+})^{n-1} D^{+}_{\mu}|0> = (E_n + V_2 - V_0)(S^{+})^{n-1} D^{+}_{\mu}|0> . \qquad (29)$$

Provided conditions (26) and (28) are also satisfied, the J=2 states in (29) are eigenstates of the shell model Hamiltonian. Their structure leads directly to the form of their eigenvalues given by (29). Hence, the positions of these J=2 states (with generalized seniority

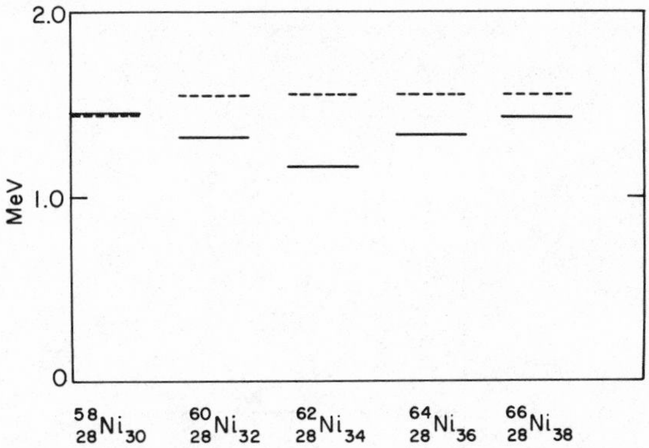

Fig. 12. Experimental (solid lines) and calculated (dashed lines) positions of J=2 levels in Ni isotopes. The 0-2 spacings are fairly independent of neutron number.

v=2) above the ground states is independent of n. This nice feature of seniority is preserved also in generalized seniority but only for a selected set of J=0 and J=2 states. This agrees well with experimental data and shell model calculations shown in Fig. 12 for nickel isotopes. The resulting α_j values however are quite different. In that case other states need not have constant spacings as clearly seen in Fig. 13. The fact that 0-2 spacings are independent of nucleon number is clearly seen in semi-magic tin isotopes. No reliable shell model calculations are available for that region but this feature holds throughout the major shell (Fig. 14). This is manifestly not the case for levels of odd tin isotopes shown in Fig. 15.

Seniority and generalized seniority give a very simple prescription for some low lying levels in semi-magic nuclei. Ground states are very accurately given by (21). Provided the effective shell model Hamiltonian satisfies conditions (22) and (23) it is sufficient to solve the two-nucleon problem to obtain V_0 and the α_j, and then to solve the four-nucleon problem to obtain Δ. Even in the absence of information on all two-body matrix elements, only two parameters V_0 and Δ are necessary to compute all ground state energies and another parameter V_2 determines the constant positions of the first excited J=2 states. Any shell model Hamiltonian must yield the eigenvalues in (24) and (29) to agree with experiment. These are very stringent and numerous conditions on both single nucleon energies and two-body matrix elements. The set of conditions (22), (23), (26) and (28) turn out to be the best way to reproduce the data. Very simple interactions do not lead to the experimentally observed features if the single nucleon energies are not degenerate. The pairing interaction and the nicer surface delta interaction satisfy those conditions only for degenerate orbits in which case all α_j are essentially equal. The effective interaction could not be as

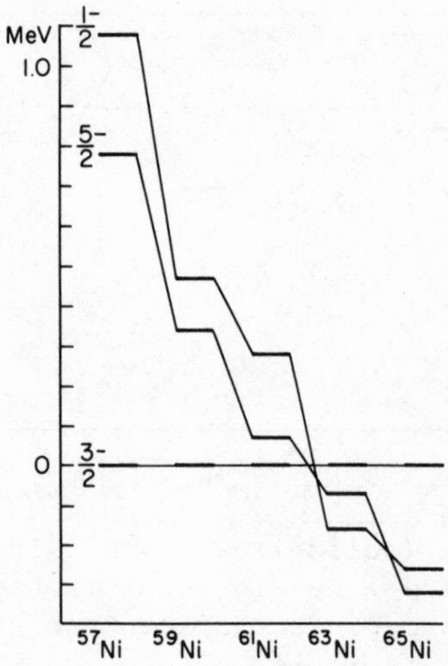

Fig. 13. Experimental (and calculated) levels in odd Ni isotopes. Level spacings change drastically unlike the 0-2 spacings.

general and schematic. It could be non-local and strongly dependent on the shell model subspace used. If we want to use the shell model to obtain good agreement with the data for semi-magic nuclei we must use an effective two-body interaction which has some eigenstates with generalized seniority v=0 and v=2.

It is now time to consider the majority of nuclei—those with both valence protons and valence neutrons. There the situation is much more complex and the emerging regularities are rather different.

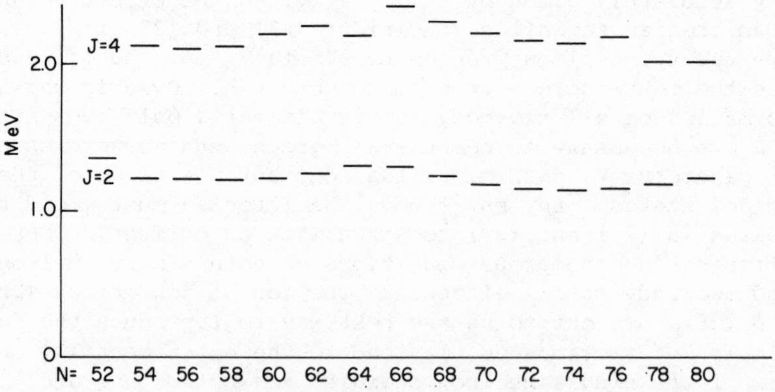

Fig. 14. Lowest levels in even Sn isotopes. The 0-2 spacings are remarkably constant throughout the major shell.

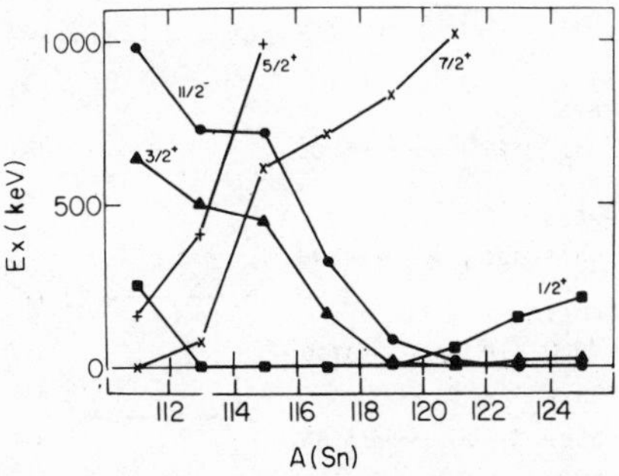

Fig. 15. Lowest levels in odd Sn isotopes. Level spacings change dramatically as functions of neutron number unlike the 0-2 spacings.

COLLECTIVE BANDS AND THE INTERACTING BOSON MODEL (IBA-1)

Whenever there are several protons and neutrons outside closed shells the shell model problem increases tremendously. Spectra of such nuclei become more complex but with increasing numbers of both valence protons and valence neutrons the experimental situation becomes simpler. The multitude of levels can be grouped into several bands. In many cases, J(J+1) dependence of energy levels within bands is observed and strong intraband E2 transitions have been measured. Such level schemes are naturally described as rotational spectra. A striking example of such a situation is shown in Fig. 16.

Such rotational bands were successfully described by the well known collective model.[10,11] In it the nucleus is described by a small number of collective variables associated with a surface (encompassing a constant volume) with quadrupole deformations. These could be the five α_μ which are the coefficients of the quadrupole mass tensor in a fixed frame in space. An alternative description utilizes a body-fixed frame whose orientation in space is given by the three Euler angles Ω_i. The amount of deformation is given by the variable β and deviations from axial symmetry by an angle γ. If the surface has a stable deformation at $\beta_0 \neq 0$, it can rotate and, for $\gamma_0 = 0$, give rise to a rotational band (the ground state band). Vibrations involving the β and γ dynamical variables can also give rise to rotational bands (β-band $J = 0,2,4,\ldots$, γ-band $J = 2,3,4,\ldots$ and higher ones).

This phenomenological collective model has been very useful in the analysis of rotational spectra. Naturally, attempts have been made to derive the collective variables from the nucleon degrees of freedom. In spite of a rather large effort no satisfactory results have been so far obtained. A more successful approach to a microscopic description of the collective model makes use of a

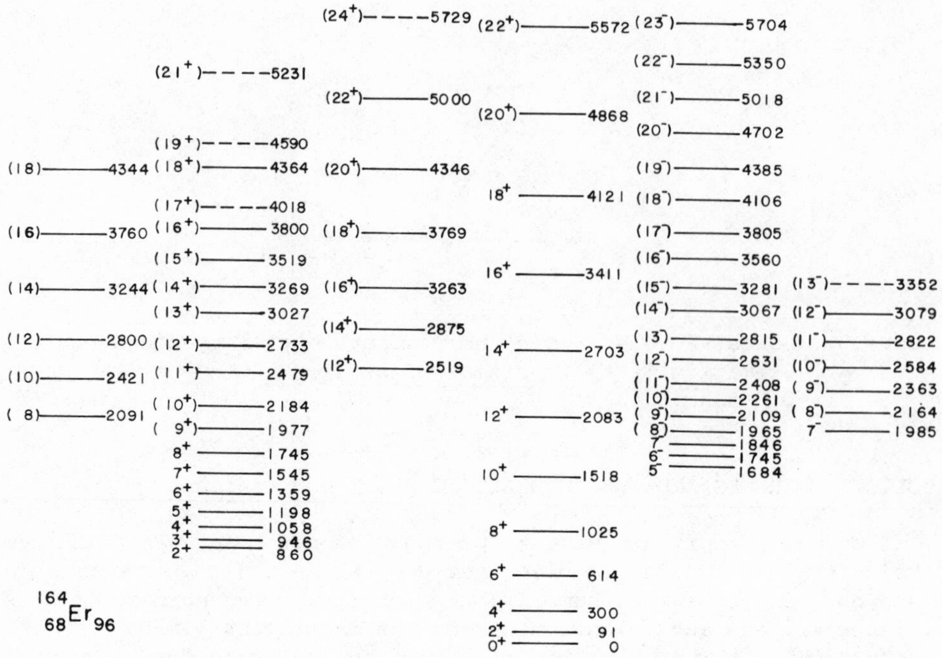

Fig. 16. Six rotational bands observed in the ^{164}Er spectrum.

deformed potential well in which nucleons move independently. In principle, such a field should be determined by a Hartree-Fock calculation. In practice, the very successful Nilsson scheme[12] is determined by minimizing the sum of single nucleon energies in an axially symmetric deformed oscillator potential well. The eigenstates of the nuclear Hamiltonian have good angular momenta J. Such states can, in principle, be projected from the wave functions of the individual nucleons in the deformed potential well. In practice, they are obtained by "cranking" the nuclear wave function around an axis perpendicular to the axis of symmetry. This procedure may become exact in the limit of strong deformations and many nucleons. Only then will wave functions with different spatial orientations become approximately orthogonal. In many cases of small or moderate deformation this procedure is very doubtful and the resulting moments of inertia totally unreliable.

In any case this "unified model" leaves unanswered many questions of physical interest. Which nuclei are strongly deformed? Which are the nuclear interactions that lead to strong deformations and rotational spectra? How does the transition between simple shell model spectra and collective bands take place? All these questions could be answered if we could obtain a shell model description of collective bands. Apart from these fundamental problems, the greatest practical problem of the collective or unified model is that there is no general method to apply it to the whole range of nuclei with collective bands. Different approaches must be applied to different types of collective spectra like the vibrational and rotational regions. A most attractive feature of the *interacting boson model* is its unified approach to all types of collective bands.

The use of bosons is simply connected to the collective model. The kinetic energy can be written down in terms of the collective variables α_μ and their conjugate momenta. We can expand all collective eigenstates in states obtained by using harmonic oscillator potential wells in the five variables α_μ for the potential energy. The scheme of states obtained in this way is that of harmonic vibrations around a spherical equilibrium shape. These states can be conveniently expressed in terms of the five oscillator bosons. These bosons are the 5 components of a boson with angular momentum $\ell = 2$. The creation operator for this d-boson is $d_\mu^\dagger$ where $\mu = -2,-1,0,1,2$. The operators $d_\mu^\dagger$ and their hermitian conjugates satisfy the Bose commutation relations $[d_\mu, d_{\mu'}^\dagger] = \delta_{\mu\mu'}$. Using the annihilation operator $\tilde{d}_\mu = (-1)^\mu d_{-\mu}$, the oscillator Hamiltonian is simply

$$\hbar\omega \sum_\mu d_\mu^\dagger d_\mu + \frac{5}{2}\hbar\omega = \hbar\omega(d^\dagger \cdot \tilde{d}) + \frac{5}{2}\hbar\omega \quad . \tag{30}$$

The collective Hamiltonian with *any* potential can be expressed in terms of the operators $d_\mu^\dagger$, d_μ and its eigenstates expressed as d-boson states.

The use of d-bosons is very convenient in describing vibrations around a spherical shape. However, this description becomes very complicated for rotations and vibrations of a deformed stable shape $(\beta_0 \neq 0)$. Eigenstates in such cases may be linear combinations of states with arbitrarily high numbers n_d of d-bosons. A boson Hamiltonian whose eigenstates, characterized by integers N, have only boson numbers satisfying $n_d \leqslant N$ was introduced by Janssen, Jolos and Dönau.[13] The eigenstates of that Hamiltonian form bases of symmetric irreducible representations of the U(6) group characterized by N. It turned out that such states may yield a simple description of rotational spectra as shown by Arima and Iachello[14] as well as by Jolos, Dönau and Janssen.[15]

Arima and Iachello introduced in addition to d-bosons also s-bosons[14] with $\ell = 0$. This way a six-dimensional space is obtained in which to the five components of a d-boson the single component of s-boson is added. This greatly simplifies the construction of U(6) Hamiltonians and their eigenstates. The latter are characterized by N which is simply the total number of s-bosons and d-bosons.

$$N = n_s + n_d \quad . \tag{31}$$

The number operator (31) commutes with the Arima-Iachello Hamiltonian. The most general rotationally invariant and hermitian Hamiltonian, commuting with N, which includes single boson energies and boson-boson interactions can be written as

$$H = \varepsilon_s\ s^\dagger s + \varepsilon_d(d^\dagger\cdot\tilde{d}) + \frac{1}{2}\sum_{L=0,2,4} c_L(d^\dagger\times d^\dagger)^{(L)}\cdot(\tilde{d}\times\tilde{d})^{(L)} +$$

$$+ \frac{1}{2\sqrt{5}}\tilde{v}_0\left\{(d^\dagger\cdot d^\dagger)s^2 + (s^\dagger)^2\ (\tilde{d}\cdot\tilde{d})\right\} +$$

$$+ \frac{1}{\sqrt{10}}\tilde{v}_2\left\{(d^\dagger\times d^\dagger)^{(2)}\cdot\tilde{d}s + s^\dagger d^\dagger\cdot(\tilde{d}\times\tilde{d})^{(2)}\right\} +$$

$$+ \frac{1}{2}u_0(s^\dagger)^2 s^2 + \frac{1}{\sqrt{5}}u_2\ s^\dagger s(d^\dagger\cdot\tilde{d}) \quad . \tag{32}$$

This Hamiltonian contains 9 parameters, two coefficients of the single boson terms and 7 of boson-boson interactions (the notation used conforms to original definitions).

The Hamiltonian (32) can be simplified, using explicitly the fact that it commutes with N, into the form

$$H = \varepsilon_s N + \frac{1}{2}u_0 N(N-1) + \varepsilon'(d^\dagger\cdot\tilde{d}) +$$

$$+ \frac{1}{2}\sum_{L=0,2,4} c'_L(d^\dagger\times d^\dagger)^{(L)}\cdot(\tilde{d}\times\tilde{d})^{(L)} + \frac{1}{2\sqrt{5}}\tilde{v}_0\Big\{(d^\dagger\cdot d^\dagger)s^2 +$$

$$+ (s^\dagger)^2(\tilde{d}\cdot\tilde{d})\Big\} + \frac{1}{\sqrt{10}}\tilde{v}_2\left\{(d^\dagger\times d^\dagger)^{(2)}\cdot\tilde{d}s + s^\dagger d^\dagger\cdot(\tilde{d}\times\tilde{d})^{(2)}\right\} . \tag{33}$$

This Hamiltonian has only 8 independent coefficients. The first two terms, however, contribute equally to all states with the same N value. Spacings of energy levels are thus determined by 6 parameters only. The coefficients in (33) are related to those in (32) by

$$\varepsilon' = \varepsilon_d - \varepsilon_s + (N-1)\left(\frac{u_2}{\sqrt{5}} - u_0\right)$$

$$c'_L = c_L + u_0 - \frac{2u_2}{\sqrt{5}} \quad . \tag{34}$$

For any given number N and specified values of the coefficients, the Hamiltonian matrix can be calculated in any convenient basis. It can then be diagonalized yielding eigenstates and correspoinding eigenvalues. Matrix elements of E2 transitions are obtained in each case by using the resulting eigenstates and the single boson hermitian operator

$$Q_\mu = \alpha_2(d^\dagger_\mu s + s^\dagger\tilde{d}_\mu) + \beta_2(d^\dagger\times\tilde{d})^{(2)}_\mu \quad . \tag{35}$$

The quadrupole operator (35) is determined by the two parameters α_2 and β_2. In general, the diagonalization must be carried out numerically but this is rather trivial compared to diagonalizing the shell model Hamiltonian or solving the differential equation of the collective model. By changing smoothly the parameters in (33) various types of collective states are obtained. It turns out that for special values of the coeffients in (33) it is possible to obtain analytical solutions of eigenvalues and eigenstates. The spectra and transitions in such cases resemble very much the results of the collective model in certain limits. To consider these special cases it is necessary to examine various sub-groups of the U(6) group.

The Hamiltonian (32) can be constructed from the generators of the Lie algebra of U(6). There are 36 such generators [k^2 for U(k)] and can be chosen as $d_\mu^\dagger d_{\mu'}$, $d_\mu^\dagger s$, $s^\dagger d_\mu$ and $s^\dagger s$. Any Hamiltonian constructed from these generators will have eigenstates that belong to irreducible representations of U(6). In order to construct a rotationally invariant Hamiltonian it is more convenient to group the generators into components of irreducible tensors. This results in the following expressions

$$\text{25 components} \qquad (d^\dagger \times \tilde{d})_\kappa^{(k)} \qquad k=0,1,2,3,4 \tag{36}$$

$$\text{two rank } k=2 \text{ tensors} \qquad d_\mu^\dagger s \qquad s^\dagger \tilde{d}_\mu \tag{37}$$

$$\text{a scalar} \qquad s^\dagger s \quad . \tag{38}$$

Any rotationally invariant U(6) Hamiltonian is a linear combination of the scalar generators in (36) and (38) and scalar products of all generators.

If we consider a subgroup of U(6) irreducible U(6) representations may contain several irreducible representations of the subgroup. Any state with given N may have then a better characterization—the irreducible representation of the subgroup to which it belongs. This process can be carried out further yielding a set of labels or quantum numbers which are the irreducible representations of the chain of subgroups. In order to have a physically meaningful classification of states the last subgroup in a chain must be O(3), the group of transformations induced in the given space by three-dimensional rotations. The generators of O(3) are the well known components of the total angular momentum operator. In the case of s- and d-bosons this vector must be proportional to the only rank k=1 tensor among the U(6) generators. The proper normalization, given by the commutation relations, yields for the angular momentum vector the form

$$L_\kappa = \sqrt{10}\,(d^\dagger \times \tilde{d})_\kappa^{(1)} \quad . \tag{39}$$

In some cases this procedure leads to complete labelling of states.

In the case of the U(6) subgroups this procedure leads to simple algebraic expressions for eigenvalues and eigenstates.

The simplest set of special values of the coefficients of the general U(6) Hamiltonian (32) is to keep only terms obtained with the 25 generators (36). They form the Lie algebra U(5) of unitary transformations in the five-dimensional space of d-bosons.[14,16] All these generators commute with $n_d = (d^\dagger \cdot d)$ which uniquely characterizes the fully symmetric irreducible representations of U(5). The U(5) subgroup of U(6) can be used for the procedure described above since it includes O(3) as a subgroup; the three generators (39) are among the generators (36). There is, however, another subgroup of U(5) which includes O(3) as a subgroup. This is O(5), the group of real orthogonal transformations in the five-dimensional space of d-bosons. Its generators are the components of irreducible tensors in (36) with odd values of k, k=1 and k=3 [the number of O(n) generators is n(n-1)].

The Hamiltonian (32) with U(5) symmetry has three independent parameters for the boson-boson interactions [since there are only three allowed $(d^\dagger d)^2|0\rangle$ states with $J=0,2,4$]. It is therefore possible to use the quantum numbers (labels) supplied by the irreducible representations of U(5), O(5) and O(3) to obtain an algebraic expression for the eigenvalues. Every Lie group has a (quadratic) Casimir operator which is a quadratic function of the generators. It commutes with all the generators and has therefore the same eigenvalue in all states of a given irreducible representation. The eigenvalues of the Casimir operator can thus be used to label the irreducible representations of the group. It turns out that it is possible to express the two-body part of the most general U(5) Hamiltonian as a linear combination of the Casimir operators of U(5), O(5) and O(3). By their construction they commute with each other and thus the eigenvalues of the Hamiltonian will be linear combinations of their eigenvalues.

The quadratic Casimir operator of U(5) can be written as

$$\sum_{k=0}^{4} (d^\dagger \times \tilde{d})^{(k)} \cdot (d^\dagger \times \tilde{d})^{(k)} \quad . \tag{40}$$

Its eigenvalues are given by $n_d(n_d+4)$. The Casimir operator of O(5) is defined by

$$C_5 = 2 \sum_{k=1,3} (d^\dagger \times \tilde{d})^{(k)} \cdot (d^\dagger \times \tilde{d})^{(k)} \quad . \tag{41}$$

Its eigenvalues are given by $v(v+3)$ where v is the seniority quantum number for states with n_d bosons. The fully symmetric irreducible representations of O(5) are completely characterized by n_d and the seniority v which can assume the values $v = n_d, n_d-2, \ldots, 1$ or 0. Instead of the Casimir operator C_5 one can use the pairing interaction operator $(1/2)(d^\dagger \cdot d^\dagger)(\tilde{d} \cdot \tilde{d})$ with eigenvalues 5 for L=0 and 0 for L=2,4 two d-boson states. For n_d boson states with seniority v, the eigenvalues of this pairing operator are

$$\frac{1}{2}\left[n_d(n_d+3) - v(v+3)\right] = \frac{n_d - v}{2}(3 + n_d + v) \quad .$$

Comparing this expression with (12) in the fermion case we see the effect of the Pauli principle in the latter case. The Casimir operator of O(3) is the usual (L•L) expression.

The U(5) Hamiltonian, apart from the linear term in n_d, can be written as the linear combination of the Casimir operator of U(5), the O(5) pairing interaction and (L•L) with coefficients $\alpha/2$, 2β and γ respectively. The quadratic Casimir operators of U(5) and O(3) contain single boson terms which we combine with the linear term in n_d. We thus obtain for the eigenvalues of the most general U(5) Hamiltonian the expression[14,16]

$$\varepsilon n_d + \alpha n_d(n_d-1)/2 + \beta\left\{n_d(n_d+3) - v(v+3)\right\} + \gamma\left\{L(L+1) - 6n_d\right\} \quad . \tag{42}$$

To this expression terms linear and quadratic in N may be added.

The eigenstates of the U(5) Hamiltonian are fully characterized by N, n_d, v and L. The corresponding eigenvalues depend only on these quantum numbers. The U(5) eigenvalues given by (42) are typical of a vibrational spectrum. An example is shown in Fig. 17. For ε large compared with other coefficients the energies are essentially given by n_d but are not proportional to it. There are anharmonicities in the spectrum which are due to the boson-boson interactions. These anharmonicities are given by the dependence of (42) on v and L. The characterization of eigenstates by the quantum numbers listed above is, however, not complete. In the reduction of irreducible representations of O(5) into those of O(3), the same O(3) representation may occur more than once. For a given v, there may be more than one state with a certain L. To distinguish between such states another quantum number is necessary. It has been defined[16] as n_Δ— the number of d-boson tryads coupled to L=0. As shown above, the energy eigenvalues (42) are independent of it.

Another special case of the U(6) Hamiltonian is obtained by using in its constructions only the generators of another subgroup, namely O(6).[17,18] This is the group of real orthogonal transformations in the six-dimensional space of s- and d-bosons. The group O(5) considered above can be augmented to O(6) by adding 15-10 = 5 generators which must be the components of an irreducible operator of rank 2. The components of the quadrupole operator

$$Q' = s^\dagger \tilde{d} + d^\dagger s \tag{43}$$

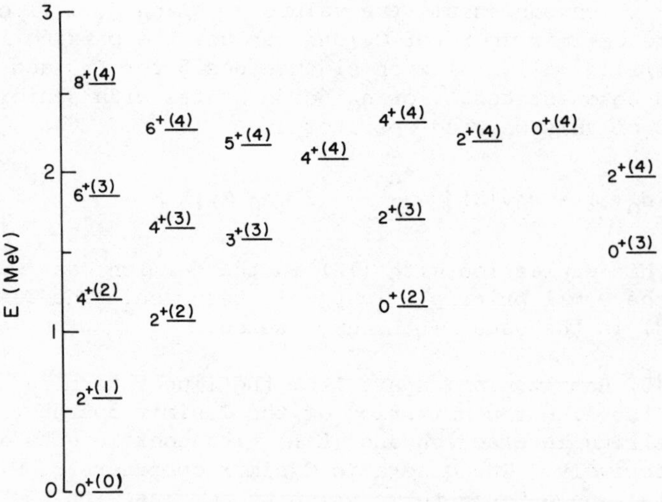

Fig. 17. A typical vibrational [U(5)] spectrum of IBA-1. Numbers of d-bosons are given in brackets.

satisfy the necessary commutation relations with the components of $(d^\dagger \times \tilde{d})^{(k)}$, $k=1,3$.

Also in this case there are only three independent boson-boson interactions obtained from the scalar products of the tensors with $k=1,2,3$. The most general O(6) boson-boson interaction can be written down as a linear combination of the Casimir operators of O(6), O(5) and O(3), [C_6, C_5 and (L·L) respectively]. To these the Casimir operator of U(6) can be added as well as any term linear in N. Apart from the single boson terms included in C_6, C_5 and (L·L) only terms proportional to $N=n_s+n_d$ can be added to a O(6) Hamiltonian. The number operators n_s and n_d do not commute with the Casimir operator of O(6).

The Casimir operator of O(6) is given by

$$C_6 = Q'\cdot Q' - 2\sum_{k=1,3} (d^\dagger \times \tilde{d})^{(k)} \cdot (d^\dagger \times \tilde{d})^{(k)} = Q'\cdot Q' - C_5 \quad . \qquad (44)$$

It has the eigenvalues $\sigma(\sigma+4)$ where the integer σ, together with N, completely characterizes the fully symmetric irreducible O(6) representations. The possible values of σ are $\sigma=N, N-2, N-4, \ldots, 1$ or 0. Instead of the operator C_6 it is possible to use the pairing interaction in the six-dimensional space of s- and d-bosons. It is defined as

$$P^\dagger P = \frac{1}{4}\,(d^\dagger\cdot d^\dagger - s^\dagger s^\dagger)(\tilde{d}\cdot\tilde{d} - ss) = \frac{1}{4}\,N(N+4) - C_6 \tag{45}$$

and is thus simply related to C_6. It has the eigenvalue 3 in the state $(d^\dagger\cdot d^\dagger - s^\dagger s^\dagger)|0>$ and eigenvalue 0 in all other two-boson states. Its eigenvalues are given accordingly by

$$\frac{1}{4}\big(N(N+4) - \sigma(\sigma+4)\big) = \frac{1}{4}\,(N-\sigma)(4+N+\sigma) \quad .$$

If now the most general O(6) Hamiltonian is taken to be a linear combination of $P^\dagger P$, C_5 and $(L\cdot L)$ with coefficients A, B/6 and C, its eigenvalues are given by[17,18]

$$\frac{A}{4}\,(N-\sigma)(N+\sigma+4) + \frac{B}{6}\,\tau(\tau+3) + CL(L+1) \quad . \tag{46}$$

In (46) the quantum number τ determines the eigenvalues of C_5 and, as the number v used above, is the seniority of d-boson states. The change in notation is customary and is useful in stressing the differences between the U(5) and O(6) schemes. In the present case n_d is not a good quantum number since $Q'\cdot Q'$ does not commute with it. However, $Q'\cdot Q'$ admixes only *pairs* of s-bosons into states of d-bosons. Since s-boson pairs have L=0, adding or removing them does not change the seniority of the d-bosons. Hence, the seniority τ is a good quantum number and its possible values are $\tau = \sigma, \sigma-1, \ldots, 0$.

The quantum numbers N, σ, τ and L completely characterize the irreducible representations of O(6) and the eigenstates of the O(6) Hamiltonian. The corresponding eigenvalues are quadratic (and linear) functions of these quantum numbers. The resulting spectrum (46) corresponds to the γ-unstable limit of the collective model. That spectrum occurs whenever there is a strong minimum at $\beta_0 \neq 0$ of the potential energy but it is completely independent of γ. No eigenstates of (32) correspond to a stable minimum with $\gamma_0 \neq 0$. Higher order terms are needed to produce such states. An example of an O(6) spectrum is given in Fig. 18. Such spectra give[19] a good approximation to some actual nuclei like ^{196}Pt. Also in the present case another quantum number is necessary, in analogy with n_Δ, to distinguish between states with a certain value of τ and the same value of L. It is denoted in the present case[18] by ν_μ. The eigenvalues (46) are independent of it.

There is an even more interesting special case of the U(6) Hamiltonian obtained by using the SU(3) subgroup of U(6).[14,20] The latter was introduced into nuclear physics many years ago in order to obtain rotational bands in the shell model.[21] The applications of the fermion SU(3) to actual nuclei were severely limited by the strong spin-orbit interaction destroying the required degeneracy between single nucleon levels in oscillator shells. In the boson model, however, there is no intrinsic spin to interact with the ℓ=2

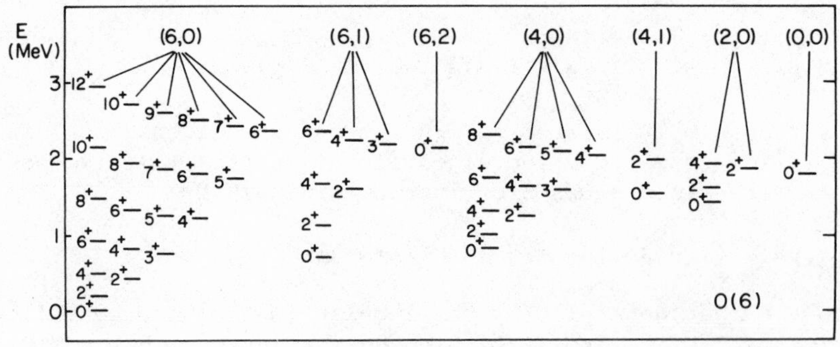

Fig. 18. A typical O(6) (γ-unstable) spectrum of IBA-1. The boson number is N=6 and the quantum numbers (σ,ν_Δ) are given in brackets.

angular momentum of a d-boson. There is no *a priori* information on the splitting between s and d single boson energies and boson Hamiltonians can have, in certain cases of physical interest, SU(3) symmetry.

The group U(3) has 9 generators which can be grouped into scalar, vector and quadrupole tensor operators. The requirement that the determinant of the unitary transformation be equal to 1 removes the scalar operator (being $N = n_s + n_d$ in the present case). The vector must be the angular momentum (39) since we require O(3) to be a subgroup of SU(3). The commutation relations of the angular momentum components with components of any irreducible tensor operator are linear combinations of those components. Hence the quadrupole tensor can be determined by the condition that commutation relations between its components are equal to angular momentum components. It can be easily varified that the five generators of SU(3), besides those in (39), are given by

$$Q_\mu = d_\mu^\dagger s + s^\dagger \tilde{d}_\mu - \left(\sqrt{7}/2\right)\left(d^\dagger \times \tilde{d}\right)_\mu^{(2)} \quad . \tag{47}$$

A positive sign in front of the second term in (47) leads to the same commutation relations. The negative sign was adopted here since it corresponds to prolate deformation.

The most general SU(3) Hamiltonian has only two independent terms, the scalar products (L·L) and (Q·Q). Any single boson term, other than those included in the scalar products must be proportional to $N = n_s + n_d$. The operator Q does not commute with either n_s or n_d. The Hamiltonian can be written as a linear combination of those scalar products with coefficients $-\kappa'$ and $-\kappa$ respectively. Alternatively, the Casimir operators of O(3) and of SU(3) can be used. The latter is given by

$$2(Q\cdot Q) + \frac{3}{4}(L\cdot L) \quad . \tag{48}$$

The two operators commute and an eigenvalue of the SU(3) Hamiltonian is the sum of their eigenvalues. The irreducible representations of SU(3) are completely characterized by pairs of integers (λ,μ). The eigenvalues of the Casimir operator are quadratic functions of the λ and μ quantum numbers. The eigenvalues of the most general SU(3) Hamiltonian can thus be obtained as[14,20]

$$\left(\frac{3}{4}\kappa-\kappa'\right) L(L+1) - \kappa\left[\lambda^2 + \mu^2 + \lambda\mu + 3(\lambda+\mu)\right] \quad . \tag{49}$$

The eigenvalues (49) are naturally arranged into rotational bands. Energies of states in any possible irreducible representation are proportional to L(L+1) like the energy levels of the ideal axially symmetric rigid rotor. The irreducible representations of SU(3) are completely determined by N,λ,μ and L and the eigenvalues completely determined by them [to (49) terms linear and quadratic in N may be added]. Still, as in the SU(3) case, these quantum numbers are not sufficient to uniquely label the eigenstates. Several identical O(3) irreducible representations may be included in one irreducible representation of SU(3). In order to distinguish among states with the same values of λ,μ and L an additional quantum number K is needed.[14,20,21] Spectra of SU(3) Hamiltonians give a very good description of actual rotational nuclei like ^{156}Gd.

The lowest state belongs to the fully symmetric irreducible representation of SU(3) given by (2N,0). In the cases discussed above, the irreducible representations of U(6) and U(5) were characterized by a single number N and n_d which was equal to the number of bosons. The situation here is different since we are not dealing with p-bosons from which the U(3) irreducible representations could be naturally constructed. The s- and d-bosons may be considered as *pairs* of p-bosons coupled symmetrically to L=0 and L=2. The number which determines the U(3) or SU(3) irreducible representation fully symmetric in the p-bosons is then 2N. The SU(3) irreducible representations which are fully symmetric with respect to s- and d-bosons are, however, not necessarily fully symmetric with respect to the fictious p-bosons. Therefore, the two numbers λ and μ are necessary and μ may be different from zero for fully symmetric states of s- and d-bosons.

A typical SU(3) spectrum is presented in Fig. 19. The lowest band has K=0 and J = 0,2,4,...,2N and, its states belong to the (2N,0) irreducible representation. The next representation is characterized by (2N-4,2) and contains two bands. One band can be assigned K=0 and contains states with J = 0,2,4,... The other band is characterized by K=2 and its states have J = 2,3,4,... These

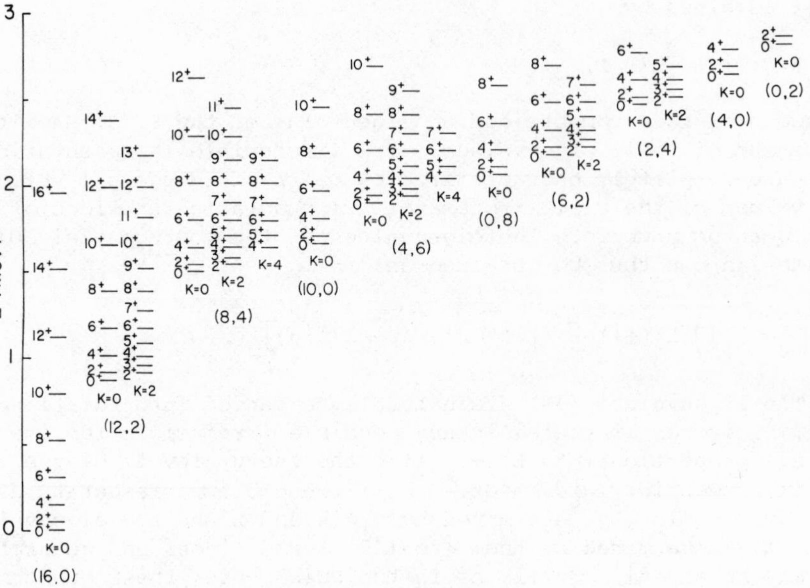

Fig. 19. A typical rotational [SU(3)] spectrum of IBA-1. The boson number is N=8, the quantum numbers (λ,μ) are given in brackets.

higher K=0 and K=2 bands closely resemble the rotational band based on a β-vibration and γ-vibration respectively.

The interacting boson model IBA-1 presented above is well defined mathematically and simple to work with. It utilizes a unified description for various kinds of collective motion and gives good agreement with the experimental data. Most actual nuclei have spectra which are between the special limits described above and have been successfully calculated by adjusting a few parameters of the general Hamiltonian (33).[22] Still, the special cases discussed above have a very simple structure and correspond to simple limits of the collective model [it can be easily verified that there are no other subgroups of U(6) which contain O(3) as a subgroup]. The simplicity of the boson model also gives a new approach to the problem of shell model structure of rotational (and other collective) states. It is now sufficient to look for the shell model description of the s- and d-bosons. This is the problem which we will now address.

THE BOSON MODEL BASED ON THE SHELL MODEL (IBA-2)

As mentioned above, d-bosons have had a long association with the collective model of the nucleus. Also there have been many attempts to reformulate the shell model in terms of bosons. Operators creating pairs of valence nucleons as well as pairs made of a valence nucleon and a hole in closed shells were expanded in terms of boson operators. Similarly, attempts were made to expand the shell model Hamiltonian in terms of boson operators. Such "boson expansions" involved infinite series of terms, the conditions on whose convergence could not be stated apart from cases of schematic models. It was not clear which terms should be kept to obtain a meaningful approximation. Hence, there was no way to make use of such expansions for the description of collective motion in actual nuclei. On the other hand, the attractive feature of the phenomenological IBA-1 model is simplicity and yet sufficient versatility to give a good description of the experimental data. The question arises how could such a simple model substitute for the extremely complicated problem of the shell model.

Let us look at the immensity of the straightforward approach of the shell model. Collective bands occur in nuclei with both valence protons and valence neutrons. Consider the case of the Z = 62 samarium isotopes. The 12 valence protons are distributed over the five orbits in the major shell, namely $2d_{5/2}$, $1g_{7/2}$, $3s_{1/2}$, $2d_{3/2}$ and $1h_{11/2}$. The configuration of the 10 valence neutrons in ^{154}Sm have components from the six orbits $1h_{9/2}$, $2f_{7/2}$, $2f_{5/2}$, $3p_{3/2}$, $3p_{1/2}$ and $1i_{13/2}$. There are millions of allowed proton-states and millions of neutron states. As a result, the numbers of states due to valence protons and neutrons in ^{154}Sm are

41,654,193,516,797	J = 0 states with positive parity
346,132,052,934,889	J = 2 states with positive parity
530,897,397,260,575	J = 4 states with positive parity .

In order to calculate eigenstates and eigenvalues with given J, matrices of this order should be constructed and diagonalized. Clearly this is an impossible task and even if performed the results would mean very little to us. We are mainly interested in the low lying collective bands in Sm isotopes which show remarkable simplicity (Fig. 20). The IBA-1 model can reproduce those bands very successfully. We are looking for a similar coupling scheme which will very drastically truncate the huge shell model space and yet give meaningful results.

Guided by the boson model we should look for analogous building blocks among the nucleons. The simplest candidates are pairs of nucleons coupled to J=0 and J=2 states with positive parity. This implies that we consider particle-particle pairs rather than particle-hole pairs (which are the favorite choice in various boson expansions). Nucleons must be excited into the major shell above the valence shell in order to couple with the hole to a J=0 state with positive pairty. This is a high excitation which could not play an important role for low lying levels. Also the conservation of the

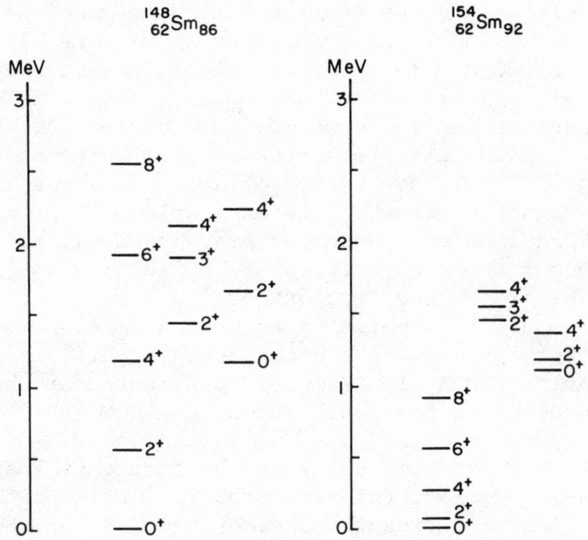

Fig. 20. Low-lying levels in Sm isotopes. There are only three collective bands. In ^{148}Sm they are quasi-vibrational whereas in ^{154}Sm they are the rotational ground state band, β-band and γ-band.

total boson number corresponds to fermion conservation of particle-particle pairs. There is little sense in no mixing of different numbers of particle-hole excitations. The next question is which particle-particle pairs should be considered, proton-proton, neutron-neutron or perhaps proton-neutron pairs. In most nuclei where collective bands occur valence protons and valence neutrons are in different major shells [all states of such valence nucleons have good isospin given by (N-Z)/2]. Hence, they cannot couple to positive parity pair states with J=0. We should thus consider for building blocks pairs with J=0 and J=2 of valence protons and pairs of valence neutrons.[23,24] As we saw above, such pairs of identical nucleons have a simple description in terms of generalized seniority. We can now see the connection between such pair states and boson states.

Let us recall the eigenvalues (24) and (29) obtained for the states $(S^+)^n|0>$ and $(S^+)^{n-1} D^+|0>$ respectively. The same eigenvalues can be obtained for *boson* eigenstates $(s^\dagger)^n|0>$ and $(s^\dagger)^{n-1} d^\dagger|0>$ from the boson Hamiltonian[25]

$$V_0 s^\dagger s + V_2(d^\dagger \cdot \tilde{d}) + \frac{1}{2}\Delta(s^\dagger)^2 s^2 + \Delta s^\dagger s(d^\dagger \cdot \tilde{d}) \quad . \tag{50}$$

The boson Hamiltonian (50) is not the first term in a "boson expansion" of the shell model Hamiltonian. Similarly, the boson operators $s^\dagger$ and $d^\dagger$ are not approximate creation operators of fermion pairs. They satisfy boson commutation relations where the commutators $[(S^+)^\dagger, S^+]$ and $[(D_\mu^+)^\dagger, D_{\mu'}^+]$ are extremely complicated. The last two terms in (50) do not represent four-fermion interactions which are not present in the shell model Hamiltonian.

The boson Hamiltonian (50) is just a model which can replace the shell model Hamiltonian only for a very special set of the lowest shell model states. It contains only single boson terms and boson-boson interactions. No higher-order terms are needed to reproduce exactly the eigenvalues of the shell model Hamiltonian where the Pauli principle is strictly obeyed. This is, of course, true only as long as $n \leqslant \Omega$. For $n > \Omega$ the shell model states (21) and (29) vanish because of the Pauli principle whereas corresponding boson states may have arbitrarily large n values. The correspondence thus established between nucleon and boson states should be simply discontinued beyond $n=\Omega$. In fact, the correspondence should be modified beyond $\Omega/2$, as we shall presently see.

The boson Hamiltonian (50) does not include interactions between d-bosons. Hence, states of the form

$$(s^\dagger)^{n_s} (d^\dagger)^{n_d}_{\gamma JM} |0\rangle \tag{51}$$

are eigenstates of (50) with the eigenvalues

$$V_0 \, n_s + V_2 \, n_d + \frac{1}{2} \Delta n_s (n_s - 1) + \Delta n_s \, n_d \quad . \tag{52}$$

In (51) the n_d operators $d^\dagger$ are coupled to a state with definite J and M. The additional quantum number γ distinguishes between the various orthogonal states with the same values of J and M. The situation for fermion states is more complex. States with even two operators D^+ like $(D^+ \times D^+)^{(J)}_M |0\rangle$ are not orthogonal to states with generalized seniorities v=0 and v=2. The state $(D^+ \cdot D^+)|0\rangle$ is generally not orthogonal to $(S^+)^2|0\rangle$ and $(D^+ \times D^+)^{(2)}|0\rangle$ is not orthogonal to $S^+ D^+_M |0\rangle$. Such states could not correspond to $(d^\dagger \times d^\dagger)^{(J)}_M |0\rangle$ boson states which are orthogonal to $(s^\dagger)^2|0\rangle$ and $s^\dagger d^\dagger_M |0\rangle$. Orthogonalization of fermion states is thus necessary and is carried out by removing from each state $(S^+)^{n-v/2} (D^+)^{v/2}_{JM} |0\rangle$ components with lower seniorities.[23,24,25] In other words, we project out all components of the form $(S^+)^{n-v/2+1} B^+_{JM} |0\rangle$ where B^+_{JM} creates a state with v-2 fermions. The resulting states are orthogonal and are now made to correspond to boson states (51) where

$$n_d = v/2 \qquad n_s = n - v/2 \quad . \tag{53}$$

If all α_j coefficients are equal, such fermion states are eigenstates of the pairing Hamiltonian

$$\varepsilon \sum_{jm} a_{jm}^{\dagger} a_{jm} + 2GS^{+}S^{-} \tag{54}$$

where $S^{-} = (S^{+})^{\dagger}$. The eigenvalues are given by the expression (52) with the coefficients

$$V_0 = \varepsilon + 2G\Omega \; ; \quad V_2 = \varepsilon \; ; \quad \Delta = -4G \; .$$

Beyond the middle of the shell, for $n > \Omega/2$, all fermion states can be put into the form $S^{+}B_{JM}^{+}|0>$ where B_{JM}^{+} creates n-2 nucleons.[27] Hence, they will not survive the projection described above. In order to establish a reasonable correspondence between fermion and boson states we use as building blocks beyond the middle of the shell J=0 and J=2 *hole* pairs rather than pairs of identical nucleons.[23,24]

In the case of (even-even) semi-magic nuclei generalized seniority gives an accurate and simple description of a very selected set of states of identical valence nucleons. It yields, for J=0 and J=2, eigenstates of matrices of a very high order. We see that we can also construct a very simple boson model for such states of semi-magic nuclei. The boson description is unnecessary, however, since the eigenvalues were first calculated from the shell model Hamiltonian in a simple way. The boson model yields its full power when we consider nuclei with both valence protons and neutrons. For such nuclei we need a very powerful coupling scheme which will drastically truncate the $10^{14} \times 10^{14}$ matrices into a manageable problem.

For semi-magic nuclei an appropriate coupling scheme was obtained by the observation that in simple configurations the T=1 part of the nuclear interaction is diagonal in the seniority scheme. Obviously, the coupling scheme must depend on the nature of the nuclear interaction. To check whether seniority is also a good scheme for the T=0 part of the nuclear interaction we can look at configurations with both protons and neutrons in the $1g_{9/2}$ orbit.[25] If the nuclear interaction is diagonal in the seniority scheme, spacings of levels with the same seniority quantum number should be independent of nucleon number and total isospin. These quantum numbers are the seniority v and reduced isospin t. In Fig. 21 we compare levels with J = 2,4,6,8 with v=2 and t=1 in ^{88}Y and ^{92}Mo. The spectra are very different, the levels in the two nuclei are almost exactly inverted. Comparing this situation with Fig. 8 we conclude that seniority is badly broken by the proton-neutron interaction (i.e. by the T=0 part of the nuclear interaction). Such strong breaking can be caused by interactions which have scalar products of even tensors in their expansion.

The simplest seniority breaking interaction is the quadrupole-quadrupole interaction. The coefficient of the quadrupole term in simple two-body systems is rather large and attractive. In the collective model, as well as in IBA-1, the quadrupole degree of freedom

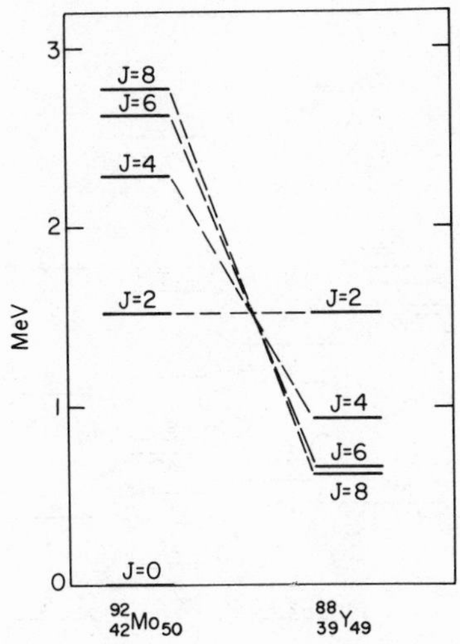

Fig. 21. Experimental levels of $(1g_{9/2})^n$ configurations of protons and neutrons. Level spacings of v=2, t=1 level spacings are almost reversed indicating major breakdown of seniority.

plays a dominant role. We can thus try a very drastic simplification of the interaction between protons and neutrons. We consider only a quadrupole-quadrupole interaction term between protons and neutrons. The strong and attractive monopole term will not strongly affect level spacings and will be omitted from the discussion. The seniority breaking effect of the quadrupole interaction is the decisive factor which determines the nature of nuclear spectra. Had the nuclear interaction been diagonal in the seniority scheme, level spacings would have been independent of nucleon number just as in semi-magic nuclei. When nucleons of the other kind are added the situation changes dramatically, the first excited J=2 level becomes much closer to the J=0 ground state. Other levels also change their positions and when there are enough valence nucleons *of both kinds* levels form rotational bands. The crucial factor which determines the appearance of the spectrum is the presence of both valence protons and valence neutrons. It is not just "the number of valence nucleons". A clear demonstration of this trend is given in Fig. 22.

Since the quadrupole interaction has an overriding importance in the T=0 part of the nuclear interaction, one may wonder what role it plays in the T=1 part. Sometimes estimates are even given for the strength of the quadrupole interaction between identical nucleons. The precise situation is that this question is irrelevant and estimates of strength are actually meaningless. The space of T=1 states in the j^2 configuration is limited to those with even values of J. Hence, the expansion of the interaction in scalar products of

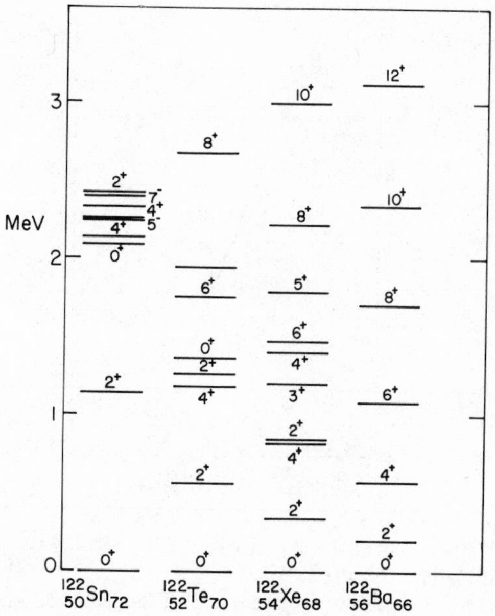

Fig. 22. Levels of A = 122 nuclei with 12 valence nucleons. The spectrum depends critically on the number of protons and neutrons. ^{122}Sn is semi-magic, ^{122}Te quasi-vibrational, whereas in ^{122}Ba a quasi-rotational ground state band is evident.

irreducible tensor operators in that space is not unique.[6] The coefficients of this expansion cannot be determined from the $V(j^2\ T{=}1\ J)$ energies. Certain two-body interactions with even tensors in their expansion can be expressed in terms of odd tensors only (like the extreme short range δ-potential). In fact, it was shown[6,9] that any interaction diagonal in the seniority scheme can be expressed as a linear combination with odd tensors and a monopole term. The T=1 interaction in actual nuclei is diagonal in the seniority scheme. It may well contain a quadrupole term but its seniority-breaking effect is completely cancelled by the effects of other even tensor interactions. This feature persists also in semi-magic nuclei with several active valence j orbits. The situation is rather different in the case of valence protons and neutrons. The expansion of the interaction (including the T=0 as well as the T=1 parts) in scalar products of irreducible tensors is unique. In such cases the quadrupole operator has a strong effect on level spacings and may lead to strong deformations and rotational spectra.

A complete set of shell model states is obtained by coupling all states of valence protons with all states of valence neutrons. A complete orthonormal set of the valence protons is $|\alpha_\pi\ J_\pi\ M_\pi >$

where α_π are the additional quantum numbers (labels) necessary to distinguish between states with the same value of J_π. The neutron set of states is similarly denoted by $|\alpha_\nu\ J_\nu\ M_\nu\ >$. To obtain the complete set of shell model states of the combined system we use Clebsch-Gordan coefficients and obtain

$$|\alpha_\pi J_\pi \alpha_\nu J_\nu JM > \sum_{\mu_\pi \mu_\nu} (J_\pi \mu_\pi J_\nu M_\nu | J_\pi J_\nu JM) |\alpha_\pi J_\pi M_\pi > |\alpha_\nu J_\nu M_\nu > . \quad (55)$$

This looks like a simple formula but it may express as many as 10^{14} states in actual cases. In this set of states matrix elements of the quadrupole interaction can be expressed by

$$\langle \alpha_\pi J_\pi \alpha_\nu J_\nu JM | Q_\pi \cdot Q_\nu | \alpha'_\pi J'_\pi \alpha'_\nu J'_\nu JM \rangle$$

$$= (-1)^{J_\nu + J'_\pi + J} \begin{Bmatrix} J_\pi & J_\nu & J \\ J'_\nu & J'_\pi & 2 \end{Bmatrix} (\alpha_\pi J_\pi \| Q_\pi \| \alpha'_\pi J'_\pi)(\alpha_\nu J_\nu \| Q_\nu \| \alpha'_\nu J'_\nu) . \quad (56)$$

The form of the matrix elements (56) is indeed simple. They are products of reduced matrix elements of the proton and neutron quadrupole operators multiplied by a Racah coefficient (6-j symbol). Still, they are elements of a matrix of order 10^{14}. Some way of drastically truncating these matrices must be found.

The truncation should yield a set of states so that matrix elements (56) between them should be large. We adopt the simple shell model Hamiltonian

$$H_\pi + H_\nu - Q_\pi \cdot Q_\nu \quad . \quad (57)$$

In the absence of proton-neutron interaction, the lowest J=0 eigenstate would be simply $(S_\pi^+)^{n_\pi}(S_\nu^+)^{n_\nu}|0>$. We take it as the first member of the truncated set and then let the Hamiltonian (57) operate on it. The resulting state will be taken as the second member of the truncated set and the procedure is then continued. If Q_π and Q_ν in (57) are approximately given by[28,29]

$$D_\pi^+ = \frac{1}{2}[Q_\pi, S_\pi^+]$$

$$D_\nu^+ = \frac{1}{2}[Q_\nu, S_\nu^+] \quad (58)$$

then the resulting state will be $(D_\pi^+ \cdot D_\nu^+)(S_\pi^+)^{n_\pi - 1}(S_\nu^+)^{n_\nu - 1}|0>$. It can be argued[25] that other states of an appropriately truncated set generated in this way for J=0 as well as for J=2,4,... are obtained by taking all proton (neutron) states obtained by using S_π^+ and D_π^+ (S_ν^+ and D_ν^+) in their construction (S-D space). In this way, the giant shell model space is reduced to the much smaller S-D space. The resulting matrices have reasonable sizes and can be easily diagonalized. In certain theoretical models this truncation is exact.[30] In actual cases it is expected to be a good approximation.

Although the size of the shell model matrices has been drastically and effectively reduced, there are still great difficulties in the actual evaluation of the matrix elements (56) in the S-D space. The effect of the Pauli principle makes the calculations very complicated, especially in actual cases with unequal α_j coefficients in S_π^+ and S_ν^+. It is here that the boson picture of the fermion S-D states can show its full strength. We saw above how certain eigenstates of H_π can correspond to states of s_π and d_π bosons and the corresponding eigenvalues obtained from an equivalent boson Hamiltonian. The same relation holds between eigenstates of H_ν and corresponding states of s_ν and d_ν bosons. In order to obtain a boson picture for the Hamiltonian (57) we should construct boson quadrupole operators equivalent to Q_π and Q_ν. This means that matrix elements of $Q_\pi(Q_\nu)$ between proton (neutron) S-D states should be equal to those of a proton (neutron) boson quadrupole operator between corresponding s_π and d_π (s_ν and d_ν) boson states. In certain theoretical models this program can be carried out exactly.[29,30] In actual cases it is expected to give a good approximation.

We started by looking for a shell model basis for IBA-1 with one kind of s- and d-boson. Shell model considerations led us to a boson model which has four kinds of bosons: s_π, d_π, s_ν and d_ν bosons.[23,24] This more detailed model is referred to as IBA-2, IBM-2 or simply IBM. It is more complicated than IBA-1 but it is more detailed. It is capable of demonstrating the transitions among the various types of collective spectra as functions of the valence proton and valence neutron numbers. Still, IBA-1 has a simpler group structure and can exhibit the various limits of collective motion explicitly and elegantly. The question arises whether IBA-1 could be obtained from IBA-2. It turns out that in certain IBA-2 cases states are completely symmetric in the proton and neutron bosons. Such states could be expressed in terms of one kind of unspecified s- and d-boson.[23]

It is worthwhile pointing out that there is no shell model approach which directly yields IBA-1. Without specifying the proton and neutron structure of boson states it is impossible to obtain a meaningful correspondence with proton and neutron states. The IBA-1 Hamiltonian is therefore not directly related to the fermion interactions. The coefficients of IBA-2 Hamiltonians are directly related, at least in principle, to the effective interactions between nucleons. This is not so in the IBA-1 Hamiltonian. The quadrupole interaction appearing in it is not directly related to the quadrupole interaction between protons and neutrons.[31] The latter is responsible for the development of rotational bands and one of its strong effects is the reduction of the value of ε. The latter no longer has the fixed large value of the fermion V_2-V_0. It is much smaller and is even put to zero in SU(3) and O(6) Hamiltonians as well as in actual cases which lie between these limits.

The actual construction of equivalent Q_π and Q_ν boson operators is not an easy task. For equal α_j values, there are simple formulae of the seniority scheme. In that case, an equivalent boson operator for protons and for neutrons has the form (35) but with coefficients α_2 and β_2 which are well defined functions of n_π and $n_{d\pi}$ (n_ν and $n_{d\nu}$).[24,26] For unequal α_j values there are only preliminary calculations which indicate that the dependence on n and n_d (for protons and neutrons) may be much more complicated.[32,27] The IBA-2 model has been successfully applied to many nuclei. In these applications the adopted form of H_π is very simply $\varepsilon(d_\pi^\dagger \cdot \tilde{d}_\pi)$ and the equivalent H_ν is taken to be $\varepsilon(d_\nu^\dagger \cdot \tilde{d}_\nu)$. It is further assumed that α_2 and β_2 for proton (neutron) bosons depend only on $n_\pi(n_\nu)$. With these simplifications only a few parameters remain and they are obtained from the experimental data. In cases where there are more data than parameters the fit obtained is quantitatively meaningful.[33] An example of such agreement is given in Fig. 23.

In arriving at the boson model from the shell model we made many approximations and assumptions. They were not seriously considered before the phenomenological success of IBA-1 became apparent. That made it possible to take the necessary step that would lead to it from the shell model. These approximations are expected to hold only for a very limited set of states which are members of the low lying collective bands. Properties of these levels change smoothly when going from one nucleus to the other throughout a major shell. Such behavior is determined by a tiny fraction of the information required to completely determine the shell model Hamiltonian and its eigenstates. Out of the 10^{14} J=0 states we are calculating only 10^2 states or fewer. In Sn isotopes there are 165 one-body and two-body matrix elements of the valence neutron Hamiltonian. Still, the constant energy separation between the J=2 and J=0 ground states, with generalized seniorities v=2 and v=0 respectively, is given by *one* parameter V_2-V_0. This simple behavior makes it possible to describe it by a simple boson model with Hamiltonian (50) where $\varepsilon = V_2 - V_0$. In nuclei with valence protons and neutrons there is smooth and regular behavior of the low lying collective bands. This makes the boson model much more useful and practical than the full shell model. Once the form of the boson Hamiltonian is given, possible contributions of fermion states excluded from the S-D space to energies and transitions can be still incorporated by renormalization of the boson model parameters.

The interacting boson model is simple and easy to work with. Yet it is versatile enough to give good agreement with experiment for collective bands of various kinds. Its success is due to the fact that it incorporates two main ingredients of the nuclear interaction. These are generalized seniority ("pairing") for identical

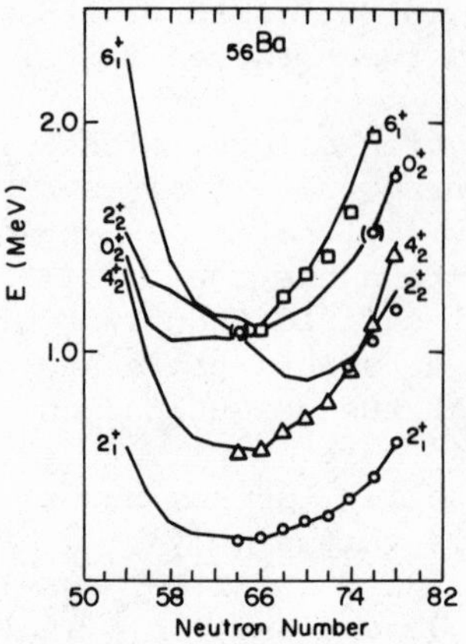

Fig. 23. Levels of Ba isotopes. The level spacings calculated from the proton-neutron boson model IBA-2 are connected by solid lines. Good agreement is obtained with the experimental values (circles, triangles and squares).

nucleons and the strong attractive quadrupole interaction between protons and neutrons. It has a close relationship with the collective model even if it is not exactly equivalent to it. On the other hand it has a shell model basis and could be considered as a good approximation to complex shell model calculations. It should be stressed that this is a good model for collective bands in even-even nuclei. Coupling with other degrees of freedom is necessary even for such nuclei. For odd mass nuclei, coupling of a fermion to the systems of bosons has been considered with very interesting results. As with any good model, the interacting boson model does not give complete answers to all problems of nuclear structure. It gives us the understanding and methods to deal with other interesting problems.

REFERENCES

1. M. G. Mayer and J. H. D. Jensen, "Elementary Theory of Nuclear Shell Structure," New York, N.Y. (1955).
2. I. Talmi, Rev. Mod. Phys. 34:704 (1962).
3. I. Talmi, Phys. Rev. 126:2116 (1962).
4. G. Racah, Phys. Rev. 63:367 (1943).
5. A. K. Kerman, Ann. Phys. (N.Y.), 12:300 (1961).
6. Most spectroscopic formula are found in: A. de-Shalit and I. Talmi, "Nuclear Shell Theory," New York, N.Y. (1963).
7. I. Talmi, in: "Elementary Modes of Excitation in Nuclei," A. Bohr and R. Broglia, eds., North Holland, Amsterdam (1977).
8. I. Talmi, Phys. Rev. 107:326 (1957).
9. I. Talmi, Nucl. Phys. 172:1 (1971).

10. A. Bohr and B. R. Mottelson, Mat. Fys. Med., Dan. Vid. Selsk. 27:No. 16 (1953).
11. A. Bohr and B. R. Mottelson, "Nuclear Structure Vol. II," Benjamin, Reading, Mass. (1975).
12. S. G. Nilsson, Mat. Fys. Med., Dan. Vid. Selsk. 29:No. 16 (1955).
13. D. Janssen, R. V. Jolos and F. Dönau, Nucl. Phys. A224:93 (1974).
14. A. Arima and F. Iachello, Phys. Rev. Lett. 35:1069 (1975).
15. R. V. Jolos, F. Dönau and D. Janssen, Yad. Fiz. 22:965 (1975).
16. A. Arima and F. Iachello, Ann. Phys. (N.Y.) 99:253 (1976).
17. A. Arima and F. Iachello, Phys. Rev. Lett. 40:385 (1978).
18. A. Arima and F. Iachello, Ann. Phys. (N.Y.) 123:468 (1979).
19. J. A. Cizewski, R. F. Casten, G. J. Smith, M. L. Stelts, W. R. Kane, H. G. Börner and W. F. Davidson, Phys. Rev. Lett. 40: 167 (1978).
20. A. Arima and F. Iachello, Ann. Phys. (N.Y.) 111:201 (1978).
21. J. P. Elliott, Proc. Roy. Soc. A245:562 (1958).
22. D. D. Warner, R. F. Casten and W. F. Davidson, Phys. Rev. Lett. 45:1761 (1980), and Phys. Rev. C24:1713 (1981); D. D. Warner and R. F. Casten, Phys. Rev. Lett. 48:1385 (1982).
23. A. Arima, T. Otsuka, F. Iachello and I. Talmi, Phys. Lett. B66: 205 (1977).
24. T. Otsuka, A. Arima, F. Iachello and I. Talmi, Phys. Lett. B76: 141 (1978).
25. I. Talmi, in: "From Nuclei to Particles," A. Molinari, ed., North Holland, Amsterdam (1982).
26. T. Otsuka, A. Arima and F. Iachello, Nucl. Phys. A309:1 (1978).
27. I. Talmi, Phys. Rev. C25:3189 (1982).
28. S. Shlomo and I. Talmi, Nucl. Phys. A198:81 (1972) and I. Talmi, Riv. Nuovo Cimento 3:85 (1973).
29. J. N. Ginocchio and I. Talmi, Nucl. Phys. A337:431 (1980).
30. J. N. Ginocchio, Phys. Lett. B79:173 (1978); 85:9 (1979); Ann. Phys. (N.Y.) 126:234 (1980).
31. A. Bohr and B. R. Mottelson, Physica Scripta 25:28 (1982) Post-script.
32. S. Pittel, P. D. Duval and B. R. Barrett, Phys. Rev. C25:2834 (1982).
33. "Interacting Bosons in Nuclear Physics," F. Iachello, ed., Plenum, New York (1979).

ROLE OF PIONS AND ISOBARS IN NUCLEI

F.C. Khanna and I.S. Towner

Atomic Energy of Canada Limited
Chalk River Nuclear Laboratories
Chalk River, Ontario, Canada K0J 1J0

INTRODUCTION

Nuclear dynamics should be based on QCD with quarks and gluons interacting via a gauge field. But as we have learned during this summer school, a complete treatment of interacting quarks and gluons is hard even for a simple system like a nucleon. In order to calculate properties of a nuclear many-body system, a quark and gluon model appears prohibitively difficult. One may be able to gain some insight from non-relativistic quark models and the use of symmetry properties to relate some relevant quantities like coupling constants and masses of various particles.

Nuclear dynamics can also be discussed within the framework of the classical treatment of strong interactions in which there are nucleons, isobars and mesons. The mesons mediate the interaction between baryons. The meson-baryon-baryon coupling may be deduced from experiment with the relationship of coupling to baryons within an octet being given by a quark model. Such a treatment may be viewed as similar to that in solid state physics where all interactions are Coulombic in nature but in order to calculate the phonons in crystals it is sufficient to use the Lennard-Jones potential. In fact there are relatively few microscopic attempts to calculate the properties of crystals.

The potential between two nucleons as mediated by mesons is short-ranged with a strong repulsion at short distances so that the two nucleons are inhibited from staying close together. However, the force at larger distances is primarily given by the one pion exchange potential (OPEP). These general features are reproduced both by phenomenological[1] as well as by meson-exchange potentials.[2] The ground state properties of nuclear matter and of the deuteron can be explained reasonably well by these models.

One should really ask the question: "What do we learn from a study of the excitation modes of a nuclear many-body system?" It is immediately apparent that a study of the nuclear response to external probes can provide rich and useful information about the constituents and structure of the nuclear system. Of course, the many-body system has some complications of its own. But a careful consideration of the various contributions may yield a wealth of information about the role of pions and isobars in a system that in its ground state consists mostly of a certain number of nucleons. Only a careful analysis will yield the relative importance of nucleons, isobars and pions in the nuclear response. It should be emphasized that among the mesons, pions have been singled out since the effects from all other mesons are relatively much shorter ranged and are damped out by short-range correlations between two nucleons. In addition there is a great deal of information available about soft pions that allows some non-perturbative statements to be made about the pionic contributions. It should be emphasized that we are interested in calculating low-energy properties at small momentum transfers and the soft-pion theorems do prove useful.

There are four independent categories of contributions:

(i) Configuration mixing and core polarization
(ii) Meson-exchange currents
(iii) Internal structure of nucleons i.e. isobars
(iv) Relativistic effects.

We relegate some of the technical discussion to the appendix and give here the physics of the problem.

Configuration Mixing and Core Polarization

In the shell model the nucleons sequentially fill the lowest-energy single-particle orbitals. Some of the orbitals are partially filled. The relative probability for the various occupancies in the valence shell may be obtained by configuration mixing. For example to consider the case of ^{48}Ca, configuration mixing involves $(fp)^8$ with ^{40}Ca being considered as a closed-shell nucleus. For simplicity one may consider cases with one particle or one hole in the LS-closed shell nuclei. This gets around the problem of configuration mixing. However, the question of core polarization remains and may be an important contributor to the nuclear response or to the transition rates. Core polarization involves the excitation of one (or more) of the particles in the core to an excited orbital through the two-body force. In practice core polarization may be evaluated in perturbation theory taken to a certain order. Here we will calculate contributions up to second order of perturbation theory.

In considering core-polarization effects it is very important to pick a two-body interaction such that the role of meson currents and isobars is not masked in any particular way. Care is taken that there is no double counting of the various effects. This puts limitations on the calculational procedures adopted. For example we use an N-N potential that is derived from π, ρ, ω and σ-exchange but we do not form a reaction matrix lest the ladder sum of various mesons simulates the effects of other mesons.

Meson-exchange Currents

Qualitatively it is clear that, due to the presence of short-range correlations, the meson-exchange currents can be calculated reliably if the pion effects dominate. In the soft-pion limit,[3] the electro-production of pions from a nucleon is described by a nucleon current: $\overline{\psi}(p')\gamma_\mu \gamma_5 \psi(p)$, where $\psi(p)$ is a Dirac spinor for the nucleon of momentum p. In the non-relativistic limit, the space component of the current reduces to $\phi^+ \vec{\sigma}\phi$ which is of order unity, while the time component reduces to $\phi^+ \vec{\sigma}\cdot(\vec{p}+\vec{p}\,')/2M\ \phi$ which is of order p/M. Here ϕ is the Pauli spinor for a nucleon. Similarly, weak production of pions from a nucleon is described by a nucleon current $\overline{\psi}(p')\gamma_\mu\cdot\psi(p)$, which non-relativistically reduces to the following: $\phi^+\left\{(\vec{\sigma}\cdot\vec{p}\,')\vec{\sigma} + \vec{\sigma}(\vec{\sigma}\cdot\vec{p})\right\}/2M\ \phi$ which is of order p/M for the space component, and $\phi^+\phi$ of order unity for the time component.

Combining the response to vector and axial-vector currents, we get Table 1. These qualitative results suggest that for Mλ transitions or moments and for axial charge the soft-pion effects may dominate and hence the role of meson-exchange currents may be calculated reliably. But for vector charge and for Gamow-Teller transitions pion effects may not dominate and the role of mesonic currents may be quite difficult to assess. In fact it may be that other mesons like ρ, ω, A_1 etc. will play a significant role in the response of the nuclear system.

Table 1. Non-relativistic Limit of Vector and Axial Vector Currents

External Probe	$\mu = 0$ Time	$\mu = 1,2,3$ Space
Vector current	0(p/M) Charge	0(1) Mλ
Axial-vector current	0(1) Axial charge	0(p/M) Gamow-Teller

We shall come back and consider nuclear response to each of these four external probes separately to assess the validity of qualitative arguments presented above.

Internal Structure of the Nucleon

The Δ-isobar (1232 MeV) is the lowest excited state of the nucleon and is related to the nucleon in the quark model by spin-isospin flip of one of the quarks. In this respect the isobar is akin to a Gamow-Teller resonance of the nucleon or to an isovector M1 transition. This resemblance would suggest that the isobar may have a role to play in Gamow-Teller transitions or in isovector M1-transitions. However, will the large energy separation (294 MeV) between the isobar and the nucleon tend to inhibit the isobar contrition? It may have done this except that the isobar has a width of 110 MeV for decay to the (N+π) system. This suggests a very strong NΔπ coupling.

In our work, the isobar is included as though it were a stable particle. The nucleon particle-hole (ph) calculations are extended to include isobar particle-nucleon hole states with the ph interaction given by meson exchanges. We assume NΔπ-coupling to be related to the NNπ-coupling through the non-relativistic quark model ($g_{N\Delta\pi} = 6\sqrt{2}/5\ g_{NN\pi}$). In practice these quark-model predictions are similar to those of Chew-Low theory ($g_{N\Delta\pi} \simeq 2g_{NN\pi}$).

THE MODEL

Many of the features mentioned above can be tested within a simple model that incorporates the important dynamical aspects of nuclear structure, meson exchange currents and isobars. In order to be able to isolate the role of nucleons, isobars and mesons the model has the following features:

(i) The NN→NN interaction is given by π, ρ, ω and σ exchange. The coupling constants of π, ρ and ω are taken from experiment, the main uncertainty being in the tensor coupling of ρ and ω. The mass of σ is fixed at 500 MeV and the coupling to nucleons is fixed by calculating the ground-state energy of closed-shell nuclei in the Hartree-Fock approximation. The cut-off parameter in the vertex form-factor is fixed at $\Lambda_\pi = 1$ GeV and $\Lambda_\rho = \Lambda_\omega = \Lambda_\sigma = 1.44$ GeV.

(ii) The NΔ→ΔN interaction is given by π and ρ exchange. The coupling constants are related to those of the NN system by the quark model.

(iii) The NΔ→NΔ interaction is given by π, ρ, ω and σ exchange and the coupling constants are again related to those of the NN system by the quark model.

(iv) All wave functions are harmonic oscillator functions.
(v) All radial integrals are cut off at 0.5 $\hbar/m_\pi c$ to simulate the effect of short-range correlations.

It should be stressed that the model has limitations but it has the distinction that there is no double-counting and therefore the effects of pions and isobars can be distinguished from those of the nucleons.

Let us consider the nuclear response to electromagnetic and axial-vector external probes.

Axial Charge

An excellent example for the study of axial charge is the β-decay of $^{16}N(0^-)$:

$$^{16}N(0^-) \rightarrow {}^{16}O(0^+) + e^- + \bar{\nu}_e \quad .$$

In this case the μ^--capture rate[5] on ^{16}O is also known. In the impulse approximation, the β-decay is described in terms of two one-body operators: (a) $\vec{\sigma}\cdot\vec{r}$ arising from the space component of axial current, and (b) $\vec{\sigma}\cdot\vec{p}/M$ arising from the time component of axial current. The matrix elements of these two operators have opposite sign and the net result is a small β-transition rate. Core polarization further decreases the decay rate. Mesonic effects are very large and enhance the decay rate by a factor of 3.5. In this particular problem isobars play a very small role. The final result reflects an interesting interplay between core-polarization and mesonic effects.

The calculated numbers[6] are compared in Table 2 with the experimental rates[4,5] for β-decay and for μ-capture. Overall agreement is quite reasonable.

The case of the A = 12 triplet has been considered in great detail by Guichon and Samour.[7] In this case the angular correlation of the $\beta^\pm$ decays, the $1^+ \rightarrow 0^+$ M1-decay and the μ-capture processes are known. By a careful analysis Guichon and Samour have isolated the time component of the axial current and have established that pion effects enhance it by a factor of about 3. These results along with those of A = 16 system make the strong dominance of pionic effects in the time component of the axial current quite convincing.

It is quite likely that 0^- excitations will also be observed in strong interactions such as (p,p′) and (p,n) experiments. However, it is not obvious what the signature will be in these reactions for strong pion enhancements that are the analogues of the enhancements predicted by soft-pion theorems for weak processes such as β-decay and μ-capture. Certainly this will be a field of future endeavor since it is clear that pion-like excitations have much to teach us.

Table 2. Calculated and Measured β-decay and μ-capture Rates

	Λ_β	$\Lambda_\mu(10^3)$
Impulse approximation	0.30 s^{-1}	2.20 s^{-1}
+ core polarization	0.12 s^{-1}	1.48 s^{-1}
+ mesonic effects	0.42 s^{-1}	2.00 s^{-1}
Experiment:		
Palffy et al.[4]	0.43 ± 0.10 s^{-1}	
Gagliardi et al.[4]	0.40 ± 0.05 s^{-1}	
Guichon et al.[5]		1.56 ± 0.11 s^{-1}

Magnetic Moments and M1 Transitions

The presence of mesonic effects in magnetic transitions has been confirmed for the elementary process[8] $n+p \rightarrow d+\gamma$. It is an interesting problem to investigate the role of mesonic effects in heavier systems and we choose to examine the magnetic moments of the simple configurations such as that of LS-closed shell plus (and minus) one particle.

Isoscalar magnetic moments. The magnitude of isoscalar magnetic moments depends[9] mainly on nuclear structure, namely wave function and core-polarization phenomena. Isobars have no effect and the dependence on mesonic effects is very weak. The calculated changes in the single-particle magnetic moments are compared with the experimental values in Table 3. The overall agreement is quite good suggesting that the treatment of the nuclear structure and the strength of the tensor force in the model one-Boson-exchange potential are reasonable.

Isovector magentic moments. With the agreement of the theoretical and experimental isoscalar magnetic moments (Table 3), it is reasonable to assume that the nuclear structure effects are well in hand. Now we can explore the role of isobars and mesonic effects. Within the model the results for isovector magnetic moments are compared with experiment in Table 4.

The calculated results are quite reasonable except for the case of the A = 41 system. In this case the large cancellation among various large contributions indicates the critical balance and interplay of core polarization, isobar and mesonic effects. Each of these quantities is individually large but the final theoretical prediction is small.

Table 3. Calculated Corrections to the Isoscalar Magnetic Moment Expressed as a Percentage of the Schmidt Value

Nucleus, A J^{π}	3 $1/2^+$	15 $1/2^-$	17 $5/2^+$	39 $3/2^+$	41 $7/2^-$
Theory	-8.7	16.5	-1.7	8.8	-1.5
Experiment	-3.2	16.7	-1.8	11.1	-1.0

The isovector M1 transition rates[9] between the single-particle states $j = \ell + 1/2$ and $j = \ell - 1/2$ are given in Table 5. In this case we find the cancellation between the core-polarization and mesonic effects is less severe. The isobar effects contribute roughly half of the decrease in the calculated M1 matrix element.

There are no experimental results on isovector transitions for these simple systems. It is possible that future (p,p′) experiments will be able to elaborate both on the structure of low-lying 1^+ states in even-even nuclei and on spin-flip transitions in odd-mass nuclei and determine the extent to which the M1-strength is quenched at low energies. At higher energies in the region of isobar excitation it is even possible that an enhancement will be seen. The sum rule for the total M1 strength is much larger in isobar particle-nucleon hole states than in the nucleon particle-hole states seen at low energies.

It should be remarked that a strong 1^+ state is known[10] in ^{48}Ca and the summed strength to this and other 1^+ states in the vicinity is of order $B(M1) \sim 5.2\ \mu_N^2$. A pure neutron $f_{7/2} \rightarrow f_{5/2}$ single-particle spin-flip transition gives a $B(M1) = 12\ \mu_N^2$. Configuration-mixing calculations of McGrory and Wildenthal[11] within the fp-shell

Table 4. Calculated Corrections to the Isovector Magnetic Moment Expressed as Percentage of the Schmidt Value

Nucleus, A J^{π}	3 $1/2^+$	15 $1/2^-$	17 $5/2^+$	39 $3/2^+$	41 $7/2^-$
Core polarization	-9.8	9.1	-14.1	37.3	-19.5
Isobars	1.7	17.5	- 0.6	3.5	- 1.2
Mesonic	15.0	-17.7	14.3	-64.8	19.8
Theory	7.0	8.9	- 0.5	-23.9	- 0.9
Experiment	8.5	11.1	- 1.3	-38.4	- 8.9

reduce the strength to 8.96 μ_N^2. Assuming that the quenching in ^{48}Ca is similar in magnitude to that in isovector transitions in ^{41}Ca (see Table 5), the strength will be reduced further to ~6 μ_N^2 which is to be compared with the experimental value of 5.2 μ_N^2. Therefore the quenching of M1-strength in ^{48}Ca arises from configuration mixing, core polarization and the isobaric current, and no one effect is the dominant cause of the reduction in M1 strength. Mesonic effects seem to be small in these transition rates and lead to enhancements rather than quenching.

There is no doubt that the excitation and understanding of unnatural parity states in finite nuclei will help in understanding the relative importance of mesonic and isobar effects. A larger body of data will help to establish clearly the nature of nuclear response to external electromagnetic probes.

Gamow-Teller Giant Resonances

Development of high-energy proton beams has revived an interest in the study of new excitation modes of the nucleus. Experiments with (p,n) reactions at small angles have preferentially excited Gamow-Teller resonances in nuclei.[12] These resonances are spin-flip isospin-flip states in nuclei. Within a simple quark model these resonances are the analogues of the excitation of isobar states in a single nucleon. Intuitively it is believed that isobar particle-nucleon hole states will dominate the structure of Gamow-Teller states.

Transition operators for the excitation of Gamow-Teller resonances derived from the space component of the axial-vector current and soft-pion theorems, which gave useful and model-independent predictions for the mesonic effects in axial charge and Mλ transitions, indicate in this case that pionic effects will be suppressed. Thus mesonic effects now become a delicate interplay of short-range phenomena and as such are much more difficult to compute reliably. For this reason, we follow the arguments of Rho[13] to be discussed below, and regroup the contributions to the Gamow-Teller matrix element not

Table 5. Calculated Corrections to the Isovector M1 Spin-flip Matrix Element Expressed as Percentage of the Schmidt Value

Nucleus, A	15	17	39	41
Core polarization	-12.4	-10.0	-15.4	-12.7
Isobars	- 3.6	- 4.8	- 5.6	- 5.9
Mesonic	6.8	2.2	6.3	3.1
Theory	- 9.3	-12.6	-14.6	-15.6

simply as core polarization, mesonic and isobars as used for the axial charge and $M\lambda$ transitions, but as those categories used by Oset and Rho[14] in which core polarization from low-energy excited states, isobar currents involving direct matrix elements and relativistic contributions are separated from the rest. In Table 6, we give our calculated results[15] for the diagonal Gamow-Teller matrix element and those deduced experimentally from β-decay between mirror states.

The contribution of the isobars is roughly one third of the total quenching of the Gamow-Teller strength. However, if we had calculated the isobar contribution by including both the direct and the exchange part of the particle-hole matrix element, the isobar contribution would have been considerably reduced. The overall agreement with experiment is quite reasonable.

An alternative explanation of the quenching of Gamow-Teller strength can be formulated in terms of the Migdal theory[14] of nuclear systems. In such a theory the short-range part of the particle-hole interaction is parametrized as

$$C\, g_0' \, \vec{\sigma}_1 \cdot \vec{\sigma}_2 \; \vec{\tau}_1 \cdot \vec{\tau}_2 \; \delta(r)$$

with $C \sim 400$ MeV fm^3 and $0.5 \leq g_0' \leq 0.7$. This interaction is obtained after summation over complicated multiparticle-multihole states and is to be used to calculate RPA series with only the direct part of the matrix elements. Such an interaction can lead to very large quenching depending on the magnitude of g_0'. However, one may ask what happens to core polarization and to relativistic effects etc.? Earlier calculations based on the potential model[15] described

Table 6. Corrections to the GT Matrix Element as a Percentage of the Single-particle Value

Nucleus, A	3	15	17	39	41
Core polarization (2 ħω)	-4.4	-4.4	-3.5	-6.3	-3.9
Isobars (direct only)	-4.7	0.3	-3.5	-5.2	-5.0
Relativistic	-1.5	-4.4	-2.7	-3.7	-3.4
Rest[a]	2.6	0.3	-2.3	-5.1	-4.4
Theory	-8.1	-8.2	-12.2	-20.3	-16.6
Experiment[16]	-4.4	-13.2	-13.8	-33.7	-26.3

[a]Includes core polarization (≥4 ħω), isobars (exchange diagrams) and mesonic effects from ρ-π diagram.

above and a model[14] based on Migdal theory yielded similar results only after due account is taken of core polarization (up to 2 $\hbar\omega$) and relativistic effects. At that time it was noticed that contributions from the diagrams that involved high excitations [the exchange part of isobaric current, core polarization (with energy >2 $\hbar\omega$), exchange currents etc.] showed large cancellation. This result was largely based on extensive computations.

Now it appears[13] that such large cancellations may be based on a gauge invariance argument similar to the case of QED, where current conservation leads to some stringent limitations on the renormalization of the charge. The argument is due to Rho[13] and is based on the dominance of the A_1-meson (1150 MeV). The axial-vector current may be written as

$$A_\mu = A_\mu^{(1)} + f_\pi \partial_\mu \phi \quad ,$$

where the first term arises from the A_1-dominance and the second term is the usual pion term: f_π is the pion-decay constant, ϕ the pion field. Since A_μ satisfies PCAC:

$$\partial_\mu A_\mu = f_\pi \, m_\pi^2 \, \phi$$

we have

$$\partial_\mu A_\mu^{(1)} = 0$$

and as far as the A_1-meson is concerned, the axial-vector current is conserved.

In this picture, the Gamow-Teller operator, being obtained from the space component of A_μ, derives principally from the first term $A_\mu^{(1)}$. That is, the weak decay is dominated by the A_1-meson and the only normalization of the weak vertex is due to the vacuum polarization due to the external field mediated by the A_1-meson. This will have the implication that a class of diagrams, which correspond to vertex renormalization and wave function renormalization, should sum to zero. It is likely that the large cancellation of terms arising from core-polarization, exchange terms of the isobar RPA series and meson exchange currents, is due to A_1-dominance of the axial current. Full implications of A_1-dominance in a nuclear system are not understood as yet.

Charge Density

The extent of mesonic contributions to the charge form-factors of light nuclei remains a serious problem. The most important example is the charge form-factor[17] of ^{3}He, which defies explanation. There are indications[17] that three-body forces may be responsible for the discrepancy, but this problem is far from settled.

CONCLUSIONS

These calculations within the model used here suggest that mesonic effects are large and should be included in studies of axial charge and Mλ moments and transitions. In fact pion effects dominate and hence provide a clue to the validity of chiral symmetry in nuclei.

But for Gamow-Teller giant resonances isobars appear to play a significant role. There is an interesting interplay in the roles of nucleon and isobar degrees of freedom in nuclei and this provides a great deal of richness to the study of nuclear physics.

It should be stressed that the coupling of mesons to nucleons is well understood and empirical data help to establish the coupling constants. But meson-nucleon-isobar and meson-isobar-isobar couplings are obtained by using the constituent quark model. It will be useful to find independent experimental support for the values of the coupling constants used in isobar-hole calculations.

Experimental study of Gamow-Teller resonances in (p,n) and (n,p) reactions and of Mλ resonances in (e,e′) and (p,p′) will help in understanding the nature of nuclear response to external probes. Data at higher energies and higher momentum transfer are needed.

But the best lesson to be learned from these studies is that before mesonic and isobar degrees of freedom can be isolated, a careful study of the more mundane nuclear-structure phenomena has to be made. The role of nucleons and isobars has to be considered in a model where there is no double counting. The next challenge is to isolate effects that may be attributed specifically to quarks i.e. that require a model that has to go beyond a separate treatment of nucleons and isobars.

The authors would like to express their thanks to M. Rho, M. Harvey, W. Weise, P. Guichon and J. Delorme for conversations at various stages of this work.

APPENDIX

All the technical details of the problem have been included in this appendix.

Baryon-Baryon Potential

The one-boson-exchange potential between nucleons used in the present calculation contains central, spin-spin, tensor and spin-orbit terms with the exchanged bosons being π, ρ, ω and σ; the latter is the phenomenological scalar boson providing intermediate-range attraction to the NN interaction arising from 2π-exchange processes.

The form of the potential is

$$V(r) = \frac{1}{3} m_\pi f_\pi^2(\vec{\tau}_1\cdot\vec{\tau}_2)\left\{(\vec{\sigma}_1\cdot\vec{\sigma}_2)V_{02}(x_\pi) + S_{12}V_{22}(x_\pi)\right\}$$
$$+ \frac{1}{4\pi} m_\sigma g_\sigma^2\left\{-V_{00}(x_\sigma) + \frac{m_\sigma{}^2}{4M^2} V_{02}(x_\sigma) - \frac{m_\sigma{}^2}{2M^2} \vec{L}\cdot\vec{S}\, V_{12}(x_\sigma)\right\}$$
$$+ \frac{1}{3} m_\rho f_\rho^2(\vec{\tau}_1\cdot\vec{\tau}_2)\left\{2(\vec{\sigma}_1\cdot\vec{\sigma}_2)\, V_{02}(x_\rho) - S_{12}V_{22}(x_\rho)\right\}$$
$$+ \frac{1}{4\pi} m_\rho g_\rho^2 (\tau_1\cdot\tau_2)\left\{V_{00}(x_\rho) + \frac{m_\rho{}^2}{2M^2} K\, V_{02}(x_\rho) - \frac{m_\rho{}^2}{2M^2}(3+4K)\vec{L}\cdot\vec{S}\, V_{12}(x_\rho)\right\}$$
$$+ \omega\text{-exchange} \qquad (A.1)$$

where the form of the ω-exchange potential is the same as that for ρ-exchange but without the isospin $\vec{\tau}_1\cdot\vec{\tau}_2$ factor. Here the tensor operator is $S_{12} = 3(\vec{\sigma}_1\cdot\hat{\vec{r}})(\vec{\sigma}_2\cdot\hat{\vec{r}}) - (\vec{\sigma}_1\cdot\vec{\sigma}_2)$. Monopole form factors $[(\Lambda^2-m^2)/(\Lambda^2+\vec{q}^2)]$ are introduced at the boson-nucleon-nucleon vertices in momentum space which lead to the following radial dependences in the coordinate space representation:

$$V_{00}(x) = Y_0(x) - \frac{\Lambda}{m} Y_0(x_\Lambda) - \frac{\Lambda^2-m^2}{2\Lambda m} x_\Lambda\, Y_0(x_\Lambda) \qquad (A.2)$$

$$V_{02}(x) = Y_0(x) - \frac{\Lambda^3}{m^3} Y_0(x_\Lambda) + \frac{\Lambda(\Lambda^2-m^2)}{2m^3}(2-x_\Lambda)\, Y_0(x_\Lambda) \qquad (A.3)$$

$$V_{12}(x) = \frac{1}{x} Y_1(x) - \frac{\Lambda^3}{m^3}\frac{1}{x_\Lambda} Y_1(x_\Lambda) - \frac{\Lambda(\Lambda^2-m^2)}{2m^3} Y_0(x_\Lambda) \qquad (A.4)$$

$$V_{22}(x) = Y_2(x) - \frac{\Lambda^3}{m^3} Y_2(x_\Lambda) - \frac{\Lambda(\Lambda^2-m^2)}{2m^3}(1+x_\Lambda)\, Y_0(x_\Lambda) \qquad (A.5)$$

where $Y_0(x) = e^{-x}/x$, $Y_1(x) = (1 + 1/x)\, Y_0(x)$, $Y_2(x) = (1 + 3/x + 3/x^2)\, Y_0(x)$, $x = mr$ and $x_\Lambda = \Lambda r$. The ranges of the form factors have been set at $\Lambda_\pi = 1$ GeV (consistent with NN interaction results) and $\Lambda_\rho = \Lambda_\omega = \Lambda_\sigma = 1.44$ GeV (from the monopole fit[18] to the nucleon form factor). Short-range correlations are approximately accommodated by introducing a lower cut-off in all radial integrals at $d = 0.5\ \hbar/m_\pi c$.

The coupling constants used are listed in Table 7. The constants $g_{\rho NN}$ and $g_{\omega NN}$ can be deduced from analyses of exclusive 2π and 3π production in e^+e^- annihilation[19] and vector-meson dominance: $g_{\rho NN} = 2.84$ and $g_{\omega NN} = 7.60$. Strict SU(3) would require $g_{\omega NN} = 3g_{\rho NN}$. The vector-meson-dominance model identifies K for ρ-exchange with the isovector anomalous moment, K = 3.7, and for ω-exchange with the isoscalar anomalous moment, K = -0.12. Recent developments[20] suggest stronger ρ-coupling with K = 6.6 (and $g_{\rho NN} = 2.63$)

and we adopt these values. The mass of the fictitious σ-meson has been fixed at a typically used value of m_σ = 500 MeV and its coupling constant adjusted so that the correct binding energy of the closed-shell nucleus is obtained in a Hartree-Fock calculation in a large oscillator basis. For A = 4 (with the oscillator energy fixed at $\hbar\omega$ = 20.4 MeV) we obtain $g_{\sigma NN}$ = 7.15, for A = 16 ($\hbar\omega$ = 13.3 MeV) $g_{\sigma NN}$ = 5.83 and for A = 40 ($\hbar\omega$ = 10.8 MeV) $g_{\sigma NN}$ = 5.54. Without the vertex form factors, smaller values of $g_{\sigma NN}$ would have led to the same binding.

The $N\Delta \to \Delta N$ potential is obtained from above by retaining only the π- and ρ-exchange components with the $N\Delta B$-coupling (B = boson) obtained from NNB-coupling by the transformation

$$g_{N\Delta B} = \frac{6\sqrt{2}}{5}\, g_{NNB} \; . \tag{A.6}$$

The spin (σ) and isospin (τ) operators are changed to transition spin and isospin operators.

The $N\Delta \to N\Delta$ potential has the same structure as $NN \to NN$ potential with coupling constants transformed as

$$g_{\Delta\Delta B} = \frac{1}{5}\, g_{NNB} \; . \tag{A.7}$$

The spin (σ) and isospin (τ) operators between Δ-states have to be interpreted as spin-3/2 operators.

The NNB coupling constants are given in Table 7.

Table 7. Masses and Coupling Constants for Meson-nucleon-nucleon vertices where $f^2 = \frac{1}{4\pi}\left(\frac{gm}{2M}\right)^2 (1+K)^2$ and M = nucleon mass

	m (MeV)	g	K	$g^2/4\pi$	f^2
π	138	13.49	—	14.48	0.078
ρ	766	2.63	6.6	0.55	5.30
ω	783	7.60	-0.12	4.60	0.62
σ	500	~5.5[a]	—	~2.5[a]	~0.2[a]

[a]Adjusted at mass A = 4,16,40 to reproduce binding energy of closed-shell nucleus in Hartree-Fock calculation. See text.

Meson-Exchange Current Operators

In this section we write down the non-relativistic two-body operators for magnetic moments and Gamow-Teller β-decay transitions as classifed for use in coordinate space by Chemtob and Rho.[22] The translationally invariant operators (spin terms) have the following form:

$$\begin{aligned}\vec{\mu}_2 = \frac{1}{2} g \Big\{ & (\vec{\tau}_1 \times \vec{\tau}_2)\left[(\vec{\sigma}_1 \times \vec{\sigma}_2) g_I + T_{12}^{(x)} \; g_{II}\right] \\ & + (\vec{\tau}_1 - \vec{\tau}_2)\left[(\vec{\sigma}_1 - \vec{\sigma}_2) h_I + T_{12}^{(-)} \; h_{II}\right] \\ & + (\vec{\tau}_1 + \vec{\tau}_2)\left[(\vec{\sigma}_1 + \vec{\sigma}_2) j_I + T_{12}^{(+)} \; j_{II}\right] \\ & + \left[(\vec{\sigma}_1 + \vec{\sigma}_2) \ell_I + T_{12}^{(+)} \; \ell_{II}\right] \\ & + (\vec{\tau}_1 \cdot \vec{\tau}_2)\left[(\vec{\sigma}_1 + \vec{\sigma}_2) m_I + T_{12}^{(+)} \; m_{II}\right] \Big\} \end{aligned} \tag{A.8}$$

where $g = e/2M$ the nuclear magneton for magnetic moment calculations and $g = g_A$ the axial-vector coupling constant for β-decay. (The one-body β-decay operator is $1/2 \; g\vec{\sigma}\vec{\tau}$). The operators $T_{12}^{\boxdot}$ are the D-wave parts of $(\vec{\sigma}_1 \boxdot \vec{\sigma}_2) \cdot \hat{\vec{r}}\hat{\vec{r}}$:

$$\begin{aligned} T_{12}^{\boxdot} &= \left[(\vec{\sigma}_1 \boxdot \vec{\sigma}_2) \cdot \hat{\vec{r}}\hat{\vec{r}} - \frac{1}{3} (\vec{\sigma}_1 \boxdot \vec{\sigma}_2)\right] , \qquad \boxdot = \pm, \times \\ &= - \frac{1}{3} \sqrt{8\pi} \left[Y_2(\hat{\vec{r}}), \; (\vec{\sigma}_1 \boxdot \vec{\sigma}_2)\right]^{(1)} . \end{aligned} \tag{A.9}$$

The functions g_I, g_{II}, h_I etc. are real scalar functions of $|\vec{r}|$, $\vec{r} = \vec{r}_1 - \vec{r}_2$ and are detailed below. The classification of the translationally non-invariant operators (orbital terms) we take from Hyuga, Arima and Shimizu:[6]

$$\begin{aligned}\vec{\mu}_2 = g \Big\{ & (\vec{\tau}_1 \times \vec{\tau}_2)(\hat{\vec{r}} \times \hat{\vec{R}}) \left[F_I + (\vec{\sigma}_1 \cdot \vec{\sigma}_2) F_{II} + S_{12} \, F_{III}\right] \\ & + (\vec{\tau}_1 \times \vec{\tau}_2)\left[(\vec{\sigma}_1 \cdot \hat{\vec{r}})(\vec{\sigma}_2 \times \hat{\vec{R}}) + (\vec{\sigma}_1 \times \hat{\vec{R}})(\vec{\sigma}_2 \cdot \hat{\vec{r}})\right] G \\ & + (\vec{\tau}_1 - \vec{\tau}_2)\left[\hat{\vec{R}} \times (\hat{\vec{r}} \times (\vec{\sigma}_1 + \vec{\sigma}_2))\right] J \\ & + \left[(\vec{\tau}_1 + \vec{\tau}_2) H + L + (\vec{\tau}_1 \cdot \vec{\tau}_2) M\right] \hat{\vec{R}} \times (\hat{\vec{r}} \times (\vec{\sigma}_1 - \vec{\sigma}_2)) \Big\} \end{aligned} \tag{A.10}$$

where $\vec{R} = 1/2 \; (\vec{r}_1 + \vec{r}_2)$ and F,G,H. etc. are real scalar functions of $|\vec{r}|$.

One-pion pair excitation current.

$$g_I = \frac{2}{3} \frac{M}{m_\pi} f_{\pi NN}^2 \; x_\pi^2 \; V_{12}(x_\pi) \tag{A.11}$$

$$g_{II} = -\frac{3}{2} g_I \tag{A.12}$$

$$G = \frac{M}{m_\pi} f^2_{\pi NN} X_\pi x_\pi V_{12}(x_\pi) \tag{A.13}$$

where $x_\pi = m_\pi r$, $X_\pi = m_\pi R$ and $V_{12}(x)$ is defined in Eq. (A.4). The coupling constants and meson masses are listed in Table 7.

One π-mesonic current.

$$g_I = \frac{2}{3}\frac{M}{m_\pi} f^2_{\pi NN} x^2_\pi W_0(x_\pi) \tag{A.14}$$

$$g_{II} = -\frac{M}{m_\pi} f^2_{\pi NN} x^2_\pi W_{12}(x_\pi) \tag{A.15}$$

$$G = -\frac{M}{m_\pi} f^2_{\pi NN} X_\pi x_\pi W_{12}(x_\pi) \tag{A.16}$$

$$F_{II} = \frac{1}{3}\frac{M}{m_\pi} f^2_{\pi NN} X_\pi x_\pi W_{02}(x_\pi) \tag{A.17}$$

$$F_{III} = \frac{1}{3}\frac{M}{m_\pi} f^2_{\pi NN} X_\pi x_\pi W_{22}(x_\pi) \tag{A.18}$$

where

$$W_0(x) = \frac{1}{x^2}(x-2)Y_0(x) + \frac{\Lambda^3}{m^3}\frac{1}{{x_\Lambda}^2}(x_\Lambda-2)Y_0(x_\Lambda) - \frac{4m^2}{\Lambda^2-m^2}\left\{\frac{1}{x^2}Y_0(x) - \frac{\Lambda^5}{m^5}\frac{1}{{x_\Lambda}^2}Y_0(x_\Lambda)\right\} \tag{A.19}$$

$$W_{12}(x) = \frac{1}{x}Y_1(x) + \frac{\Lambda^3}{m^3}\frac{1}{x_\Lambda}Y_1(x_\Lambda) - \frac{4m^2}{\Lambda^2-m^2}\left\{\frac{1}{x^2}Y_2(x) - \frac{\Lambda^5}{m^5}\frac{1}{{x_\Lambda}^2}Y_2(x_\Lambda)\right\} \tag{A.20}$$

$$W_{02}(x) = Y_0(x) + \frac{\Lambda^3}{m^3}Y_0(x_\Lambda) - \frac{4m^2}{\Lambda^2-m^2}\left\{\frac{1}{x}Y_3(x) - \frac{\Lambda^5}{m^5}\frac{1}{x_\Lambda}Y_3(x_\Lambda)\right\} \tag{A.21}$$

$$W_{22}(x) = Y_2(x) + \frac{\Lambda^3}{m^3}Y_2(x_\Lambda) - \frac{4m^2}{\Lambda^2-m^2}\left\{\frac{1}{x}Y_4(x) - \frac{\Lambda^5}{m^5}\frac{1}{x_\Lambda}Y_4(x_\Lambda)\right\} \tag{A.22}$$

with $Y_3(x) = (1 + 3/x + 6/x^2 + 6/x^3)\ Y_0(x)$ and $Y_4(x) = (1 + 6/x + 15/x^2 + 15/x^3)\ Y_0(x)$. Note that without form factors ($\Lambda \to \infty$) the radial functions $W_{12}(x)$ and $V_{12}(x)$ both reduce to $Y_1(x)/x$ and the translationally noninvariant term G cancels between the π-pair and π-mesonic currents.

One ρ-meson pair excitation current and gauge term.

$$g_I = \frac{2}{3}\frac{M}{m_\rho} f_{\rho NN}^2 x_\rho^2 V_{12}(x_\rho) \tag{A.23}$$

$$g_{II} = -\frac{3}{2} g_I \tag{A.24}$$

$$m_I = 2j_I = -\frac{1}{3}\frac{m_\rho}{M}\frac{g_{\rho NN}^2}{4\pi}(1+K)\, x_\rho^2 V_{12}(x_\rho) \tag{A.25}$$

$$m_{II} = 2j_{II} = -\frac{3}{2} m_I \tag{A.26}$$

$$F_I = \frac{1}{2}\frac{m_\rho}{M}\frac{g_{\rho NN}^2}{4\pi}(1+2K)\, X_\rho x_\rho V_{12}(x_\rho) \tag{A.27}$$

$$F_{II} = -2G = 2\frac{M}{m_\rho} f_{\rho NN}^2 X_\rho x_\rho V_{12}(x_\rho) \tag{A.28}$$

$$J = -\frac{1}{4}\frac{m_\rho}{M}\frac{g_{\rho NN}^2}{4\pi}(1+K)\, X_\rho x_\rho V_{12}(x_\rho)\,. \tag{A.29}$$

One ρ-mesonic current.

$$g_I = \frac{4}{3}\frac{M}{m_\rho} f_{\rho NN}^2 \left(1+\frac{1}{2}\kappa_\rho\right) x_\rho^2 W_0(x_\rho) \tag{A.30}$$

$$g_{II} = \frac{M}{m_\rho} f_{\rho NN}^2 (1+2\kappa_\rho)\, x_\rho^2 W_{12}(x_\rho) \tag{A.31}$$

$$G = \frac{M}{m_\rho} f_{\rho NN}^2 X_\rho x_\rho W_{12}(x_\rho) \tag{A.32}$$

$$F_I = \frac{1}{4}\frac{m_\rho}{M}\frac{g_{\rho NN}^2}{4\pi} X_\rho x_\rho \left[\left\{\left(\frac{2M}{m_\rho}\right)^2 + 2K\right\} W_{02}(x_\rho) - 4K\, W_{12}(x_\rho)\right] \tag{A.33}$$

$$F_{II} = \frac{M}{m_\rho} f_{\rho NN}^2 X_\rho x_\rho \left[\frac{2}{3} W_{02}(x_\rho) - 2W_{12}(x_\rho)\right] \tag{A.34}$$

$$F_{III} = -\frac{1}{3}\frac{M}{m_\rho} f_{\rho NN}^2 X_\rho x_\rho W_{22}(x_\rho) \tag{A.35}$$

where κ_ρ is the anomalous magnetic moment of the ρ-meson, $\kappa_\rho \sim 0.14$ in the static quark model.

The ρ-π current. For vector currents, the results are:

$$m_I = 2\zeta_\rho \frac{m_\pi^3}{m_\rho^3} Z_0(x_\pi, x_\rho) \tag{A.36}$$

$$m_{II} = 6\zeta_\rho \frac{m_\pi^3}{m_\rho^3} Z_2(x_\pi, x_\rho) \tag{A.37}$$

where $\zeta_\rho = -g_{\rho NN} g_{\rho\pi\gamma} g_{\pi NN}/12\pi$ and $g_{\rho\pi\gamma} = 0.406$ from Nagels et al.[21]

The radial functions are

$$Z_{0,2}(x_\pi,x_\rho) = a_1Y_{0,2}(x_\pi) - a_2Y_{0,2}(x_\rho) + a_3Y_{0,2}(x_{\Lambda\pi}) - a_4Y_{0,2}(x_{\Lambda\rho}) \tag{A.38}$$

with

$$a_1 = \frac{\left(\Lambda_\rho^2-m_\rho^2\right)m_\rho^2}{\left(m_\rho^2-m_\pi^2\right)\left(\Lambda_\rho^2-m_\pi^2\right)} \tag{A.39}$$

$$a_2 = \left(\frac{m_\rho}{m_\pi}\right)^3 \frac{\left(\Lambda_\pi^2-m_\pi^2\right)m_\rho^2}{\left(m_\rho^2-m_\pi^2\right)\left(\Lambda_\pi^2-m_\rho^2\right)} \tag{A.40}$$

$$a_3 = \left(\frac{\Lambda_\pi}{m_\pi}\right)^3 \frac{\left(\Lambda_\rho^2-m_\rho^2\right)m_\rho^2}{\left(\Lambda_\rho^2-\Lambda_\pi^2\right)\left(\Lambda_\pi^2-m_\rho^2\right)} \tag{A.41}$$

$$a_4 = \left(\frac{\Lambda_\rho}{m_\pi}\right)^3 \frac{\left(\Lambda_\pi^2-m_\pi^2\right)m_\rho^2}{\left(\Lambda_\rho^2-m_\pi^2\right)\left(\Lambda_\rho^2-\Lambda_\pi^2\right)} \tag{A.42}$$

For axial vector currents, the results are:

$$g_I = \frac{2}{3}\,\alpha\xi\; Z_0(x_\pi,x_\rho) \tag{A.43}$$

$$g_{II} = -\alpha\xi\; Z_2(x_\pi,x_\rho) \tag{A.44}$$

where $\xi = g_{\pi NN}\, m_\pi^3/(8\pi g_A M)$ and $\alpha = g_{\rho NN}\, g_{\rho\pi A}(1+K)/(Mm_\rho^2)$. The ρ-π-A coupling constant is given according to Chemtob and Rho[22] by the KSFR relation: $g_{\rho\pi A} = g_{\pi NN}\, m_\rho^2/(2Mg_A g_{\rho NN})$.

One ω-meson pair excitation current.

$$\ell_I = 2j_I = -\frac{1}{3}\frac{m_\omega}{M}\frac{g_{\omega NN}^2}{4\pi}\, x_\omega^2\, V_{12}(x_\omega) \tag{A.45}$$

$$\ell_{II} = 2j_{II} = -\frac{3}{2}\,\ell_I \tag{A.46}$$

$$J = \frac{1}{4}\frac{m_\omega}{M}\frac{g_{\omega NN}^2}{4\pi}\,(1+K)\; X_\omega x_\omega\; V_{12}(x_\omega) \tag{A.47}$$

The ω-π current.

$$h_I = j_I = \zeta_\omega \frac{m_\pi^3}{m_\omega^3}\, Z_0(x_\pi,x_\omega) \tag{A.48}$$

$$h_{II} = j_{II} = 3\zeta_\omega \frac{m_\pi^3}{m_\omega^3} Z_2(x_\pi, x_\omega) \tag{A.49}$$

where $\zeta_\omega = -g_{\omega NN}\, g_{\omega\pi\gamma}\, g_{\pi NN}/12\pi$ and $g_{\omega\pi\gamma} = 2.03$ from Nagels et al.[21]

All two-body matrix elements are evaluated with harmonic oscillator wave functions transformed into relative and center-of-mass coordinates. Short-range correlations are introduced, as was the case for matrix elements of the residual interaction, by introducing a lower limit cut-off at $d = 0.5\ \hbar/m_\pi c$ in the radial integral in the relative coordinate, but not in the center-of-mass coordinate.

Core Polarization

Core polarization effects are included up to second order in the two-body potential V. The role of Hartree-Fock diagrams is quite small.

Core polarization with isobars in the particle states is considered up to second-order perturbation theory with the restriction that intermediate states have no more than one isobar particle.

Relativistic Effects

The Gamow-Teller operator is corrected for relativistic effects by including terms of order $1/M^2$; M is the nucleon mass.

For magnetic moments and M1-transitions the relativistic corrections are of order $1/M^3$ and are not included.

REFERENCES

1. T. Hamada and I. D. Johnston, Nucl. Phys. 34:383 (1962).
2. K. Erkelenz, Phys. Rep. 13:193 (1974); K. Hohlinde, Phys. Rep. 68:121 (1981); M. Lacombe, B. Loiseau, J. M. Richard, R. Vinh Mau, J. Cote, P. Pires and R. de Tourreil, Phys. Rev. C21:861 (1980).
3. K. Kubodera, J. Delorme and M. Rho, Phys. Rev. Lett. 40:755 (1978).
4. L. Palffy, J. P. Deutsch, L. Grenacs, J. Lehmann and M. Steels, Phys. Rev. Lett. 34:212 (1975); C. A. Gagliardi, G. T. Garvey, J. R. Wrobel and S. J. Freedman, Phys. Rev. Lett. 48:914 (1982).
5. P. Guichon, B. Bihoreau, M. Giffon, A. Goncalves, J. Julien, L. Roussel and C. Samour, Phys. Rev. C19:987 (1979).
6. I. S. Towner and F. C. Khanna, Nucl. Phys. A372:331 (1981); P. Guichon, M. Giffon, J. Joseph, R. Laverriere and C. Samour, Z. Phys. A285:183 (1978); P. Guichon and C. Samour, Phys.

Lett. 82B:28 (1979); K. Koshigiri, H. Ohtsubo and M. Morita, Prog. Theor. Phys. 62:706 (1979); W. K. Cheng, B. Lorazo and B. Goulard, Phys. Rev. C21:374 (1980).

7. P. A. M. Guichon and C. Samour, Nucl. Phys. A382:461 (1982).
8. D. O. Riska and G. E. Brown, Phys. Lett. 38B:193 (1972).
9. For more details on these aspects see I. S. Towner and F. C. Khanna, to be published.
10. W. Steffen, H. D. Gräf, W. Cross, D. Meuer, A. Richter, E. Spamer, O. Tilze and W. Knüpfer, Phys. Lett. 95B:23 (1980).
11. J. B. McGrory and B. H. Wildenthal, Phys. Lett. 103B:173 (1981).
12. C. D. Goodman, Nucl. Phys. A374:241c (1982) and the references quoted therein.
13. M. Rho, private communication; talk at Amsterdam Conference (1982).
14. E. Oset and M. Rho, Phys. Rev. Lett. 42:47 (1979).
15. I. S. Towner and F. C. Khanna, Phys. Rev. Lett. 42:51 (1979). This contains an earlier version of the calculation of the quenching of Gamow-Teller transitions. For the latest results see I. S. Towner and F. C. Khanna, to be published.
16. G. Azuelos and J. E. Kitching, Nucl. Phys. A285:19 (1977); S. Raman, C. A. Houser, T. A. Walkiewicz and I. S. Towner, Atomic Data and Nuclear Data Tables 21:567 (1978).
17. E. Hadjimichael, R. Bornais and B. Goulard, Phys. Rev. Lett. 48:583 (1982).
18. F. Iachello, A. Jackson and A. Lande, Phys. Lett. 43B:191 (1973).
19. A. Silverman, in: "Proc. of the 7th Int. Symposium on Lepton and Photon Interactions at High Energies, Stanford, 1975," W. T. Kirk, ed., SLAC, Stanford (1975); M. Chemtob, in: "Mesons in Nuclei," D. H. Wilkinson and M. Rho, eds., North-Holland, Amsterdam (1979).
20. G. Höhler and E. Pietarinen, Nucl. Phys. B95:210 (1975).
21. M. M. Nagels et al., Nucl. Phys. B147:189 (1979).
22. M. Chemtob and M. Rho, Nucl. Phys. A163:1 (1971).

NUCLEAR STRUCTURE, DOUBLE BETA DECAY AND GIANT RESONANCES

L. Zamick

Physics Department
Rutgers University
Piscataway, N.J. 08854, USA

SHELL MODEL

I will start by talking about the nuclear shell model. To keep things interesting I will apply this to a topic of current interest—double beta decay.

At a conference such as this we look for signals that the nucleus does not merely consist of neutrons and protons. Such signals have up to now been surprisingly sparce, but they are there.

But why look at large complicated nuclei, where the wave functions are ill understood? Let me cite a few examples to justify the study of large nuclei.

Take as an example magnetic moments, please. I think it was Dieter Kurath who remarked that below mass 24 there are no anomalous magnetic moments. More precisely the deviations from shell model calculations are very small. But in heavier nuclei the deviations can become very large. For example for the ground state of ^{209}Bi the measured magnetic moment is 4.080 μ_N, while the single particle model (Schmidt) gives a value of 2.657 μ_N. Ironically ^{208}Pb is considered to be the best closed shell nucleus in the periodic table.

(We now understand that because of meson exchange currents and core polarization the orbital g factor changes from 1 to about 1.1. In ^{209}Bi the correction is $\delta g_e \ell$. The orbit is $h_{9/2}$ with $\ell = 5$. This large ℓ value provides the amplification of the effect.)

As another example, the clearest signal for parity violation in nuclear wave functions occurs in heavy nuclei, especially in

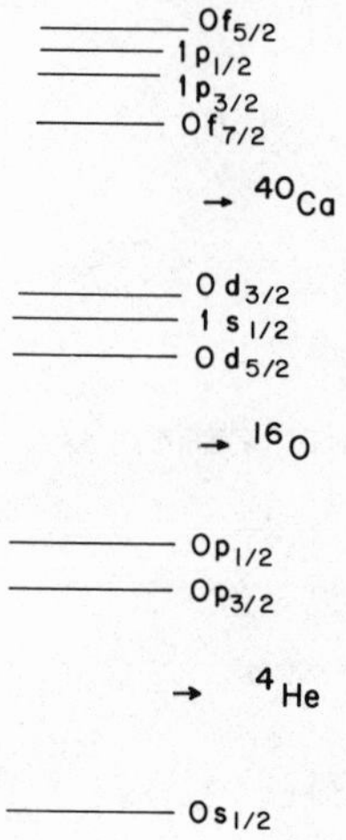

Fig. 1. Single particle levels in a nuclear shell model potential.

^{180}Hf—a strongly deformed nucleus. The relevant states are $J = 8^+$ and $J = 8^-$. The calculated mixing of these two basis states is very difficult because the 8^+ state belongs to a K=0 rotational band and the 8^- to a K=8 rotational band.

I also hope to show that the double beta decay process is enhanced in large nuclei where pairing correlations are important.

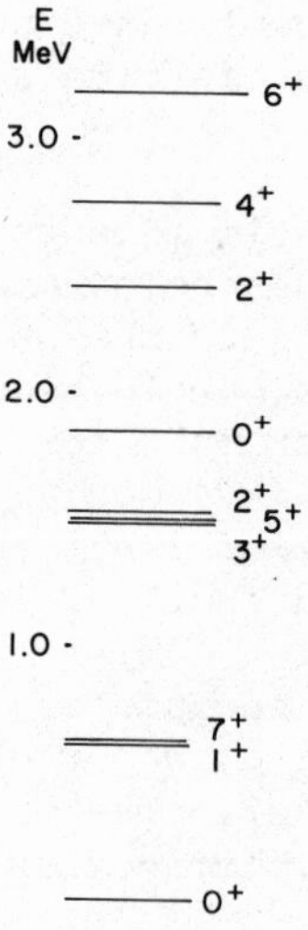

Fig. 2. Spectrum of ^{42}Sc (not all levels shown).

Example of a Shell Model Calculation

Since this is a school it is appropriate to discuss some of the techniques of shell model calculations. I will therefore discuss an old calculation performed by John McCullen, Ben Bayman and myself[1] (MBZ). This calculation was also performed by Ginocchio and French[2] at about the same time.

We classify the single particle levels in j-j coupling as $n\ell j$ e.g. $0s_{1/2}$, $0p_{3/2}$, $0p_{1/2}$, $0d_{5/2}$, $1s_{1/2}$, $0d_{3/2}$, $0f_{7/2}$, $1p_{3/2}$, $0f_{5/2}$, $1p_{1/2}$ etc. Here n is the number of nodes in the radial wave function, ℓ the orbital angular momentum and j the total angular momentum of the single particle orbit. See Fig. 1.

We can put 2 neutrons and 2 protons in the $0s_{1/2}$ orbit forming the closed shell or magic nucleus ^{4}He. We next fill the $0p_{3/2}$ and $0p_{1/2}$ shells forming the next magic nucleus ^{16}O. (We can fill just the $0p_{3/2}$ shell to form ^{12}C. But there is a great deal of configuration mixing in ^{12}C, i.e. excitations from $p_{3/2}$ to $p_{1/2}$, so it cannot be regarded as a good closed shell.)

We next fill the 1s,0d shell and reach the next magic nucleus ^{40}Ca.

The aim of the MBZ calculation was to try to describe many low lying levels beyond ^{40}Ca in terms of neutrons and protons in the $f_{7/2}$ shell. The nuclei covered range from ^{40}Ca to ^{56}Ni.

The prediction for the ground state of ^{41}Ca is simple. We put a particle into an $f_{7/2}$ orbit and therefore predict that the ground state spin is $J = 7/2$. This is correct. The magnetic moment of a neutron in an $f_{7/2}$ orbit should be the same as that of a free neutron $-1.91\ \mu_N$. In fact the measured value is -1.59. In calculating the magnetic moments of more complex nuclei we use the effective magnetic moments of ^{41}Ca and ^{41}Sc. (That the $f_{7/2}$ moment should be equal to the free moment is evident from the fact that, by definition, the moment is taken in a state with $M = j = 7/2$. The wave function for this state factorizes as $Y_{3,3}\ \chi\uparrow$ i.e. a pure spin-up neutron state.)

When we add two nucleons to the $f_{7/2}$ shell we quickly run into an ambiguity. We can form a wave function $f_{7/2,m_1}(1)\ f_{7/2,m_2}(2)$. At this level the ground state of ^{42}Sc (one neutron and one proton beyond ^{40}Ca) is 64-fold degenerate.

This degeneracy is removed by a residual interaction. Just as the high energy speakers have told us that their criteria for the correct Lagrangian for QCD are those that fit experiment (they are not derived from fundamentals, as many people seem to think), so in the nuclear case we chose a Lagrangian or Hamiltonian to fit

Table 1. Matrix Elements of the Effective Interaction

	J	V^J	
T=1	0	0	
	2	1.5246	(1.5864 in ^{42}S)
	4	2.7523	
	6	3.1893	
T=0	1	0.611	
	3	1.490	
	5	1.511	
	7	0.617	

experiment. Indeed, in this old calculation, we followed Talmi,[3] who had suggested taking the matrix elements to fit the spectrum of the two-body system ^{42}Ca or ^{42}Sc.

Since the interaction is a scalar under rotations, the resulting eigenfunctions will have good total angular momentum. We can construct states of good angular momentum using Clebsch-Gordan or Wigner coefficients

$$[f_{7/2}(1)\ f_{7/2}(2)]^J_M = \sum_{m_1 m_2} (jjm_1m_2|JM) f_{7/2,m_1}(1)\ f_{7/2,m_2}(2)\ .$$

We determine the matrix elements of the interaction from the spectrum of ^{42}Sc or ^{42}Ca. See Fig. 2.

$$V^J = \langle (f^2_{7/2})^J\ V(12)\ (f^2_{7/2})^J \rangle\ .$$

The values are shown in Table 1. Some comments on the Table are in order. Using the property of a Clebsch-Gordan coefficient

$$(j_1j_2m_1m_2|JM) = (-1)^{J-j_1-j_2}\ (j_2j_1m_2m_1|JM)$$

we find that

$$[f_{7/2}(1)f_{7/2}(2)]^{JT} = (-1)^{J+T}\ [f_{7/2}(2)f_{7/2}(1)]^{JT}\ .$$

The state must be antisymmetric under the interchange of the two nucleons. Hence $(-1)^{J+T} = -1$. Thus, if J is even, the isospin T=1, if J is odd, T=0. For the two neutron system $T_z = (N-Z)/2 = +1$. Since $T \geq |T_z|$ we must have T=1. Hence for ^{42}Ca we can only have states of even angular momentum for the $f^2_{7/2}$ configuration J = 0,2,4 and 6. For ^{42}Sc because of charge independence we expect to have these even states as well, with the same energy splitting as in ^{42}Ca, but we also expect additional states of odd angular momentum i.e. $J = 1,3,5$ and 7.

Indeed there is a striking difference between the spectra of ^{42}Sc and ^{42}Ca; ^{42}Sc does have these additional low lying odd spin states. (When the MBZ calculations were done the energies of the $J = 3^+, 5^+$ and 7^+ were wrongly taken to be 2.25, 1.95 and 1.008 MeV respectively. This is because the experimental situation was not clear in 1963. Remarkably this misassignment did not seem to make very much difference; as borne out by subsequent calculations with correct matrix elements.)

In the above we set the matrix element of the J=0 state to zero energy. We can add a constant to all the matrix elements without affecting the wave functions or relative spacings. This constant is the true interaction of two nucleons in the J=0 ground state of ^{42}Ca. It is obtained from the masses or binding energies of various calcium isotopes.

$$E^0 = -[BE(^{42}Ca) + BE(^{40}Ca) - 2\,BE(^{41}Ca)] = -3.1 \text{ MeV} \quad .$$

We can then use $\overline{V}^J = V^J + E^0$ instead of V^J. We would then obtain predictions for the binding energies of the heavier nuclei.

In examining the spectrum of ^{42}Ca we see that there is some ambiguity in taking the matrix elements from experiment.[1] Between the first 2^+ state at 2.42 MeV and the first 4^+ state at 2.75 MeV there are two so-called intruder states, the 0_2^+ state at 1.84 MeV and the 2_2^+ state at 2.42 MeV. They have been associated by Gerace and Green[4] (in analogy with earlier work by Brown and Green[5] in the oxygen region) with four particle-two hole deformed states, with the four particles in the 0f-1p shell and two holes in the 0d-1s shell. Indeed, Gerace and Green believe the 2_1^+ and 2_2^+ state have approximately equal admixtures of two particle and four particle-two hole states. There is also evidence of considerable mixing of the 4p-2h and 2p 0^+ states, coming from the observed rather strong E(0) transition from the 0_2^+ state at 1.84 MeV to the ground state. Since the operator is $\sum_i r^2(i)$ i.e. a one-body operator, there would be no E(0) transition if the ground state were (2p) and the excited 0^+ were pure 4p-2h.

We will not be talking too much more about deformed states, except to remark that they could have a large effect on the currently interesting problem of the quenching of Gamow-Teller transitions e.g. the $^{41}Si \rightarrow {}^{41}Ca\ 7/2_i \rightarrow 7/2_i$ transition, considered by Khanna and Towner.[6] The introduction of deformed components will be a source of additional quenching, over and above what the above authors calculate due to other mechanisms. Since the authors obtain too little quenching, the overall result of the deformed states will be to bring the calculated results closer to experiment.

More than Two Particles in the $f_{7/2}$ Shell

Consider first identical nucleons i.e. neutrons. The Pauli principle is important. We already saw that for the $f^2_{7/2}$ configuration only even J's are allowed.

In the ^{43}Ca $f^3_{7/2}$ configuration only states with J = 3/2,5/2,7/2, 9/2,11/2 and 15/2, each occurring only once, can satisfy the Pauli principle. It is easy, for example, to see why one cannot have J=21/2. The J=21/2 M=21/2 state is $f_{7/2,7/2}(1)\ f_{7/2,7/2}(2)\ f_{7/2,7/2}(3)$. This is clearly symmetric.

For n=4 we have J=0, v=0, J=2,4,6 v=2 and J=2,4,5,8 v=4. Here we have introduced the seniority quantum number v which is roughly the number of particles not coupled to J=0 (see Prof. Talmi's lectures for more details). Thus for ^{41}Ca v=1, and in ^{42}Ca v=0 for J=0 and v=2 for J=2,4 and 6. In ^{43}Ca we can form the J=7/2$^-$ state by coupling two nucleons to J=0 and adding the 3rd; thus v=1. For J = 3/2,5/2,9/2,11/2 and 15/2 we cannot couple two particles to J=0; hence v=3.

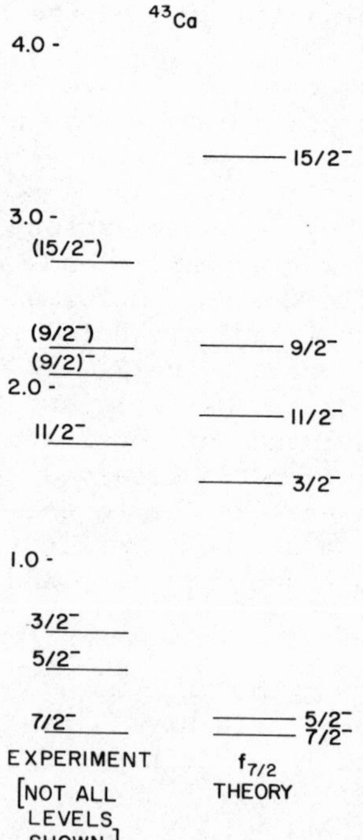

Fig. 3. Levels of ^{43}Ca compared with the $f^3_{7/2}$ theory.

In ^{44}Ca, where states of a given angular momentum can occur more than once, seniority is still a good quantum number e.g. $\langle J=2\ v=2|V(12)|J=2\ v=4\rangle = 0$. For $j > 7/2$ this would no longer be true.

The states and spectrum of ^{45}Ca, three $f_{7/2}$ holes, is identical to that of ^{43}Ca three $f_{7/2}$ particles. (Likewise ^{46}Ca and ^{42}Ca.)

The spectrum of the calcium isotopes was obtained by Talmi.[3] If we examine the spectrum of ^{43}Ca, see Fig. 3, we see a fairly good fit with the glaring exception of the $J = 3/2$ state being predicted to lie almost 1 MeV higher than it actually does. This could be due to the importance of the $p_{3/2}$ admixture.

This example affords an opportunity to study the strengths and weaknesses of the approach being taken. When we take the two-body matrix elements from experiment we are incorporating all sorts of complicated effects i.e., the effects of hard-core correlations, the exchange of bosons between two nucleons, the exchange of a phonon between two nucleons (Bertsch-Kuo-Brown bubble) etc. See Fig. 4(a). When we come to three particles we still take a tremendous amount of renormalization into account. But some things are left out, for example the diagram, Fig. 4(b). This looks like an effective three-body interaction, and indeed it is—it describes the admixture of $f^2_{7/2}\ p_{3/2}$ into a basic $f^3_{7/2}$ configuration.

This is not a fundamental three-body interaction. If we replace the $p_{3/2}$ nucleon by a $\Delta(J = 3/2,\ T = 3/2)$ we would have a relatively more fundamental three-body interaction.

This example illustrates the difficulty of getting evidence of three-body interactions in nuclei. When we cannot get the spectrum of a three-body system from that of a two-body system, the explanation could be that we are suppressing *nucleon* degrees of freedom (likely to be Δ), but it could also be, and is more likely, that we are suppressing *nuclear* degrees of freedom by using too small a model space.

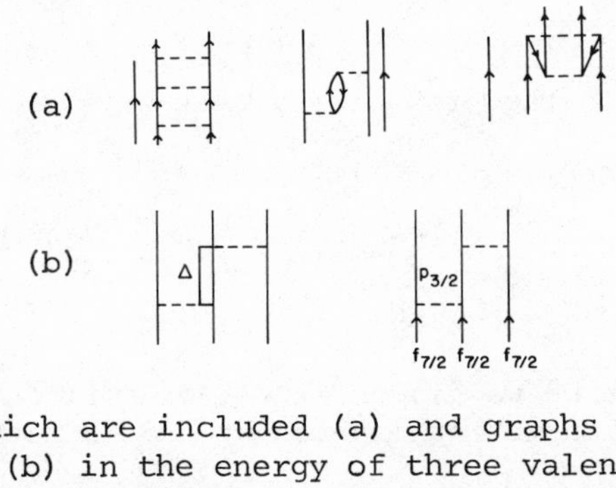

Fig. 4. Graphs which are included (a) and graphs which are not included (b) in the energy of three valence nucleons when two-body matrix elements are taken from experiment.

In the $f_{7/2}$ shell for neutrons only, i.e. the calcium isotopes, each state of given angular momentum and seniority occurs only once. The wave function for given J,v is unique and can be constructed without knowledge of the residual interaction.

A convenient way of characterizing the wave functions is in terms of fractional parentage coefficients (see Bayman and Lande).[8]

$$\psi(j^n\ Jv) = \sum_{J_0 v_0} (j^{n-1}\ J_0 v_0 j|\}j^n\ Jv)\ \left[(j^{n-1})^{J_0 v_0}\ j(n)\right]^{Jv}$$

We expand the wave function, which is antisymmetric in n nucleons, in terms of a larger class of basis states which are necessarily antisymmetric only in (n-1) nucleons. This is useful for evaluating the expectation value of one-body operators. Despite the fact that the wave function is antisymmetric in all n nucleons we manage to single out the nth nucleon in this expression.

Neutrons and Protons in the $f_{7/2}$ Shell

When we have several neutrons and protons in the $f_{7/2}$ shell we can get several states of given angular momentum J. We now need the residual interaction to obtain the wave functions. In the $f_{7/2}$ model, the residual interaction is just the eight numbers $\overline{V}^J$ (or $\overline{V}^J$) obtained from the spectrum of ^{42}Sc and ^{42}Ca.

Take as an example the J=2 states in $^{44}_{21}Sc_{23}$. We have 3 valence $f_{7/2}$ neutrons and one $f_{7/2}$ proton. We can use as basis states $(L_n\ j)^{J=2}$ where L_N is the angular momentum of the 3 neutrons and $j = 7/2$ the angular momentum of the proton. The allowed basis states are obtained from the triangle relation $\vec{L}_N + \vec{7}/2 = 2$, i.e. $|L_N - 7/2| \leq 2 \leq (L_N + 7/2)$. Thus the allowed basis states are $(3/2\ 7/2)^2$ $(5/2\ 7/2)^2$ $(7/2\ 7/2)^2$ $(9/2\ 7/2)^2$ and $(11/2\ 7/2)^2$.

We write our wave function as

$$\psi^\alpha = \sum_{L_N} D^\alpha(L_N,j)\ (L_N,j)^{J=2}\ .$$

Here $|D^\alpha(L_N,j)|^2$ is the probability that in the $\alpha + h$ J=2 state the three neutrons couple to angular momentum L_N. The D^α satisfy the normalization conditions

$$\sum_{L_N} D^\alpha(L_N,j)|^2 = 1\ ; \qquad \sum_{L_N} D^\alpha(L_N,j)\ D^{\alpha'}(L_N,j) = \delta_{\alpha\alpha'}\ .$$

To obtain the D^α we diagonalize the Hamiltonian H. In this case the Hamiltonian is represented by a 5×5 matrix with matrix elements

$$H_{(L_N,j),(L_N',j)} = \left\langle \left[\left(f^3_{7/2_\nu}\right)^{L_N j}\right]^{J=2} \sum_{i<j} V(ij) \left[\left(f^3_{7/2_\nu}\right)^{L_N' j}\right]^{J=2} \right\rangle .$$

The wave function is represented by a column vector whose elements are the $D(L_N,j)$. The equation reads

$$5 \times 5 \quad \left[H_{(L_N j),(L_N' j)} - E^\alpha \, \delta_{L_N L_N'} \right] \begin{bmatrix} D^\alpha\ (3/2,j) \\ D^\alpha\ (5/2,j) \\ D^\alpha\ (7/2,j) \\ D^\alpha\ (9/2,j) \\ D^\alpha (11/2,j) \end{bmatrix} = 0 \ .$$

The E^α are the eigenvalues. There will be five eigenvalues and correspondingly 5 eigenfunctions.

The many-body matrix element $H_{(L_N j),(L_N' j)}$ can be expressed as a linear combination of the two-body matrix elements $\langle (f^2_{7/2})^L\ V(12)\ (f^2_{7/2})^L \rangle$ which, we recall we identify with V^L, the energy levels of ^{42}Sc (^{42}Ca). That is

$$H_{(L_N,j),(L_N',j)} = \sum_L f^{J=2}\ (L_N,L_N',L) V^L \ .$$

The technique of evaluating f will not be given here; it is outlined clearly in MBZ. The evaluation is facilitated by the use of the (normalized) Racah coefficients defined by

$$\left[(j_A j_B)^{J_{AB}}\ j_C\right]^J = \sum_{J_{BC}} U(j_A j_B J j_C; J_{AB} J_{BC}) \left[j_A (j_B j_C)^{J_{BC}}\right]^J \ .$$

That is, if we have three particles in various j orbits we can form basis states of total J in different ways. We can couple the first two particles to angular momentum J_{AB} and then couple the third to the first two to form a total J; or we can first couple the second and third particle to J_{BC}, and then couple the first one to this combination to form the same total J. The normalized Racah coefficients U are elements of a unitary transformation which relates one set of basis states to the other.

But the important thing to remember is that we can express the many-body matrix element of the interaction in terms of the spectrum of ^{42}Sc (^{42}Ca).

We list in Table 2 the wave functions of the J=2 states of ^{44}Sc, as well as the energies.

Some comments are in order. Previous to these calculations there had been some naive ideas that since we worked in the $f_{7/2}$ shell the lowest state in ^{44}Sc would correspond to three neutrons

Table 2. $D(L_P = j, L_N)$ ^{44}Sc J = 2 States

E(MeV)		0.0	2.986	4.317	4.854	6.661
L_P	L_N			T = 2		T = 2
7/2	3/2	-0.144	0.146	-0.254	0.761	-0.561
7/2	5/2	0.685	0.470	0.524	0.098	-0.158
7/2	7/2	0.713	-0.398	-0.577	0.019	0.000
7/2	9/2	-0.020	-0.456	0.278	-0.380	-0.755
7/2	11/2	0.042	-0.625	0.500	0.516	0.302

coupled to $L_N = 7/2$. We see that for the lowest state (at 0.0 MeV) the wave function is approximately 0.685 $\left[f_{7/2}(f^3_{7/2})^{7/2}\right]^{J=2}$. There is considerable mixing. The seniority of the neutrons is not preserved when neutrons and protons are present.

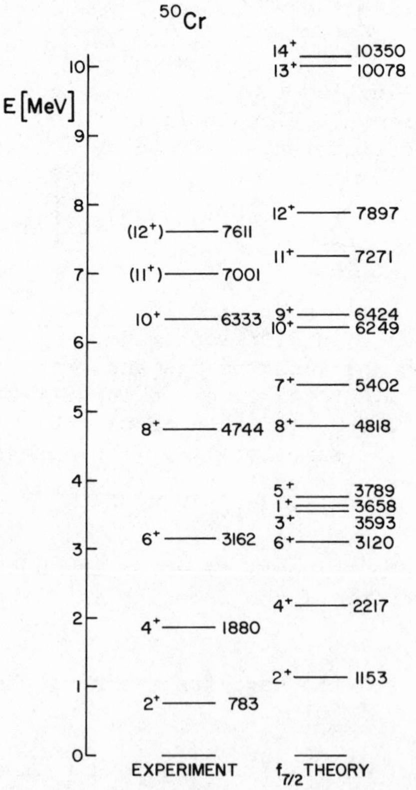

Fig. 5. Levels of $^{50}_{24}Cr_{26}$ compared with the $f_{7/2}$ model.

For ^{44}Sc $T_Z = (N-Z)/2 = 1$. Hence for four nucleons the possible values of isospin are 1 and 2. There are two T=2 J=2 states, at 4.317 MeV and 6.661 MeV. We know that there must be two because these $\psi^{T=2}_{T_Z=1}$ states are analogues of states in ^{44}Ca i.e. $\psi^{T=2}_{T_Z=2}$. We know that in ^{44}Ca there are two J=2 states with seniorities v=2 and v=4 respectively. The T=2 wave functions in ^{44}Sc can be obtained from those in ^{44}Ca by the isospin lowering operator $\psi(^{44}\mathrm{Sc})^{T=2}_{M_T=1} = T_-\psi(^{44}\mathrm{Ca})^{T=2}_{M_T=2}$. Actually a little thought will convince us that (up to a phase) the $D(L_P=j, L_N)$ for the T=2 states in ^{44}Sc are simply the fractional parentage coefficients $[(j^3)^{L_N}\, j\,|\}\, j^4\, J=2]$.

Just to show that things hang together we give the experimental and calculated spectrum of ^{50}Cr whose configuration is $(f^4_{7/2})_\pi(f^6_{7/2})_\nu$, see Fig. 5. Note that the agreement between theory and experiment is quite good, except for the fact that the 2^+_1 level does not come down low enough. Not all levels have been seen, especially those of odd-angular momentum. The highest level seen experimentally is $J=12^+$. The $f_{7/2}$ model also predicts states with angular momentum 13 and 14 at higher energies.

Application to Magnetic Moments

The magnetic moment of a free neutron is $-1.913\ \mu_N$ and of a proton $2.793\ \mu_N$. The magnetic moment of a nuclear state of angular momentum J is defined as

$$\mu = \langle \psi^J_{M=J}\ \hat{\mu}_z\ \psi^J_{M=J} \rangle$$

where

$$\hat{\mu}_z = g_\ell\ \ell_z + g_s\ s_z$$

$$g_\ell = \begin{matrix} 1 \\ 0 \end{matrix} \qquad g_s = \begin{matrix} 2\times(\ 2.793) & \text{proton} \\ 2\times(-1.913) & \text{neutron} \end{matrix} \quad .$$

For an $f_{7/2}$ neutron (^{41}Ca)

$$\mu = \langle f_{7/2,7/2}\ \hat{\mu}_z\ f_{7/2,7/2} \rangle \quad .$$

But $f_{7/2,7/2}$ factorizes to $Y_{3,M=3}\ \chi\uparrow$ i.e. is a pure spin-up wave function. We thus get the prediction $\mu(^{41}\mathrm{Ca})_{\text{ground state}} = \mu_{\text{free neutron}} = -1.913\ \mu_N$. The measured value is $-1.59\ \mu_N$, close but nevertheless significantly different from the free value.

The magnetic moment of an $f_{7/2}$ proton is predicted to be $\ell + 2.793 = 5.793\ \mu_N$. The measured value is $5.43\ \mu_N$.

We use experimental moments as input for calculations in the $f_{7/2}$ region. We define

$$g_n = \frac{\mu(^{41}\text{Ca})}{j} \qquad j = 7/2$$

$$g_p = \frac{\mu(^{41}\text{Sc})}{j} \quad .$$

Thus the magnetic moment operator for the $f_{7/2}$ shell can be written

$$\vec{\mu} = g_n \vec{L}_N + g_p \vec{L}_P \, ,$$

we can use the Lande trick $\vec{J} = \vec{L}_N + \vec{L}_P$ for a basis state $(L_P L_N)^J$

$$\vec{\mu} = g \vec{J} \qquad \vec{\mu}\cdot\vec{J} = g \, \vec{J}\cdot\vec{J}$$

$$g = \frac{1}{J(J+1)} \left(g_n \, \vec{L}_N \cdot \vec{J} + g_p \, \vec{L}_P \cdot \vec{J} \right) \; .$$

Thus for a wave function

$$\psi = \sum_{L_P L_N} D(L_P L_N) \, (L_P L_N)^J$$

the magnetic moment is

$$\mu = \frac{g_p + g_n}{2} J + \sum_{L_P L_N} \frac{|D(L_P L_N)|^2 [L_P(L_P+1) - L_N(L_N+1)]}{(J+1)} \times \frac{(g_p - g_n)}{2} \; .$$

This is a remarkably simple formula, involving no Racah or fractional parentage coefficients. Recall that $|D(L_P L_N)|^2$ is the probability that protons couple to L_P and neutrons to L_N.

For the $J = 1^+_1$ state of ^{44}Sc the measured moment is $2.56 \pm 0.04 \; \mu_N$. The MBZ calculation gives $2.21 \; \mu_N$. Had we used naive wave functions where the three neutrons couple only to $L_N = 7/2$ we would have obtained $\mu = (g_p + g_n) = 1.1 \; \mu_N$ about a factor of two *smaller* than what the more exact wave functions furnish.

Spectroscopic Strengths

To test whether the wave functions make any sense, the most useful tools have been the stripping reaction (d,p) and the pickup reactions (p,d). I will here discuss these rather briefly and uncritically (as far as the reaction mechanism goes).

We regard the above as direct reactions. In the (d,p) we drop a neutron into the $f_{7/2}$ shell; in (p,d) we pick up a neutron from the $f_{7/2}$ shell. We define the maximum (or sum rule) strength for (d,p) as the number of neutron holes in the $f_{7/2}$ shell; for (p,d) it is the number of neutron particles in the $f_{7/2}$ shell.

Here are some examples. Consider the reaction

$$p + {}^{48}Ca \rightarrow d + {}^{47}Ca \quad J = 7/2^- \ .$$

Clearly $\mathcal{S}$ is eight because there are 8 $f_{7/2}$ neutrons to be picked up. If the shell model picture is good all the strength $\mathcal{S}$ would be concentrated in the $J = 7/2$ ground state of ^{47}Ca. This seems to be the case.

In the reaction $d + {}^{40}Ca \rightarrow p + {}^{41}Ca$ the strength for $f_{7/2}$ is also eight, since there are eight neutron holes in ^{40}Ca. Although one reaches the $J = 7/2^-$ ground state of ^{41}Ca with most of the strength, there is some strength to a second $7/2^-$ state at about 3 MeV excitation. This would be in accord with the Gerace-Green picture in which low-lying highly deformed states are admixed with the basic shell model states.

In the reaction $d + {}^{42}Ca \rightarrow {}^{43}Ca + p$ in the $f_{7/2}$ picture we are dropping an $f_{7/2}$ nucleon into a $J=0$ ^{42}C core. Clearly then for the states of the configuration $(f^2_{7/2})_\nu$ we can only reach the $J = 7/2^-$ state. The strength $\mathcal{S}$ is zero for $J = 3/2, 5/2, 9/2, 11/2$ and $15/2$. This seems to be borne out by experiment.

In the reaction $^{43}Ca(p,d)^{42}Ca$ (I don't remember if this can be done—is ^{43}Ca stable? [*Ed. note: Yes*]) one can reach the states of ^{42}Ca with all spins $J = 0,2,4,6$; in fact one can easily show that a state of angular momentum J is reached with a probability (2J+1). We expect $\mathcal{S}^J = C(2J+1)$ where C is such that $\sum \mathcal{S}^J = 3$.

An interesting test of the shell model is afforded by the reaction $p + {}^{47}Ti \rightarrow d + {}^{46}Ti$. The ground state spin of ^{47}Ti is $J = 5/2$. If we take out an $f_{7/2}$ neutron from ^{47}Ti the states in ^{46}Ti we can reach will have $J = \vec{5}/2 + \vec{7}/2$. Hence we will not be able to reach the $J=0$ ground state of ^{46}Ti. This again seems to be borne out by experiment.

We see then that stripping and pickup experiments, in the region where the direct reaction assumption is valid give us a good test of our shell model assumptions. The MBZ[1] wave functions are too simple and this manifests itself in a larger fragmentation of strength than is predicted by this single j shell model. Nevertheless things do hang together; the gross predictions of the theory seem to be borne out reasonably well.

BETA DECAY

Single Beta Decay

The Fermi operator for allowed transitions is $O_F = {}_{i}\, t_{-}(i)$

where $t_-|\text{neutron}\rangle = |\text{proton}\rangle$. Note that $O_F = T_-$, the lowering operator of isospin. Recall the rules

$$T_z\, \psi^T_{M_T} = M_T\, \psi^T_{M_T}$$

$$T_+\, \psi^T_{M_T} = \sqrt{T(T+1) - M_T(M_T+1)}\;\psi^T_{M_T+1}$$

$$T_-\, \psi^T_{M_T} = \sqrt{T(T+1) - M_T(M_T-1)}\;\psi^T_{M_T-1}$$

We see that the Fermi operator cannot change the isospin of a state i.e. we have the selection rule $\Delta T = 0$ (also $\Delta J = 0$). The Fermi operator only connects a state to its analogue in the neighboring nucleus.

With *very* few exceptions, though, the ground state of a nucleus has for its isospin $T = T_Z = (N-Z)/2$. Hence a typical transition involves states of different isospin.

The allowed Gamow-Teller operator is

$$O_{GT} = \sum_i \sigma_\mu(i)\; t_-(i) \quad .$$

This can change the isospin by one unit. The Gamow-Teller matrix element is given by

$$M^2_{GT} = \frac{1}{2J_i+1} \sum_{M_F M_i \mu} |\langle \psi^{J_F}_{M_F} \sum_i \sigma_\mu(i) t_-(i)\; \psi^{J_i}_{M_i}\rangle|^2 \quad .$$

The log ft value is given by

$$\log ft \simeq 3.65 - \log M^2_{GT} \ .$$

Thus if $M_{GT} = 1$ we would have $\log ft = 3.65$. Looking casually through the nuclear data sheets we find that usually log ft is of the order of 5, 5.5 or 6 with of course some exceptions. A value of $\log ft = 5.65$ would imply $M_{GT} = 0.1$. This is a not untypical situation. Somehow the many-body aspects of a nuclear wave function conspire to suppress the single beta decay matrix element.

The reason that I emphasize this here is because later I will try to convince you that despite the fact that single beta decay matrix elements are suppressed, *double* beta decay matrix elements from $J=0^+$ to $J=0^+$ can be large.

A table of calculated log ft values, compared with experiment is given in Table 3.

In performing the calculations we wrote our wave functions in isospin notation introducing $\chi_{1/2}$ for a neutron and $\chi_{-1/2}$ for a proton. Thus a wave function for n neutrons and z protons in the

Table 3. log ft Values

Nuclear Decay	J_i	J_F	Expt.	Calc.	Expt. Cross Conjugate	Nuclear Decay Cross Conjugate
$Sc^{43} \rightarrow Ca^{43}$	7/2	7/2	4.9	4.58	5.4	$Fe^{53} \rightarrow Mn^{53}$
	7/2	5/2	5.0	4.45	5.2	
$Sc^{44} \rightarrow Ca^{44}$	2	2	5.3	5.11	5.5	$Mn^{52} \rightarrow Cr^{52}$
	6	6	5.8	5.40	5.5	
$Ca^{45} \rightarrow Sc^{45}$	7/2	7/2	6.0	5.28	5.4	$Cr^{51} \rightarrow V^{51}$
$Ti^{45} \rightarrow Sc^{45}$	7/2	7/2	4.6	4.55	5.1	$Mn^{51} \rightarrow Cr^{51}$
$Sc^{46} \rightarrow Ti^{46}$	4	4	6.2	5.94	—	
$Ca^{47} \rightarrow Sc^{47}$	7/2	7/2	8.5	5.85	5.7	$Sc^{49} \rightarrow Ti^{49}$
	7/2	5/2	6.0	4.99	—	
$Sc^{47} \rightarrow Ti^{47}$	7/2	7/2	5.3	5.41	6.2	$V^{49} \rightarrow Ti^{49}$
	7/2	5/2	6.1	5.62	—	
$Sc^{48} \rightarrow Ti^{48}$	6	6	5.5	5.45		
	6	6	6.4	forbidden		

$f_{7/2}$ shell, originally written as

$$\sum D(L_P L_N) \left[f^n_{7/2}(1\ldots n)^{L_N} \; f^z_{7/2}(n+1\ldots n+z)^{L_P} \right]^J$$

now becomes

$$\sum D(L_P L_N) \frac{1}{\sqrt{\frac{(n+z)!}{n!z!}}} \underset{(1\ldots n)(n+1\ldots n+z)}{\mathcal{A}} [L_P L_N]^J \times$$

$$\times \, \chi_{1/2}(1)\ldots\chi_{1/2}(n) \; \chi_{-1/2}(n+1)\ldots\chi_{-1/2}(n+z) \; .$$

Double Beta Decay: $^{48}Ca \rightarrow {}^{48}Ti \;\; 0^+ \rightarrow 0^+$

Having discussed the shell model at some length we can now apply it to the problem of double beta decay. The single beta decay process ^{48}Ca $J=0^+ \rightarrow Sc^{48}$ $J=1^+$ is energetically forbidden. The double beta decay process

$$Ca^{48} \rightarrow e^- + e^- + \bar{\nu} + \bar{\nu} + {}^{48}Ti$$

is energetically allowed. (There is also the possibility of

neutrinoless emission, which is one of the prime points of interest in this subject.)

The transition is second order in the weak interaction and the relevant matrix element is

$$\sum_{INT} \frac{\langle F\ O_{GT}\ INT\rangle \langle INT\ O_{GT}\ I\rangle}{E_I - E_{INT}}$$

where I, INT and F represent respectively the initial, the intermediate and the final states. In our case I = ^{48}Ca J=0 ground state; INT = 1^+ state in ^{48}Sc and F is the J=0^+ ground state of ^{48}Ti. The isospin of the ^{48}Ca ground state is T=4 (consistent with eight neutrons in the $f_{7/2}$ shell); that of ^{48}Ti g.s. is T=2. Hence the Fermi transition would contribute only if there were isospin impurities. These are small so we limit the discussion to Gamow-Teller transitions.

Following Haxton, Stephenson and Strottmann[9] one can use closure in the sum over intermediate states by assuming the relevant intermediate states are localized in energy. The matrix element can be written as $M_{GT,\beta\beta}/\Delta\bar{E}$ where

$$M_{GT,\beta\beta} = \left\langle F\ \frac{1}{2} \sum_{ij} \vec{\sigma}(i)\cdot\vec{\sigma}(j)\ t_-(i)\ t_-(j) I \right\rangle \ .$$

The wave function of Ca^{48} is $(f^8_{7/2})_\nu\ \chi_{1/2}(1) \ldots \chi_{1/2}(8)$. The wave function of ^{48}Ti is

$$\sqrt{\frac{2}{N(N-1)}} \sum_{L\ \mathrm{even}} D(L,L) \underset{(12)(3\ldots 8)}{\mathcal{A}} \left[(f^2_{7/2})^L_\pi\ (f^6_{7/2})^L_\nu\right]^{J=0} \times$$
$$\times\ \chi_{-1/2}(1)\ \chi_{-1/2}(2)\ \chi_{1/2}(3) \ldots \chi_{1/2}(8)$$

where N = 8. Using various tricks, described by N. Auerbach and the author, we find

$$M_{GT,\beta\beta} = \frac{g^2}{2}\left(\frac{N(N-1)}{2}\right)^{1/2} \times$$
$$\times \sum_{L\ \mathrm{even}} D(L,L) \frac{\sqrt{(2L+1)}}{\left[\sum_{L'\mathrm{even}} (2L'+1)\right]^{1/2}} \left[L(L+1)-2j(j+1)\right]$$

where $j = 7/2$, $g = \vec{\sigma}\cdot\vec{j}/j(j+1) = 2/7$ for the $f_{7/2}$ shell.

The values of D(L,L) for the J=0 ground state of Ti^{48} obtained in the MBZ diagonalization are

$$D(0,0) = -0.9136 \qquad D(2,2) = 0.4058$$
$$D(4,4) = 0.0196 \qquad D(6,6) = -0.0146\ .$$

The calculated value of $M_{GT,\beta\beta}$ is 0.18. Thus this matrix element is suppressed.

This result is at first surprising. Note the factor in $M^2_{GT,\beta\beta}$ of N(N-1)/2 (N=8). This would suggest that ^{48}Ca should be a nice juicy case to consider. By the way the results here are consistent with limits set by the Columbia group, Wu et al.[10] (Note also that Haxton et al.[9] got a factor of seven larger for the matrix element when they used renormalized Kuo-Brown matrix elements than when they used the bare ones. The larger value was $M_{GT,\beta\beta} = 0.2$). In the next section we discuss reasons for this. Note that previous work on the calcium isotopes was done by Khodel.[11]

The K selection rule of Lawson. Lawson[12] suggested another way of constructing wave functions in the $f_{7/2}$ shell, following the idea of the Nilsson model. The idea is to stick with a one-body potential as long as possible.

For the closed shell ^{40}Ca it is appropriate to use a spherical one-body potential. As we add nucleons to the nucleus, however, we find a very large degeneracy which we showed could be removed by using a residual interaction.

Another way of removing the degeneracy is to use a deformed one-body potential. Indeed, Hartree-Fock calculations of open shell nuclei do yield deformed solutions.

The simplest deformed potential is axially symmetric. The Nilsson[13] one-body Hamiltonian is

$$H_N = \frac{p^2}{2m} + \frac{1}{2} m \omega^2 r^2 + C\vec{\ell}\cdot\vec{s} + \beta r^2 Y_{2,0} + D\vec{\ell}\cdot\vec{\ell} .$$

Angular momentum is not conserved with this interaction. Note however, that the projection of the total angular momentum along the body-fixed three-axis commutes with this Hamiltonian. We can therefore assign a quantum number K for this projection.

There is an interesting systematic that can be associated with this scheme. Consider the nuclei ^{45}Ti, ^{47}Ti and ^{49}Ti. Following Lawson[12] we consider the weak deformation limit, in which the wave function of the intrinsic state is expressed in terms of $f_{7/2}$ orbitals; see Fig. 6. The intrinsic state for the protons will be $f_{7/2,1/2}(1)\ f_{7/2,-1/2}(2)$ in all three nuclei. The intrinsic states of the neutrons are:

^{45}Ti	$f_{7/2,1/2}\ f_{7/2,-1/2}\ f_{7/2,3/2}$	K = 3/2
^{47}Ti	$f_{7/2,1/2}\ f_{7/2,-1/2}\ f_{7/2,3/2}\ f_{7/2,-3/2}\ f_{7/2,5/2}$	K = 5/2
^{49}Ti	$f_{7/2,1/2}\ f_{7/2,-1/2}\ f_{7/2,3/2}\ f_{7/2,-3/2}\ f_{7/2,5/2}$	
	$f_{7/2,-5/2}\ f_{7/2,7/2}$	K = 7/2

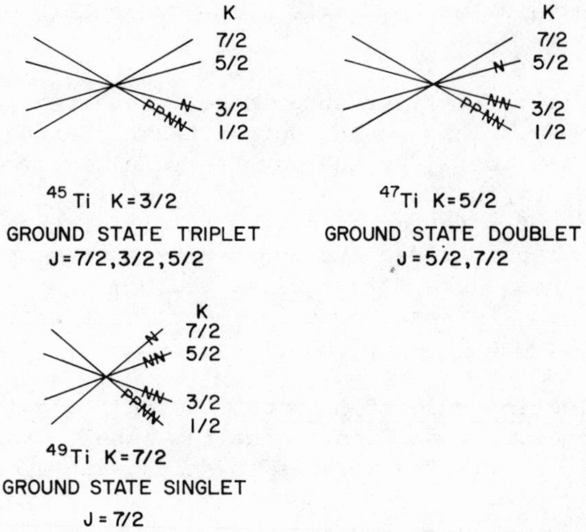

Fig. 6. Nilsson intrinsic states for titanium isotopes.

If we look at the low-lying spectra of these nuclei, shown in Fig. 7, we find that there is a near degenerate triplet with $J = 3/2^-$, $5/2^-$ and $7/2^-$ near the ground state of ^{45}Ti; in ^{47}Ti the $J = 5/2^-$ is slightly lower than $J = 7/2^-$, but the $J = 3/2^-$ is at a much higher energy; in ^{49}Ti the $J = 7/2^-$ ground state is alone, much lower than $J = 3/2^-$ or $5/2^-$. We thus formulate the rule that if $K = 3/2$ there will be a low-lying state with $J = K = 3/2$; if $K = 5/2$ there will be a

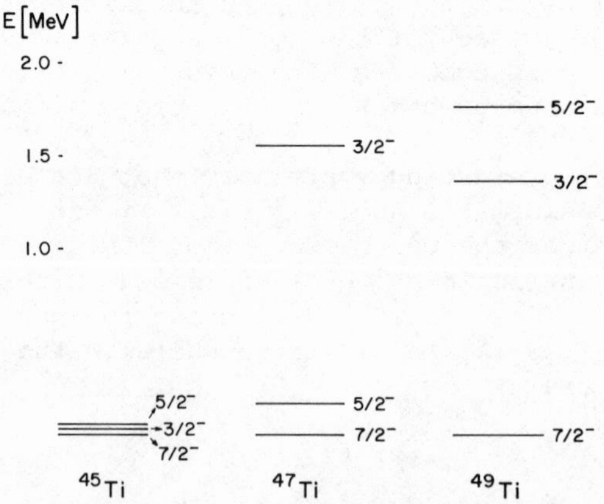

Fig. 7. The $J = 3/2, 5/2$ and $7/2$ states of odd titanium isotopes.

low-lying J = K = 5/2 state etc. This seems to work throughout the $f_{7/2}$ region, not just the titaniums.

It should be added that one of the successes of MBZ was to show that we could correctly predict the low-lying J = 5/2 state, e.g., the near doublet $J = 5/2^-$ and $J = 7/2^-$ in ^{47}Ti. But the calculation cannot get the low lying $J = 3/2^-$ states. This was already mentioned in the context of ^{43}Ca (which by the way would have K = 3/2 and which does have an anomalously low $J = 3/2^-$ state), and could be due to the importance of $p_{3/2}$ admixtures.

Getting back to double beta decay we can easily see why the ^{48}Ca to ^{48}Ti transition would be inhibited. The intrinsic state for ^{48}Ti is

$$(1-P_{12})(f_{7/2,1/2}\ f_{7/2,-1/2})_\pi\ (1-P_{12})(f^{-1}_{7/2,7/2}\ f^{-1}_{7/2,-7/2})_\nu$$

i.e. two protons in K = 1/2 and two neutron holes in K = 7/2. In this picture we see why there is an inhibition. In order to go from ground state to ground state we have to remove two neutrons from the top of the Fermi sea i.e. from K = 7/2, and change them into two protons in the lowest state i.e. K = 1/2. But the operator σt can change K by at most one unit. Hence we have a selection rule working against this process.

The above argument is rough, and somewhat unsatisfactory. Obviously ^{48}Ca is spherical and the Nilsson model[13] does not apply. To make the argument more correctly, we note that the intrinsic state can be regarded as a wave packet containing several states of different angular momentum. We must project out a state of good total J. The projected state is

$$N \sum_{\substack{L\ \text{even} \\ j=7/2}} (jj\tfrac{1}{2} - \tfrac{1}{2}|L0)(jjj-j|L0)(LL00|00)\left[(f^2_{7/2\pi})^L\ (f^{-2}_{7/2})^L\right]^0$$

where N is a normalization factor. [Note $(LL00|00) = 1/\sqrt{2L+1}$.] Thus, instead of using the MBZ wave functions in ^{48}Ti we use the projected ones i.e.

$$D(L,L) \propto (jj\tfrac{1}{2} - \tfrac{1}{2}|L0)(jjj-j|L0)\ \frac{1}{\sqrt{(2L+1)}}\ .$$

When these are inserted into the expression for $M_{GT,\beta\beta}$ it is easy to show that $M_{GT,\beta\beta}$ vanishes.

Thus ^{48}Ca, which at first looked like an excellent case for studying double beta decay now becomes questionable. We should add that earlier work by Khodel reached the same conclusion.

Table 4. Double Beta Matrix Elements. Here, T_o is the Kinetic Energy carried off by the Leptons.

Reaction	$T_o(m_ec^2)$	$\|M_{GT}\|$ theory	$\|M_{GT}\|$ exp
$^{130}Te \rightarrow {}^{130}Xe$	5.0	1.48	0.10 - 0.13[a]
$^{128}Te \rightarrow {}^{128}Xe$	1.7	1.47	0.18 - 0.23
$^{82}Se \rightarrow {}^{82}Kr$	5.9	0.94	{1.43 / 0.27[a]
$^{76}Ge \rightarrow {}^{76}Se$	4.0	1.28	
$^{48}Ca \rightarrow {}^{48}Ti$	8.4	0.22	$\lesssim$0.19

[a]Maximum values determined from total geochemical rates.

Double beta decay in heavier nuclei. But do not despair. There are other nuclei. The most encouraging work is that of Haxton, Stephenson and Strottmann[9] who have done large shell model calculations in several nuclei, and who have found that in many cases the value of $M_{GT,\beta\beta}$ is quite large, of the order of 1 or even larger. (Although, as Talmi points out, the number of configurations become astronomical, for $J=0^+$ states the number of configurations though large, are still small enough to make the calculations feasible.)

The results of Table 4 have been presented by Haxton[14] at various conferences. Note that the theoretical values are much larger than the values or limits obtained from geochemical rates. Laboratory rates are however much larger than the geochemical, as is seen above in the case of ^{82}Se decay. It is difficult at present to assess the experimental situation.

Coherent pairing effects. We wish to show that in nuclei where pairing correlations are important the double beta decay matrix element will be large. Here work by N. Auerbach and the author[15] is discussed.

Consider as an example the double beta decay $^{76}Ge \rightarrow {}^{76}Se$ shown in Fig. 8. We will consider a somewhat crude calculation in which the single particle wave functions are asymptotic Nilsson states $|Nn_z\Lambda\Sigma>$ where Λ is the projection of orbital and Σ of spin angular momentum on the body-fixed z-axis (axis of symmetry).

In the Nilsson model[13] each energy has a twofold 'Kramers' degeneracy. If one state is ψ_K the other is $\psi_{\bar{K}}$.

The time reversed state of $|Nn_z\Lambda\Sigma>$ is

$$(-1)^{\ell+1/2+\Lambda+\Sigma}\ |Nn_z - \Lambda - \Sigma>$$

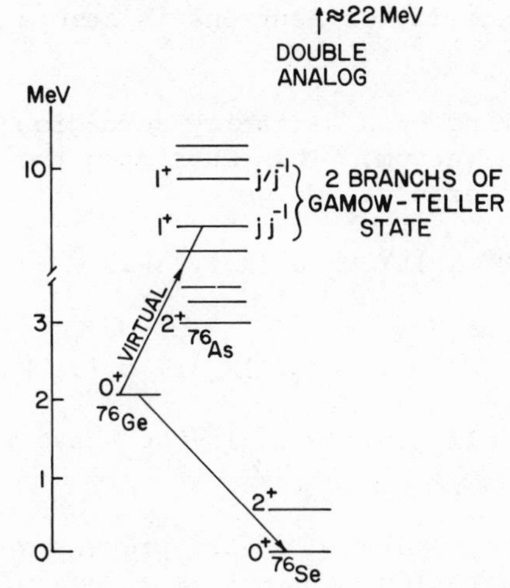

Fig. 8. Double beta decay of ^{76}Ge to ^{76}Se.

where $(-1)^{\ell}$ is the parity of the Nilsson orbit.

If we did not have any pairing then the transition would have to take place by 2 neutrons in the #26 orbit $|3\ 0\ 1\ -1/2>$ changing into 2 protons in the #16 orbit $|3\ 1\ 2\ -1/2>$. But these orbits have different values of n_z i.e. the radial wave functions are different. But the operator $\sigma_1 \cdot \sigma_2\ t_-(1)\ t_-(2)$ has no radial dependence and therefore cannot change the radial quantum numbers. At this level $M_{GT,\beta\beta}$ would be zero.

By turning on pairing correlations we allow partial occupation of several orbits for the neutrons, and likewise for the protons. Hence there will be favorable orbits i.e. ones for which the radial quantum numbers n_z and Λ are the same for the neutrons and the protons—Σ may or may not be the same. We thus see qualitatively how pairing can enhance this process.

In second quantized form the t_- operator is $a_\pi^\dagger a_\nu$ i.e. it destroys a neutron and creates a proton. The quasi-particle operators in the pairing theory are

$$\alpha^\dagger(\nu) = U(\nu)\ a^\dagger(\nu) + V(\nu)\ a(\bar{\nu})$$

$$\alpha^\dagger(\bar{\nu}) = U(\nu)\ a^\dagger(\bar{\nu}) - V(\nu)\ a(\nu)$$

where $V(\nu)^2$ is the probability that the state ν is occupied and $U(\nu)^2 = 1 - V(\nu)^2$. We write the double beta decay operator $\sigma_1 \cdot \sigma_2\ a_\pi^\dagger a_\nu\ a_{\bar{\pi}}^\dagger a_{\bar{\nu}}^\dagger$ in terms of the quasi-particle creation and destruction operators $\alpha^\dagger$ and α. We assume that both for the initial and final states the seniority of neutrons is zero and the seniority of protons is also zero.

Note that α acting on a seniority zero ground state gives zero i.e. annihilates the vacuum. This cuts down the number of terms in the expression considerably. We get

$$M_{GT,\beta\beta} = \sum_{K_\nu K_\pi} U^\nu(K_\nu)V^\nu(K_\nu)U^\pi(K_\pi)V^\pi(K_\pi) \times$$

$$\times \left\langle (1-P_{12})\ K_\pi(1)\bar{K}_\pi(2)\ \sigma_1\cdot\sigma_2\ K_\nu(1)\bar{K}_\nu(2)\right\rangle \ .$$

The U's and V's are all positive. In the above we characterize the orbit by $K = \Lambda+\Sigma$.

To establish the *coherence* of the process we must convince ourselves that the expectation of $\sigma_1\cdot\sigma_2$ is always of one sign. When

$$|nn_z\Lambda\Sigma >_\nu = |nn_z\Lambda\Sigma >_\pi \quad \Lambda \neq 0 \quad \sigma_1\cdot\sigma_2 = -1 \ ,$$

when

$$|nn_z\Lambda\Sigma >_\nu = |nn_z\Lambda\Sigma >_\pi \quad \Lambda = 0 \quad \sigma_1\cdot\sigma_2 = -3 \ ,$$

when

$$|nn_z\Lambda\Sigma >_\nu = |nn_z\Lambda\Sigma \pm 1 >_\pi \quad \sigma_1\cdot\sigma_2 = -2 \ .$$

Thus $\sigma_1\cdot\sigma_2$ is always negative. (Of course in a spin triple state $\sigma_1\cdot\sigma_2 = +1$, but we never have a pure triplet state in this scheme.)

We express our results in terms of the strength of the pairing gap Δ, which is related to the U's and V's as follows:

$$U(\nu) = \frac{1}{\sqrt{2}}\left(1 + \frac{\varepsilon(\nu)-\lambda}{E(\nu)}\right)^{1/2}$$

$$V(\nu) = \frac{1}{\sqrt{2}}\left(1 - \frac{\varepsilon(\nu)-\lambda}{E(\nu)}\right)^{1/2}$$

$$E(\nu) = \left[\left(\varepsilon(\nu)-\lambda\right)^2 + \Delta^2\right]^{1/2} \ .$$

In the above λ is the Fermi energy. The values we obtained for $M_{GT,\beta\beta}$ were 0,-1.61 and -3.05 for Δ equal to 0, 1 and 2 respectively. The favored value of Δ in this region is about 1.4 MeV. Thus our results are supportive to the findings of Haxton et al.

We mentioned before that, in general, single beta decay in odd-even or even-odd nuclei is suppressed. Part of the suppression arises in the pairing theory. As a result of pairing the matrix element is multiplied by $U^\nu U^\pi$ (or $V^\nu V^\pi$) of the *odd* nucleon undergoing the transition. This reduces the matrix element by about a factor of two and the transition rate by a factor of four.

The Effect of Delta Particle-Nucleon Hole Admixtures on Double Beta Decay

The single beta decay. I believe that Faqir Khanna is going to talk about Δ-hole admixtures for single beta decay, (p,n) and magnetic dipole transitions. So I will be brief.

There is currently a lot of excitement about quenching of Gamow-Teller strength due to Δ particle-nucleon hole admixtures.[16] This subject has been stimulated by the (p,n) experiments of Goodman et al.[17] as well as the long standing problem of missing M1 strength (is it really missing or is there experimental confusion?).

The Gamow-Teller state, obtained by acting with the operator $\sum_i \sigma t_-$ ($t_+|\text{proton}\rangle = |\text{neutron}\rangle$) on a closed (but spin unsaturated) shell involves very few particle-hole configurations. For example with ^{48}Ca as the closed shell the configurations are $f_{7/2\pi}\, f^{-1}_{7/2\nu}$ and $f_{5/2\pi}\, f^{-1}_{7/2\nu}$. The reason for this is of course the Pauli principle. We cannot for example change a $0d_{5/2}$ nucleon into a $0d_{3/2}$ because the $0d_{3/2}$ shell is fully occupied.

On the other hand we can extend the Gamow-Teller operator so that it can excite internal degrees of freedom of the nucleon i.e. change a nucleon into a delta. We now have,

$$O_{GT} = O_N + O_\Delta$$

where

$$O_N = \sum_i \sigma(i)t_-(i) \quad \text{and} \quad O_\Delta = \sum_i S(i)T_-(i) \quad .$$

The Δ particle-nucleon hole state created by the operator O_Δ is at a very high energy ~300 MeV, whereas the nucleon Gamow-Teller state is at an energy of about 10 MeV in a medium-heavy mass nucleus.

We call the two states $|GT\rangle_N$ and $|GT\rangle_\Delta$. Because of residual interaction the two states can be admixed, yielding 'physical' Gamow-Teller states

$$|1\rangle = |GT\rangle_N - \alpha\,|GT\rangle_\Delta$$

$$|2\rangle = \alpha|GT\rangle_N + |GT\rangle_\Delta$$

with

$$\alpha = \frac{{}_N\langle GT|V_{\Delta N}|GT\rangle_\Delta}{E_\Delta - E_N} \quad .$$

Despite the fact that the energy denominator is very large $(E_\Delta - E_N) \approx 300$ MeV, the effect of the admixture can be very important for the low lying GT transition i.e. involving the state $|1>$. There are two reasons for this:

(a) The basic M1 transition matrix element for changing a nucleon to a delta is about the same as for a nucleon remaining a nucleon. [In the simple quark model a nucleon is an admixture of two spin-up and one spin-down quark, all in radial s-states; in the Δ state all three spins are up. The M1 or GT operator either leaves the spin alone (nucleon to nucleon), or it flips the spin from down to up (nucleon to delta).]

(b) There is (to a first approximation) no inhibition due to the Pauli principle for the Δ-h state. Whereas in ^{48}Ca there are only two nucleon particle-hole states ($f_{7/2\pi}\ f^{-1}_{7/2\nu}$ and $f_{5/2\pi}\ f^{-1}_{7/2\nu}$), there are many Δ-h states. For example the states $(\Delta)_{0s1/2}\ 0s^{-1}_{1/2\text{nucleon}}$, $\Delta_{0p3/2}\ 0p^{-1}_{3/2\text{nucleon}}$ etc. are allowed. There is nothing to prevent the Δ particle from being in a $0s_{1/2}$ or $0p_{3/2}$ state. This would not be the case for the nucleon. This fact enhances the importance of Δ-h admixtures by a factor of A, the number of nucleons, in its effect on quenching of low-lying Gamow-Teller resonances.

When this is put together with other factors of A one finds that the effective coupling strength for Gamow-Teller resonances gets renormalized, and to a first approximation this renormalization is, as shown by Bohr and Mottelson,[16] independent of A. Different calculations differ somewhat on the amount of renormalization. In the Bohr-Mottelson calculation the Gamow-Teller transition rate is cut down by a factor of two due to a combination of Δ-h admixtures and RPA correlations.

Most calculations obtain large effects due to Δ-h admixtures. There are, however, some which do not, e.g. those of Khanna and Towner (KT). I am sure Khanna will discuss this. Amongst other things the KT calculation includes Δ-h exchange terms which in part cancel the direct terms. The exchange terms are more important for orbits $j \to j$ than for orbits $j \to j' = j \pm 1$. Hence this is more important for ground state magnetic moments than for GT transitions. There are currently arguments as to whether Δ-h exchange terms should or should not be included.

In this vein, a calculation by Bertsch and Hamamoto[18] considers an alternate mechanism—simply the robbing of the one particle-hole Gamow-Teller strength by higher lying 2 particle-2 hole states. They find considerable admixture (~56%) of one particle-one hole strength in the basic 2 particle-2 hole region.

It should be remarked that not all of this 56% necessarily comes from the main peak. As was shown by Halemane, Abbas and I,[19] when one introduces ground state correlations one can generate new strength e.g. in closed L-S shell nuclei there is no M1 strength unless ground state correlations are present.

Still, some of the depletion of M1 or GT strength can come from the above mechanism, thus making the Δ-h admixture, though still significant, of somewhat lesser importance.

Double beta decay and Δ-h admixtures. What is the effect of Δ-h admixtures on double beta decay? If all that happens is that the effective coupling constant for single beta decay gets renormalized from say 1 to 0.8, then, since double beta decay is a second order process the matrix element would be changed by $(0.8)^2$ and the transition rate by $(0.8)^4$.

There is, however, one point which Auerbach and I worried about. In the second order expression, before the closure approximation is made, we sum over intermediate states

$$M_{GT,\beta\beta} = \sum_M \frac{<FOM><MOI>}{E_M - E_I}$$

$$O = O_N + O_\Delta \ .$$

What happens when we extend the sum to the Δ region i.e. $E_M - E_I \approx$ 300 MeV. If we just extended the sum to the low-lying Gamow-Teller region, ~10 MeV but used the experimental (quenched) matrix elements, for <IOM> and <MOF> then we would clearly get the simple-minded reduction of the double beta decay mentioned above.

Does the inclusion of the high intermediate states change the picture? The answer is no, and we are indebted to Bohr, Mottelson and Hamamoto for discussions on this point. In Fig. 9, there are four graphs depicting double beta decay from a closed shell nucleus like ^{48}Ca. Graph (a) involves nucleons only, whereas (b), (c) and (d) involve Δ-h admixtures.

We should compare, in particular, graphs (b) and (c). In (b) the first decay excites the Δ-h state, which then de-excites to the GT state before the second decay takes place. The intermediate state energy here is about 10 MeV, and this graph corresponds to a renormalization of the coupling constant for single beta decay.

In graph (c), the first decay also excites the Δ-h state. The second decay occurs *before* the Δ-h state in the first decay de-excites into a GT state.

We see however, that (c) is identical to (b) in every respect except for the fact that the energy denominator is now ~300 MeV instead of 10 MeV. (As Brodsky comments, the Δ-h state lives a very

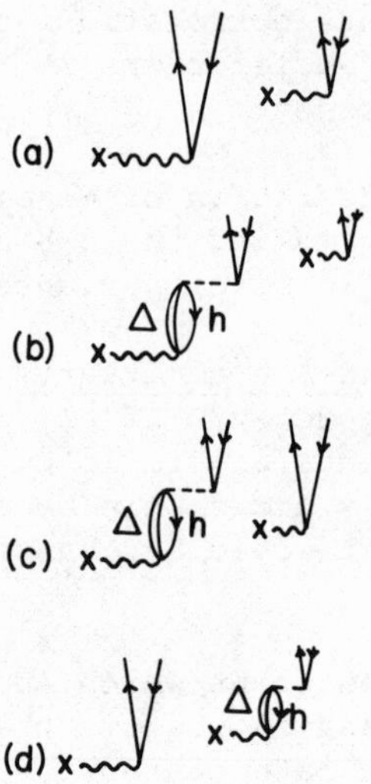

Fig. 9. Contributions of Δ particle-nucleon hole admixtures to double beta decay.

short time.) Thus graph (c), which corresponds to the high energy intermediate state in the Δ-h region is down by a factor of about 10/300 relative to (b) and so can be neglected.

GIANT RESONANCES

Soluble Model-Monopole-Monopole Interaction

Just for fun we will consider as an example of collective giant resonance state, the breathing mode state, in a soluble model.[20]

$$H = \sum_i P_i^2/2m + \lambda \sum_{i<j} r^2(i)\ r^2(j) \ .$$

A model like this was considered several years ago by Goshen and Lipkin.[21] See also the work of Kirson.[22]

We remind ourselves that in the 3-dimensional harmonic oscillator the levels are separated by $1\hbar\omega$ and the orbit is characterized by $n\ell j$ $(N = 2n+\ell)$ e.g.

N = 0   0s
N = 1   0p
N = 2   1s,0d
N = 3   1p,0f
N = 4   2s,1d,0g .

In the above, n, the radial quantum number, equals the number of nodes in the radial wave function. There are two matrix elements of r^2: $<nr^2n> = (2n+\ell+3/2)b^2$ where $b^2 = \hbar^2/m\omega$,

$$< nr^2n+1 > = -\sqrt{(n+1)(n+\ell+3/2)}\ b^2 \ .$$

Thus if we put several nucleons in the oscillator states the mean square *mass* radius (R^2_{MS}) is 1/A Sum$(2n+\ell+3/2)b^2 = \Sigma b^2/A$, i.e.

$$\begin{aligned} \Sigma &= 6 \quad \text{for} \quad {}^4\text{He} \\ &= 36 \quad \text{for} \quad {}^{16}\text{O} \\ &= 120 \quad \text{for} \quad {}^{40}\text{Ca} \quad \text{etc.} \end{aligned}$$

For example, $R^2_{MS}(^{40}Ca) = 3b^2$. (N.B. $b \sim A^{1/6}$ fm, $b \simeq 1.75$ fm for ^{16}O, $\simeq 1.95$ fm for ^{40}Ca.)

With the above monopole-monopole Hamiltonian we can obtain the binding energy of a closed shell nucleus. To do this it is convenient to replace λ by another parameter $\hbar\omega_0$ such that

$$\left(\frac{\hbar^2}{m}\right)^2 \lambda = \frac{\hbar\omega_0{}^3}{2\Sigma} \ .$$

In that case, the energy becomes

$$E(\hbar\omega) = \hbar\omega\ \frac{\Sigma}{2} + \frac{(\hbar\omega_0)^3}{4(\hbar\omega)^2}\ \Sigma \ .$$

The Hartree condition $dE/d\hbar\omega = 0$ leads to $\hbar\omega = \hbar\omega_0$ i.e. the nucleus will have a minimum energy at a mean square radius $R^2_{MS} = \Sigma b^2/A$ with $b_0^2 = \hbar^2/m\omega_0$.

We can get the energy of the breathing mode state in two different ways. First the Tamm-Dancoff approximation. We define this state as

$$\psi = \Sigma\ r^2|0>$$

i.e. the r^2 operator acting on the ground state, but it is understood that we keep only the part of the wave function orthogonal to the ground state. This state clearly has J=0, T=0 (and in L-S coupling approximation S=0).

For example in ^{16}O we would take a linear combination of 1s $0s^{-1}$ and 1p $0p^{-1}$

$$\psi_B = \Sigma \left\langle (ph^{-1})^{J=0,T=0}\ r^2 0\right\rangle \left| (ph^{-1})^{00}\right\rangle \ .$$

The energy of the state ψ_B can be written as

$$E_B = 2\hbar\omega + V_{\text{particle-hole}} \quad .$$

When we work this out we obtain

$$E_B = 2\hbar\omega + \frac{\hbar\omega}{2} = 2.5\ \hbar\omega \quad .$$

The calculation is facilitated by an energy weighted sum rule. An alternative way of getting the energy is to note that there is a relation between the compressibility and the breathing mode energy

$$E_B = \hbar \sqrt{\frac{KA}{mA\langle r^2\rangle}}$$

where

$$K = \frac{1}{A}\,(\hbar\omega)^2\,\frac{d^2E}{d\hbar\omega^2} \qquad\qquad E = \hbar\omega\,\frac{\Sigma}{2} + \frac{(\hbar\omega_0)^3}{4(\hbar\omega)^2}\,\Sigma \quad .$$

$A\langle r^2\rangle$ is the mean square radius.

One obtains $E_B = \sqrt{6}\ \hbar\omega$. This is very close to 2.5 $\hbar\omega$, and indeed if one extends the Tamm-Dancoff calculation by allowing backward going graphs, i.e. one does an RPA calculation, one also obtains $\sqrt{6}\ \hbar\omega$. (It was pointed out by Kirson[22] that Suzuki[23] obtained a different result $\sqrt{2}\ \hbar\omega$ because he used an interaction with the opposite sign. Thus my result is $\sqrt{4+2}\ \hbar\omega$ while Suzuki's is $\sqrt{4-2}\ \hbar\omega$. Suzuki has done valuable work in obtaining simple formulae for the energies of giant resonances of various multipolarities.)

Core Polarization with the Monopole-Monopole Interaction

When we add a valence nucleon to a closed shell it polarizes the core.[24] We then can define a polarization charge. The most familiar example is the E2 core where we replace

$$\sum_{\text{protons}} r^2\, Y_2$$

by

$$(1 + e_p) \sum_{\text{protons}} r^2\, Y_2 + e_N \sum_{\text{neutrons}} r^2\, Y_2 \quad .$$

A popular choice (though not necessarily correct) is to take $e_p = 1/2$, $e_N = 1/2$. In general the polarization charge depends on the multipole, the spin and the isospin $e = e(L,S,T)$.

In our monopole example the polarization charge corresponds to an isotope shift—the change in radius of the core of a nucleus when we add a valence nucleon.

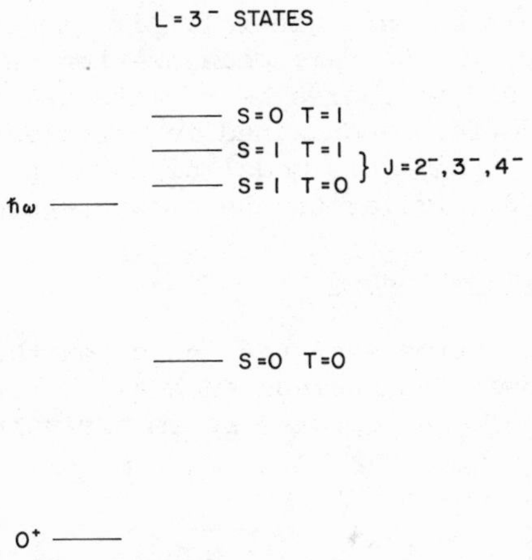

Fig. 10. The spin-isospin energy dependence of giant resonances, e.g. $L=3^-$ states in ^{16}O.

With the simple interaction we can easily calculate the polarization charge (see Fig. 10) in first order, TDA and RPA. The first order result is $e = -1/2$. The TDA result is $e_{TDA} = -2/5$. We can understand this result as being due to replacing the energy denominator 1/2 by 1/2.5 $\hbar\omega$, where 2.5 $\hbar\omega$ is the energy of the breathing mode state in TDA.

Another way of seeing this is to note that with schematic interactions, if the first order result is e^0, then the TDA result is $e^0/(1-e^0/2)$ and the RPA result is $e^0/(1-e^0)$. Thus, the RPA result is $-1/3$. We can do the calculation in yet another way—a Hartree calculation for the odd system. We consider the difference in square radius between that of a core plus one valence nucleon and that of a core. The square radius goes as Σ/ω. Thus

$$\Sigma/\omega = \frac{\Sigma_C}{\omega_C} + (e_{HF}+1)\left(\frac{\Sigma-\Sigma_C}{\omega_C}\right)$$

This reads: the square radius of the A+1 system Σ/ω is equal to the square radius of the core Σ/ω plus the square radius of the valence particle assuming an effective charge $(e_{HF}+1)$. We find

$$(e_{HF}+1) = \frac{(\Sigma^{2/3} - \Sigma_C{}^{2/3})\Sigma^{1/3}}{(\Sigma-\Sigma_C)} \quad .$$

E.g. for ^{16}O $\Sigma_C = 36$, for ^{17}O $\Sigma = 36+7/2$. This result looks quite different than the others. But if we linearize the above expression we find $e_{HF} = -1/3$. This is the same as the RPA result.

The above result is more general: if we linearize the Hartree-Fock equation we obtain the same polarization charge as in an RPA calculation. It would of course be of interest to see how important are the non-linear effects contained in Hartree-Fock. (N.B. Bhaduri points out amusingly that this model gives us a fractional effective charge $(1+e_{HF}) = 2/3$, similar to the up quark charge.)

Isoscalar Quadrupole Resonances

We can make an estimate of the energy of the isoscalar quadrupole state in a closed L-S nucleus such as ^{16}O or ^{40}Ca. Let us use a hybrid model where we accept the schematic results for the polarization charge

$$e_{TDA} = \frac{e_0}{(1-e_0/2)} \quad ; \qquad e_{RPA} = \frac{e_0}{(1-e_0)}$$

where e_0 is the first-order polarization charge. The renormalized E2 operator is

$$(1+e_p)\sum_p r^2\, Y_2 + e_n \sum_n r^2\, Y_2 \quad .$$

We accept the popular choice $e_n = 0.5$ and $e_p = 0.5$. This leads to an isoscalar polarization charge $e_{SCL} \equiv e_p + e_n = 1$. We identify this with the RPA value. In that case $(e_0)_{SCL} = 0.5$ and $(e_{TDA})_{SCL} = 2/3$. But e_{TDA} is obtained by replacing the energy denominator $1/2\ \hbar\omega$ by $1/E_{TDA}$. This leads to the result $E_{TDA} = 3/2\ \hbar\omega$ for the isoscalar quadrupole state. If we went over to the RPA we would get $E_{RPA} = \sqrt{2}\ \hbar\omega$ which is very close to this result.

Schematic versus delta interaction—degenerate model. This work was done by A. Abbas and myself.[25] We can also obtain the $L=2^+$ $S=0$, $T=0$ states by a matrix diagonalization. In ^{16}O the particle-hole states which enter in L-S coupling are:

$$(0f\ 0p^{-1})^{L=2} \quad (1p\ 0p^{-1})^{L=2} \quad \text{and} \quad (0d\ 0s^{-1})^{L=2} \quad .$$

We neglect the one-body spin-orbit interaction and assume a central two-body interaction, so L-S is good.)

We will consider a degenerate model, where the unperturbed energy of these states is $2\ \hbar\omega$. We now turn on the particle-hole interaction. If we use a schematic interaction, as was used by Brown and Bolsterli[26] we know what will happen: one collective state will come down in energy and two will remain at the unperturbed position $2\ \hbar\omega$. [If we choose the strength correctly then the position of the collective state can be made to coincide with $\sqrt{2}\ \hbar\omega$. The collective state carries all the B(E2) strength.]

When Abbas[25] did the calculation using a delta interaction rather than a schematic one, an amusing thing happened—two collective states came down in energy and one state remained at 2 $\hbar\omega$. (In ^{40}Ca where there are 6 particle-hole configurations with L=2, S=0, T=0, three will come down and 3 will be unperturbed.) These results depend on the use of harmonic oscillator radial integrals. Why is this so?

The delta interaction is nearly schematic but not quite

$$\langle (P_A H_A^{-1})^L \, V (P_B H_B^{-1})^L \rangle = \text{Constant} \begin{bmatrix} P_A H_A L \\ 0\;0\;0 \end{bmatrix} \begin{bmatrix} P_B H_B L \\ 0\;0\;0 \end{bmatrix} \times$$

$$\times \int R_{PA} \, R_{HA} \, R_{PB} \, R_{HB} \, r^2 dr$$

Everything factorizes into $(P_A H_A)$ and $(P_B H_B)$ except the radial integral. We can write the matrix element of the delta interaction

$$\int f_{P_A H_A}(r) \, f_{P_B H_B}(r) dr \quad .$$

The schematic interaction, on the other hand, has the structure $g_{P_A H_A} \, g_{P_B H_B}$.

Details about the number of states which do and which do not come down are given by Abbas and me so I won't discuss this any further. It is interesting to note however, that the two interactions give quite different results.

It should also be noted that the density dependent Skyrme[27] interactions give results that are closer to the schematic than the delta interaction—that is they tend to concentrate the collectivity into one state.

Skyrme interaction for the isoscalar quadrupole state—also isoscalar. We are going progressively to more detailed interactions. We next consider the Skyrme interaction,[27] which, although not realistic, is able to correlate a good deal of data over a wide range of the periodic table. The interaction is:

$$V = -t_o(1 + xP^\sigma)\delta(\vec{r}_1 - \vec{r}_2) + \frac{t_1}{2}\left[\delta(\vec{r}_1 - \vec{r}_2)k^2 + k^2\delta(\vec{r}_1 - \vec{r}_2)\right] +$$

$$+ t_2 \, \vec{k}\cdot\delta(\vec{r}_1 - \vec{r}_2)\vec{k} + \frac{t_3}{6}\,\rho^\sigma\left(\frac{\vec{r}_1+\vec{r}_2}{2}\right)(1 + x_3 P^\sigma)\,\delta(\vec{r}_1 - \vec{r}_2)$$

where

$$\vec{k} = \frac{\vec{\nabla}_1 - \vec{\nabla}_2}{2} \quad .$$

We have generalized the interaction somewhat by raising the density to a power σ. The original Skyrme interaction had a linear dependence on ρ.

In the context of the baryon and meson spectrum, M. Harvey noted that a delta interaction can cause the pion and nucleon masses to 'go through the floor', i.e. the energy goes to $-\infty$ if the delta interaction is used in a self-consistent way. Hence good results obtained with *restricted* use of the delta interaction are suspect.

One encounters the same problem with nuclei: it can be argued that a delta interaction could be used as a residual interaction between valence nucleons in a single major shell. But an attractive delta interaction in a Hartree-Fock calculation would lead to a nuclear collapse. In the context of the Skyrme interaction that is why we introduce a repulsive density dependent term $t_3\rho^\sigma\,\delta(\vec{r}_1-\vec{r}_2)$, (and/or a repulsive $k^2\delta$ term).

We see this more explicitly by using as a trial wave function a Slater determinant with harmonic oscillator radial wave functions which are characterized by a single length parameter $b\,(b^2=\hbar/m\omega)$. The energy will be of the form

$$E = \frac{\Sigma'}{b^2} - \frac{At_0}{b^3} + \frac{Bt_1}{b^5} + \frac{Ct_3}{b^{(3+3\sigma)}} \quad .$$

Here, in the kinetic energy term, we have $\Sigma' = \hbar^2/2m\ \Sigma$ $[\Sigma = \mathrm{sum}(2n+\ell+3/2)]$. The matrix element of $\delta(\vec{r}_1-\vec{r}_2)$ goes as $1/b^3$. This leads to the attractive $-At_0/b^3$ term. If we only had this plus the kinetic energy term then we would clearly get a collapse ($E \to -\omega$) as $b \to 0$.

Repulsive $k^2\delta(\vec{r}_1-\vec{r}_2)$ and $\rho^\sigma\delta(\vec{r}_1-\vec{r}_2)$ terms ($\sigma > 0$) will clearly stabilize the situation. That is, there will now be a value $b = b_0$ where $dE/db = 0$. At this point we can evaluate the compressibility

$$K = \frac{1}{A}\, b^2 \left.\frac{d^2E}{db^2}\right|_{b_0} \quad .$$

In the simple case where $t_1 = t_2 = 0$, i.e. we have no velocity dependence but only an energy dependence, one finds[28]

$$K = \frac{1}{A}\left[\langle T\rangle + 9E_B + \sigma(3\langle T\rangle + 9E_B)\right] \quad .$$

Here $\langle T\rangle/A$ is the mean kinetic energy per nucleon and E_B/A the binding energy per nucleon. Using $\langle T\rangle/A = 20$ MeV, $E_B/A = 8$ MeV we get

$$K = (92 + \sigma 132)\ \mathrm{MeV} \quad .$$

Note that K is linear in σ. It should be emphasized that as we change σ, we are keeping the binding energy and radius fixed. Other gross properties, especially single particle energies do not change in any significant way as we change σ. This shows that we cannot claim to really know the compressibility merely because our interaction gives us good radii, binding energies and single particle energies.

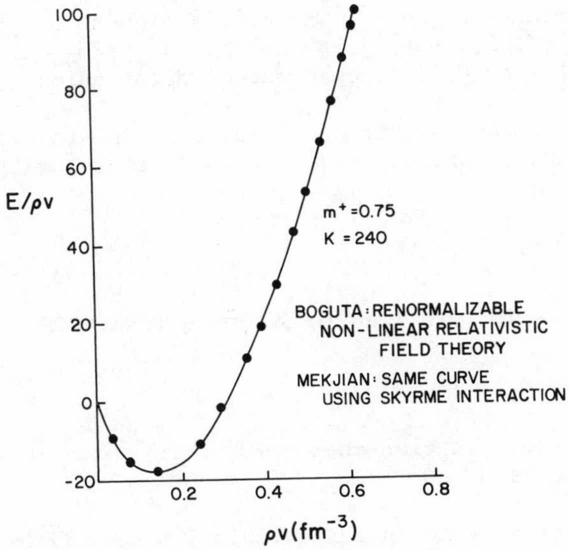

Fig. 11. The equation of state-energy versus density.

Amusingly, Mekjian,[29] using one version of the Skyrme interaction gets an identical curve for the energy as a function of density, as does Boguta who used a renormalizable non-linear relativistic field theory, (Fig. 11). As mentioned in the context of the schematic model, the energy of the breathing mode state gives us information about K. Current experiments favor $E_B \approx 2\hbar\omega$. This would imply a somewhat smaller K than most Skyrme interactions give.

We should also mention work of Jennings and Jackson.[30] They point out that the breathing mode strength is not located in one state but is distributed over a wide energy interval. Therefore different definitions of K can lead to different results. The K we have been talking about so far is K_{sc} where sc stands for scaling. Another definition is K_A—the compressibility obtained by doing a Hartree-Fock calculation in which there is a constraint on the mean square radius of the nucleus. In infinite nuclear matter $K_A \simeq 0.7\, K_{sc}$. The quantity K_{sc} is obtained from the energy $\sqrt{m_3/m_1}$ whereas K_A is from $\sqrt{m_1/m_{-1}}$. Here $m_K = \sum_n (E_n - E_o)^K\, B(EL\ 0 \rightarrow n)$ where $(E_n - E_o)$ is the excitation energy and B(EL) the transition rate to the state n.

We now get back to the quadrupole properties with Skyrme interactions. We use deformed oscillator wave functions in our Slater determinant. $\psi = \psi(x/b_x, y/b_y, z/b_z)$ with $b_x b_y b_z = b_0^3$. We set $b_x = b_y \neq b_z$. The kinetic energy now becomes

$$\frac{\Sigma_x' + \Sigma_y'}{b_x^{\,2}} + \Sigma_z'/b_z^2 = \frac{(\Sigma_x' + \Sigma_y')}{b_0^{\,3}}\, b_z + \Sigma_z'/b_z^2$$

where $\Sigma_x' = \text{Sum}(N_x + 1/2)$ etc. The matrix element of $\delta(\vec{r}_1-\vec{r}_2)$ i.e. $\delta(x_1-x_2)\ \delta(y_1-y_2)\ \delta(z_1-z_2)$ goes as $1/b_x\ 1/b_y\ 1/b_z = 1/b_0{}^3$. Hence this potential energy is independent of deformation.

Hence, if we do a variational calculation with just a delta interaction, the condition $\partial E/\partial b_z = 0$ leads to the well-known Mottelson conditions[31]

$$(b_z/b_0)^3 = \frac{2\Sigma_z}{(\Sigma_x+\Sigma_y)} \quad .$$

These conditions in turn lead to a simple result for the isoscalar E2 charge

$$e_{sc} = e_p + e_n = 1$$

(e_p and e_n are polarization charges). This agrees with the popular prescription $e_p = 1/2$, $e_n = 1/2$.

One also obtains in an RPA calculation $E_{2+} = \sqrt{2}\ \hbar\omega$. We see that all these nice results stem from the fact that the potential energy does not depend on deformation for a zero range interaction.

The effect of the finite range term is interesting, we can write

$$k^2\delta(\vec{r}_1-\vec{r}_2) \quad \text{as} \quad (k_x^2 + k_y^2 + k_z^2)\ \delta(\vec{r}_1-\vec{r}_2) \quad .$$

Clearly for a deformation of a closed shell nucleus we will get a contribution of the form

$$\left(\frac{1}{b_x{}^2} + \frac{1}{b_y{}^2} + \frac{1}{b_z{}^2}\right)\frac{1}{b^3} \quad .$$

This looks something like the kinetic energy term and indeed can be lumped together with it, in terms of an effective mass. Indeed, the effective mass, in the context of the Skyrme interactions[27] can be defined as

$$\frac{m}{m^*} = \text{KINETIC ENERGY + FINITE RANGE ENERGY} = \frac{\Sigma'}{b^2} + Bt_1/b^2$$

because $t_1 > 0$, m^*/m is less than one.

An effective mass $m^*/m < 1$ leads to more widely spaced single particle energies. In nuclear matter the level spacing is $(p_1{}^2-p_2{}^2)/2m^*$. In a finite nucleus we have approximately that if a level spacing for $m^*/m = 1$ is $\hbar\omega$, it now becomes $\hbar\omega\ m/m^*$. The energy of the isoscalar quadrupole state can be shown to become

$$E_2{}^+ = \sqrt{2}\ \hbar\omega/\sqrt{m^*/m} \quad .$$

The work described here was done by M. Golin and myself.[32] (The

Skyrme interaction[27] does not have the feature of nature, that $m^*/m \simeq 2$ for levels near the Fermi surface and $m^*/m < 1$ for levels far away from the Fermi surface. As noted by Brown and Speth[33] this could lead to a small modification of the formula.)

Overview: What we have shown up to now is that the energies of isoscalar modes depend on some gross bulk properties of nuclei, e.g., effective mass, compressibility. They are not directly sensitive to exotic processes like meson exchange etc.

Spin Dependent Modes

In the L-S coupling scheme there are 4 types of giant resonances: S=0 T=0, S=1 T=0, S=0 T=1, S=1 T=1.

In discussing the breathing mode state and isoscalar quadrupole state, we have been focussing on the S=0 T=0 mode. As we shall soon see these modes lie lowest in energy. We should remind ourselves, however, that the spin admixtures will cause these states to come down in energy. (Also because of the Coulomb interaction, there can be weak isospin admixtures.)

To see how these states arrange themselves in energy we consider a calculation of Towner, Castel and Zamick.[34] We use a delta interaction of the form

$$-G(1 + xP^{\sigma})\ \delta(\vec{r}_1-\vec{r}_2)\,; \qquad P^{\sigma} = \left(\frac{1+\sigma_1\cdot\sigma_2}{2}\right) .$$

Actually, we set all the radial matrix elements equal so it becomes a schematic interaction.

In a Tamm-Dancoff calculation the states move from their unperturbed positions by amounts proportional to $-3G$ for S=0 T=0, $G(1-2x)$ for S=1 T=0, G for S=1 T=1, and $G(1+2x)$ for S=0 T=1, see Fig. 12. Common values of x and G are 1/3 and 380 MeV-fm^3. Hence, we see that only one mode gets a collective downward shift, S=0 T=0; the other three modes go up in energies.

Amusingly, the mode which has caused the most interest in recent years, the pionic mode with S=1 T=1 goes up in energy by an amount proportional to G, i.e. independent of the spin dependent parameter x. Current experimental evidence is in agreement with the fact that these pionic mode states are not very low in energy. (If they were low we would have strong precritical phenomena, or if they actually collapsed below the ground state in self-consistent RPA calculations, we would have a new pion condensed ground state.)

The fact that our simple calculation gets the pionic modes at about the right energy is probably an accident; a cancellation of two effects—a tensor interaction which wants to bring these pionic

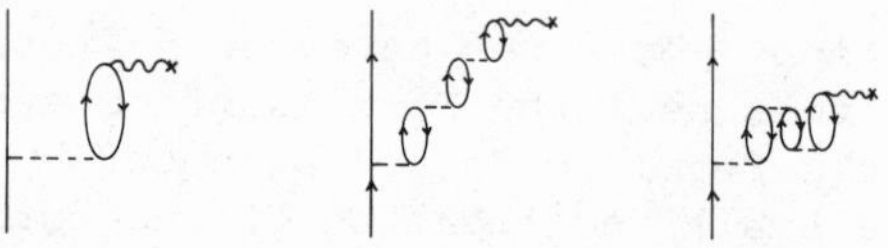

Fig. 12. Polarization charges calculated in first order, TDA and RPA.

modes down in energy, and a stronger interaction of the type that we have considered i.e., delta which pushes the states up in energy.[35] This same tensor interaction will not contribute in first order to the binding energy of a closed shell nucleus, or indeed to the energies of the S=0 T=0 modes like the isoscalar quadrupole, octupole (which, by the way, is the most *familiar* isoscalar mode because it lies lowest in energy) or monopole modes. (However, the tensor interaction acting in second order, i.e. with correlated ground state wave functions, will contribute to all of these modes.) Even to this day the tensor interaction has proven to be surprisingly elusive. There are less than a handful of cases where the interaction seems to be needed in day to day nuclear spectroscopy. One example is the suppressed beta decay of ^{14}C (J=0 T=1) to (or is it from?) ^{14}N (J=1 T=0). [*Ed. note: come now.*] Of course there is the deuteron itself, if you want to stretch things. But can you think of any other examples?

A possible new region of importance of the tensor interaction is in core polarization at intermediate to large momentum transfer magnetic electron scattering.[35] For example in the inelastic scattering from the J=0 T=0 ground state of ^{12}C to the 15.11 MeV state J=1, T=1 there is a large second maximum in the cross section at about $q \simeq 2.5\ fm^{-1}$, which cannot be explained by using wave functions consisting of 8 nucleons in the p-shell. Calculations in which core polarization is included, especially with a strong tensor interaction, more or less succeed in obtaining a large second maximum.

Evidently, this tensor interaction, although strong, is not strong enough to overcome the repulsion due to a short range delta interaction, and therefore does not lead to pion condensation or even to precritical phenomena. The importance of the two-pion exchange part of the nucleon-nucleon interaction has been emphasized by Toki and Weise. There is, however, in my opinion, still enough uncertainty about the strength of the tensor interaction, and about other things, so that we should still keep the matter open.

Let us lastly show a connection between the delta force used here and the Migdal parameters. In the Migdal theory one takes only the direct matrix elements of the particle hole interaction. The exchange operator is $P^M P^\sigma P^\tau$. If we use a delta interaction only relative $\ell=0$ states contribute so $P^M = 1$. Thus, to form an antisymmetric state we operate with $(1 - P^\sigma P^\tau)$ on an antisymmetrized state. But we can also lump this in with the interaction

$$-G \langle (1 + xP^\sigma)\ \delta(\vec{r}_1-\vec{r}_2)\rangle_a = -G \langle (1 + xP^\sigma)(1-P^\sigma P^\tau)\ \delta(\vec{r}_1-\vec{r}_2)\rangle_{na}$$

where a and na denote antisymmetrized and non-antisymmetrized states, respectively. Thus the effective interaction is

$$\frac{G}{4}\left[-3 + (1-2x)\sigma_1\cdot\sigma_2 + (1+2x)\tau_1\cdot\tau_2 + \sigma_1\cdot\sigma_2\ \tau_1\cdot\tau_2\right] .$$

Note that the coefficients of 1, $\sigma_1\cdot\sigma_2$, $\tau_1\cdot\tau_2$ and $\sigma_1\cdot\sigma_2\ \tau_1\cdot\tau_2$ are the same as the energy shifts for the four different ST modes. Indeed it can be proven that if one uses direct matrix elements, then the interaction, say $\tau_1\cdot\tau_2\ \delta(\vec{r}_1-\vec{r}_2)$ affects only the S=0 T=1 mode etc.

In the Migdal theory the interaction is written

$$C_0\left[f_0 + g_0\ \sigma_1\cdot\sigma_2 + f_0'\ \tau_1\cdot\tau_2 + g_0'\ \sigma_1\cdot\sigma_2\ \tau_1\cdot\tau_2\right]$$

with $C_0 \simeq 380$ MeV-fm^3 (the same choice we took for G). This has four independent parameters whereas the delta interaction has only 2 (G and x). The four parameters can be justified by the fact that the Migdal interaction is an effective interaction, which contains the effects of high order perturbative correlations.

With $C_0' = 380$ MeV a currently favored value of g_0' is about 0.7. Note, however, with our delta interaction with $G = 380$ MeV we would get $g_0' = 1/4$. In our 2-parameter delta theory we could increase G by a large factor—2.5 to 3—but this would destroy the fit to the well studied T=0 S=0 modes. In the Migdal theory we can change g_0' without making a change in f_0. (Strictly speaking, there are sum rules which the f's and g's must obey, but they are often ignored. The physical origin of the sum rules can be seen from our previous discussion of two particle states e.g. ^{42}Ca. We noted that for the configuration j^2 if J is even T is odd and if J is odd T is even. Using the Migdal interaction, only for certain combinations of the parameters do we get no contribution from the forbidden particle-particle modes.) The small value of g_0' that we get ($g_0' = 1/4$) is due, as previously mentioned, to the fact that we are using the delta interaction to mock up an attractive tensor interaction and a stronger delta interaction.

Higher Order Effects in Core Polarization

The popular way of calculating core polarization is to do an RPA calculation. This was discussed in the context of the soluble monopole-monopole interaction.

In the usual application, the polarization charge, calculated when one has one nucleon beyond a closed shell is used also for several nucleons beyond a closed shell. Is this procedure correct? There are examples where this seems to work—in large portions of the 0d-1S shell—as shown by Wildenthal, one seems able to get by with one set of E2 and M1 polarization charges. For E2, I recall, the values are $e_n = 0.5$ $e_p = 0.2$. But we can find other models for which this does not work. At the last NATO conference (Banff, 1978) I discussed E0 polarization charges, i.e. isotope shifts.[36] My only excuse for discussing this again is that there are more experimental data available, and that these data seem to bear out the idea of an *effective* two-body radius operator.

Consider as an example the calcium isotopes. If one adds a neutron to ^{40}Ca the charge radius of the ground state of ^{41}Ca will change. Since the neutron has no charge (although it does have a charge distribution which could affect things) we expect the isotope shift to arise from the fact that the valence neutron polarizes the ^{40}Ca core.

If we calculate an effective E(0) charge, that is, if we write the operator as $e(E0) = \sum_n r^2$, then we would expect a linear behavior for the change in square charge radius of the calcium isotopes. This is not observed.

I had suggested that if we postulate that the effective radius operator has a two-body part

$$\Sigma r^2 + \sum_{i<j} V(ij)$$

then the formula for the square radius of the calcium isotopes would have the same form as the formula for binding energies that Talmi[3] has obtained

$$r^2(A+n) - r^2(A) = nC + \frac{n(n-1)}{2}\alpha + \left[\frac{n}{2}\right]\beta$$

where $[n/2] = n/2$ for even n and $(n-1)/2$ for odd n. In the proceedings of the last NATO conference[36] the radii of some calcium isotopes were not known so the formula given above had some predictions.

In Table 5 I show the *proton* (as opposed to charge) radii. The experimental analysis is due to Wohlfahrt et al.[37]

There is a clear prediction that the charge radius of ^{41}Ca would not lie half way between ^{40}Ca and ^{42}Ca but would be much closer to ^{40}Ca. This has been borne out by experiments of Träger et al.[38] who determined the rms radii of ^{41}Ca, ^{43}Ca and ^{45}Ca by laser spectroscopy techniques. They obtain $^{41}\langle r^2\rangle^{1/2} = {}^{40}\langle r^2\rangle^{1/2} \pm 0.006$ fm. The results of Träger et al. are presented in Fig. 13. The ordinate is

Table 5. Proton Radii for Calcium Isotopes

n	$\sqrt{r^2(A+n)_p}$ - $\sqrt{r^2(A)_p}$ ($\times 10^{-3}$ fm) Expt.	Our Formula
1		0.45
2	36	34.5
3	25	25.0
4	53	49.1
5		29.6
6	37	43.7
7		14.2
8	21	18.3

$$\frac{\delta\langle r^2\rangle\ (A-40)}{\delta\langle r^2\rangle (42-40)}\ .$$

This quantity is of course *one* for ^{42}Ca. The even-odd staggering is clearly shown.

E6 Suppression: Open Shell Nucleus Effect?

I here report on the work of Castel, Satchler and myself. The motivation is that E6 transitions in the 0f-1p shell were observed and their rates were much lower than one would expect by doing the usual perturbation calculations. The experiments were in ^{53}Fe where a $19/2^-$ isomeric state (MBZ predicts this state is isomeric because the $15/2^-$ and $17/2^-$ states are at higher energies than the $19/2^-$ so that M1 and E2 decay channels are not open) decays by E6 to the $7/2^-$ ground state. Also in ^{50}Cr inelastic scattering from the 0^+ ground state to a 6^+ excited state was strongly suppressed.

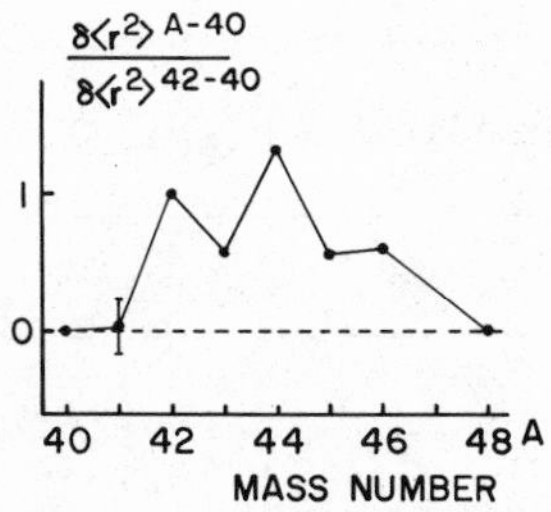

Fig. 13. Relative changes in the mean square nuclear charge radii of the calcium isotopes, (normalized to the ^{42}Ca-^{40}Ca change).

Table 6. Comparison of Theory and Experiment for E6 Polarization Charges

Force	Schematic	S.T.O.	Reid	Expt.
δe_p	-0.10	0.10	0.21	-0.44[a]
δe_n	0.46	0.52	0.55	-0.50[b]

[a]McHarris et al. ^{53}Fe.
[b](e,e′) Saclay [50]

A comparison of theory and experiment is given in terms of E6 polarization charges in Table 6. In the above theories, one couples a valence nucleon to an RPA vibration. This vibration must have L=6.

We try to take into account higher order effects in a physical way. We ask, what are the effects of *quadrupole* distortions on the E6 moments? The P_6 operator is

$$\frac{-5}{16}\, r^6 \left[1-21\cos^2\theta + 63\cos^4\theta - \frac{231}{5}\cos^6\theta\right] .$$

We may imagine using deformed oscillator wave functions $\psi(x/b_x, y/b_y, z/b_z)$. By making the transformation $\bar{x}/b = x/b_x$ etc. we restore the wave function to sphericity in the coordinates $\bar{x},\bar{y},\bar{z}$ but the operator gets transformed. To first order in the familiar deformation parameter δ

$$P_6 \to \bar{P}_6 = P_6 + \frac{5}{16}\,\delta r^6 \left[1-3\cos^2\theta - 21\cos^4\theta + \frac{147}{5}\cos^6\theta\right] .$$

We now evaluate $\bar{P}_6$ in the Nilsson intrinsic state for the appropriate deformation. [Recall that although the Nilsson Hamiltonian allows for $\Delta N = 2$ mixing, the actual calculations were limited to $\Delta N = 0$. By replacing P_6 by $\bar{P}_6$ we are approximately (though by no means perfectly) allowing for $\Delta N = 2$ mixing.] The results are shown in Table 7, in which the term proportional to δ comes from replacing P_6 by $\bar{P}_6$. Note

Table 7. Predictions for $<\bar{P}_6>$ for ^{50}Ca

δ	$<\bar{P}_6>$
0	-9.3 − 14.7δ = -9.3
0.1	-9.4 + 31.3δ = -6.3
0.2	-8.1 + 46.5δ = 2.1
0.3	-7.4 + 62.3δ = 11.3

Table 8. Predictions for $<\overline{P}_6>$ for ^{53}Fe

δ	$<\overline{P}_6>$
0	2.3 - 23.3δ = 2.33
0.1	3.1 - 11.0δ = 1.97
0.2	5.1 + 1.0δ = 5.27
0.3	5.9 + 16.1δ = 10.74

that if we had used P_6 we would have obtained a negligible variation as a function of δ, from -9.3 to -7.4. But the inclusion of the quadrupole term causes as sign change (from -9.3 at δ=0 to +11.3 at δ=0.3). This result is then encouraging for our thesis that open-shell effects can be important.

The ^{53}Fe results shown in Table 8 are not so clear cut. Only for a very narrow band of δ does the matrix element decrease. For $\delta \gtrsim 0.2$ the quadrupole distortion increases the matrix element.

Clearly for this nucleus the calculation will have to be done with greater care, i.e. a complete $\Delta N = 2$ Nilsson calculation, together with the inclusion of hexadecapole deformations.

We have at least demonstrated that open shell effects *can* be very important and that they should be thoroughly investigated before more exotic mechanisms are proposed.

ACKNOWLEDGMENT

The author wishes to thank the U.S. National Science Foundation for financial support.

REFERENCES

1. J. D. McCullen, B. F. Bayman and L. Zamick, Phys. Rev. 134B:515 (1964).
2. J. N. Ginocchio and J. B. French, Phys. Lett. 7:137 (1963).
3. I. Talmi, in: "Proceedings of the Rehovot Conference on Nuclear Structure," North-Holand, Amsterdam (1958); I. Talmi and I. Urna, Ann. Rev. Nucl. Sci. 10:353 (1960); I. Talmi, Rev. Mod. Phys. 34:704 (1962).
4. W. J. Gerace and A. M. Green, Nucl. Phys. 93A:110 (1967).
5. G. E. Brown and A. M. Green, Phys. Lett. 15:168 (1965); Nucl. Phys. 75:401 (1966).
6. I. S. Towner and F. C. Khanna, Phys. Rev. Lett. 42:51 (1979).

7. G. F. Bertsch, Nucl. Phys. 79:209 (1966); T. T. S. Kuo and G. E. Brown, Nucl. Phys. 85:40 (1966).
8. B. F. Bayman and A. Lande, Nucl. Phys. 77:1 (1966).
9. W. C. Haxton, G. J. Stephenson Jr. and D. Strottman, Phys. Rev. Lett. 47:153 (1981); see also H. Primakoff and S. P. Rosen, Ann. Rev. Nucl. Part. Sci. 31:145 (1981).
10. R. D. Bardin, P. J. Gollon, J. D. Ullman and C. S. Wu, Nucl. Phys. A158:337 (1970).
11. V. A. Khodel, Phys. Lett. 32B:583 (1970).
12. R. D. Lawson, Phys. Rev. 124:1500 (1981).
13. S. G. Nilsson, K. Dan. Vidensk. Selsk. Mat.-Fys. Medd. 29:16 (1955).
14. e.g. W. C. Haxton, lecture at "Conference of Spin Modes in Nuclei," (1982).
15. L. Zamick and N. Auerbach, Phys. Rev., to be published.
16. W. Knüpfer et al., Phys. Lett.77B:367 (1979); I. S. Towner and F. C. Khanna, Ref. 6; H. Toki and W. Weise, Phys. Lett. 97B:12 (1980); A. Bohr and B. R. Mottelson, *ibid.* 100B:10 (1981); G. E. Brown and M. Rho, Nucl. Phys. A372:397 (1981).
17. C. D. Goodman et al., Phys. Rev. Lett. 44:1755 (1980); C. D. Gaarde et al., Nucl. Phys. A369:258 (1981).
18. G. F. Bertsch and I. Hamamoto, to be published.
19. L. Zamick, A. Abbas and T. Halemane, Phys. Lett. 103B:87 (1981).
20. L. Zamick, Nucl. Phys. A232:13 (1974).
21. S. Goshen and H. J. Lipkin, Ann. Phys. 6:301 (1959).
22. M. W. Kirson, Nucl. Phys. A317:388 (1979).
23. T. Suzuki, Phys. Rev. C8:2111 (1973); Nucl. Phys. A220:569 (1974).
24. L. Zamick, Nucl. Phys. A249:63 (1975); Nucl. Phys. A254:544 (1975) (E).
25. A. Abbas and L. Zamick, Phys. Rev. C22:1755 (1980).
26. G. E. Brown and M. Bolsterli, Phys. Rev. Lett. 3:472 (1959).
27. D. Vautherin and D. M. Brink, Phys. Rev. C5:626 (1972).
28. L. Zamick, Phys. Lett. 47B:119 (1973).
29. A. Mekjian, private communication.
30. B. K. Jennings and A. D. Jackson, Nucl. Phys. A342:23 (1980).
31. B. R. Mottelson, "The Many-Body Problem," Wilson, New York (1958).
32. M. Golin and L. Zamick, Nucl. Phys. A249:320 (1975).
33. G. E. Brown, J. S. Dehsa and J. Speth, Nucl. Phys. A330:290 (1979).
34. I. Towner, B. Castel and L. Zamick, Nucl. Phys. A365:189 (1981).
35. J. Meyer-ter-Vehn, Phys. Rep. C74:323 (1981); E. Oset, H. Toki and W. Weise, Phys. Rep. C83:281 (1982); T. Suzuki, H. Hyuga, A. Arima and K. Yazaki, Phys. Lett. 106B:19 (1981); H. Toki and W. Weise, Phys. Lett. 92B:265 (1980).
36. L. Zamick, in: "Common Problems in Low and Medium Energy Nuclear Physics," B. Castel, B. Goulard and F. C. Khanna, eds., Plenum, New York (1979); Ann. Phys. 66:784 (1971).
37. H. D. Wohlfahrt, E. B. Shera, M. V. Hoehn, Y. Yamazaki,

G. Fricke and R. M. Steffen, Phys. Lett. 73B:131 (1978).
38. F. Träger et al., Z. Phys. A290:143 (1979); *ibid.* 290:345 (1979); *ibid.* 292:401 (1979).
39. B. Castel, R. Satchler and L. Zamick, submitted for publication.

UNITY IN DIVERSITY—A SUMMARY TALK

F.C. Khanna

Atomic Energy of Canada Limited
Chalk River Nuclear Laboratories
Chalk River, Ontario, Canada K0J 1J0

On the first day of this summer school the organizers asked me to summarize the proceedings. This assures you that this talk was not written before I arrived here, but there is still the possibility that it contains some of my prejudices and these should be apparent to this audience.

The organizers have been presumptious in assuming that in a 45 minute presentation the careful and eloquent talks (34) of 13 different speakers can be condensed in a capsule form. In the "thermodynamic" limit this would allow approximately one minute/talk and a sermon of 11 minutes. This will not do justice to the eloquence of the lecturers. But there is another approach to this problem that is "chirally invariant". There were six high energy talks and let me label them as "leftist" by virtue of their tendency towards revolution. There were six talks on nuclear physics and let me label them as "rightist" by virtue of their conservative approach to ideas. Let me classify one of the talks as background (the talk by F. Khanna).

One would expect that in any meeting of the two "camps" there is no yielding of any ground. But the message I would like to convey to you is the basic unity in diversity of ideas. Quantum chromodynamics (QCD), with quarks, gluons, flavor and color, has become reasonably well established as the theory of strong interactions. It is a non-Abelian gauge theory and is asymptotically free. At large Q^2 the running coupling constant $[\alpha_s(Q^2)]$ is small enough to justify use of perturbation theory.

The internal structure of nucleons and mesons is understood in terms of gluons and quarks. There is experimental support for a

Table 1. Results of Lattice Gauge Theory Calculations Using Monte Carlo Techniques

Theoretical	Experiment
$\frac{M_\Delta}{M_\rho} = 1.55 \pm 0.24$	1.58
$\frac{m_p}{m_\rho} = 1.46 \pm 0.23$	1.20
$g_p = 2.99 \pm 0.17$	2.79
$g_n = -1.91 \pm 0.08$	-1.90
$\Lambda = 75 \pm 8$ MeV	

variety of flavors. But it appears that at low Q^2 it is more profitable to consider the structure of nuclei in terms of nucleons and mesons. The color degree of freedom is masked. To establish the unity in diversity, physics has to be done in a region where the two diverse and independent aspects are needed simultaneously. This remains our task for the future.

Before going on to discuss the problems that lie ahead, let me digress and list the success stories of the rightists and the leftists.

The High Energy Viewpoint of the successes in the study of dynamics of strong interactions is that:

i. QCD is a successful theory with strong support from high energy experiments;
ii. quarks have fractional charges and five different flavors so far;
iii. gluons have spin 1 and are colorful;
iv. QCD is a non-Abelian gauge theory with a running coupling constant $\alpha_s(Q^2)$ such that $\lim \alpha_s(Q^2) \to$ small value as $Q^2 \to \infty$. This makes perturbation theory valid at large Q^2. At small Q^2, $\alpha_s(Q^2)$ increases making it imperative to consider non-perturbative approaches;
v. there is one and only one parameter, a scale parameter, Λ;
vi. a non-perturbative approach to QCD is based on a numerical study of lattice gauge theory using Monte Carlo techniques. With an input of the masses of the u, d and s quarks and of the gluon field, numerical studies at zero temperature (T=0) give masses of low-lying baryon states and the g-factors of proton and neutron, as shown in Table 1. These calculations

assume $m_u = m_d = 9.3 \pm 1.2$ MeV, $m_s = 239 \pm 14$ MeV and a lattice size $a = 0.14$ fm;

vii. critical behavior in finite temperature QCD has been obtained from lattice gauge theory by Monte Carlo techniques. There are probably two phase transitions, a deconfinement transition at $T_c \simeq 200$ MeV and a chiral phase transition at $T_{ch} \simeq 250$ MeV. These may imply that there is a confining bag with radius $\sim$1 fm and a chiral bag with a somewhat smaller radius, $\sim$0.6 fm. With the chiral phase transition there is the possibility of Goldstone pions. If further calculations confirm that $T_c \neq T_{ch}$, it will be very interesting for low-energy phenomena;

viii. there is a possibility of using light-cone perturbation theory at large Q^2. There are some very useful features of this approach. One of them is that form factors in a hadronic process may be obtained by quark counting with

$$F^2(Q^2) \sim (Q^2)^{2-n_T}$$

where n_T is the total number of quarks in the process. Such an approach will be very useful for few nucleon systems;

ix. QCD inspired potential models for the low-lying nucleon spectrum are successful;

x. potential models for NN, $\Delta\Delta$ and hidden color states as six-quark states yield some interesting results;

xi. QCD has spectacular support from the good Lord, i.e. Cosmology puts some severe constraints on the number of flavors, number of generations, masses of neutrino etc.

In Fig. 1 the nucleon spectrum is plotted in a manner that nuclear physicists are more used to. Energies are divided by a factor of 100. The spectrum looks like a typical spectrum of an odd-A nuclear system. The main differences are that the energy separation between levels in the particle spectrum is larger by a factor of 100 and most of the levels have a large width ($\sim$100 MeV).

The nuclear physics viewpoint of the successes in the study of the dynamics of strong interaction is that:

i. low-lying states are well-described as rotational or vibrational levels;

ii. the interacting Boson model is simple and is successful in correlating data. The results are spectacular in ^{196}Pt;

iii. shell model and collective models are again useful in correlating data at low energies. The computational complexity of shell model calculations is perhaps much less than that in lattice gauge theories;

iv. giant states in nuclei have been studied by considering the nuclear response to external electromagnetic and weak probes. Valuable information has been gained about the interaction of particles and holes in nuclei and about the propagation of p-h states in many-particle systems even though the surface introduces some complications;

v. anomalons have been observed but defy a logical explanation within the available information about dynamics of strongly interacting particles;

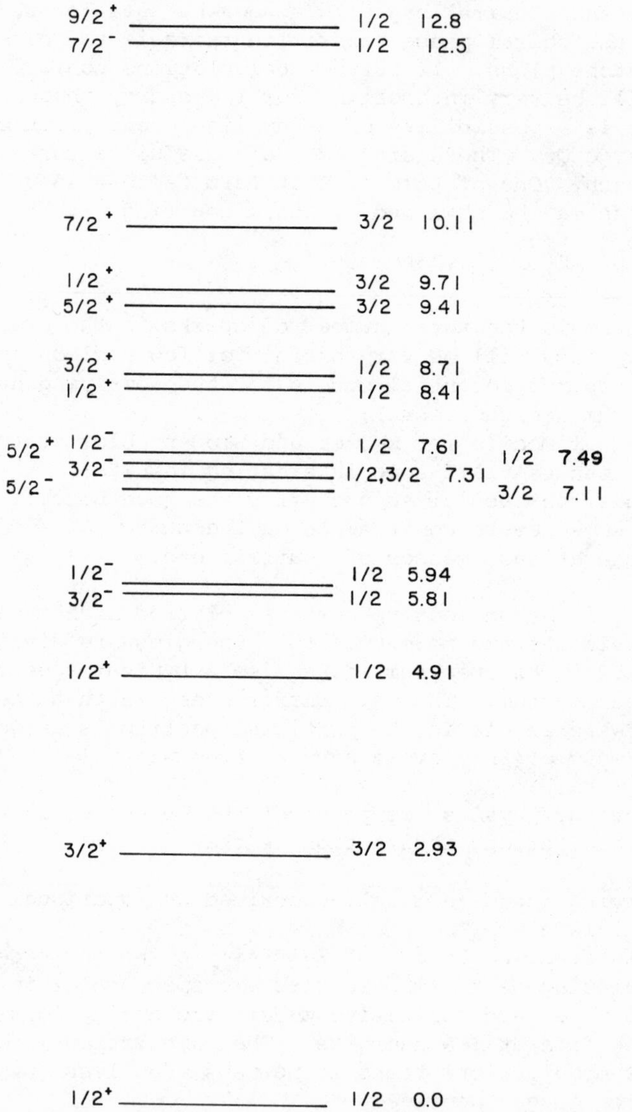

Fig. 1. Spectrum of low-lying baryon states (J^{π}-T). The energies are divided by a factor of 100 so as to provide a resemblance to the energy spectrum of a typical odd-A nuclear system.

vi. a spectacular role for pion condensates in nuclei has been thwarted by the important role played by multiple-pion exchanges;

vii. isobars play an essential role in the quenching of Gamow-Teller strength in nuclei;

viii. double β-decay provides a fundamental experiment in testing grand unified theories (GUTS). Nuclear structure plays a very important role in the understanding of this process;

ix. nuclei are useful as laboratories for testing GUTS and QCD;

x. possibility of producing a quark gluon plasma in heavy ion collisions with energies of several TeV/nucleon. Such events are rare but are found in cosmic ray studies. A single event may provide a large amount of information.

The successes of the high energy physics and nuclear physics approaches given above indicate that the two "camps" are not speaking the same language. Within each area the descriptions are largely

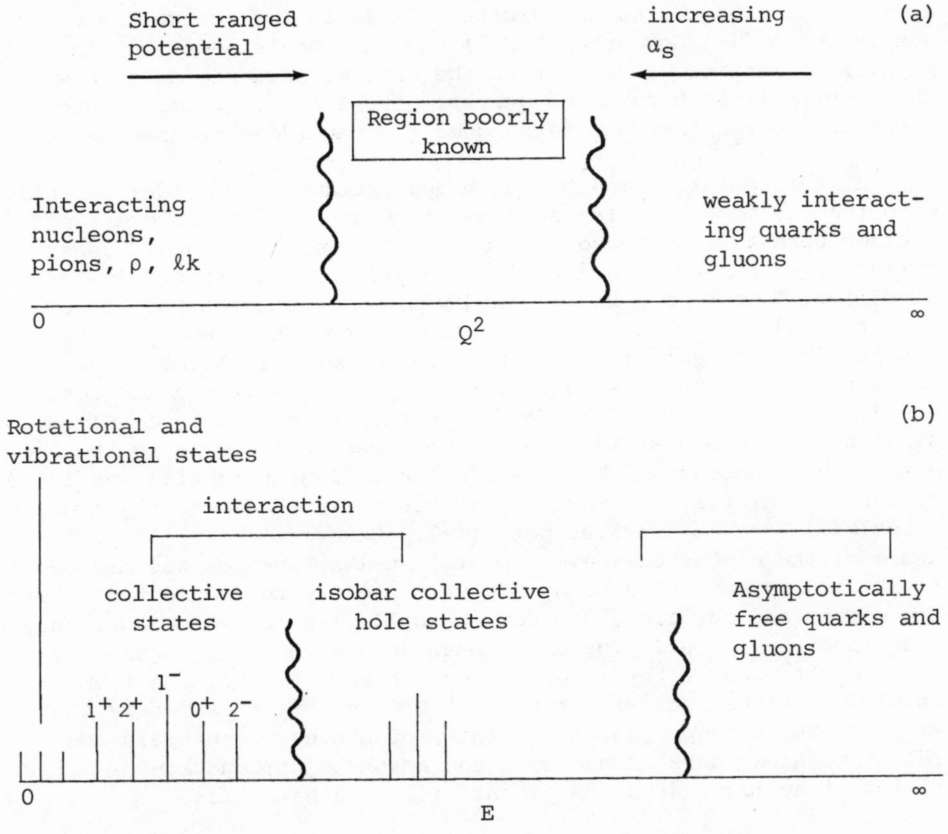

Fig. 2. Variation of physical phenomena in a nuclear system, (a) as a function of Q^2, (b) as a function of E.

adequate, but one has to be aware of discrepancies. Pictorially the areas of uncertainty are shown in Fig. 2. The Q^2 dependence is shown in Fig. 2(a). At low momentum transfers a model of nuclei as composites of nucleons with forces mediated by π, ρ ... etc. is sufficient. As Q^2 is increased the conventional description requires shorter-ranged potentials, which can be viewed as due to exchanges of heavier bosons, which brings about many uncertainties. One also knows that at asymptotic values of Q^2, nuclear systems may be viewed as weakly interacting systems of quarks and gluons. The intermediate region of Q^2 where α_s is increasing in magnitude and the potential model is becoming less and less meaningful is where the future challenge lies.

Similarly, if nuclear excitations are viewed as a function of energy, rotational and vibrational states appear at very low energies. At higher energies (10-30 MeV) there are the nucleonic giant states of various multipolarity. These are followed by isobar-hole collective states (~300 MeV) which are mixed with nucleonic giant states. Again at very high energies there is the weakly interacting plasma of quarks and gluons. It is in the intermediate energy region ($\sim$1 GeV $<$ E $<$ $\sim$5 GeV) that an understanding of the physics is lacking. Again it is the intermediate region of E where the descriptions in terms of quarks, gluons and in terms of nucleons and mesons merge, that new techniques and new ideas are needed.

At this point it may be quite appropriate to consider an analogy with the Coulomb potential as shown in Fig. 3. The discovery of Coulomb potential preceded the construction of the Lennard Jones potential which led to a detailed analysis of the properities of crystals. It may be asked: Could we have discovered the Coulomb potential and its relevance to crystals from a knowledge of the Lennard Jones potential? Probably the answer will be *no*. This is the situation in the case of nuclear physics. A logical sequence would be to discover color fields and then establish that only color singlets should appear in nature. Then one could go on to the discovery of mesons and baryons, to the nuclear potential and lastly to nuclear physics. It may be possible to proceed directly from color fields to the nuclear potential. However, we are all well aware of the historical development. Nuclear physics was discovered first, and we have been going backwards to try to deduce that there are mesons and nucleons, and now we would like to learn about quarks and gluons in nuclei. There has been an unsuccessful attempt by Zeldovich to deduce the internal constituents of nucleons from a knowledge of the nuclear potential. One can see immediately that, with the built-in prejudices in favor of nuclear potentials and reluctant acceptance of nucleons and mesons, introduction of color fields, flavors, quarks and gluons will be a hard task.

The talks during the last two weeks have raised many questions and I would like to provide a partial list of them.

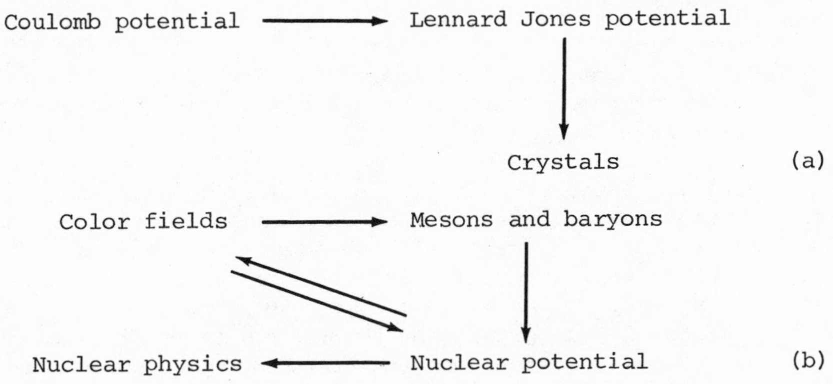

Fig. 3. (a) Description of crystals in terms of the Coulomb interaction has developed through the effective Lennard Jones potential. (b) Possible routes to the description of phenomena in nuclear physics in terms of color fields.

i. How do we bridge the gap between low and high energy and momentum transfer? It will be useful to have models, preferably solvable.

ii. Do we have to depend on a non-perturbative treatment of QCD at small Q^2? This would make life very hard and there is a danger that physics may be obscured in the process.

iii. How far can we push lattice gauge theories? Do they provide sufficient physical insight? It is not clear that inclusion of one or more $q\bar{q}$ pairs will not make the numerical task impossible or at least alter the conclusions drastically.

iv. Are lattice gauge theories a panacea, or do they provide only a hint of the physics involved? Elaboration of such physics may require some intuitively simple models.

v. Do gluons appear only as "phonons" of the lattice? Can there be glueballs?

vi. What is a 6-quark state? Why do the two-nucleon, two-isobar and hidden-color states not exhaust all possibilities for the 6-quark system?

vii. Why does nature not reveal color to us? Are there situations where color fluctuations may occur and how do we know that this has happened?

viii. What are the limits of applicability of proton being considered as a Dirac particle? Only by pushing it hard to show deviations is it possible that the internal structure of proton may be revealed.

ix. How far can the nucleon and meson picture be pushed? There may be processes like heavy-ion reactions that show departures from a composite system of elementary nucleons and mesons. We may need high-energy heavy ion collisions to find an indication of quarks and gluons. Are anomalons an indication of such a breakdown in the description of nuclei in terms of nucleons and mesons?

x. Can the light-cone perturbation theory be pushed to small values of momentum transfer and energy and still yield meaningful results?

There is no doubt that we have to learn to deal with the interesting region where quarks, gluons, nucleon and meson degrees of freedom are mixed. There is also no doubt that there is a basic unity in these diverse descriptions of the dynamics of strongly interacting system. We know that it took us almost thirty years after the postulate of pion physics by Yukawa before we were convinced that meson degrees of freedom are needed in the study of electromagnetic and weak processes in nuclei. We are well aware that quarks are the basic constituents of baryons and mesons. But we have to ask the question: how do we find nuclear processes that require explicit presence of quarks and gluons? It will require hard work, dedication and above all new ideas to establish the basic unity in the strong interaction dynamics of nucleons and nuclei. A humorous look at this is given in Fig. 4. Introduction of an elephant in the intermediate region is done to exemplify unity, but at the same time to stress the complexity of the problem at hand.

But this summary would not be complete without giving a quote from an eminent reporter who attended this summer school: "A better understanding of Nuclear Dynamics is obtainable only by replacing Dirac's Equation and Maxwell's Equations by the Maxbell Equations which were projected in the Max Bell auditorium during this NATO institute. The lectures all started with a Big Bang, but tended near the end to undergo a phase transition into an inflationary mode with exponential expansion as the speaker disappeared beyond the horizon in a balloon, leaving the audience confined within a bubble. It was impossible to get to the walls fast enough, even by following Brodsky on the light cone, to discover the fascinating and exciting physics on the other side, which even Rabbi Schramm did not understand.

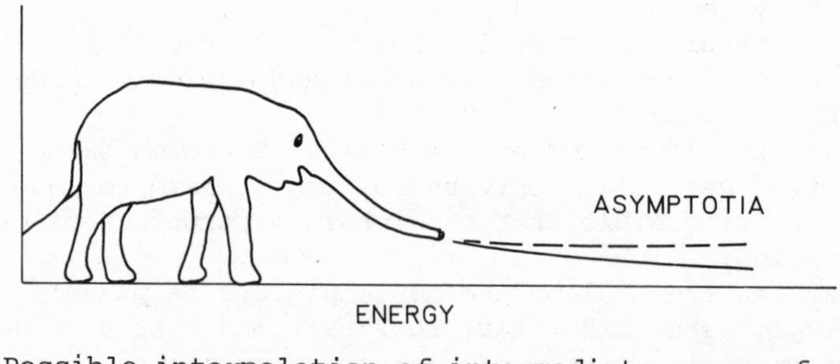

Fig. 4. Possible interpolation of intermediate range of energies. The understanding of physics in this energy region is akin to the case of four blind men and an elephant.

But there is still hope. Replace the open space-time expanding universe by a small finite imaginary-temperature space lattice and place all bets on Monte Carlo. Then we can catch up with the running coupling constant and understand why nuclei can flow like honey, peanut butter or jam, why anomalons behave like interacting bosons, and how to find the hidden color." Direct report from Pearson College of the Pacific by HARRY LIPKIN.

INDEX